DIGITAL SYSTEMS
with algorithm implementation

DIGITAL SYSTEMS
with algorithm implementation

M. Davio, J.-P. Deschamps and A. Thayse
Philips & MBLE Associated S.A., Philips Research Laboratory, Brussels

A Wiley–Interscience Publication

JOHN WILEY & SONS

Chichester · New York · Brisbane · Toronto · Singapore

Library of Congress Cataloging in Publication Data:
Davio, Marc.
Digital systems, with algorithm implementation.
'A Wiley–Interscience publication.'
Bibliography: p.
Includes index.
1. Computers—Circuits. 2. Logic circuits.
3. Algorithms. 4. Computer architecture. I. Deschamps, Jean-Pierre, 1945- II. Thayse, André, 1940-
III. Title.
TK7888.4.D38 621.3819'535 82-2710
ISBN 0 471 10413 2 (cloth) AACR2
ISBN 0 471 10414 0 (paper)

British Library Cataloguing in Publication Data:
Davio, M.
Digital systems.
1. Digital electronics
I. Title II. Deschamps, J.-P.
III. Thayse, A
621.3815 TK7868.D5
ISBN 0 471 10413 2 (cloth)
ISBN 0 471 10414 0 (paper)

Filmset and printed in Northern Ireland at The Universities Press (Belfast) Ltd.

Contents

List of symbols

$\{a \mid P\}$	set of elements satisfying a binary property P
$\subseteq$	set inclusion
$\cup$	set union
$\cap$	set intersection
$\lvert E \rvert$	number of elements of E
$\in$	is an element of
$\notin$	is not an element of
$\forall$	for any
$\exists$	there exists
$\nexists$	there does not exist
$\emptyset$	empty set

Bold face letters like $\mathbf{x}$, $\mathbf{e}$, and $\mathbf{B}$ are used to denote a vector

pr_j	jth projection
$w(\mathbf{x})$	weight of $\mathbf{x}$
$d(\mathbf{a}, \mathbf{b})$	distance between $\mathbf{a}$ and $\mathbf{b}$
$\oplus$	exclusive or
$\vee$	disjunction
$\wedge$	conjunction
$\bar{}$	complementation
$\langle f; g\rangle$	P-function
∇	is a P-function of
$[f, h]$	maximum element of $\{g_i \mid \langle g_i ; h\rangle \nabla f\}$
T	composition law
T^0, T^1, T^l	composition laws on P-function
$x^{(c)} = m-1,$	iff $x \in C$ (lattice exponentiation)
$= 0,$	otherwise
$\lceil a \rceil$	smallest integer greater than or equal to a
$\lfloor a \rfloor$	greatest integer smaller than or equal to a
$\simeq$	is represented by
$\lVert X \rVert$	absolute value of X
min	minimum
max	maximum
$\log_a n$	logarithm of n in base a
$\log n$	logarithm of n in the base given by the context
$\cong$	is approximately equal to

$\gg$	is much greater than
$\ll$	is much smaller than
$\binom{i}{j}$	$=\dfrac{i!}{j!\,(i-j)!}$
$\|m\|_n$	m modulo, i.e. the remainder of the division of m by n
$0(n)$	$\leqslant Cn$ for some constant value C
Σ, +	real sum
GF	Galois field
Π	real product
mod n	modulo n
Z	set of integers
Z_n	ring of integers modulo n
:=	is replaced by
[A]	contents of register A
[	chronology
$\Rightarrow$	implies
$\Leftrightarrow$	is equivalent to
$\subset$	precedes
$\not\subset$	does not precede
$\equiv$	is equivalent to
λ	empty command, empty word
t	true condition

Introduction

During the last decade the applications of digital techniques have pervaded everyday life: the microprocessor, a 'computer on a chip', appears in pocket calculators, toys and domestic appliances; entire technical fields, such as audio and video broadcasting, traditionally using analog techniques, are progressively moving to digital techniques. These major changes are obviously due to our ever-increasing capabilities in electronic integration and the trend outlined above is likely to maintain its accelerating pace during the next years.

It thus appears worthwhile to spend some effort in measuring the impact and the nature of that evolution and in trying to delineate its characteristic features. The authors of this book attempt to present a synthesis of the main technical aspects of the evolution which digital circuits have undergone since the early Shannon papers. The purpose is ambitious and, to have some chance of success, the project should focus on a small number of central ideas. Fortunately enough, a single concept may serve to unify the whole discussion: this concept is that of algorithm, i.e. of computation method. The central point to get is that, every time an appropriate and rigorously defined computation method is discovered, whether in storing and processing huge information files, or in setting up assisted education programs, it is possible to commit that task to automated processsing systems.

The introduction of the algorithm concept has furthermore the advantage of linking tightly the hardware and software domains. The present book indeed takes the algorithm concept as the basis for hardware design; once one has accepted that this very concept spans all digital design, one is immediately faced with the need for some language in which to express one's algorithms. Classical switching theory offers a number of these languages: for example, a normal form in Boolean algebra expresses a method for computing the local values of a binary function, an automaton represents an iteration, and so on. The representation of an algorithm in an appropriate language may be seen as a program and this observation clearly provides us with the expected link from, say, the gate level to the microprocessor.

The program concept as such is, however, still insufficient to cover the needs of hardware design. One has indeed to tie in some formal way a program and its implementation. In some sense, one has to complete the program description with a hardware interpretation. One of the immediate conclusions one reaches from that observation is that, with a given algorithm, it is generally possible to associate, by varying the interpretation, a number of distinct implementations. For example, the well known serial addition algorithm may be interpreted both as a combinational and as a sequential circuit. The design problem will thus appear as a choice or as a trade-off problem, and the discussion of the possible trade-offs will obviously be a central concern.

The concept of algorithm is introduced in Chapter I, after a short introduction to digital systems. Simultaneously, one describes the discrete components upon which will be based all the later designs. Chapter II covers the classical algebraic structures of lattice and of Boolean algebra. Simultaneously, it introduces the quality criteria, or performance measures, to be used afterwards to evaluate and compare functionally equivalent designs.

The study of combinational components is spread over Chapters III and IV. Chapter III is devoted to general purpose combinational components: multiplexers, demultiplexers, read only memories and programmable logic arrays; Chapter IV is devoted to special purpose combinational components: adders, subtractors, amplitude comparators, arithmetic and logic units, radix conversion networks and permutation networks. If the circuits described here most often correspond to commercial components, the chapter in itself has a second purpose, viz. to describe a family of algorithms that will receive in later chapters a number of different implementations.

The study of synchronous sequential circuits is started in Chapter V, where one observes for the first time that the use of additional memories for the storage of intermediate results and of some form of synchronization allows an interesting saving on the amount of circuitry involved in the implementation of an iterative process. The classical methods of designing synchronous sequential circuits by means of flip-flops are reviewed in Chapter VI, and are applied in Chapter VII to the synthesis of various sequential components, such as registers, counters, and frequency dividers.

Chapter VIII generalizes the observation made about iterations to a larger class of algorithms, called computation schemes: these algorithms result from the loop free functional composition of combinational functions; roughly speaking, the family of computation schemes contains all the algorithms, the number of steps of which is fixed. Here again, one discovers a number of possible implementations, as combinational, sequential or pipeline circuits. Chapter IX applies these methods to the design of multipliers and of dividers.

In Chapter X, we reach the most general class of algorithms, including those algorithms the execution time of which varies according to the values of the processed numerical data. The relevant model is that of algorithmic state machine, in which every computing process is modelled by the cooperation of two automata, the control unit and the processing unit.

Chapter XI introduces the new important idea of sequentialization (or serialization): the point is that two independent partial computations may be carried out either simultaneously (in parallel) or one after the other (sequentially) in an arbitrary order; the expectation is obviously that the sequentialization process will yield a corresponding return on the cost of the required circuitry. When one applies that idea to the processing unit of an algorithmic state machine, one reaches the classical architectures of familiar processing units and of bit-slice processors. The same ideas are applied in Chapters XII and XIII to the control units of algorithmic state machines. On the one hand, one describes here a number of algorithm transformations which allow the designer to associate with a given algorithm a number of control unit architectures; on the other hand, one discusses the associated optimization problems.

Chapter XIII is also devoted to the control unit; essentially, it studies the situations where the computation of the label of the next instruction to be executed is performed by some specialized circuits. Here are introduced the concepts of program counter, of stack automaton and of sequencer.

A last concept is required before reaching a microprocessor architecture, i.e. that of interpretation algorithm: this is the algorithm by which a command expressed in some upper level language is executed as a sequence of elementary operations of the processing unit level. That idea is explored in Chapter XIV where one furthermore develops an elementary microprocessor.

While writing this book, the authors have contracted important debts of gratitude. They wish to thank particularly Professor Vitold Belevitch, Director of the Philips Research Laboratory, both for his constant encouragement and for giving them the possibility of materializing this work within the framework of the Philips Research Laboratory activities.

Mrs Edith Moes typed the manuscript with amiability and competence. Mr Clause Semaille executed the numerous drawings with his usual care and Mr Heinz Georgi provided us with his kind editorial assistance.

Brussells
January 1982

M. Davio
J.-P. Deschamps
A. Thayse

Chapter I
Digital systems

This introductory chapter gives the main basic concepts. Emphasis is put on the relation between algorithms and digital systems. This relation is the most important underlying idea: a digital system implements an algorithm, and the performance of the designed circuit strongly depends on decisions which are made at the algorithm level.

1 INPUT SIGNALS, OUTPUT SIGNALS

1.1 Definitions

Physical systems establish causal relations between excitations, which are called *input signals*, and responses to these excitations, which are called *output signals*. The set of relations that the system establishes between input and output signals is designated by the general term *input–output behaviour*. The starting point of the design of a physical system is its input–output behaviour.

It frequently arises that the sets

$$\Sigma = \{\sigma_1, \sigma_2, \ldots, \sigma_N\} \quad \text{and} \quad \Omega = \{\omega_1, \omega_2, \ldots, \omega_M\}$$

of input and output signals are both finite. One then says that the system is a *digital system*. Equivalent terms are: *discrete system, logical system.*

Let us introduce some definitions. The sets Σ and Ω of input and output signals are called *input* and *output alphabets*. Their elements are the *input* and *output letters*. An ordered string of letters is a *word*. The sets of input and output words are denoted by Σ^* and Ω^*. These infinite sets are the *input* and *output dictionaries*.

1.2 Encoding of the signals: generalities

One must distinguish between a particular signal (input, output or internal signal) and its actual representation in the digital system. Most often, the elements of a given alphabet A are represented by means of symbols belonging to a set S, such that

$$|S| \ll |A|,$$

where $\ll$ stands for 'much smaller than'. The elements of A are represented by n-tuples of symbols belonging to S. An *encoding of the alphabet* A is a one-to-one mapping:

$$\psi : A \to S^n.$$

Hence

$$|S^n| = |S|^n \geqslant |A|$$

and

$$n \geqslant \lceil \log_{|S|} |A| \rceil,$$

where $\log_{|S|}$ stands for the logarithm in base $|S|$.

The choice of an appropriate encoding is an important problem: it can strongly act upon the performances of the corresponding system implementation. The two following sections give two practical examples: the encoding of the natural integers and the encoding of the alphanumeric characters.

1.3 Mixed radix number representations

1.3.1 Definition

Let us recall, for the sake of completeness, the fundamental lemma of Euclidean division.

Lemma 1. *Given two natural integers m and n, there exist two natural integers, the quotient q and the remainder r, such that*

$$m = nq + r \quad \textit{and} \quad 0 \leqslant r < b. \tag{1}$$

This familiar Lemma is mentioned without proof. As an immediate consequence, we may write:

$$q = \lfloor m/n \rfloor \quad \text{and} \quad r = [\![m]\!]_n. \tag{2}$$

Let us now consider a vector

$$\mathbf{b} = [b_{n-1}, \ldots, b_1, b_0]$$

having n integer components $b_i \geqslant 2$, for every $i \in Z_n = \{0, 1, \ldots, n-1\}$. This vector will be called *basis vector.* From this vector, we build up a second vector

$$\mathbf{w} = [w_n, w_{n-1}, \ldots, w_1, w_0]$$

having $(n+1)$ integer valued components defined by

$$w_0 = 1, \qquad \forall i \in Z_n : w_{i+1} = w_i b_i = \prod_{j=0}^{i} b_j. \tag{3}$$

This vector is called *weight vector.* To simplify the notations, we also put

$$N = w_n = \prod_{j=0}^{n-1} b_j. \tag{4}$$

These notations allow the mixed radix representation theorem to be stated.

Theorem 1. (*Mixed radix representation*). *Any natural integer* $X \in Z_N = \{0, 1, \ldots, N-1\}$ *has a unique representation* $[x_{n-1}, \ldots, x_1, x_0]$ *satisfying the two following conditions*:

(i)
$$X = \sum_{i=0}^{n-1} x_i w_i; \tag{5}$$

(ii)
$$\forall i \in Z_n : 0 \leqslant x_i < b_i. \tag{6}$$

Proof. Consider the following sequence of Euclidean divisions

$$X = b_0 q_1 + x_0; \qquad 0 \leqslant x_0 < b_0; \tag{7.1}$$

$$q_1 = b_1 q_2 + x_1; \qquad 0 \leqslant x_1 < b_1; \tag{7.2}$$

$$q_i = b_i q_{i+1} + x_i; \qquad 0 \leqslant x_i < b_i; \tag{7.i}$$

$$q_{n-1} = b_{n-1} q_n + x_{n-1}; \qquad 0 \leqslant x_{n-1} < b_{n-1}. \tag{7.n-1}$$

The n quotients $q_0, q_1, \ldots, q_{n-1}$ are easily eliminated by substitution. In a first step, we obtain:

$$X = ((\ldots((q_n b_{n-1} + x_{n-1}) b_{n-2} + x_{n-2}) b_{n-3} + \ldots) b_1 + x_1) b_0 + x_0. \tag{8}$$

Expanding the latter equation, and using (3) we obtain:

$$X = q_n w_n + \sum_{i=0}^{n-1} x_i w_i. \tag{9}$$

The hypothesis $X \in Z_N$ means in particular $X < N = w_n$. This in turn implies in (9), and thus in (7.n-1), that $q_n = 0$. In this case, (9) reduces to (5). □

The vector $\mathbf{x} = [x_{n-1}, \ldots, x_1, x_0]$ obtained in Theorem 1 is the mixed radix representation (*m.r.r.*) of X with respect to the basis vector $\mathbf{b} = [b_{n-1}, \ldots, b_1, b_0]$.

Remark. The mixed radix numeration systems play a key role in several fields. Formulae (7) and (8) are used in numeration system transformations. They also appear in residue arithmetic (see Exercise 5) and in the definition of some interconnection networks, namely the perfect shuffles (see Chapter IV).

The homogeneous systems are studied in Section 1.3.3.

1.3.2 Examples

(1) The time measure system in

hours, minutes, seconds

is a mixed radix numeration system based on the vector [24, 60, 60].

Z_{30}	**b**: **w**: 30	5 6	3 2	2 1
0		0	0	0
1		0	0	1
2		0	1	0
3		0	1	1
4		0	2	0
5		0	2	1
6		1	0	0
7		1	0	1
8		1	1	0
9		1	1	1
10		1	2	0
11		1	2	1
12		2	0	0
13		2	0	1
14		2	1	0
15		2	1	1
16		2	2	0
17		2	2	1
18		3	0	0
19		3	0	1
20		3	1	0
21		3	1	1
22		3	2	0
23		3	2	1
24		4	0	0
25		4	0	1
26		4	1	0
27		4	1	1
28		4	2	0
29		4	2	1

Figure 1. Mixed radix representation

(2) Let us consider the basis vector $[b_2, b_1, b_0]=[5, 3, 2]$. The corresponding weight vector is $[w_3, w_2, w_1, w_0]=[30, 6, 3, 1]$. It allows the representation of the elements of Z_{30} according to the table of Figure 1.

For instance,

$$19=3\cdot w_1+1\cdot w_0=3\cdot 6+1.$$

1.3.3 Radix B numeration system

Let us suppose that

$$b_0=b_1=\ldots=b_{n-1}=B.$$

Hence, according to (3) we may define

$$\forall i\in Z_{n+1}: w_i=B^i. \tag{10}$$

Corollary 1. (*Radix B representation*). *Any natural integer* $X \in Z_{B^n} = \{0, 1, \ldots, B^n - 1\}$ *has a unique representation* $[x_{n-1}, \ldots, x_1, x_0]$ *satisfying the two following conditions:*

(i)
$$X = \sum_{i=0}^{n-1} x_i B^i, \tag{11}$$

(ii)
$$\forall i \in Z_n : 0 \leqslant x_i < B. \tag{12}$$

Expression (11) is called the *radix B representation* of X. An alternative form is

$$X = (\ldots((x_{n-1}B + x_{n-2})B + x_{n-3}) \ldots x_1)B + x_0. \tag{13}$$

The numeration systems in bases 2, 8, 10, and 16 are called *binary, octal, decimal,* and *hexadecimal* numeration systems. In the last case, the sixteen digits are

$$0 \quad 1 \quad 2 \quad 3 \quad 4 \quad 5 \quad 6 \quad 7 \quad 8 \quad 9 \quad A \quad B \quad C \quad D \quad E \quad F;$$

for instance, the hexadecimal number $A09$ corresponds to the decimal number

$$10.16^2 + 0.16 + 9 = 2569.$$

The digits $0, 1, \ldots, B-1$ appearing in the radix B numeration system can themselves be represented in the binary system by means of $\lceil \log_2 B \rceil$ bits. This allows the definition of *binary coded radix B representations.* For instance, in the *binary coded decimal* (BCD) *system the decimal number* 197 is written under the form

$$0001 \quad 1001 \quad 0111.$$

1.3.4 Lexicographic order

Let us consider the Cartesian product

$$\mathbf{Z} = Z_{b_{n-1}} \times Z_{b_{n-2}} \times \ldots \times Z_{b_1} \times Z_{b_0}$$

in which every set $Z_{b_i} = \{0, 1, \ldots, b_i - 1\}$ is a totally ordered set:

$$0 < 1 < \ldots < b_i - 1.$$

The set $\mathbf{Z}$ itself can be totally ordered according to the following rule:

$$\mathbf{x} = [x_{n-1}, \ldots, x_1, x_0] < [y_{n-1}, \ldots, y_1, y_0] = \mathbf{y},$$

iff there exists an index j such that

$$x_j < y_j \quad \text{and} \quad x_k = y_k, \quad \forall k > j.$$

It is the *lexicographic order* over $\mathbf{Z}$.

Theorem 1 defines a mapping

$$\psi : Z_N \rightarrow \mathbf{Z};$$

this mapping associates with every $X \in Z$ its representation $\mathbf{x} = [x_{n-1}, \ldots, x_1, x_0]$. Obviously ψ is a one-to-one and onto mapping. Furthermore, it preserves the order relation, i.e. if $\mathbf{x}$ and $\mathbf{y}$ are the m.r.r. of X and Y, then

$$X \leqslant Y \Leftrightarrow \mathbf{x} \leqslant \mathbf{y}:$$

(1) according to Theorem 1

$$X = Y \Leftrightarrow \mathbf{x} = \mathbf{y};$$

(2) if $\mathbf{x} < \mathbf{y}$ there exists an index j such that $x_j < y_j$ and $x_k = y_k, \forall k > j$.

Bit positions	Bit positions 654							
3210	000	001	010	011	100	101	110	111
0000	NULL	DC0	blank	0		P		p
0001	SOM	DC1	!	1	A	Q	a	q
0010	EOA	DC2	"	2	B	R	b	r
0011	EOM	DC3	#	3	C	S	c	s
0100	EOT	DC4	$	4	D	T	d	t
0101	WRU	ERR	%	5	E	U	e	u
0110	RU	SYNC	&	6	F	V	f	v
0111	BELL	LEM	'	7	G	W	g	w
1000	BKSP	SO	(	8	H	X	h	x
1001	HT	S1	)	9	I	Y	i	y
1010	LF	S2	*	:	J	Z	j	z
1011	VT	S3	+	;	K	[	k	{
1100	FF	S4	,	<	L	\	1	|
1101	CR	S5	−	=	M	]	m	}
1110	SO	S6	.	>	N	↑	n	ESC
1111	SI	S7	/	?	O	←	o	DEL

NULL	Null/Idle	BKSP	Backspace	DC0–DC4	Device control
SOM	Start of message	HT	Horizontal tab	ERR	Error
EOA	End of address	LF	Line feed	SYNC	Synchronous idle
EOM	End of message	VT	Vertical tab	LEM	Logical end of media
EOT	End of transmission	FF	Form feed	S0–S7	Separator information
WRU	'Who are You?'	CR	Carriage return		Blank
RU	'Are you. . .?'	SO	Shut out	ESC	Escape
BELL	Audible signal	SI	Shift in	DEL	Delete/Idle

Bit positions of code format = | 6 5 4 3 2 1 0 |

Figure 2. ASCII code

Hence,

$$Y-X=\sum_{i=0}^{j-1}(y_i-x_i)w_i+(y_j-x_j)w_j$$

$$\geqslant w_j-\sum_{i=0}^{j-1}(b_i-1)w_i=1.$$

Therefore

$$\mathbf{x}<\mathbf{y}\Rightarrow X<Y.$$

(3) One similarly proves that

$$\mathbf{x}>\mathbf{y}\Rightarrow X>Y.$$

1.4 Encoding of the alphanumeric symbols

A common requirement of data processing is the representation of the alphabetic, numerical and punctuation symbols. Such codes are called alphanumeric codes. One of the very commonly used codes is the ASCII code. It is shown in Figure 2.

Remarks.

(1) The integers $0, 1, \ldots, 9$, are represented in the binary numeration system, with prefix 011.

(2) The lexicographic order of the representations of the letters $A, B, \ldots, Z, a, b, \ldots, z$ corresponds to the alphabetic order.

2 EXPLICIT DESCRIPTION OF THE INPUT–OUTPUT BEHAVIOUR

In the simplest case, the correspondence between the input and output letters is described by a mapping

$$\phi:\Sigma\rightarrow\Omega. \tag{14}$$

The input alphabet is finite and it is possible to define entirely the mapping ϕ by means of a table; it is called the *value table* or the *truth table.* The knowledge of the table gives the *explicit description of the input–output behaviour.* Let us give an example.

Example 1. Two integers a and b, lying in the interval $0\leqslant a, b<2^n-1$, are given by their binary representations $[a_{n-1}, \ldots, a_1, a_0]$ and $[b_{n-1}, \ldots, b_1, b_0]$. Define a digital system which computes the binary representation $[c_n, c_{n-1}, \ldots, c_1, c_0]$ of the sum $c=a+b$. Figure 3 describes the input–output behaviour of such an adder in the particular case where $n=2$.

a_1	a_0	b_1	b_0	c_2	c_1	c_0
0	0	0	0	0	0	0
0	0	0	1	0	0	1
0	0	1	0	0	1	0
0	0	1	1	0	1	1
0	1	0	0	0	0	1
0	1	0	1	0	1	0
0	1	1	0	0	1	1
0	1	1	1	1	0	0
1	0	0	0	0	1	0
1	0	0	1	0	1	1
1	0	1	0	1	0	0
1	0	1	1	1	0	1
1	1	0	0	0	1	1
1	1	0	1	1	0	0
1	1	1	0	1	0	1
1	1	1	1	1	1	0

Figure 3. Explicit description of the input–output behaviour of an adder

Comments. Example 1 suggests several comments.

(1) The construction of the table of Figure 3 allowed us to translate the informal description of the problem, namely 'to build an adder', into a formal one. Such a transformation is very important for the synthesis problem: the existence of formal synthesis techniques obviously depends on the existence of formal descriptions. These formal techniques play a key role in the design of digital systems.

(2) The adder defined by the table of Figure 3 is a system with four binary input and three binary outputs. It works as follows: if one applies the binary representations of a and b on the four inputs, then the system will generate the binary representation of the sum $c=a+b$; it appears on the three outputs after some *response time.* Hence the system may be considered as a *permanent memory* which stores the results of the addition: the input letters are the *addresses,* the output letters are the *data,* the computation of a sum consists in the *readout* of the datum selected by the address, and the response time is called *access time.* Such a storing device may be called *read only memory.*

(3) A quantitative argument, namely the size of an explicit description by means of a table, immediately shows the necessity of other description methods: the truth table of an adder which sums integers lying between 0 and 1023 would contain $2^{20}(\cong 10^6)$ rows. One thus has to look for more condensed description methods. The use of appropriate *algebraic structures* can be useful in this respect.

(4) Table description also presents a qualitative drawback. One cannot get any 'original' result from a system defined by an explicit table. All the

expected results, namely the sums c, are computed during the table construction. Hence, as it has been seen above, the system must be considered as a memory which stores the solutions previously computed by the designer. This leads us to look for implicit descriptions, which do not imply the *a priori* knowledge of the results. In implicit descriptions, a digital system is defined as a memory which stores the solutions previously computed by the designer. relation between digital systems and other branches of computer sciences like the theory of algorithms and the computability theory.

3 ALGORITHMS

3.1 Computation primitives, admissible constructs

We concluded the last section by showing the necessity of implicit description methods of the input–output behaviour. This does not mean that explicit description methods are never used. As a matter of fact they play an important role, but at a different level.

We thus have to look for new descriptions of the input–output behaviour. For that purpose we introduce the notion of algorithm. We adopt an intuitive definition: an *algorithm* results from the assembly of some computation primitives, according to a series of construction rules.

The *computation primitives* correspond to partial computations whose results are known in advance and stored. For instance, in the classical paper and pencil multiplication algorithm, the multiplication tables are computation primitives. The computation primitives are *discrete functions,* that is functions defining a mapping from a finite set to a finite set, whose explicit description is known.

The *admissible constructs* are a set of rules according to which the computation primitives may be combined in order to obtain more complex results.

Very familiar examples of constructs are given by the classical programming instructions:

if then else,

for step until do,

while do.

The next section gives a concrete example.

3.2 Example: radix B serial addition

We describe an algorithm of addition in the radix B numeration system. Given two integers X and Y, belonging to the interval $0 \leq X, Y < B^n$, and their radix B representations $[x_{n-1}, \ldots, x_1, x_0]$ and $[y_{n-1}, \ldots, z_1, z_0]$, one computes the radix B representation $[z_n, z_{n-1}, \ldots, z_1, z_0]$ of

$$Z = X + Y + c_0, \tag{15}$$

where c_0 is an initial carry belonging to $\{0, 1\}$.

Note that

$$0 \leqslant X + Y + c_0 \leqslant (B^n - 1) + (B^n - 1) + 1 = 2B^n - 1, \tag{16}$$

and thus $z_n \in \{0, 1\}$.

Let us write a system of equalities based on the Euclidean division (see Lemma 1):

$$x_0 + y_0 + c_0 = Bc_1 + s_0; \quad 0 \leqslant s_0 < B; \tag{17.0}$$

$$x_1 + y_1 + c_1 = Bc_2 + s_1; \quad 0 \leqslant s_1 < B; \tag{17.1}$$

$$x_i + y_i + c_i = Bc_{i+1} + s_i; \quad 0 \leqslant s_i < B; \tag{17.i}$$

$$x_{n-1} + y_{n-1} + c_{n-1} = Bc_n + s_{n-1}; \quad 0 \leqslant s_{n-1} < B. \tag{17.n-1}$$

By multiplying both members of (17.i) by B^i, for every index $i = 0, 1, \ldots, n-1$, and summing up their equalities, one obtains:

$$\sum_{i=0}^{n-1} x_i B^i + \sum_{i=0}^{n-1} y_i B^i + \sum_{i=0}^{n-1} c_i B^i = \sum_{i=1}^{n} c_i B^i + \sum_{i=0}^{n-1} s_i B^i; \tag{18}$$

hence,

$$X + Y + c_0 = c_n B^n + \sum_{i=0}^{n-1} s_i B^i. \tag{19}$$

According to (16) and (19), $c_n < 2 \leqslant B$, and

$$[z_n, z_{n-1}, \ldots, z_1, z_0] = [c_n, s_{n-1}, \ldots, s_1, s_0]. \tag{20}$$

The result is generated by an iterative algorithm that can be written as follows:

for $i = 0$ *step* 1 *until* $n - 1$ *do*
begin
$c_{i+1} = \lfloor (x_i + y_i + c_i)/B \rfloor;$ (21)
$s_i = [\![x_i + y_i + c_i]\!]_B$ (22)
end.

This algorithm is based on the possibility to compute c_{i+1} and s_i according to (21) and (22). This partial computation constitutes the only computation primitive of the algorithm. The subsystem which actually computes c_{i+1} and s_i is symbolized by the black box of Figure 4. It is a *full adder in base B*. The hardware counterpart of a computation primitive is often called a *computation resource.*

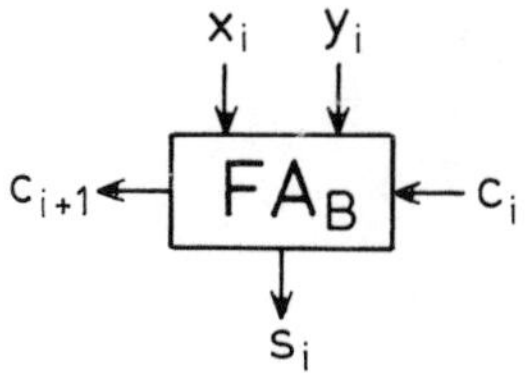

Figure 4. Full adder in base B

Remark. The above constructs can be extended to the case of m.r.r. systems.

The algorithm is composed of n identical steps, each of which performs (21) and (22). One of the two values obtained at step i, namely c_{i+1}, is used at step $i+1$; in algebraic terms this corresponds to *functional composition*: if

$$f(a, b, c) = \lfloor (a+b+c)/B \rfloor,$$

and

$$g(a, b, c) = [\![a+b+c]\!]_B,$$

then

$$\begin{aligned}
s_0 &= f(x_0, y_0, c_0), \\
s_1 &= f(x_1, y_1, g(x_0, y_0, c_0)), \\
s_2 &= f(x_2, y_2, g(x_1, y_1, g(x_0, y_0, c_0))), \\
&\text{-----------------------------------} \\
s_{n-1} &= f(x_{n-1}, y_{n-1}, g(\ldots, g(x_0, y_0, c_0)\ldots)), \\
c_{n-1} &= g(x_{n-1}, y_{n-1}, g(\ldots, g(x_0, y_0, c_0)\ldots)).
\end{aligned} \tag{23}$$

The circuit counterpart of the functional composition is a series of connections between computation resources. This is shown in Figure 5 which describes a radix B serial adder.

Comments.

(1) The choice of an algorithm strongly influences the performances of the corresponding digital circuit. In what concerns the choice of the computation resources, rather few constraints will be imposed in the next sections and chapters. On the other hand, we shall first consider the algorithms in which the only admissible construct is the functional composition. Afterwards we shall progressively enlarge the category of admissible constructs, and study the influence of this enrichment on the circuit performances.

(2) In the design of a digital system, the concept of algorithm often appears at several different levels. The serial adder of Figure 5 implements a classical addition algorithm based on an iterative use of (21) and (22). However, the computation resource performing these operations must itself be realized by means of smaller components, for instance NAND gates (see Section 4.3). Hence, one must look for an algorithm performing the function of a full adder and whose computation primitives are NAND functions. On the other hand, the serial adder may be part of a larger system performing

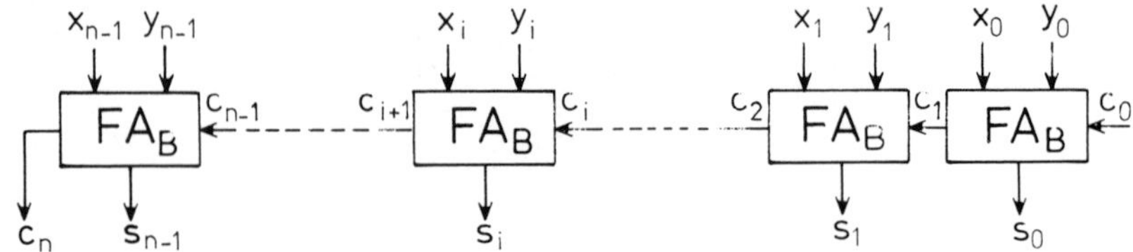

Figure 5. Radix B serial adder

multiplications and the addition could be a computation primitive of the multiplication algorithm. Hence, a digital system often implements a series of nested and hierarchically organized algorithms. An algorithm lying at some hierarchy level is a computation primitive of algorithms lying at upper levels and uses itself computation primitives which are algorithms lying at lower levels. For instance, the discrete Fourier transform algorithm uses multiplications and additions, the multiplication algorithm uses additions, the addition algorithm uses full adder function, and the full adder function can be generated by NAND functions.

3.3 Computation schemes

This section is devoted to a particular class of algorithms which are very often encountered in practical applications. They may be roughly characterized as follows: they consist of loop free compositions of basic functions, namely the computation primitives.

3.3.1 Definitions

Let S be a set of symbols. One wishes to compute the value taken by a *logic function*

$$f: S^n \to S^m$$

at point

$$\mathbf{x} = (x_{n-1}, \ldots, x_1, x_0) \in S^n.$$

For that purpose, one disposes of a set

$$\Pi = \{\mathbf{\Pi}_i \mid i \in I \text{ and } \mathbf{\Pi}_i: S^{n_i} \to S^{m_i}\}$$

of *computation primitives*, where I is some indexing set. One also disposes of two construction rules:

(1) the *jth projection* pr_j associates with an element $[s_{m_i-1}, \ldots, s_1, s_0]$ of S^{m_i} its jth component s_j, for every $j = 0, 1, \ldots, m_{i-1}$. This allows us to define the jth projection of a computation primitive

$$\text{pr}_j\, \mathbf{\Pi}_i: S^{n_i} \to S.$$

It is simply defined as follows:

$$\forall \mathbf{x}_i \in S^{n_i}: (\mathrm{pr}_j\ \mathbf{\Pi}_i)_{\mathbf{x}_i} = \mathrm{pr}_j\ (\mathbf{\Pi}_i(\mathbf{x}_i)).$$

(2) The *composition operation* associates with the (n_i+1) functions

$$\mathbf{\Pi}_i: S^{n_i} \to S^{m_i}; \qquad g_0, g_1, \ldots, g_{n_i-1}: S^n \to S$$

the function h which is defined by

$$\forall \mathbf{x} \in S^n: h(\mathbf{x}) = \mathbf{\Pi}_i(g_{n_i-1}(\mathbf{x}), \ldots, g_1(\mathbf{x}), g_0(\mathbf{x})).$$

A *computation scheme* is a set of functions

$$A = \{N_0, N_1, \ldots, N_R\}$$

where, for every k, either one of the following four assertions holds:
(1) N_k is a constant function, that is a member of S:

$$\forall \mathbf{x} \in S^n: N_k(\mathbf{x}) = s \in S. \tag{24}$$

(2) N_k is a variable x_l:

$$\forall \mathbf{x} \in S^n: N_k(\mathbf{x}) = x_l. \tag{25}$$

(3) N_k is the jth projection of function $N_{k'}$, where $k' < k$:

$$\forall \mathbf{x} \in S^n: N_k(\mathbf{x}) = \mathrm{pr}_j\ (N_{k'}(\mathbf{x})). \tag{26}$$

(4) N_k is obtained by function composition of $\mathbf{\Pi}_i, N_{k_0}, N_{k_1}, \ldots, N_{k_{n_i-1}}$, where $\max(k_0, k_1, \ldots, k_{n_i-1}) < k$:

$$\forall \mathbf{x} \in S^n: N_k(\mathbf{x}) = \mathbf{\Pi}_i(N_{k_{n_i-1}}(\mathbf{x}), \ldots, N_{k_1}(\mathbf{x}), N_{k_0}(\mathbf{x})). \tag{27}$$

3.3.2 Example: radix B serial addition (continued)

The radix B serial addition described in Section 3.2 is a computation scheme. It uses the set of symbols

$$Z_B = \{0, 1, \ldots, B-1\},$$

and the primitive

$$\mathbf{\Pi}: Z_B \times Z_B \times Z_2 \to Z_2 \times Z_B,$$

defined by

$$\forall (x_i, y_i, c_i) \in Z_B \times Z_B \times Z_2: \mathrm{pr}_0\ (\mathbf{\Pi}(x_i, y_i, c_i)) = z_i = [\![x_i + y_i + c_i]\!]_B,$$
$$\mathrm{pr}_1\ (\mathbf{\Pi}(x_i, y_i, c_i)) = c_i = \lfloor (x_i + y_i + c_i)/B \rfloor.$$

For $n = 4$, the corresponding scheme is

$$\mathrm{ADD} = \{N_0, N_1, \ldots, N_{20}\},$$

where

$$
\begin{array}{ll}
N = 0, & \\
N_1 = x_0, & N_{11} = x_2, \\
N_2 = y_0, & N_{12} = y_2, \\
N_3 = \mathbf{\Pi}(N_1, N_2, N_0), & N_{13} = \mathbf{\Pi}(N_{11}, N_{12}, N_{10}), \\
N_4 = \mathrm{pr}_1\,(N_3), & N_{14} = \mathrm{pr}_1\, N_{13} \\
N_5 = \mathrm{pr}_0\,(N_3) & N_{15} = \mathrm{pr}_0\, N_{13} \\
N_6 = x_1, & N_{16} = x_3, \\
N_7 = y_1, & N_{17} = y_3, \\
N_8 = \mathbf{\Pi}(N_6, N_7, N_5), & N_{18} = \mathbf{\Pi}(N_{16}, N_{17}, N_{15}), \\
N_9 = \mathrm{pr}_1\, N_8, & N_{19} = \mathrm{pr}_1\, N_{18}, \\
N_{10} = \mathrm{pr}_0\, N_8, & N_{20} = \mathrm{pr}_0\, N_{18}.
\end{array}
$$

3.3.3 Precedence relation. Precedence graph

With every computation scheme $A = \{N_0, N_1, \ldots, N_R\}$ is associated a directed graph:

(1) Its vertices are in one-to-one correspondence with the functons N_k belonging to types (24), (25) or (27). Let us denote by V the set of vertices.

(2) The binary relation $\mathcal{R}$ defining the graph is

$$\mathcal{R} = \{(N_k, N_l) \in V^2 \mid \exists N_{i_1} = N_k, N_{i_2}, \ldots, N_{i_r} = N_l;\ N_{i_{j+1}} = \mathbf{\Pi}_t(\ldots, \mathrm{pr}_s\, N_{i_j}, \ldots)\}. \tag{28}$$

One writes

$$N_k \subset N_l. \tag{29}$$

According to (28) this means that during the execution of algorithm A, the evaluation of N_k *precedes* the evaluation of N_l. Relation $\mathcal{R}$ is called the *precedence relation* of algorithm A.

The precedence relation is both transitive ($N_i \subset N_j$ and $N_j \subset N_k \Rightarrow N_i \subset N_k$) and irreflexive ($N_k \not\subset N_k$). Hence, its graph is acyclic. It is represented by its *Hasse diagram* (see Chapter II, Section 2.1), that is the set of edges

$$E = \{(N_k, N_l) \mid N_k \subset N_l \text{ and } \nexists N_t\colon N_k \subset N_t \subset N_l\}.$$

The Hasse diagram of relation $\mathcal{R}$ is also an acyclic directed graph. It is called the *precedence graph* of algorithm A.

If N_k is either a constant or a variable (see (24) and (25)), then the corresponding vertex has no incident edge; it is a *source node.* Similarly, a vertex which has no successor is a *terminal node.*

The precedence graph of the radix B serial addition (Section 3.3.2) is shown in Figure 6.

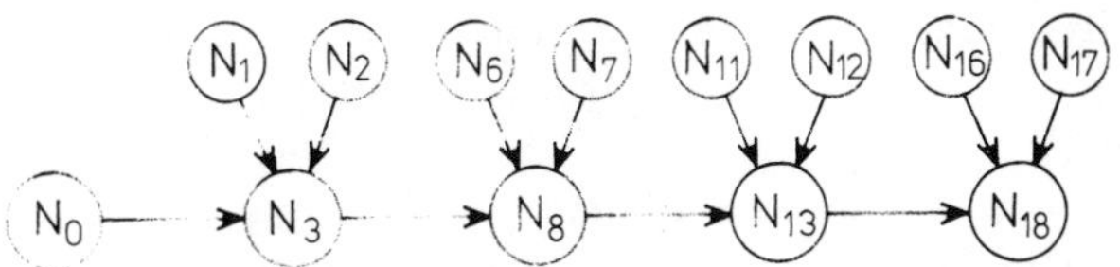

Figure 6. Precedence graph of the radix B serial addition

The precedence graph gives prominence to the skeleton of the computation scheme and, particularly to the succession constraints it implies.

Let us define two features of a computation scheme which are measured on its precedence graph:

(1) The *cost* of a computation scheme is the number of nodes (vertices) which are not source nodes.

(2) The *computation time* of a scheme is the length of the longest path in the graph.

3.3.4 Combinational circuits

The computation scheme and the corresponding precedence graph give an abstract representation of an algorithm. One may associate several implementations with this algorithm. The simplest way is to consider the computation scheme as a structure description:

(1) To every function N_k of type (27), that is

$$N_k = \Pi_i(N_{k_{n_i-1}}, \ldots, N_{k_1}, N_{k_0}),$$

corresponds an operator with n_i input wires and m_i output wires, each of them being able to carry the symbols of S.

(2) The computation scheme describes the set of connections linking the operators.

One so obtains a *combinational logic circuit.* The implementation of a computation scheme is thus reduced to the implementation of the computation primitives and of the interconnections. The notions of cost and computation have a very obvious practical meaning.

4 ELEMENTARY COMPONENTS

4.1 Introduction

Digital systems are mainly characterized by the finite number of input and output signals. On the other hand, many physical devices have the following property: they can stand in only a finite number of different stable states (exclude transients). For instance, the following devices have two stable states: a relay contact can be open or closed, a magnetic material with rectangular hysteresis cycle can present either a positive or a negative remanent flux density, a switching transistor may be in the cutoff or in the saturation state.

These two-state elementary digital systems may obviously be used as *building blocks* in order to construct larger systems. We shall only describe electronic components which are superior as regards items such as speed, dimension, cost and reliability.

In Section 3.2 we made a comment about the hierarchical organization of algorithms and of digital systems which implement algorithms. The two-state elementary systems described above constitute the bottom level of this hierarchy. Larger systems are constructed by step by step process: given the digital systems belonging to some hierarchical level, they are combined, in a rather heuristic way, in order to obtain 'useful' systems belonging to the next level.

Here is a list of the main circuits which are described in this book. They are (arbitrarily) grouped into five levels.

(1) Devices: resistors, diodes, bipolar transistors, and MOS field effect transistors (Section 4.2 and 4.3).

(2) Minimum computation or storing ability: logic gates (Sections 4.2 and 4.3) and flip-flops (Chapter VI).

(3) Small size combinational and sequential circuits: multiplexers, demultiplexers, address decoders (Chapter III), registers, counters, scalers, and sequence generators (Chapter VII).

(4) Memories and medium size combinational circuits: arithmetic and logic units, comparators (Chapters IV), read only memories, programmable logic arrays (Chapter III), and random access memories (Chapter VII).

(5) Large size circuits: bit slice processors (Chapter XI), sequencers (Chapter XIII), and microprocessors (Chapter XIV).

The basic building blocks are two-state devices. It is thus very natural to study Boolean functions which map $\{0, 1\}^n$ into $\{0, 1\}^m$ (see Chapter II, Section 3.1).

4.2 Logic gates

We aim at designing operators (see Figure 7), which implement Boolean functions by means of components such as resistors, diodes, and transistors. In order to reach this goal, we proceed as follows: we first determine the functions which are realized by interconnection of a few components. After,

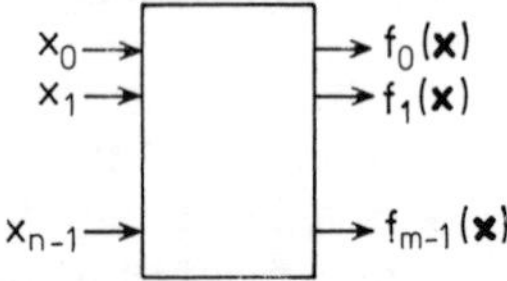

Figure 7. Boolean operator

we try to express the function we have to implement in terms of these basic functions.

The determination of the basic functions, that is the functions which are easily implemented by a few components, will guide us in the choice of the appropriate algebraic tools.

Figure 8a illustrates this approach by three particular examples. The three circuits are analysed under the assumption that diodes and transistors are ideal switches (no leakage currents, no voltage drop in the forward bias or saturation region). The voltage level of every input wire may be either 0(V) or E(V). The static behaviour analysis of these circuits is summarized in Figure 8b. It is more convenient to use binary digits (in short: *bits*) in order

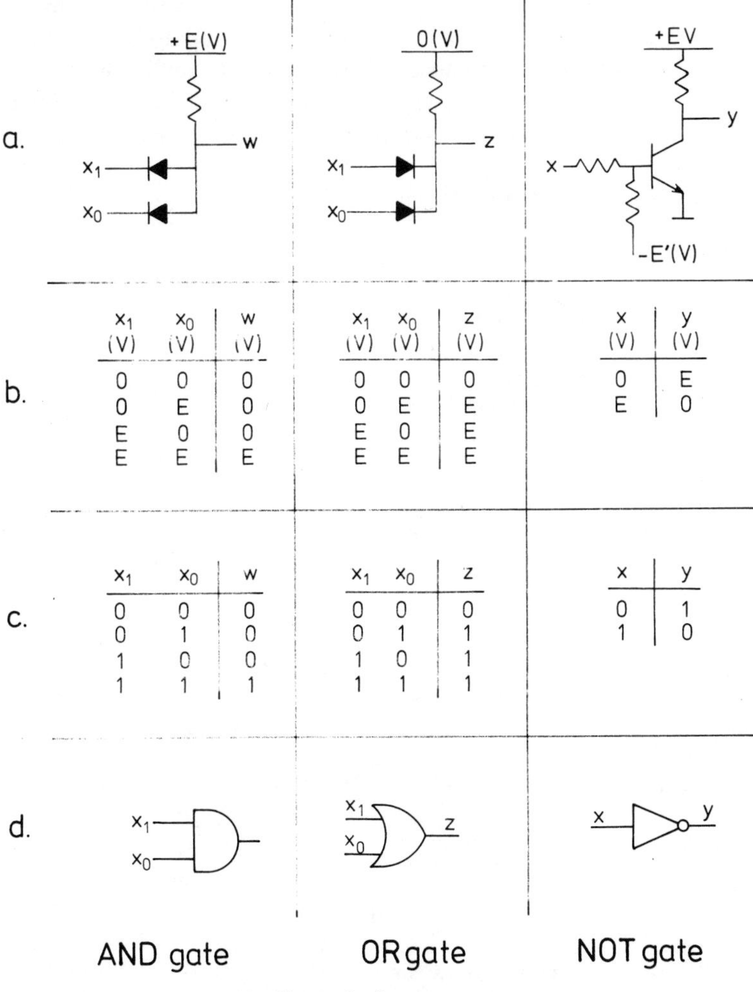

Figure 8. Logic gates

to represent the two possible voltage levels. There are two different assignments. Let us choose the following correspondence:

$$0(V) \leftrightarrow 0; \qquad E(V) \leftrightarrow 1. \tag{30}$$

It is said to be a *positive logic assignment*: 1 represents the highest voltage. In a *negative logic assignment*, 1 represents the lowest voltage.

Thanks to the chosen logic assignment (30) we obtain the truth tables of the basic functions generated by the three circuits (Figure 8c). These circuits are called *logic gates*. The left-hand circuit is an AND gate: its output w takes the value 1 if both inputs x_0 *and* x_1 take the value 1. The middle circuit is an OR gate, and the right-hand circuit is a NOT gate (or inverting gate). The corresponding MIL symbols are shown in Figure 8d.

The tables of Figure 8b and Figure 8c are logic models of the corresponding circuits. They have been derived under some approximations and, for instance, do not take into account the fact that the input signals actually can take all values between 0(V) and E(V). If one has to realize a digital system by an interconnection of logic gates, the knowledge of the logic model of the gates is generally not sufficient, and other parameters must be known. Let us define some of them.

(1) *Static behaviour.* Figure 9a shows a typical input–output characteristic

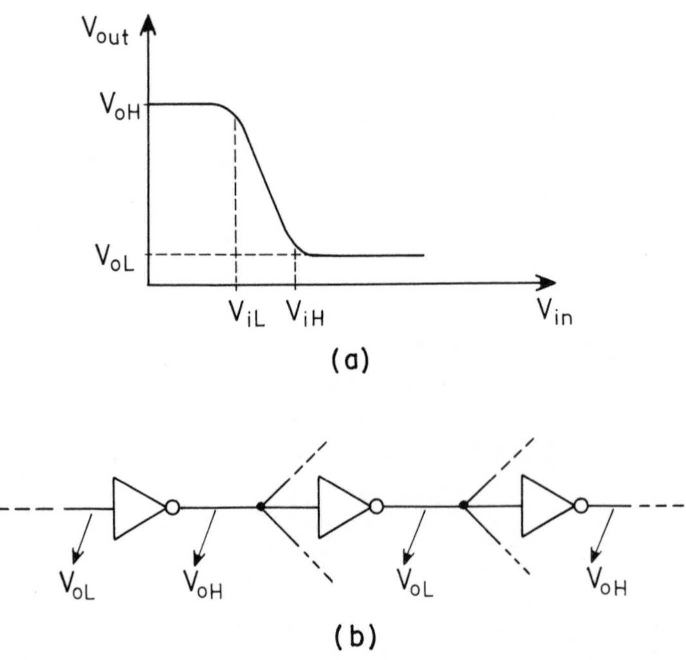

Figure 9. (a) Input–output characteristic of a NOT gate. (b) Cascade connection of NOT gates

of NOT gate. It gives prominence to four voltage levels: V_{oL}, V_{oH}, V_{iL} and V_{iH}. If V_{oL} and V_{oH} are considered as the nominal voltage levels corresponding to the logic levels 0 and 1, then some margin is allowed. Consider the cascade connection of several NOT gates (Figure 9b) and suppose that Figure 9a gives the input–output characteristic of a NOT gate whose output is connected to the inputs of a given number of similar gates. A necessary condition of correct working is

$$V_{oL} < V_{iL} \quad \text{and} \quad V_{oH} > V_{iH}. \tag{31}$$

The difference $V_{iL} - V_{oL}$ is the *noise margin for an input corresponding to logic level* 0 and the difference $V_{oH} - V_{iH}$ is the *noise margin for an input corresponding to logic level* 1.

(2) *Dynamic behaviour.* Figure 10b shows the waveform appearing at the output of a NOT gate in response to a change of voltage level on its input (Figure 10a). It is not possible to give an account of the continuous

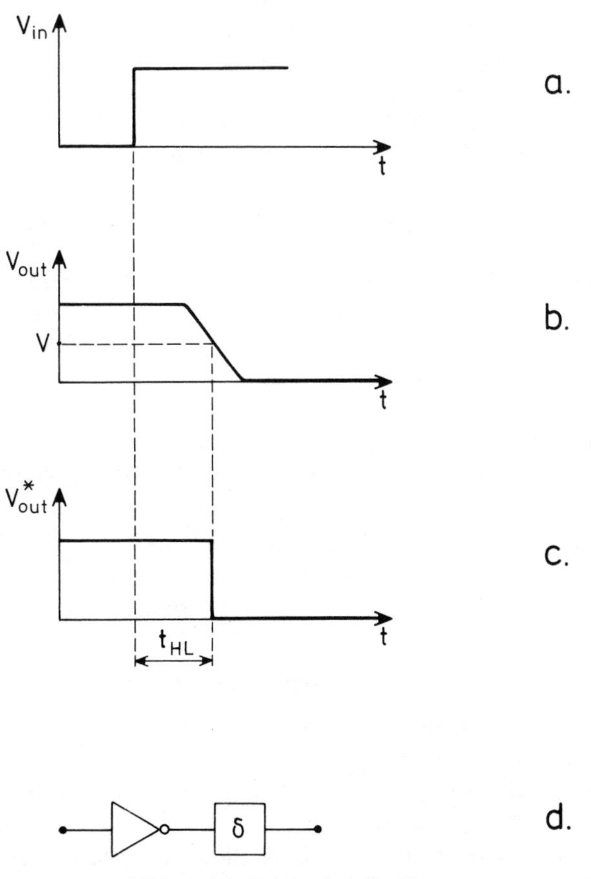

Figure 10. Dynamic behaviour

evolution depicted by Figure 10b in the frame of a logic model which is discrete by nature. Hence, in order to shorten the discussion we make the next simplifying assumption

$$V_{iL} = V_{iH} = V,$$

where V is the input voltage corresponding to the output voltage $(V_{oH} + V_{oL})/2$. One obtains the output waveform of Figure 10c. The NOT gate model is shown in Figure 10d. It includes an ideal NOT gate followed by a delay element δ. Notice that the delays can be different depending on the direction of the voltage level change: positive-going or negative-going transition.

It has been seen that the output of a given logic gate may be connected to the inputs of several similar gates. Nevertheless, there is a maximum number of gate inputs which can be driven by one gate output. For instance, if the output of the NOT gate of Figure 8a is connected to the inputs of several other NOT gates, and if the input voltage of the first NOT gate is 0(V), then the corresponding transistor is cut off. Nevertheless, its output voltage is not equal to E(V) since the voltage source supplies the input currents of the other gates and this produces a voltage drop across the collector resistor. The maximum number of input gates that a gate can drive is called its *fan-out.* Similarly, the number of inputs to a logic gate is often termed the *fan-in.* For instance the fan-in of the AND gate shown in Figure 8 is equal to 2.

To sum up, a logic gate is characterized by the following factors:

(1) The Boolean *function* it implements.

(2) The *cost* γ. Examples of cost measures are the weighted sum of the component costs, the power dissipation, and the silicon surface area.

(3) The *delay* δ.

(4) The *noise margin* (s).

(5) The *fan-out.*

(6) The *fan-in.*

Notice that if the cost measure is the power dissipation, then the *cost/performance ratio* is the delay time-power product:

$$\eta(\text{pJ}) = \gamma(\text{mW}) \cdot \delta(\text{ns}). \tag{32}$$

It is an energy.

4.3 Logic families

The diode circuits of Figure 8a have actually been used in the past but are now completely obsolete. They simply served as introductory examples.

Digital electronics makes use of several technologies which are called *logic families.* Let us describe some of them.

(1) *Diode–transistor–logic* (*DTL*). A diode–transistor–logic gate implements the NAND operation, that is the negation of the AND operation. As

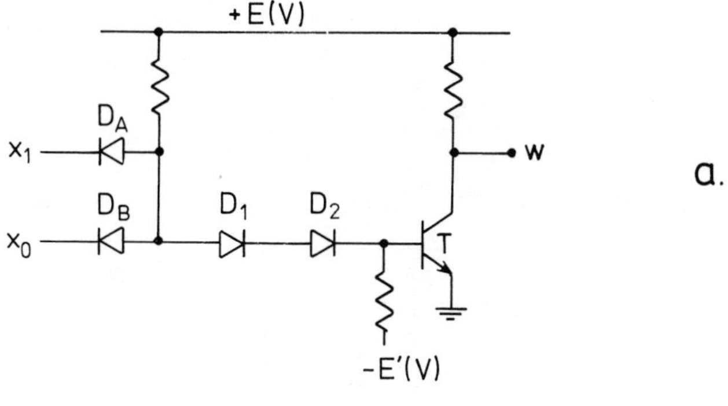

x_1 (V)	x_0 (V)	w (V)
0	0	E
0	E	E
E	0	E
E	E	0

b.

Figure 11. DTL gate

shown in Figure 11a, it consists of a diode AND gate (D_A, D_B), in cascade with a diode-level shifter (D_1, D_2) and a NOT gate (T).

(2) *Transistor–transistor–logic* (*TTL or* T^2L). A two-input TTL gate, implementing the NAND function, is shown in Figure 12. It is similar to the DTL gate of Figure 11. The base-emitter junctions of the input transistor T_1 are equivalent to the input diodes of the DTL gate. Similarly, the collector-base junction of T_1 performs the level shifting function. The advantages of TTL over DTL are that it requires less silicon surface area and is faster because the capacitances are smaller and the transistor action of T_1 provides active turn-off of T_2.

TTL circuits deserve a special mention because of their popularity: series 54, 74, 54L, 74L, 54S, and 74S, where L stands for low power and S for Schottky diode.

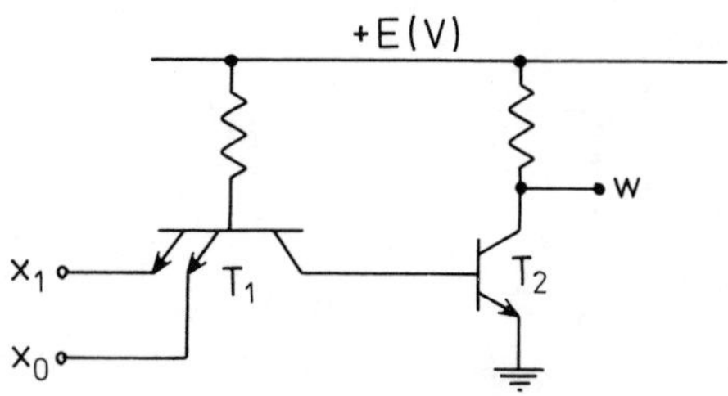

Figure 12. TTL gate

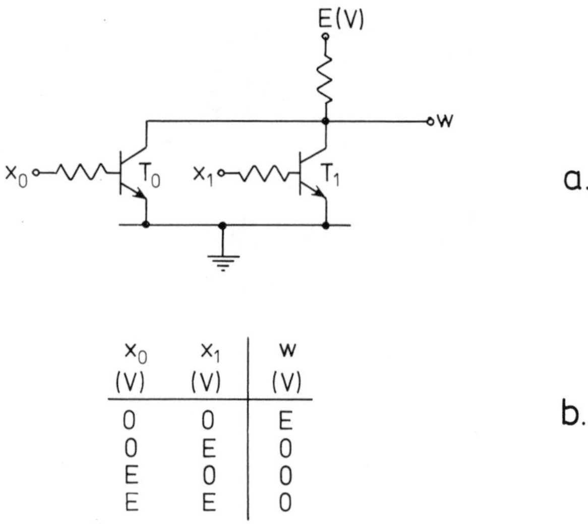

x_0 (V)	x_1 (V)	w (V)
0	0	E
0	E	0
E	0	0
E	E	0

Figure 13. RTL gate

(3) *Resistor–transistor–logic* (*RTL*). A resistor–transistor–logic gate implements the NOR operation, that is the negation of the OR operation. A two-input RTL gate is shown in Figure 13a. The output voltage is high if and only if both T_0 and T_1 are cut off, that is if and only if both input voltages are low.

The earliest gates of the type shown in Figure 13a were more economically constructed by omitting the base resistors. Such gates are called direct-coupled-transistor-logic gates (DCTL).

(4) *Integrated–Injection–Logic* (*IIL or I^2L*). Integrated-injection-logic also known as merged-transistor-logic (MTL), is closely related to direct-coupled-transistor-logic. An interconnection of DCTL gates is shown in Figure 14. Each gate consists of several transistors whose bases form the

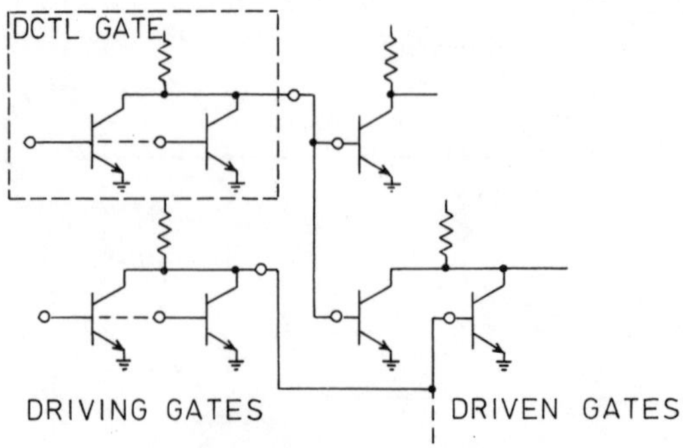

Figure 14. Interconnection of DCTL gates

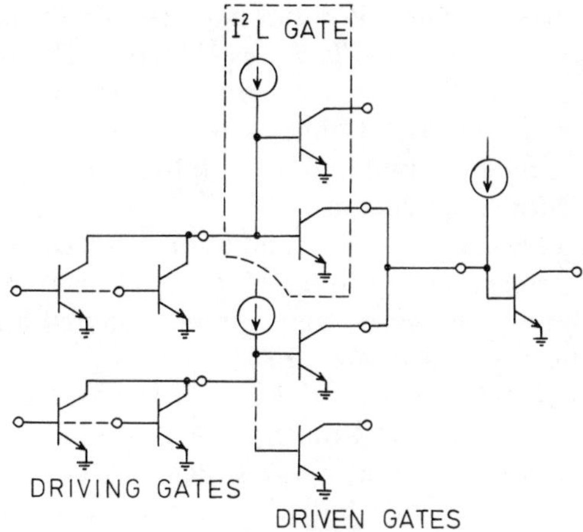

Figure 15. Circuit of Figure 14 redrawn to show $I^2$1 gate configuration

gate inputs and whose collectors are tied to form the gate output. Another way of looking at the interconnection of DCTL gates is that the collector resistor and the collector voltage supply form a current source which supplies sufficient base current to the driven input in order to saturate the transistor. The base current of the driven input can be diverted by saturating any of the transistors in the driving state. The circuit is redrawn in Figure 15 to emphasize this interpretation. The basic gate now consists of a single base, multiple collector transistor, with a current source in the base. It is shown in Figure 16a. The current source T_1 is a p–n–p transistor consisting of a p–n junction called the *injector* and the base of T_2. The n-type base of T_1 is the emitter of T_2; they are grounded. Furthermore the p-diffused region of the injector is shared by several gates.

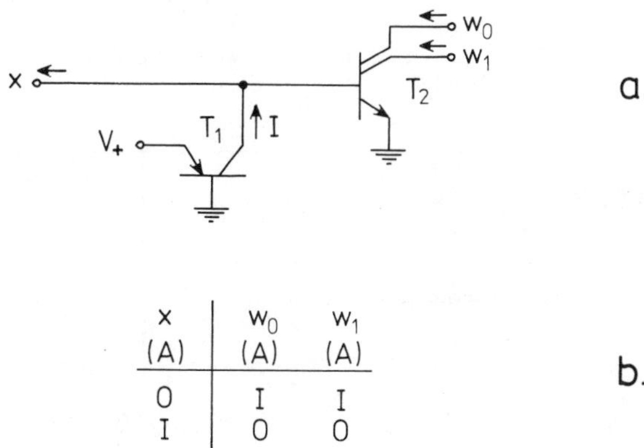

X (A)	W0 (A)	W1 (A)
0	I	I
I	0	0

Figure 16. Basic $I^2$1 gate

The working of the basic gate is better expressed in terms of current than in terms of voltage levels (*current logic*) as shown by Figure 16b in which I stands for the current supplied by the current source T_1.

I^2L circuits have a good delay time–power product and a high packing density. Most often they are buffered to look like TTL at the terminals and are thus only used for large circuits.

Notice that it is very easy to perform additions of currents: one simply has to connect the corresponding wires. Hence, I^2L is very convenient to implement threshold-logic gates and multiple-valued-logic gates (Dao, 1977; Dao, McCluskey, and Russel, 1977).

(5) *Emitter-Coupled-Logic* (*ECL*). An emitter-coupled-logic gate implements both the OR and the NOR function. A two-input ECL gate is shown in Figure 17. It includes a current mode switch (T_0, T_1, T_2) and two emitter followers (T_3, T_4) which provide buffering and low output impedance. The current mode switch works as follows: as long as all the input voltages are sufficiently negative with respect to the reference voltage, then the current I flows through T_2 and R_2. If one (or several) of the input voltages is sufficiently positive with respect to the reference voltage, then the current I flows through the corresponding transistor(s) and R_1. Typical values for the 10,000 series circuits are: high voltage = −0.9 V, low voltage = −1.75 V, and reference voltage = −1.29 V.

ECL is a very fast form of bipolar transistor logic because the transistors are never allowed to saturate. Consequently there is no excess charge to be removed from the base region before the turn-off process can begin.

(6) *MOS Logic*. The MOS field effect transistor is an attractive element for logic circuits for several reaons: MOS circuits have a high packing density and a low power consumption; their relatively simple processing results in high chip yields.

Figure 18 shows a NAND gate realized in nMOS technology. It consists of two n-channel enhancement mode transistors (T_0, T_1) and one n-channel

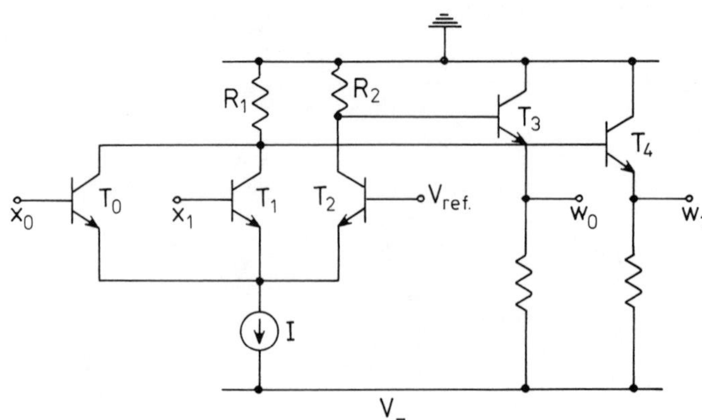

Figure 17. ECL gate

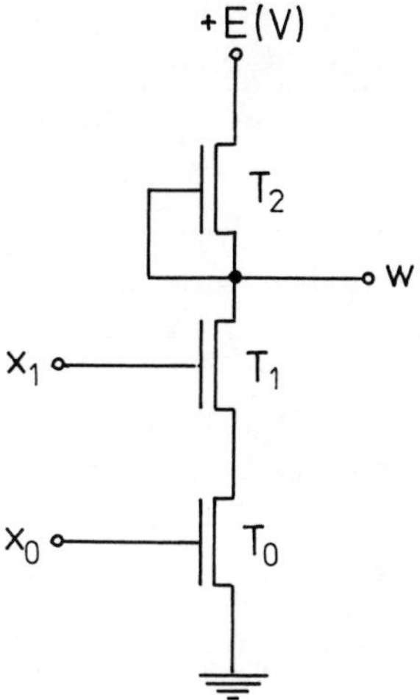

Figure 18. nMOS gate

depletion mode load transistor (T_2). Other combinations are possible: one may use p-channel transistors and one may use enhancement mode load transistors.

Figure 19 shows a NOR gate realized in CMOS technology. It includes two n-channel (T_0, T_1) and two p-channel (T_2, T_3) enhancement mode transistors. It is faster than a single channel MOS circuit and its static power consumption is very low since it is essentially determined by the leakage currents.

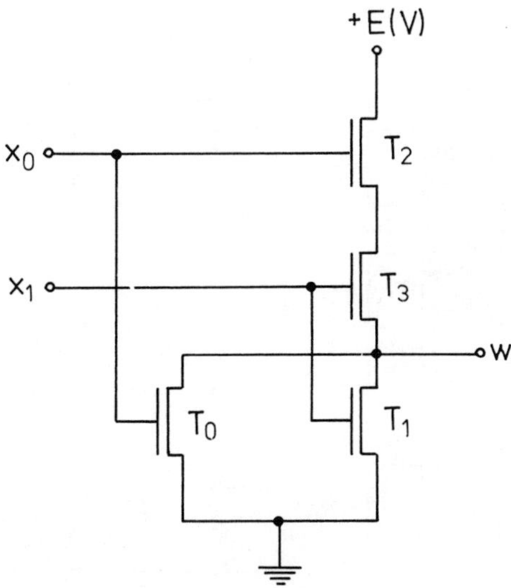

Figure 19. CMOS gate

Most of the material contained in this section may be found in the two following books: Glaser and Subak-Sharpe (1979) and Taub and Schilling (1977).

Comments.

(1) There are four main modern logic families, namely TTL, ECL, I^2L and MOS. DTL and RTL gates were presented because they help to understand the working of TTL and I^2L gates. Let us give some general characteristics of these logic families:

TTL is a medium speed, medium power dissipation, and medium cost technology. It is very popular and offers by far the largest number of

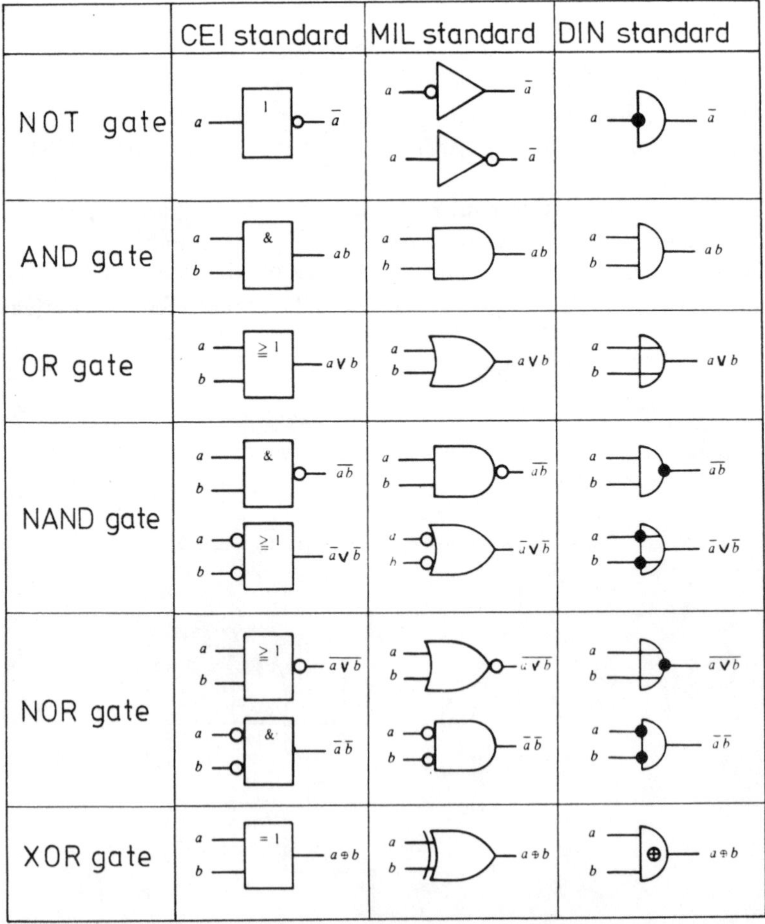

Figure 20. Standard representations of logic gates

standard SSI (*small-scale-integration* = a few gates) and MSI (*medium-scale-integration* = complex functions requiring up to about a hundred gates) circuits. TTL is also used in some LSI (*large-scale-integration* = memories, logic arrays, and processors) circuits.

ECL is a high speed, high power dissipation, and high cost technology. It offers the highest speed and is the ultimate choice for very fast systems.

MOS logic is a low speed, low power dissipation, and low cost technology. It has a high packing density. The defect densities for MOS processes are low, and the chip yields are relatively high. pMOS circuits are cheaper and slower than nMOS and CMOS circuits. CMOS circuits are faster than nMOS circuits, and have a very low power consumption.

I^2L is a medium speed, low power dissipation, and low cost technology. Among its advantages are a good delay time–power product (as low as that for CMOS), high packing density, and simple processing.

I^2L and MOS technologies are mainly used in LSI circuits.

In the future, it will be possible to fabricate chips containing some hundreds of thousands of transistors. These VLSI (*very-large-scale-integration*) circuits will be able to implement complete digital computer systems. In this respect, three technologies have emerged to date: nMOS silicon gate process, CMOS silicon gate process, and I^2L process.

(2) The description of the main logic families gave prominence to the AND, OR, NOT, NOR, and NAND functions. They can be easily implemented by electronic circuits, and this observation must guide us in the choice of the algebraic tools. Three sets of standard symbols are shown in Figure 20. The XOR gate implements the exclusive OR function, i.e. the modulo 2 sum (see Chapter II).

(3) In this section, logic gates are presented as an intermediary level between physical devices and systems. This intermediary step is not always essential for the design of digital systems. On the contrary, in some cases one can obtain more economical circuits by reasoning in terms of components rather than in terms of gates. Examples are CMOS circuits (Forbes, 1977), multiple-valued I^2L circuits (Davio and Deschamps, 1981), and Emitter Function Logic (Skokan, 1973). The component interconnection rules then determine the algebraic model. Another well-known example is given by the relay circuits.

(4) Several times we used the term multiple-valued logic, by which we understood the possibility to encode information by means of more than two symbols along with the feasibility of physical devices having more than two stable states. The main advantage of multiple-valued logic is the increase of information density. On the other hand, multiple-valued logic circuits probably have lower noise margins than binary circuits. Anyway, multiple-valued logic circuits often ought to be compatible with existing circuits and look like binary circuits at the terminals. Circuits relating binary and multiple-valued signals have been described by Etiemble (1979). Multiple-valued logic can also be a simple description tool for coping with

large discrete systems. For instance, ternary logic may help to design binary arithmetic circuits (Davio and Deschamps, 1977).

4.4 Example

Let us implement a binary full adder (see Figure 4) by means of three computation primitives: the AND function, the OR function, and the NOT function. The next computation scheme performs the full adder operations:

$$
\begin{aligned}
&N_0 = x_i, && N_6 = \mathrm{AND}(N_4, N_5),\\
&N_1 = y_i, && N_7 = \mathrm{AND}(N_2, N_6),\\
&N_2 = c_i, && N_8 = \mathrm{OR}(N_2, N_6),\\
&N_3 = \mathrm{AND}(N_0, N_1), && N_9 = \mathrm{NOT}(N_7),\\
&N_4 = \mathrm{OR}(N_0, N_1), && N_{10} = \mathrm{OR}(N_3, N_7),\\
&N_5 = \mathrm{NOT}(N_3), && N_{11} = \mathrm{AND}(N_8, N_9).
\end{aligned}
$$

To this computation scheme correspond the graph of Figure 21 and the circuit of Figure 22. Figure 23 shows the result of a step-by-step analysis of the circuit.

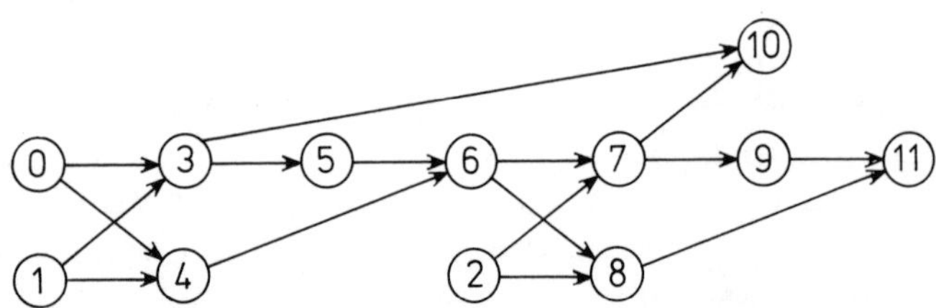

Figure 21. Full-adder: precedence graph

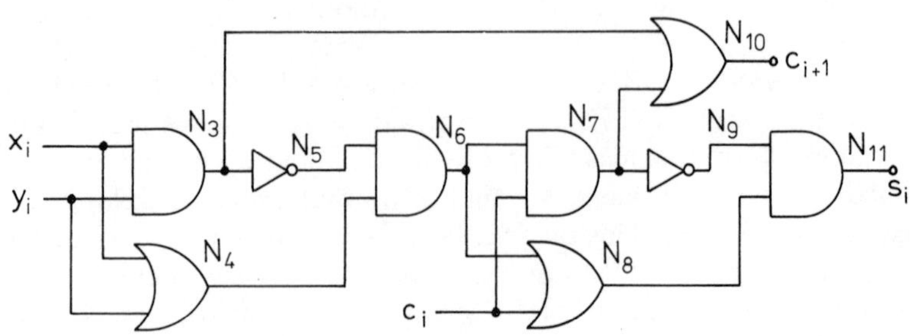

Figure 22. Full-adder

x_i	y_i	c_i	N_3	N_4	N_5	N_6	N_7	N_8	N_9	N_{10}	N_{11}
0	0	0	0	0	1	0	0	0	1	0	0
0	0	1	0	0	1	0	0	1	1	0	1
0	1	0	0	1	1	1	0	1	1	0	1
0	1	1	0	1	1	1	1	1	0	1	0
1	0	0	0	1	1	1	0	1	1	0	1
1	0	1	0	1	1	1	1	1	0	1	0
1	1	0	1	1	0	0	0	0	1	1	0
1	1	1	1	1	0	0	0	1	1	1	1

Figure 23. Circuit analysis

5 SUMMARY

This introductory chapter has presented the main concepts along with a short review of current digital circuit engineering. We saw that the input–output behaviour of digital systems can be described in an explicit way, by means of discrete functions or tables, or in an implicit way, by means of algorithms or computation schemes.

On the other hand, prominence was given to physical devices having a finite number of different stable states. They are used as building blocks to construct larger systems.

The synthesis problem may be stated as follows: design a digital circuit consisting of interconnected basic components, which implements a given input–output behaviour, and achieve a compromise solution between generally contradictory criteria such as the cost (component cost, development cost, testability, maintenability, and reliability) and the computation time (length of the precedence graph, and number of consecutive execution steps). Anyway, the solution of the problem is based on a decomposition in a series of steps, and the system will be hierarchically organized.

Though it has not been mentioned, the usefulness of storing devices is quite obvious: they are used for storing initial data and intermediary or final results. In this respect, the necessity of some kind of external control already appears (Figure 24).

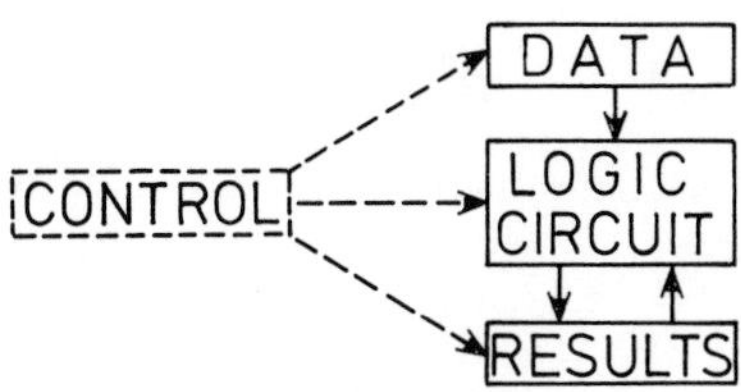

Figure 24. Basic organization

EXERCISES

E1. Extend the addition algorithm described in Section 3.2 to the case of mixed radix representations.

E2. Develop an algorithm for converting a number from a mixed radix representation to its radix B representation.

E3. *Signed digit representations* (Davio and Deschamps, 1978). Let B and K represent integers such that

$$B \geqslant 3 \quad \text{and} \quad 1 \leqslant K \leqslant B.$$

Prove the two following assertions:

(i) For any integer X, there exist unique quotient q and remainder r such that
  (a) $X = Bq + r$,
  (b) $-(B-K) \leqslant r \leqslant K-1$.

(ii) Any integer X such that

$$-(B-K)\frac{B^n-1}{B-1} \leqslant X \leqslant (K-1)\frac{B^n-1}{B-1}$$

has a unique representation $x_{n-1} \ldots x_1 x_0$ such that

(a) $X = \sum_{i=0}^{n-1} x_i B^i$,

(b) $\forall i: -(B-K) \leqslant x_i \leqslant (K-1)$.

The representation $x_{n-1} \ldots x_1 x_0$ obtained by proposition (ii) is called *signed digit representation of* X. Negative digits will be represented with an upper bar. Signed digit representations based on the integers B and K will be called (B, K)-representations.

Extend to (B, K)-systems the serial addition algorithm developed in Section 3.2. Devise the corresponding serial subtraction algorithm.

Convert a number from the $(B, K^{(1)})$-system to the $(B, K^{(2)})$-system.

Study the effect of a sign change on the (B, K) representation of a number.

E4. *Redundant signed-digit representations* (Aviziensis, 1961; Hwang, 1979). Let B represent the radix of a number representation. Assume that B is even (a similar principle would apply to odd radices). The set of allowable digit values is defined to be

$$S = \{-\alpha, \ldots, -1, 0, 1, \ldots, \alpha\}$$

with

$$\alpha = (B/2) + 1.$$

Prove that any number X such that

$$-\alpha\frac{B^n-1}{B-1} \leqslant X \leqslant \alpha\frac{B^n-1}{B-1}$$

has (at least) one representation $x^*_{n-1} \ldots x^*_1 x^*_0$ such that:

(i) $x^*_u \in S$,

(ii) $X = \sum_{i=0}^{n-1} x^*_i B^i$.

Suppose now that X is given by its radix B representation $x_{n-1} \ldots x_1 x_0$. Show that a redundant signed digit representation $x^*_{n-1} \ldots x^*_1 x^*_0$ of X is obtained by

computing the quantities b_i, d_i and x_i^* by:
(i) $b_0 = 0$;

$$b_{i+1} = 0 \quad \text{if} \quad x < \alpha,$$
$$b_{i+1} = 1 \quad \text{if} \quad x \geqslant \alpha;$$

(ii) $d_i = x_i - Bb_{i+1}$;
(iii) $x_i^* = d_i + b_i$.

Observe that this process is carried out without carry propagation. Imagine an algorithm for the back conversion process (known algorithms for this conversion all involve carry propagation). The most interesting feature of redundant signed digit representation is that they allow addition to be performed without carry propagation. To appreciate this, consider the usual equation of a radix B full-adder:

$$Br_i^* + z_i^* = x_i^* + y_i^* + r_{i-1}^*$$

and show that if:
(i) $-((B/2)+1) \leqslant x_i^*, y_i^* \leqslant (B/2)+1$,
(ii) $r_{i-1}^* \in \{-1, 0, 1\}$,
then it is possible to compute from the full adder equation a carry, r_i^*, and a result, z_i^*, satisfying the three following conditions:
(i) $-((B/2)+1) \leqslant z_i^* \leqslant (B/2)+1$,
(ii) $r_i^* \in \{-1, 0, 1\}$,
(iii) r_i^* is independent of r_{i-1}^*.
(*Hint:* split the full-adder equation into the pair of equations

$$Br_i^* + w_i = x_i^* + y_i^*,$$
$$z_i^* = w_1 + r_{i-1}^*,$$

and impose $-B/2 \leqslant w_i \leqslant B/2$).

E5. *Residue arithmetics.* Consider a radix vector $[b_{n-1}, \ldots, b_1, b_0]$ and assume that the b_j's are pairwise relatively prime (i.e.: g.c.d. $(b_i, b_j) = 1$). We accept from number theory that the latter condition implies the existence of integers $m_{i,j}$ such that

$$b_i m_{i,j} \equiv 1 (\text{mod } b_j).$$

The integer $m_{i,j}$ is the multiplicative inverse of b_i mod b_j. We assume that the multiplicative inverses $m_{i,j}$ are given once for all.

Now, it is shown in number theory that any integer X such that

$$0 \leqslant X \leqslant w_n - 1; \qquad w_n = \prod_{i=0}^{n-1} b_i$$

has a unique representation $\tilde{x}_{n-1}, \ldots, \tilde{x}_1, \tilde{x}_0$ such that

$$\tilde{x}_i = [\![X]\!]_{b_i}.$$

The proof of this result is known as the Chinese remainder theorem.

We study here an algorithm for passing from the residue representation $\tilde{x}_{n-1} \ldots \tilde{x}_1 \tilde{x}_0$ of X to its mixed radix representations $x_{n-1} \ldots x_1 x_0$. The algorithm consists of n steps. Step i may be described as follows:

Step i. Assume given
(i) The digits $x_{i-1} \ldots x_1 x_0$ of the suited mixed radix representation.

(ii) The residues $[\![q_{i-1}]\!]_{b_j}$ of the quotient q_{i-1} appearing in equation $(7, i-1)$:

$$q_{i-1} = b_{i-1}q_i + x_{i-1}$$

for $j \geqslant i$.

Compute now:

(i) $[\![q_ib_{i-1}]\!]_{b_j} = [\![q_{i-1} - x_{i-1}]\!]_{b_j}$; $j = i, i+1, \ldots, n-1$.

(ii) $[\![q_i]\!]_{b_j} = [\![[\![q_ib_{i-1}]\!]_{b_j}m_{i-1,j}]\!]_{b_j}$; $j = i, i+1, \ldots, n-1$.

(iii) $x_i = q_{i,i}$.

Show that the above algorithm, initialized by $x_0 = \tilde{x}_0$, actually provides the mixed radix representation of X. Illustrate it by a numerical example.

E6. Develop a radix B serial subtraction algorithm. Describe the corresponding combinational circuit.

Chapter II
Algebraic structures

1 INTRODUCTION

In order to describe, analyse, and design logic networks it is first necessary to become conversant with the underlying basic mathematics of the subject. In this respect Chapter I showed the importance of the concept of *discrete functions* and in particular of the concept of *Boolean functions.* These functions constitute the natural mathematical tool for describing the input–output behaviour both of the logical gates and of the computation primitives.

A first objective of this chapter is to state a set of explicit relations allowing us to describe the computation primitives in terms of logical gates. Other important applications of the subjects treated here will appear in the course of the following chapters. In addition, the mathematical concepts introduced in this chapter are currently used in operations research, graph theory, and more generally in any field connected with digital systems.

A second objective is to develop in an informal manner the most elementary (but also the most often used) properties of lattices and of Boolean algebras. In this respect this chapter is by no means a complete treatment of these subjects, but rather a survey of the results to the extent that they bear upon material developed in later chapters.

Lattices, Boolean algebras and rings are introduced in Section 2; Section 3 introduces the logic and Boolean functions and their well-formed expressions; Section 4 introduces the optimization problems that will more extensively be considered in the following chapters; finally Section 5 introduces the theory of complexity of logic functions and of their realizations.

2 LATTICES AND BOOLEAN ALGEBRA

2.1 Lattices

Lattices play an important role in characterizing various computation models. In particular, it will be shown later on that a Boolean algebra is nothing but a lattice with a few specific properties.

Definition 1. A *lattice* is an algebraic structure

$$\langle L, \vee, \wedge \rangle$$

where L is a set (the *carrier* of the lattice) and in which two binary operations:

$\vee$ called *disjunction* (or sum),

$\wedge$ called *conjunction* (or product)

are defined by the set of axioms:

$$\forall a, b, c \in L:$$

(L_1): $a \vee a = a$;	$a \wedge a = a$	(idempotency),
(L_2): $a \vee (b \vee c) = (a \vee b) \vee c$;	$a \wedge (b \wedge c) = (a \wedge b) \wedge c$	(associativity),
(L_3): $a \vee b = b \vee a$;	$a \wedge b = b \wedge a$	(commutativity),
(L_4): $a \vee (a \wedge b) = a$;	$a \wedge (a \vee b) = a$	(absorption).

Note that the set:

$$Z_m = \{0, 1, \ldots, m-1\}$$

with the operations

$$a \vee b = \max\{a, b\},$$
$$a \wedge b = \min\{a, b\},$$

is a lattice. In particular, $Z_2 = \{0, 1\}$, with the operations AND (conjunction) and OR (disjunction), is a lattice.

Note that, roughly speaking, the laws $\vee$ and $\wedge$ have the same properties as the set theoretical union and intersection, respectively.

Lemma 1. *In any lattice* $\langle L, \vee, \wedge \rangle$

$$a \vee b = b \Leftrightarrow a \wedge b = a \tag{1}$$

Proof. If $a \vee b = b$, (L_4) allows us to write:

$$a \wedge (a \vee b) = a \wedge b = a.$$

A similar proof holds for stating the reciprocal. □

An essential property that we deduce from the axioms $(L_1)-(L_4)$ is that they allow us to define two partial ordering relations on L. Let us recall (see e.g. Grätzer, 1971, p. 1) that a *partial ordering relation* R on a set L is a relation which is at the same time reflexive, antisymmetric, and transitive; $\forall a, b, c \in L$ we have:

(P_1) aRa	(reflexibility),
(P_2) aRb and bRa imply $a = b$	(antisymmetry),
(P_3) aRb and bRc imply aRc	(transitivity).

Theorem 1. *In any lattice $\langle L, \vee, \wedge\rangle$, the converse relations*

$$\leqslant : \{(a, b) \mid a \vee b = b\} \quad \text{and} \quad \geqslant : \{(a, b) \mid a \vee b = a\}$$

are partial ordering relations; moreover:

$$\forall c, d \in L : c \vee d = \textit{supremum}\ (c, d);\ c \wedge d = \textit{infimum}\ (c, d).$$

Proof. Relation $\leqslant$ is reflexive in view of (L_1) and antisymmetric in view of (L_3). It is, moreover, transitive:

$$a \leqslant b \text{ and } b \leqslant c \Rightarrow a \vee b = b \text{ and } b \vee c = c$$
$$\Rightarrow a \vee c = a \vee (b \vee c) = (a \vee b) \vee c = b \vee c = c \Rightarrow a \leqslant c.$$

We then deduce that $c \leqslant (c \vee d)$; indeed, $c \vee (c \vee d) = (c \vee c) \vee d = c \vee d$ (in view of (L_1) and (L_3)). The disjunction $c \vee d$ is larger than c and d. Let c' be an upper bound for $\{c, d\} : c \leqslant c'$ and $d \leqslant c' \Rightarrow c \wedge c' = c'$ and $d \vee c' = c' \Rightarrow (c \vee d) \vee c' = c'$ (in view of (L_1) and (L_3)) $\Rightarrow c \vee d \leqslant c'$. The property $c \wedge d =$ infimum $\{c, d\}$ is proved in a similar way. □

Theorem 2. *In any lattice, the disjunction and the conjunction are isotone, i.e.:*

$$a \leqslant b \Rightarrow a \vee c \leqslant b \vee c \quad \textit{and} \quad a \wedge c \leqslant b \wedge c, \quad \forall a, b, c \in L.$$

Proof. It suffices to write the following series of implications:

$$a \leqslant b \Rightarrow a \vee b = b \Rightarrow (a \vee b) \vee c = b \vee c \Rightarrow (a \vee b) \vee (c \vee c)$$
$$= b \vee c \Rightarrow (a \vee c) \vee (b \vee c) = b \vee c \Rightarrow a \vee c \leqslant b \vee c.$$

The second proposition is proved in a similar way. □

Corollary. *In any lattice the partial ordering relation $\leqslant$ is compatible with the disjunction and the conjunction, i.e.:*

$$a \leqslant b \quad \textit{and} \quad c \leqslant d \Rightarrow a \vee c \leqslant b \vee d \quad \textit{and} \quad a \wedge c \leqslant b \wedge d, \quad \forall a, b, c, d \in L.$$

Proof. From the preceding Theorem one deduces that

$$a \leqslant b \Rightarrow a \vee c \leqslant b \vee c$$

and

$$c \leqslant d \Rightarrow b \vee c \leqslant b \vee d;$$

therefore, by transitivity one obtains: $a \vee c \leqslant b \vee d$.

The second proposition is proved in a similar way. □

The existence of a partial ordering relation $\leqslant$ (as defined by Theorem 1) allows us to adopt the corresponding terminology: smaller than, included in, etc. (Grätzer, 1971, p. 2,3). Theorem 1 implicitly contains the statement of the *duality principle* (Grätzer, 1971, p. 7; Rudeanu, 1974, p. 5). This principle may be stated as follows:

Let Φ be a statement about lattices expressed in terms of: $\wedge$ and $\vee$. The dual of Φ is the statement we get from Φ by interchanging: $\wedge$ and $\vee$. If Φ is true for all lattices, then the dual of Φ is also true for all lattices.

This duality principle is illustrated by the symmetry of axioms (L_1)–(L_4) and of Lemma 1.

Below we shall restrict ourselves to *finite lattices.* In this case we see that any lattice has a *maximum element* (or *upper bound*) usually denoted $\mathbf{1}$:

$$\mathbf{1} = \bigvee_{x \in L} x,$$

and a *minimum element* (or *lower bound*) usually denoted $\mathbf{0}$:

$$\mathbf{0} = \bigwedge_{x \in L} x.$$

Definition 2. In a lattice $\langle L, \vee, \wedge \rangle$ with maximum and minimum elements $\mathbf{1}$ and $\mathbf{0}$ respectively, a *complement* of an element $a \in L$ is any element $b \in L$ such that

$$a \wedge b = \mathbf{0} \quad \text{and} \quad a \vee b = \mathbf{1}. \tag{2}$$

By definition of maximum element:

$$a \wedge \mathbf{1} = a.$$

It follows that the only x which satisfies $\mathbf{1} \wedge x = \mathbf{0}$ is $x = \mathbf{0}$ itself. Since

$$\mathbf{1} \vee \mathbf{0} = \mathbf{1},$$

we conclude that $\mathbf{0}$ is complement of $\mathbf{1}$. By a dual argument we equally show that $\mathbf{1}$ is a complement of $\mathbf{0}$. Note that Definition 2 specifies *a* complement, not *the* complement thereby allowing for more than one. Indeed, it is possible for a lattice element to have more than one complement satisfying relations (2). Of course there are lattices where any element has not necessarily a complement.

Any lattice L can be represented by a *Hasse diagram* in which distinct elements in L are represented by distinct vertices. If $a \leqslant b$, the vertex representing b is drawn higher than the vertex representing a. Moreover, if there does not exist an element $c : a < c < b$, an edge is drawn between a and b.

Example 1. The lattice of all subsets of the set $S \equiv \{a, b, c\}$, under the ordering relation of set inclusion, is shown in the Hasse diagram of Figure 1, where $\{a, b, c\} = \mathbf{1}$ and where the empty set $\emptyset = \mathbf{0}$.

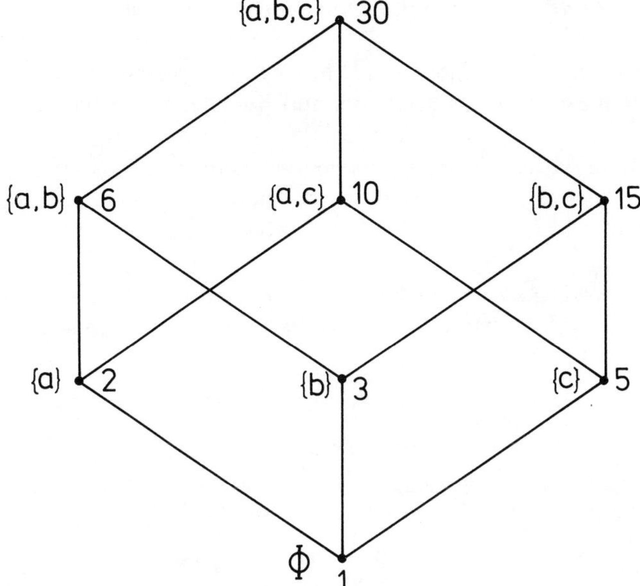

Figure 1. Hasse diagram

Example 2. Consider the set $S \equiv \{1, 2, 3, 5, 6, 10, 15, 30\}$, under the ordering relation of divisibility. We verify that S under this ordering relation is a lattice whose Hasse diagram is also given by the same Figure 1, where $30 = \mathbf{1}$ and $1 = \mathbf{0}$. Note also that the Hasse diagram of Figure 1 is the lattice L of the divisors of 30 whose ordering is the divisibility. In this respect note that if the law $\wedge$ is synonymous to greatest common divisor, then the dual law $\vee$ is synonymous with least common multiple.

2.2 Distributive and complemented lattices

Definition 3. A lattice is *distributive* if and only if (L_5) (and hence (L_5') by elementary manipulation) holds

(L_5): $a \wedge (b \vee c) = (a \wedge b) \vee (a \wedge c)$,

(L_5'): $a \vee (b \wedge c) = (a \vee b) \wedge (a \vee c)$.

Theorem 3. *In a distributive lattice, if*

$$a \wedge x = a \wedge y \quad \textit{and} \quad a \vee x = a \vee y, \quad \textit{then } x = y$$

Proof. Using repeatedly the hypotheses (L_4), (L_3) and (L_5), we have:

$$\begin{aligned} x &= x \wedge (a \vee x) = x \wedge (a \vee y) = (x \wedge a) \vee (x \wedge y) \\ &= (a \wedge y) \vee (x \wedge y) = (a \vee x) \wedge y = (a \vee y) \wedge y = y. \end{aligned}$$

□

Theorem 4. *In a distributive lattice complements are unique when they exist.*

Proof. Assume that an element $a \in L$ has two complements x_1 and x_2 in L. Then by Definition 2 of complements and by Theorem 3 we have $x_1 = x_2$.

Definition 4. A lattice is called *complemented* if all its elements have complements. Henceforth we shall use the notation $\bar{a}$ to denote the *unique* complement of an element a of a distributive lattice.

Theorem 5 (De Morgan's identities). *In a distributive lattice, if a and b have complements $\bar{a}$ and $\bar{b}$, respectively, then $a \wedge b$ and $a \vee b$ have complements $\overline{(a \wedge b)}$ and $\overline{(a \vee b)}$, respectively, and*

$$\overline{(a \wedge b)} = \bar{a} \vee \bar{b};$$
$$\overline{(a \vee b)} = \bar{a} \wedge \bar{b}. \tag{3}$$

Proof. By Theorem 4 it suffices to prove that

$$(a \wedge b) \wedge (\bar{a} \vee \bar{b}) = \mathbf{0}, \qquad (a \wedge b) \vee (\bar{a} \vee \bar{b}) = \mathbf{1},$$

to verify the first identity (3); the second is dual.

Compute:

$$(a \wedge b) \wedge (\bar{a} \vee \bar{b}) = (a \wedge b \wedge \bar{a}) \vee (a \wedge b \wedge \bar{b})$$
$$= \mathbf{0} \vee \mathbf{0} = \mathbf{0}$$

and

$$(a \wedge b) \vee (\bar{a} \vee \bar{b}) = (a \vee \bar{a} \vee \bar{b}) \wedge (b \vee \bar{a} \vee \bar{b})$$
$$= \mathbf{1} \vee \mathbf{1} = \mathbf{1}.$$

□

Theorem 6. (Involution property). *In a distributive lattice, if $a \in L$ has a complement $\bar{a}$, then $\bar{a}$ also has a complement and*

$$\overline{(\bar{a})} = a$$

Proof. By definition of a complement, $a \vee \bar{a} = \mathbf{1}$ and $a \wedge \bar{a} = \mathbf{0}$. To determine (if it exists) the complement of $\bar{a}$, we must find an element $x \in L$ which simultaneously satisfies

$$\bar{a} \vee x = \mathbf{1} \quad \text{and} \quad \bar{a} \wedge x = \mathbf{0}.$$

Since a is one such element and since complements are unique, a is the complement of $\bar{a}$. □

Example 3. Consider the set $Z_m = \{0, 1, \ldots, m-1\}$ with the usual ordering relation on the set of integers. This set is clearly a distributive lattice for the binary operations of max ($\vee$) and of min ($\wedge$); the Hasse diagram of this lattice is a chain. We verify (by means of conditions (2)) that except for

$m=2$, the lattice Z_m is not complemented. We may however define a unary operation '*', called pseudo-complementation, as follows:

$$x^* = (m-1) - x, \quad \forall x \in Z_m \tag{4}$$

(the operation '−' has here the arithmetic subtraction meaning). We verify that this pseudo-complementation operation satisfies the De Morgan's identities, i.e.:

Theorem 7.

$$\left(\bigvee_{i=0}^{r-1} a_i\right)^* = \bigwedge_{i=0}^{r-1} a_i^*, \tag{5}$$

$$\left(\bigwedge_{i=0}^{r-1} a_i\right)^* = \bigvee_{i=0}^{r-1} a_i^* \tag{6}$$

Proof. Let, for example, $a_j = \max\{a_0, a_1, \ldots, a_{r-1}\}$; we have then:

$$\left(\bigvee_{i=0}^{r-1} a_i\right)^* = (m-1) - \left(\bigvee_{i=0}^{r-1} a_i\right) = (m-1) - a_j = a_j^*$$

$$= \bigwedge_{i=0}^{r-1} a_i^* \text{ since } a_j^* \leqslant a_i^* \forall i.$$

A similar statement holds for proving (6). □

Example 4. The nth Cartesian power $Z_m^n = (Z_m \times Z_m \times \ldots \times Z_m)$ of Z_m is a distributive lattice with a pseudo-complementation operation if the disjunction, conjunction, and pseudo-complementation are defined componentwise, i.e.:

$$[a_0, \ldots, a_i, \ldots, a_{n-1}] \vee [b_0, \ldots, b_i, \ldots, b_{n-1}]$$
$$= [a_0 \vee b_0, \ldots, a_i \vee b_i, \ldots, a_{n-1} \vee b_{n-1}]$$
$$[a_0, \ldots, a_i, \ldots, a_{n-1}] \wedge [b_0, \ldots, b_i, \ldots, b_{n-1}]$$
$$= [a_0 \wedge b_0, \ldots, a_i \wedge b_i, \ldots, a_{n-1} \wedge b_{n-1}]$$
$$[a_0, \ldots, a_i, \ldots, a_{n-1}]^* = [a_0^*, \ldots, a_i^*, \ldots, a_{n-1}^*], \quad a_i, b_i \in Z_m.$$

Example 5. Let m be a positive integer; it is shown that the set L of the divisors of the integer m is a lattice if we take '∨' as the least common multiple law and '∧' as the greatest common divisor law (see also Example 2), the ordering relation being the divisibility. This lattice is known to be distributive. Consider, for example, the lattice L of the divisors of 60; the Hasse diagram of this lattice is given in Figure 2. The integers $1, 3, 4, 5, 12, 15, 20$, and 60 have a complement. For example: $20 = \bar{3}$ since l.c.m. $(20, 3) = 60$ and g.c.d. $(20, 3) = 1$. The integers $2, 6, 10$, and 30 have no complement. For example, l.c.m. $(2, x) = 60$, considered as an equation in x, has the solutions $x = 30$ and $x = 60$ whereas g.c.d. $(2, x) = 1$ has the

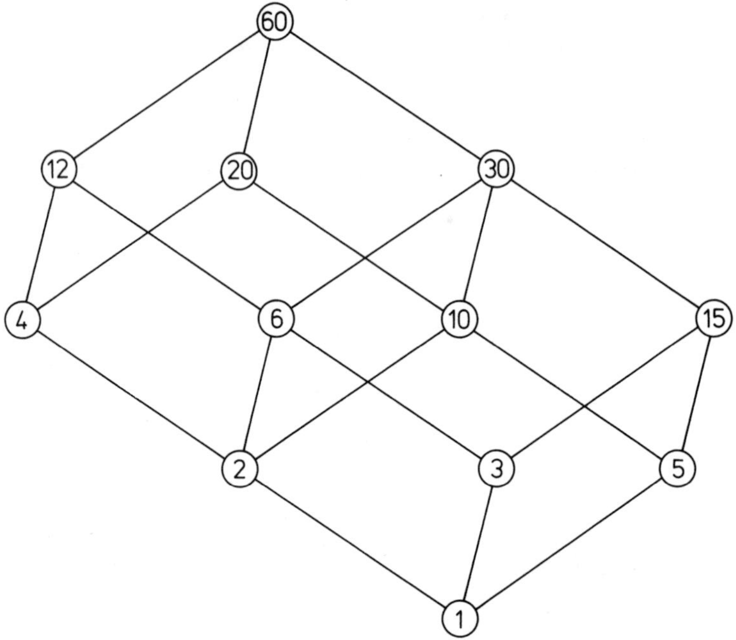

Figure 2. Hasse diagram of the divisors of 60

solutions $x=3$, $x=5$, $x=15$ and $x=1$, which proves the nonexistence of the complement of 2. The lattice L of the divisors of 60 is thus a distributive but non-complemented lattice. The lattice L of the divisors of 30 is a distributive and complemented lattice; we have indeed:

$$\bar{1}=30, \qquad \bar{2}=15, \qquad \bar{3}=10, \qquad \bar{5}=6.$$

2.3 Boolean algebras

Definition 5. A Boolean algebra (from Boole (1847)) is a distributive and complemented lattice. In a Boolean algebra we shall sometimes use the terms '*Boolean sum*' and '*Boolean product*' instead of disjunction and conjunction respectively. Furthermore, in some cases, the symbol $\wedge$ will either be replaced by the usual multiplicative notation or dropped.

In particular, $Z_2=\{0, 1\}$ is a Boolean algebra for the operations OR($\vee$), AND($\cdot$) and NOT($^-$) (see also Section I.4.2). Similarly Z_2^m is a Boolean algebra if one uses as operations the componentwise extensions of $\vee$, $\cdot$ and $^-$. A Boolean algebra is represented as an algebraic structure

$$\langle B, \vee, \mathbf{0}, \cdot, \mathbf{1}\rangle,$$

where B is the carrier set of the algebra and where each of the operations $\vee$ and $\cdot$ is followed by its neutral element. (Remember that neutral elements with respect to the operations $\vee$ and $\cdot$ satisfy the relations:

$$\forall a \in B: \; a \vee \mathbf{0} = a$$
$$a \cdot \mathbf{1} = a.)$$

As a matter of fact, it may be shown that any finite Boolean algebra is isomorphic to those algebras Z_2^m. We may thus restrict ourselves to those Boolean algebras without loss of generality.

It is now useful to summarize synoptically the formal properties of Boolean algebras. First of all, a Boolean algebra, being a lattice, satisfies the four axioms (L_1)–(L_4) of a lattice, i.e. idempotency, commutativity, associativity and absorption. Second, since a Boolean algebra is a distributive lattice, we have the two distributivity laws, i.e.:

$$a(b \vee c) = ab \vee ac, \; a \vee bc = (a \vee b)(a \vee c) \quad \text{(distributivity)}$$

Finally, since a Boolean algebra is a complemented lattice with lower bound $\mathbf{0}$ and upper bound $\mathbf{1}$, respectively, we have:

$$a \vee \mathbf{0} = a, \qquad a \cdot \mathbf{0} = \mathbf{0}, \qquad a \vee \mathbf{1} = \mathbf{1}, \qquad a \cdot \mathbf{1} = a.$$
$$\bar{\mathbf{1}} = \mathbf{0}$$
$$a \vee \bar{a} = \mathbf{1}, \qquad a\bar{a} = \mathbf{0} \qquad \text{(complementarity)}$$
$$\overline{(\bar{a})} = a \qquad \text{(involution)}$$
$$\overline{a \vee b} = \bar{a}\bar{b}, \qquad \overline{ab} = \bar{a} \vee \bar{b} \quad \text{(De Morgan's laws)}$$

2.4 Rings

Definition 6. A *ring* is a set R on which are defined two binary operations, denoted $+$(addition) and $\cdot$ (multiplication), such that the following properties are satisfied:

(a) R is a *commutative group* with respect to the addition, i.e.
  - (i) $a + (b + c) = (a + b) + c$ (associativity).
  - (ii) $a + b = b + a$ (commutativity).
  - (iii) R contains a unique *neutral element*, denoted $\mathbf{0}$, such that $\forall a \in R$: $a + \mathbf{0} = \mathbf{0} + a = a$.
  - (iv) To each a in R there corresponds a unique element $-a$ in R such that $a + (-a) = (-a) + a = \mathbf{0}$; $-a$ is called the *inverse* of a.

(b) R is a *semigroup* with respect to the multiplication, i.e.: $(a \cdot b) \cdot c = a \cdot (b \cdot c)$ (associativity).

(c) Multiplication distributes over addition, i.e.

$$a(b + c) = ab + ac \text{ and } (b + c)a = ba + ca.$$

If multiplication is also commutative, i.e. if $a \cdot b = b \cdot a$, R is a commutative ring.

Example 6. The set of integers $Z_m = \{0, 1, \ldots, m-1\}$ under the modulo m addition and multiplication operations forms a commutative ring (remember that 'modulo m' means $a \equiv b$ iff $a - b = kp$, $k \in Z$). The modulo m addition and multiplication are denoted '$\oplus$' and '$\cdot$' respectively.

The nth Cartesian power Z_m^n of Z_m is a commutative ring if the modulo m addition and multiplication are defined componentwise.

It is possible to state a one-to-one correspondence between the structures of the ring and of the lattice that we may confer to Z_m. Further on we shall restrict ourselves to the case $Z_2 = \{0, 1\}$ and to its extension Z_2^m.

Definition 7. A *Boolean ring* with neutral element is an algebraic structure

$$\langle B, \oplus, \mathbf{0}, \ldots, \mathbf{1} \rangle$$

with an additive law $\oplus$ with neutral element $\mathbf{0}$, a multiplicative law '$\cdot$' with neutral element $\mathbf{1}$, the usual axioms of the rings, and the idempotence law for multiplication, i.e.

$$a \cdot a = a.$$

Theorem 8. *A Boolean ring with neutral element satisfies the following properties*:

$$\forall a, b \in B : a \oplus a = \mathbf{0}, \tag{7}$$

$$a \oplus b = \mathbf{0} \Leftrightarrow a = b, \tag{8}$$

$$ab = ba. \tag{9}$$

Proof. The idempotence of the multiplication allows us to write:

$$(\mathbf{1} \oplus a)(\mathbf{1} \oplus a) = (\mathbf{1} \oplus a),$$

from which we deduce by distributivity:

$$(\mathbf{1} \oplus a) \oplus (a \oplus a) = \mathbf{1} \oplus a.$$

Relation (7) then follows by adding to the two members of this equality the inverse of $(\mathbf{1} \oplus a)$, i.e. $-(\mathbf{1} \oplus a)$.

From (7) we deduce:

$$a = b \Rightarrow a \oplus a = a \oplus \ = \mathbf{0}.$$

From $a \oplus b = 0$ we deduce $a = b$ by adding to $a \oplus b$ and to $\mathbf{0}$ the inverse of a; this proves (8).

Finally, starting from

$$(a \oplus b)(a \oplus b) = a \oplus b,$$

we obtain by distributivity:

$$a \oplus b \oplus ab \oplus ba = a \oplus b,$$

and hence $ab = ba$.

The central property is the one to one correspondence between a Boolean algebra and a Boolean ring.

Theorem 9 (Stone, 1935). *The Boolean algebra*

$$\langle B, \vee, \mathbf{0}, \wedge, \mathbf{1} \rangle \tag{10}$$

is a Boolean ring

$$\langle B, \oplus, \mathbf{0}, \cdot, \mathbf{1} \rangle \tag{11}$$

if we define the additive law '⊕' and the multiplicative law '·' as follows:

$$a \oplus b = (a \wedge \bar{b}) \vee (\bar{a} \wedge b),$$
$$a \cdot b = a \wedge b. \tag{12}$$

Conversely, the Boolean ring (11) *is a Boolean algebra* (10) *if we define the disjunction* $\vee$, *the conjunction* $\wedge$, *and the complementation '¯' as follows:*

$$a \vee b = a \oplus b \oplus ab,$$
$$a \wedge b = a \cdot b, \tag{13}$$
$$\bar{a} = \mathbf{1} \oplus a.$$

In both cases, the neutral elements **0** *and* **1** *of the algebra and of the ring coincide.*

Proof. The proof of the theorem reduces to a routine verification showing that the Boolean algebra (10) together with the relations (13) satisfies the axioms of the rings and that the Boolean ring (11) together with the relations (12) satisfies the axioms of the Boolean algebras. A detailed verification is left to the reader. □

Definition 8. The set F is said to be a *field* if it is a commutative ring and, in addition, satisfies the following two axioms:

(F_1) There is a unique non-zero element **1** in F such that $a \cdot \mathbf{1} = a$ for every a in .

(F_2) To each non-zero a in F corresponds a unique element a^{-1} (or $1/a$) in F such that a $a^{-1} = \mathbf{1}$.

Fields containing a finite number of elements are usually called *finite fields.*

It can be shown that if p is a prime integer, then the ring of integers, modulo p, forms a field. This finite field is called *Galois field* and is denoted GF(p). Note finally that the Boolean algebra $B_2 = \{0, 1\}$ structured in a Boolean ring according to the laws (12) is the Galois field GF(2); in this important particular case, the addition $\oplus$ is the modulo-2 sum or EXCLUSIVE-OR operation.

Consider again the Boolean algebra $\{0, 1\}^n$; we call the *weight* of the

n-tuple:

$$\mathbf{a}=(a_0, a_1, \ldots, a_{n-1}), \qquad a_i \in \{0, 1\}, \quad \forall i,$$

the number of components a_i of $\mathbf{a}$ equal to 1, i.e.

$$\text{weight of } \mathbf{a} = w(\mathbf{a}) = \sum_i a_i.$$

We verify that $\forall \mathbf{a}, \mathbf{b} \in \{0, 1\}^n$:

$$w(\mathbf{a}) + w(\mathbf{b}) = w(\mathbf{a} \vee \mathbf{b}) + w(\mathbf{a} \wedge \mathbf{b}). \tag{14}$$

The *Hamming distance* $d(\mathbf{a}, \mathbf{b})$ between the two elements $\mathbf{a}$, $\mathbf{b}$ of $\{0, 1\}^n$ can be defined as follows:

$$d(\mathbf{a}, \mathbf{b}) = w(\mathbf{a} \oplus \mathbf{b}) = w(\mathbf{a} \vee \mathbf{b}) - w(\mathbf{a} \wedge \mathbf{b}). \tag{15}$$

The distance is a non-negative integer such that:

$$d(\mathbf{a}, \mathbf{b}) = d(\mathbf{b}, \mathbf{a}) \qquad \text{(symmetry)}, \tag{16}$$

$$d(\mathbf{a}, \mathbf{b}) = 0 \Leftrightarrow \mathbf{a} = \mathbf{b} \qquad \text{(reflexivity)}, \tag{17}$$

$$d(\mathbf{a}, \mathbf{c}) \leqslant d(\mathbf{a}, \mathbf{b}) + d(\mathbf{b}, \mathbf{c}) \qquad \text{(triangular inequality)}. \tag{18}$$

3 LOGIC FUNCTIONS AND THEIR REPRESENTATION

3.1 Logic and Boolean functions

Definition 9. A *logic function* is a mapping:

$$f: \{0, 1, \ldots, m-1\}^n \to \{0, 1, \ldots, m-1\}.$$

For $m = 2$, a logic function is a *Boolean function*, i.e. a mapping:

$$f: \{0, 1\}^n \to \{0, 1\}.$$

The value of the function f at the vertex $\mathbf{x} = (x_{n-1}, \ldots, x_1, x_0) \in Z_m^n$ $(x_i \in Z_m \forall i)$ is denoted $f(x_{n-1}, \ldots, x_1, x_0)$ or $f(\mathbf{x})$.

To any vertex $\mathbf{x}$ we associate the integer X:

$$X = \sum_{i=0}^{m-1} x_i m^i,$$

which corresponds to the vector $\mathbf{x}$ written in the system of base m. The local value $f(\mathbf{x})$ will also be denoted f_X; the function f is completely described by the corresponding set of values, i.e.

$$[f_0, f_1, \ldots, f_i, \ldots, f_{(m^n-1)}].$$

This expression is often referred to as the *value vector* of f; it is a vector of m^n components, each component taking its value in Z_m. This allows us to confer to the set of logic functions

$$\{f: Z_m^n \to Z_m\}$$

	2	2	2	2	0	0	0	2	2	2
x_1	1	2	2	2	2	1	0	2	2	2
	0	2	1	0	2	1	0	2	1	0
		0	1	2	0	1	2	0	1	2	x_0
		0	0	0	1	1	1	2	2	2	x_2

Figure 3. Logic function

both the structure of lattice and of ring: one has to extend componentwise the definition of the lattice operations and of the ring operations. In this respect, to the particular set of functions:

$$\{f: Z_2^n \rightarrow Z_2\}$$

are conferred the structures of Boolean algebra and of Boolean ring.

Example 7. Consider the logic function:

$$f: \{0, 1, 2\}^3 \rightarrow \{0, 1, 2\}$$

defined by the Figure 3.
The values of the variable x_1 determine the rows of Figure 3 while the values of the variables x_0 and x_2 determine the columns of Figure 3. Observe that this tabular representation of a function derives from a two-dimensional presentation of the value vector of this function: in most cases it is easier to write a $[p > q]$ matrix than a $[1 \times pq]$ vector.

Example 8. Consider two Boolean functions:

$$f_0: \{0, 1\}^3 \rightarrow \{0, 1\} \quad \text{and} \quad f_1: \{0, 1\}^3 \rightarrow \{0, 1\}.$$

Both functions are defined by Figure 4.
The functions f_0 and f_1 constitute the sum output and the carry output of a full adder, respectively.

x_2	x_1	x_0	f_0	f_1
0	0	0	0	0
0	0	1	1	0
0	1	0	1	0
0	1	1	0	1
1	0	0	1	0
1	0	1	0	1
1	1	0	0	1
1	1	1	1	1

Figure 4. Boolean functions

Remark. We define a *discrete function* as a mapping

$$f : \mathop{\mathsf{X}}_{i=0}^{n-1} \{0, 1, \ldots, m_i - 1\} \rightarrow \{0, 1, \ldots, m - 1\}$$

A logic function is thus a discrete function where the cardinalities of the variable domain are all the same. The theorems and algorithms developed further on in this section apply for discrete functions as well as for logic functions; however only logic functions are of practical use in logic design. In the present text some illustrative theoretical examples will in some cases deal with discrete functions.

A graphical technique, the *Karnaugh map*, for representing Boolean functions was originally proposed by Veitch and later improved by Karnaugh. The Karnaugh map is a matrix representation of the value vector where every possible combination of the binary input variables is represented by an entry (or square); the entries in the matrix are coded according to a Gray code enumeration order. This ensures that there is a change in one variable only between adjacent vertical or horizontal entries. Note that entries in adjacent rows differ by one variable as do entries in adjacent columns.

The Karnaugh map for the Boolean functions of Figure 4 is shown in Figure 5.

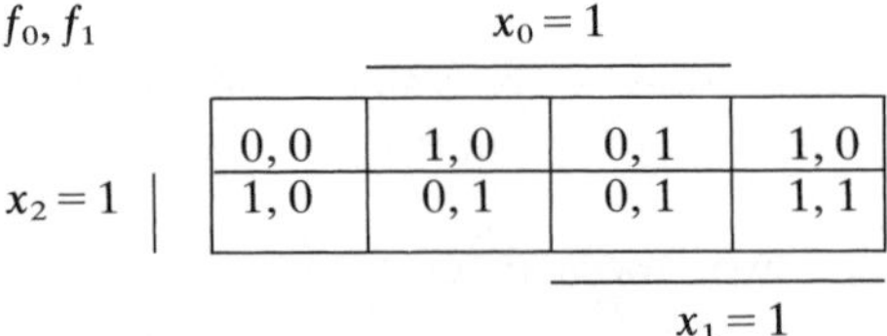

Figure 5. Karnaugh map

It is also of common use, when dealing with Boolean functions, to delete the occurrences of '0' in the Karnaugh map.

3.2 Lattice expressions

3.2.1 Lattice exponentiation

Consider a subset C of Z_m; the *lattice exponentiation function*: $Z_m \rightarrow Z_m$ whose value at the point x is denoted $x^{(C)}$, is defined as follows:

$$\begin{aligned} x^{(C)} &= m - 1, \quad \text{iff } x \in C, \\ &= 0, \qquad\quad \text{otherwise.} \end{aligned}$$

For the particular case of Boolean functions, there are four possible subsets of Z_2, i.e.

$$C=\emptyset, \{0\}, \{1\} \text{ or } \{0, 1\}.$$

By definition we have:

$$x^{(\emptyset)}=0; \qquad x^{(\{0\})}=\bar{x}, \qquad x^{(\{1\})}=x; \qquad x^{(\{0,1\})}=1. \tag{19}$$

3.2.2 Lattice expression

Let $\mathbf{x}=(x_{n-1}, \ldots, x_1, x_0)$ be a set of symbols, called variables, with $x_i \in Z_m$, $\forall i$.

Lattice expressions are defined in a recurrent way as follows:

(a) The m elements $\{0, 1, \ldots, m-1\}$ of Z_m and the variables $x_0, x_1, \ldots, x_{n-1}$ are lattice expressions.

(b) If P and Q are two lattice expressions, then $(P \vee Q)$, $(P \wedge Q)$ and $P^{(C)}$ $\forall C \subseteq Z_m$, are lattice expressions.

(c) No other lattice expressions exist that cannot be generated by a finite number of applications of rules (a) and (b).

We are able to write a lattice expression comprising, beside literals and $\vee$, $\wedge$, and $^{(C)}$, nested symmetric pairs of parentheses, ( ), enclosing the expressions operated upon according to rule (b) of the preceding definition. Therefore, for some literals $x_0, x_1, x_2, x_3 \in Z_2$

$$((x_0 \vee x_1 \vee (x_2 \vee x_3) \wedge ((\overline{x_1 \wedge x_2})))) \wedge (x_2 \vee (x_2 \wedge x_3))$$

is an expression written according to the given rule. Since it may be difficult, especially in complicated cases, to spot which parenthesis symbols form a symmetric pair, simplifications are common in the use of parentheses if no ambiguity results. For example in any lattice expression, exponentiation is performed before conjunction and conjunction before disjunction; moreover the symbol of conjunction is generally dropped.

Example 7 (continued). Three-variable lattice expression over Z_3:

$$E=1x_0^{(1)}(x_2^{(1)}x_1^{(0,1)} \vee x_1^{(0)}) \vee 2(x_0^{(0)}(x_1^{(0,1)} \vee x_2^{(0,2)}) \vee x_1^{(1,2)}x_2^{(0,2)}) \tag{20}$$

Example 8 (continued). Three-variable lattice expressions over Z_2:

(a) $E_0=(x_0\bar{x}_1 \vee \bar{x}_0 x_1)\bar{x}_2 \vee \overline{(x_0\bar{x}_1 \vee \bar{x}_0 x_1)}x_2$ (21)

(b) $E_1=x_0x_1 \vee x_0x_2 \vee x_1x_2.$ (22)

Comments.

(1) With any lattice expression is associated a logic function $f(\mathbf{x})$; the values of this function are obtained by substituting the values for the variables in the expression and by observing the computation laws in Z_m. We verify, e.g., that the lattice expressions (20) and (21, 22) compute the logic function of Example 7 and the Boolean functions of Example 8, respectively.

(2) Any lattice expression may be considered as a computation scheme using the set of primitives:

$$\mathbf{\Pi}=\{\vee, \wedge, x^{(C)} \mid C \subseteq Z_m\}.$$

Since a materialization (e.g. a logic network) is associated with each computation scheme, we may also associate a materialization with each lattice expression. Note however that the identification of a lattice expression with a computation scheme generally requires that the nested symmetric pairs of parentheses may no longer be simplified according to the rule quoted in the definition of a lattice expression.

3.2.3 Canonical conjunctive and disjunctive expansions

In the preceding section we have emphasized the distinction between logic functions and lattice expansions. Every lattice expression is clearly a logic function since its arguments x_i and values are in Z_m respectively. We will now prove that any logic function may be described by a lattice expression.

Some particular types of logic functions will first be considered.

Definition 10.

(a) A *cube function* is a logic function of the form:

$$c(\mathbf{x})=\ell \bigwedge_{i=0}^{n-1} x_i^{(C_i)}, \qquad \ell \in Z_m, \qquad C_i \subseteq Z_m,$$

(b) An *anticube function* is a logic function $d(\mathbf{x})$ of the form:

$$d(\mathbf{x})=\ell \vee \bigvee_{i=0}^{n-1} x_i^{(C_i)}, \qquad \ell \in Z_m, \qquad C_i \subseteq Z_m.$$

The cube $c(\mathbf{x})$ thus takes the value ℓ iff $x_i \in C_i$, $\forall i$, and takes the value 0 otherwise; the anticube $d(\mathbf{x})$ takes the value $(m-1)$ iff $\exists i: x_i \in C_i$ and takes the value ℓ otherwise.

Definition 11. If a logic function is expressed as a disjunction of cube functions it is called a *disjunctive normal form.* If a logic function is expressed as a conjunction of anticube functions, it is called a *conjunctive normal form.*

Every lattice expression may be expressed in a disjunctive normal form and in a conjunctive normal form by using the distributive laws in the lattice and De Morgan's laws.

Theorem 10. *Every logic function $f(\mathbf{x})$ can be written in the following disjunctive normal form*:

$$f(\mathbf{x})=\bigvee_{\mathbf{e}}\left[f(\mathbf{e}) \wedge \bigwedge_{i=0}^{n-1} x_i^{(e_i)}\right], \qquad (23)$$

$$\mathbf{e}=(e_{n-1}, \ldots, e_1, e_0), \qquad 0 \leqslant e_i \leqslant m-1.$$

Proof. As the variables $x_{n-1}, \ldots, x_1, x_0$ take on any of the m^n possible sets of values in Z_m^n, say for example $(x_{n-1}, \ldots, x_1, x_0) = (e_{n-1}, \ldots, e_1, e_0)$, only one term in the expression (23) is non-zero, namely the term

$$f(e_{n-1}, \ldots, e_1, e_0) \wedge x_0^{(e_0)} \wedge x_1^{(e_1)} \wedge \ldots \wedge x_{n-1}^{(e_{n-1})}$$

which at $x_0 = e_0$, $x_1 = e_1, \ldots, x_{n-1} = e_{n-1}$ is just $f(e_{n-1}, \ldots, e_1, e_0)$. □

The right-hand side of (23) is called the *canonical disjunctive form* of the function $f(\mathbf{x})$.

Each conjunction of the form $\ell \wedge x_0^{(e_0)} \wedge x_1^{(e_1)} \wedge \ldots \wedge x_{n-1}^{(e_{n-1})}$ is called a *minterm.*

Let $(\mathbf{x}_1, \mathbf{x}_0)$ be a partition of the set $\mathbf{x}$ of variables, with:

$$\mathbf{x}_1 = (x_{n-1}, \ldots, x_{p+1}, x_p),$$
$$\mathbf{x}_0 = (x_{p-1}, \ldots, x_1, x_0).$$

Every logic function can be written in the following disjunctive normal form:

$$f(\mathbf{x}) = \bigvee_{\mathbf{e}_0} \left[f(\mathbf{x}_1, \mathbf{e}_0) \wedge \bigwedge_{i=0}^{p-1} x_i^{(e_i)} \right] \tag{24}$$
$$\mathbf{e}_0 = (e_{p-1}, \ldots, e_1, e_0), \qquad 0 \leq e_i \leq m-1.$$

Relation (24) is called the *partial canonical disjunctive form* of $f(\mathbf{x})$ with respect to $\mathbf{x}_0$. The proof of (24) is exactly the same as (23).

Theorem 11. *Every logic function $f(\mathbf{x})$ can be written in the following conjunctive normal form*:

$$f(\mathbf{x}) = \bigwedge_{\mathbf{e}} \left[f(\mathbf{e}) \vee \bigvee_{i=0}^{n-1} x_i^{(\bar{e}_i)} \right], \qquad 0 \leq e_i \leq m-1 \tag{25}$$

($\bar{e}_i$ means the set complementation, i.e. $\bar{e}_i = \{0, 1, \ldots, m-1\} \backslash e_i$).

Proof. Similar to that of Theorem 9. □

The right-hand side of relation (25) is called the *canonical conjunctive form* of the function $f(\mathbf{x})$.

Each disjunction of the form $\ell \vee x_0^{(\bar{e}_0)} \vee x_1^{(\bar{e}_1)} \vee \ldots \vee x_{n-1}^{(\bar{e}_{n-1})}$ is called a *maxterm.*

Every logic function may also be written in the *partial canonical conjunctive form* with respect to $\mathbf{x}_0$, i.e.

$$f(\mathbf{x}_1, \mathbf{x}_0) = \bigwedge_{\mathbf{e}} \left[f(\mathbf{x}_1, \mathbf{e}_0) \vee \bigvee_{i=0}^{p-1} x_i^{(\bar{e}_i)} \right], \qquad 0 \leq e_i \leq m-1. \tag{26}$$

Example 8 (continued). The three variable Boolean functions have the following canonical disjunctive and canonical conjunctive normal forms respectively:

$$f(x_2x_1x_0) = f(000)\bar{x}_2\bar{x}_1\bar{x}_0 \vee f(001)\bar{x}_2\bar{x}_1x_0 \vee$$
$$f(010)\bar{x}_2x_1\bar{x}_0 \vee f(011)\bar{x}_2x_1x_0 \vee$$
$$f(100)x_2\bar{x}_1\bar{x}_0 \vee f(101)x_2\bar{x}_1x_0 \vee$$
$$f(110)x_2x_1\bar{x}_0 \vee f(111)x_2x_1x_0, \quad (27)$$
$$= (f(000) \vee x_2 \vee x_1 \vee x_0)(f(001) \vee x_2 \vee x_1 \vee \bar{x}_0) \wedge$$
$$(f(010) \vee x_2 \vee \bar{x}_1 \vee x_0)(f(011) \vee x_2 \vee \bar{x}_1 \vee \bar{x}_0) \wedge$$
$$(f(100) \vee \bar{x}_2 \vee x_1 \vee x_0)(f(101) \vee \bar{x}_2 \vee x_1 \vee \bar{x}_0) \wedge$$
$$(f(110) \vee \bar{x}_2 \vee \bar{x}_1 \vee x_0)(f(111) \vee \bar{x}_2 \vee \bar{x}_1 \vee \bar{x}_0). \quad (28)$$

The function f_0 of Example 8 has the following disjunctive and conjunctive forms, respectively:

$$f_0 = \bar{x}_2\bar{x}_1x_0 \vee \bar{x}_2x_1\bar{x}_0 \vee x_2\bar{x}_1\bar{x}_0 \vee x_2x_1x_0 \quad (29)$$
$$= (\bar{x}_2 \vee x_1 \vee \bar{x}_0)(\bar{x}_2 \vee \bar{x}_1 \vee x_0)(x_2 \vee x_1 \vee x_0)(x_2 \vee \bar{x}_1 \vee \bar{x}_0). \quad (30)$$

3.3 Canonical forms and their implementation

Procedures for designing logic networks corresponding to canonical disjunctive and conjunctive expansions are extensively discussed in the literature dealing with switching theory. Some of the most classical realizations will be described.

Theorems 9 and 10 allow us to state that any logic function can be represented by at least one lattice expression. This allows us to synthesize any logic function by means of the computation primitives:

$$\{\vee(\text{OR-gate}), \quad \wedge(\text{AND-gate}), x^{(c)}(\text{inverter})\}$$

The canonical disjunctive expansion may be interpreted in the following way.

(a) The minterms are in one-to-one correspondence with the AND-gates, i.e. each minterm is synthesized by an AND-gate which has thus the functions $x_i^{(e_i)}$ as inputs and produces the corresponding minterm as output.

(b) The outputs of the AND-gates are connected to the OR-gate which produces the function f at its output.

The network so obtained is called an AND–OR network; with the canonical conjunctive expansion is associated in a similar way an OR–AND network.

Example 8 (continued). The sum-function f_0 of the full-adder has the canonical expansions (29) and (30); to these expansions correspond the networks of Figure 6a and 6b respectively.

The networks of Figures 6a and 6b are referred to as *two-level AND–OR* and *two-level OR–AND* networks respectively; these networks are moreover *tree-networks* since the first level gates are fan-out free. Observe that if there is no fan-in limitation for the first level gates, any logic function may be realized by a two-level network.

As we shall see in the next section, the disjunctive and conjunctive forms of Boolean functions are the end result of a straightforward design procedure; this procedure produces in AND–OR logic the two-level networks described in Figure 6. It is important to note that this procedure and the corresponding network schemes are also useful, e.g. in NOR-logic and in NAND logic.

In Chapter I we defined both NAND and NOR operations for describing logic gates. The next theorem shows that any Boolean function can be realized solely with NAND or NOR gates. This theorem provides us also with a simple relationship between AND–OR networks and NAND or NOR networks.

Theorem 12.

(a) *A two-level AND–OR network, and a two-level NAND network obtained from the former one by substituting both AND- and OR-gates by NAND gates, realize the same Boolean function.*

(b) *A two-level OR–AND network, and a two-level NOR network obtained from the former one by substituting both OR- and AND-gates by NOR gates, realize the same Boolean function.*

Proof.

(a) From Theorem 10 we successively deduce:

$$\bar{f} = \bigwedge_{\mathbf{e}} \overline{\left[f(\mathbf{e}) \wedge \bigwedge_{i} x_i^{(e_i)} \right]},$$

$$f = \bar{\bar{f}} = \overline{\bigwedge_{\mathbf{e}} \overline{\left[f(\mathbf{e} \wedge \bigwedge_{i} x_i^{(e_i)} \right]}}.$$

The proof derives from the above relation and from the definition of the NAND gate.

(b) Similar statement as for part (a). □

Note that, if, in an AND–OR (resp. OR–AND) network an implicant (resp. implicate) reduces to only one letter, the corresponding AND-gate (resp. OR-gate) may be deleted, while in the NAND–NAND (resp. NOR–NOR) synthesis the first level NAND (resp. NOR) reduces to an inverter.

Example 8 (continued). From the AND–OR and OR–AND realizations of Figures 6a and 6b we deduce the NAND and NOR realizations of Figures 7a and 7b respectively. It should be noted that the simple relationship of Theorem 12 between AND–OR networks and NAND or NOR networks holds for two-level networks. If networks with more than two levels of gating are used, the situation is considerably more complicated.

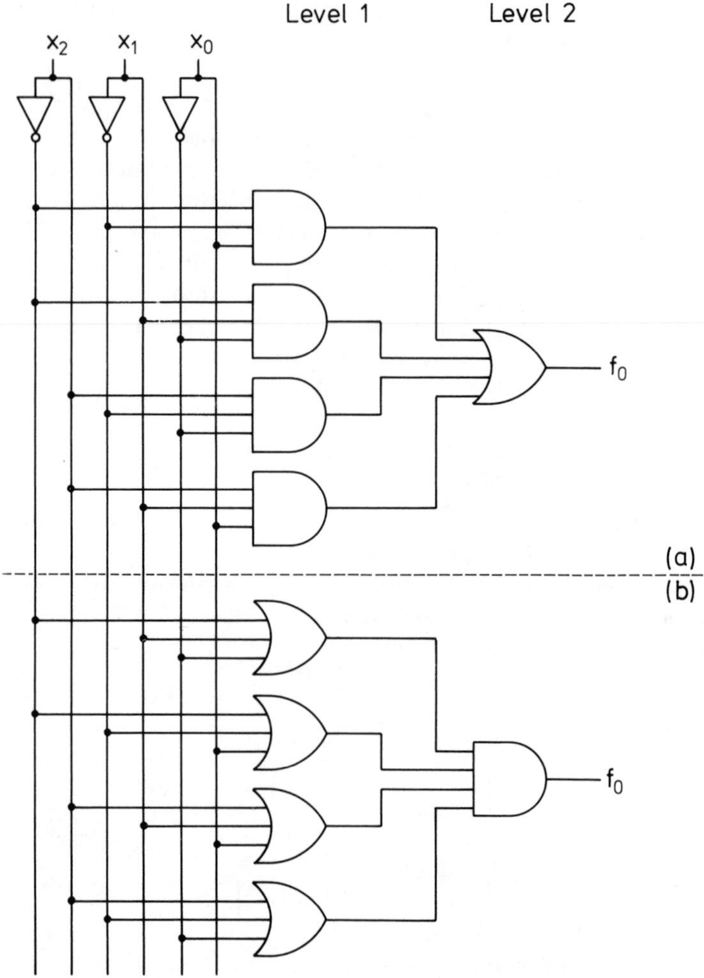

Figure 6. Two-level AND–OR and OR–AND networks

Direct design procedures for NAND or NOR networks have been developed, but they are rather limited in scope and power. By contrast, and in view of Theorem 12, the design algorithms for two-level forms are completely general, very powerful, and essentially independent of the chosen type of circuit realization. One of these algorithms will be developed extensively in Section 4.

We have seen that besides the direct implementation in AND–OR and in OR–AND networks, the disjunctive and conjunctive canonical forms provide us, by the intermediate of Theorem 12, with NAND and NOR implementations of Boolean functions. From these canonical forms we

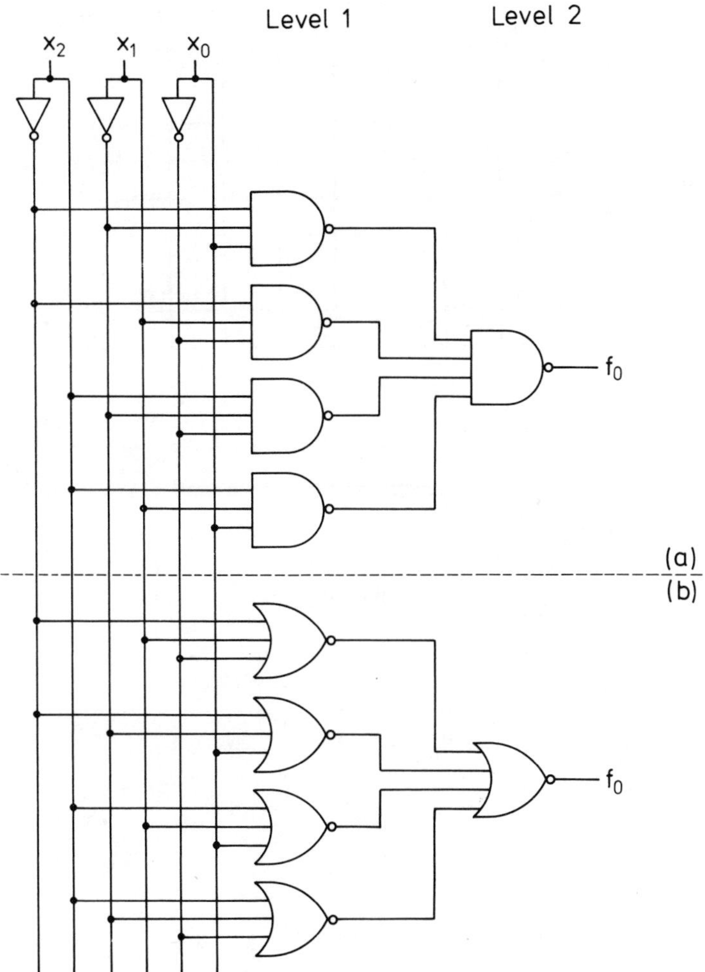

Figure 7. Two-level NAND and NOR networks

deduce also another type of implementation, namely the implementation by means of programmable logic arrays or PLA's.

Figure 8a shows the structure of a PLA, which consists of two matrices. In the first part (AND matrix) input signals and their negated signals are selectively connected to product term lines so that the outputs of the first part are capable of realizing any terms of a switching function of n variables. These outputs (or product lines) are input to the second part (OR matrix) where OR functions produce the desired sum of terms.

In Figure 8a the node structure for the coupled arrays (or matrices) is clearly shown: the dark circles within the AND matrix serve to indicate the

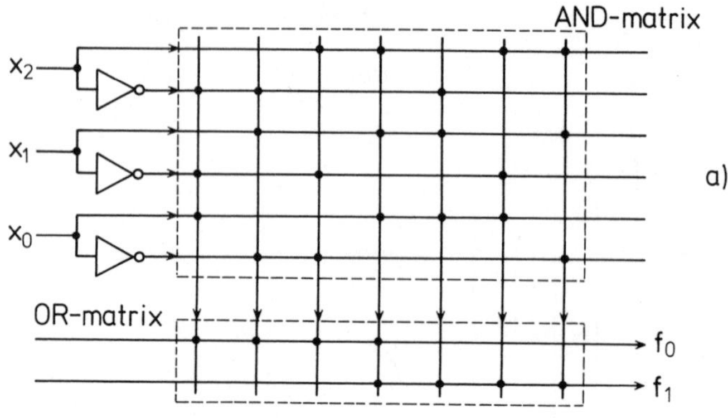

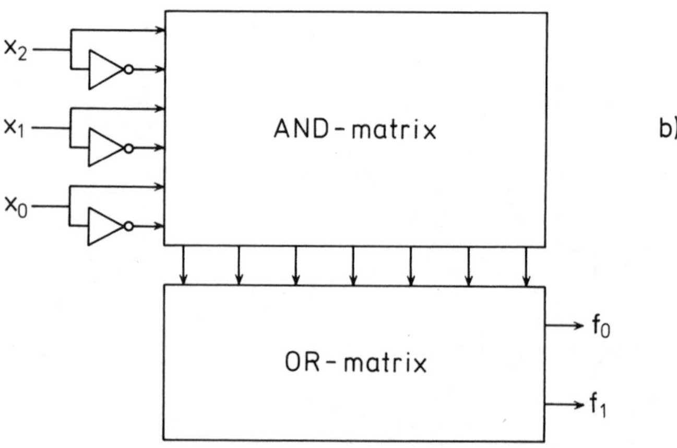

Figure 8. Programmable logic array

presence of an AND-gate while the dark circles within the OR-matrix serve to indicate the presence of an OR-gate. A block-diagram representation of this PLA is shown in Figure 8b.

Example 8 (continued) The sum f_0 and carry functions f_1 are realized by the PLA of Figure 8a; this PLA structure is deduced from the canonical disjunctive forms for f_0 and f_1, i.e.

$$f_0 = \bar{x}_2\bar{x}_1x_0 \vee \bar{x}_2x_1\bar{x}_0 \vee x_2\bar{x}_1\bar{x}_0 \vee x_2x_1x_0,$$

$$f_1 = \bar{x}_2x_1x_0 \vee x_2\bar{x}_1x_0 \vee x_2x_1\bar{x}_0 \vee x_2x_1x_0.$$

The canonical disjunctive expansion leads to a final interesting observation. We stated that the minterms are particular cube functions which are zero everywhere, except at one point of Z_m^n. Any logic function

$$f: Z_m^n \to Z_m$$

may thus be written as the disjunction of functions which are zero everywhere except at a unique point of Z_m^n. This observation allows us to write the following expansion for f:

$$f(\mathbf{x}) = \textcircled{\scriptsize $\sum$}_{\mathbf{e}} \left[f(\mathbf{e}) \wedge \bigwedge_{i=0}^{n-1} x_i^{(e_i)} \right],$$

where $\textcircled{\scriptsize $\sum$}$ means the modulo-m sum.

This suggests to us the possibility of building a logic function by means of primitives realizing the ring operations. In particular, for Boolean functions, the modulo-2 sum is realized by an EXCLUSIVE–OR gate which is a classical component in logical design.

The above examples suggest that there exists for a given logic function a large number of computation schemes evaluating this function. We are thus faced with the choice of a given computation scheme realizing a set of logic functions in an optimal way. This *optimization problem* has been extensively developed in Davio, Deschamps, and Thayse (1978); it will be introduced more briefly and discussed in the next section.

4 OPTIMIZATION PROBLEMS

4.1 Optimization criteria

The purpose of logic function synthesis is to obtain a computation scheme that, when implemented with a set Π of computation primitives, results in a low-cost network. However the criteria that determine a low-cost network must be defined if we have to evaluate the success of the achieved synthesis.

The subject of switching theory is concerned, among other things, with finding algorithms for simplifying Boolean and logic functions. Although algorithms that satisfy all possible criteria for simplification are not available, it is important for the designer to be familiar with the different requirements, restrictions and limitations of a practical situation.

Optimization criteria for logic function simplification may be divided into two broad categories:

(a) Cost optimization, which is basically concerned with the reduction in the number of components and connections, and with the reduction of component cost (power, area, etc.). However the costs of developing, maintaining, and preparing test patterns must also be taken into account in the evaluation of a more realistic cost criterion (see also Chapter I, Section 5).

(b) Time optimization, which is concerned with the reduction in propagation time.

Besides these two optimization criteria a practical limitation to be introduced in the design of logic network is a loading constraint which reduces practically to fan-in and fan-out limitations. For example, the number of inputs (or fan-in) to logic gates such as AND and OR-gates cannot be increased without limit: diodes are not ideal and they have non-zero forward resistance and finite reverse resistance. If too many diodes are included in a gate, that gate will not operate reliably. The diode characteristics determine the exact limit but 10 is a typical limit. Fan-out specifies the number of circuits the output of a gate can drive without impairing its normal operation. Sometimes the term loading is used instead of fan-out. This term is derived from the fact that the output of a gate can supply a limited amount of power, above which it ceases to operate properly. Each circuit connected to the output consumes a certain amount of power, so that each additional circuit adds to the load of a gate.

Let us now discuss the two basic optimization criteria, namely cost optimization and delay optimization.

4.1.1 Delay criterion

The signals through a logic gate take a certain amount of time to propagate from the inputs of the circuit to its output. This interval of time is defined as the propagation delay of the gate. When speed of operation is important networks must have a minimum number of series gates between inputs and outputs.

A Boolean function expressed in sum of products or products of sums can be implemented with two levels of gates provided both direct and complemented inputs are available. Obviously, if only the direct input variables are available, the inverters that generate the complements constitute a third level in the implementation. For logic functions the exponentiation operation requests in any case a third level.

The use of two-levels provides the least propagation delay and thus the fastest circuit. A classical problem in switching theory may be stated as follows: between all the two-levels (AND–OR or OR–AND), normal forms of a logic function find those having the minimum number of terms and the minimum number of literals. There exist several algorithms for solving this problem (see e.g. the Quine–McCluskey, Consensus and implicant–implicate algorithms; Davio, Deschamps, and Thayse, 1978, ch. III). The implicant–implicate algorithm will be presented in Section 4.2. It optimizes a relative cost criterion which can be stated as follows: between all the networks having a minimal propagation delay find those having a minimal cost.

4.1.2 Cost criterion

As quoted above the costs of developing, maintaining, testing, etc. logical systems must be taken into account in the evaluation of a realistic cost criterion. In this chapter we shall however limit ourselves to the following items for reducing the cost of implementing digital systems:

(1) the number of logic gates;

(2) the number of inputs to a gate; and

(3) the number of connecting wires.

More elaborated cost considerations are given in the later chapters. As proved in Section 4.2, any algorithm for obtaining the prime implicants or the prime implicates of a logic function constitutes a first step in minimizing the number of gates and inputs to a gate, provided we are satisfied with a two-level implementation. The second step in the optimization process is reached by obtaining an irredundant normal form for the logic function (see Section 4.2.3).

General algorithms for obtaining a simplified multilevel implementation are not available. The designer must resort to algebraic manipulations or to any available computer program that searches for a minimum gate implementation under certain given conditions. An example of a multilevel optimal implementation using NAND-gates for the full-adder is given in Figure 9.

It seems that at this time one of the few methods that exist for obtaining optimal multilevel networks is a method which makes use either of multiplexers or of demultiplexers as elementary gates or primitives. This method will be introduced in Chapter III, Section 4 and will extensively be developed in Chapter XII, Section 5.

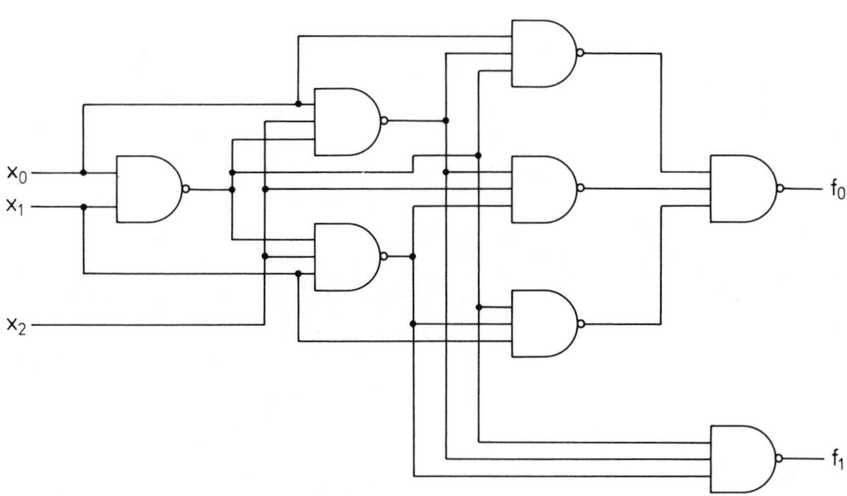

Figure 9. Full-adder realization using a minimal number of NAND-gates

4.2 The implicant–implicate method and the obtention of optimal two-level AND–OR and OR–AND networks

4.2.1 Definitions and basic theorems

A logic function $f(\mathbf{x})$ is said to *cover* another function $g(\mathbf{x})$, denoted $f \supseteq g$, if $f \geqslant g \ \forall \mathbf{x}$.

We shall also say that f contains g or is greater than g; equivalently, g is contained in f or g is smaller than f.

Cube and anticube functions were defined in Section 3.2.3; we shall first define in this section some particular types of cube and of anticube functions which will appear to be of special interest for obtaining optimal two-level AND–OR and OR–AND networks.

Remember that we called cube any conjunction of letters:

$$\ell \wedge x_i^{(C_i)} \wedge \ldots \wedge x_j^{(C_j)}.$$

Definition 12. An *implicant* of a logic function is any cube smaller than or equal to this function. A *prime implicant* is an implicant which is not contained in any other implicant of this function.

We called anticube any disjunction of letters:

$$\ell \vee x_i^{(C_i)} \vee \ldots \vee x_j^{(C_j)}.$$

Definition 13. An *implicate* of a logic function is any anticube greater than or equal to this function. A *prime implicate* is an implicate which does not contain any other implicate of this function.

Theorem 13.

(*a*) *Any logic function f is the disjunction of all its prime implicants.*

(*b*) *Any logic function f is the conjunction of all its prime implicates.*

Proof.

(a) By definition of a prime implicant $p_i(f)$ of f:

$$p_i(f) \leqslant f,$$

and thus:

$$\bigvee_i p_i(f) \leqslant f.$$

Conversely, f may be written as a disjunction of cubes $p'_j(f)$:

$$f = \bigvee_j p'_j(f),$$

each of these cubes being included in a prime implicant of f. Therefore:

$$f = \bigvee_j p'_j(f) \leqslant \bigvee_i p_i(f),$$

and thus:

$$f = \bigvee_i p_i(f).$$

(b) Similar type of statement. □

Theorem 13 provides us with two representations of a logic function, either as a disjunction of all its prime implicants, or as the conjunction of all its prime implicates. Our main problem will be the following.

A logic function being represented as a disjunction of implicants, find all its prime implicants, or, a logic function being represented as a conjunction of implicates, find all its prime implicates.

The following theorem plays a key role in the solution of this problem.

Theorem 14.

(a) Let f and g be two logic functions represented by the sets of their prime implicants $\{p_i(f)$ and $\{p_j(g)\}$, respectively. The prime implicants of the conjunction $f \wedge g$ are found by forming all the conjunctions $p_i(f) \wedge p_j(g)$ and by deleting in the obtained set the cubes smaller than other cubes.

(b) Let f and g be two logic functions represented by the sets of their prime implicates $\{q_i(f)\}$ and $\{q_j(f)\}$ respectively. The prime implicates of the disjunction $f \vee g$ are found by forming all the disjunctions $q_i(f) \vee q_j(g)$ and by deleting in the obtained set the anticubes greater than other anticubes.

Proof.

(a) Let $p(fg)$ be a prime implicant of the conjunction $f \wedge g$; $p(fg)$ is smaller than both f and g. Hence $p(fg)$ is smaller than, at least, one prime implicant of f, say $p_i(f)$, and than, at least, one prime implicate of g, say $p_i(g)$. Then:

$$p(fg) \leqslant p_i(f) \wedge p_j(g).$$

However, since the conjunction of cubes is a cube, $p_i(f) \wedge p_j(g)$ is an implicant of $f \wedge g$. Finally, as $p(fg)$ is prime:

$$p(fg) = p_i(f) \wedge p_j(g).$$

(b) Similar statement. □

4.2.2 The implicant–implicate algorithm

The implicant–implicate method for obtaining the prime implicants and the prime implicates of a logic function is grounded on the solution of an auxiliary problem stated in Theorem 15 below.

Theorem 15.

(a) Let

$$c(\mathbf{x}) = \ell \wedge \bigwedge_i x_i^{(C_i)}$$

be a cube. The prime anticubes of $c(\mathbf{x})$ are ℓ and $x_i^{(C_i)}$.

(*b*) *Let*

$$d(\mathbf{x}) = \ell \vee \bigvee_i x_i^{(C_i)}$$

be an anticube. The prime cubes of $d(\mathbf{x})$ *are* ℓ *and* $x_i^{(C_i)}$.

Proof.

(a) (1) ℓ and $x_i^{(C_i)}$ are anticubes of $c(\mathbf{x})$ since ℓ and $x_i^{(C_i)}$ are greater than $c(\mathbf{x})$.

(2) ℓ is a prime anticube of $c(\mathbf{x})$. Indeed, the only anticubes smaller than ℓ are the anticubes ℓ' such that $\ell' < \ell$. These are obviously not anticubes of $c(\mathbf{x})$. Similarly, the anticubes smaller than $x_i^{(C_i)}$ are the anticubes $x_i^{(C_i')}$, with $C_i' \subset C_i$, which are not anticubes of $C(\mathbf{x})$.

(3) Any prime anticube

$$d(\mathbf{x}) = \ell' \vee \bigvee_i x_i^{(C_i')}$$

is either ℓ or one of the $x_i^{(C_i)}$. Assume first $\ell \leqslant \ell'$. Then $d(\mathbf{x})$ has to be ℓ since $\ell \leqslant d(\mathbf{x})$ and $d(\mathbf{x})$ is prime. Assume next $\ell \not\leqslant \ell'$ (i.e. $\ell > \ell'$ or ℓ and ℓ' are not comparable). In this case, for one index at least, say k, $C_k \subseteq C_k'$ holds. In the opposite case, $d(\mathbf{x})$ would assume the value ℓ' in at least one point where $c(\mathbf{x})$ assume the value ℓ: $d(\mathbf{x})$ would not be an implicate. Then $d(\mathbf{x}) = x_k^{(C_k)}$ since $x_k^{(C_k)} \leqslant d(\mathbf{x})$ and $d(\mathbf{x})$ is prime.

(b) Similar statement as for (a). □

Algorithm. *Obtention of prime cubes and/or prime anticubes.*

(*a*) *Assume that* $f(\mathbf{x})$ *is represented as a disjunction of cubes; in view of Theorem* 15(*a*), *each of these cubes is represented as a conjunction of its prime anticubes.*

(*b*) *Compute the prime anticubes of* f *by performing disjunctions of anticubes, using Theorem* 14(*b*). *Let us recall that the disjunction of two anticubes is an anticube.*)

(*c*) *In view of Theorem* 15(*b*) *each of these anticubes is represented as a disjunction of its prime cubes.*

(*d*) *Compute the prime cubes of* $f(\mathbf{x})$ *by performing conjunctions of cubes, using Theorem* 14(*a*). (*Let us recall that the conjunction of two cubes is a cube.*)

First, it is interesting to note that the proposed algorithm provides at the same time the representation of f both by its prime cubes and by its prime anticubes. The algorithm is easily matched to partial problems, e.g. to the research of the prime anticubes from a disjunction of cubes or to other data presentation, i.e. to a representation of f as a conjunction of anticubes.

From the computational point of view, steps 2 and 4 of the algorithm are easily performed in a step-by-step way. If f is represented as the disjunction of cubes $C_0 \vee C_1 \vee C_2, \ldots,$ we will first compute the prime anticubes of $C_{01} = C_0 \vee C_1$, next the prime anticubes of $C_{012} = C_0 \vee C_1 \vee C_2 = C_{01} \vee C_2$, and so on.

Example 9. Consider a ternary function of two quaternary variables:

$$f: \{0, 1, 2, 3\}^2 \to \{0, 1, 2\}$$

and assume it is given as a disjunction of cubes, i.e.

$$f = 1x_1^{(0)}x_0^{(1)} \vee 2x_1^{(0)}x_0^{(2)} \vee 1x_1^{(0)}x_0^{(3)} \vee 1x_1^{(2)}x_0^{(0)} \vee 2x_1^{(2)}x_0^{(2)} \\ \vee 1x_1^{(3)}x_0^{(1)} 1x_1^{(3)}x_0^{(3)}.$$

The disjunction of the two first terms is:

$$x_1^{(0)}x_0^{(0,2)}(1 \vee x_0^{(2)}).$$

The result is next summed with the third term to give:

$$x_1^{(0)}x_0^{(1,2,3)}(1 \vee x_0^{(2)}).$$

The process is carried on until the last term and provides as resulting representation:

$$f = (1 \vee x_0^{(2)})(1 \vee x_1^{(0,2)})(1 \vee x_1^{(2)} \vee x_0^{(1,2,3)})(x_1^{(2)} \vee x_0^{(1,2,3)})(x_1^{(0,3)} \\ \vee x_0^{(0,2)})(x_1^{(0,2,3)} \vee x_0^{(0)})(x_1^{(0,2)} \vee x_0^{(0,1,3)})$$

(representation of f as a conjunction of prime anticubes).

Step (d) of the algorithm provides us finally with f as a disjunction of prime cubes, i.e.

$$f = 2x_1^{(0,2)}x_0^{(2)} \vee 1x_1^{(2)}x_0^{(0)} \vee 1x_1^{(0,3)}x_0^{(1,3)} \vee 1x_1^{(0)}x_0^{(1,2,3)}.$$

4.2.3 The optimal covering problem

Consider a logic function f and the set of its implicants: $p_0, p_1, \ldots, p_{q-1}$. The function f can be written as the disjunction of all its implicants. However in most cases f can be written as the disjunction of only a subset of its implicants; it may happen that some implicant, say p_k, is included in the disjunction of some other implicants. Suppose for instance that

$$p_k \leqslant p_0 \vee \ldots \vee p_{k-1},$$

then any representation of f as a disjunction of some of its implicants among which lie $p_0, \ldots, p_{k-1}$ and p_k is *redundant*, i.e. the implicant p_k may be deleted.

One of the main problems studied in switching theory is to find representations of f as the disjunction of a minimum number of its implicants; to those representations correspond indeed lower cost two-level realizations (see Theorem 17 below). The problem may be stated as follows: find a subset P' of $P=\{p_0, \ldots, p_{q-1}\}$ such that:

(1) For every point where f takes the value ℓ, there exists at least one implicant in P' that also takes the value ℓ, and

(2) the number of elements in P' is as small as possible.

This is an example of *covering–closure problem*: we must cover the points of f by means of a set of implicants while minimizing a cost function. We shall see further on that for the purpose of obtaining a minimum cost realization of f we shall delete the implicants contained in others and thus only consider the prime implicants of f: we know indeed (Theorem 13) that f may be represented as the disjunction of all its prime implicants. For that reason, in the sequel P will denote the set of prime implicants of f.

Consider the finite sets:

$$P=\{p_0, \ldots, p_{k-1}\} \equiv \text{set of the prime implicants of } f$$

and

$$M=\{m_0, \ldots, m_{s-1}\} \equiv \text{set of the minterms of } f.$$

We may associate with every element p_i of P a binary variable y_i. Every subset P' of P can thus be characterized by a k-tuple $\{y_0, \ldots, y_{k-1}\}$ defined as follows:

$$\begin{aligned} y_i &= 1, \quad \text{if } p_i \in P' \\ &= 0, \quad \text{otherwise.} \end{aligned}$$

On the other hand, define a $k \times s$ matrix $[r_{ij}]$ by

$$\begin{aligned} r_{ij} &= 1, \quad \text{if } p_i \geqslant m_j \\ &= 0, \quad \text{otherwise,} \end{aligned}$$

$\forall i=0, 1, \ldots, k-1$ and $j=0, 1, \ldots, s-1$.

The subsets $P' \subseteq P$ which cover M are those that are characterized by a k-tuple $\{y_0, \ldots, y_{k-1}\}$ such that:

$$\sum_{i=0}^{k-1} r_{ij} y_i \geqslant 1, \quad \forall j=0, 1, \ldots, s-1.$$

Indeed, this last condition means that for every m_j in M there exists at least one p_i in P' such that $r_{ij}=1$, i.e. such that $m_j \leqslant p_i$.

The covering problem can now be stated in the following form:

minimize the (cost) function:

$$C=\sum_{i=0}^{k-1} y_i$$

under the s constraints:

$$\sum_{i=0}^{k-1} r_{ij} y_i \geqslant 1, \quad \forall j=0, 1, \ldots, s-1.$$

The constraints may also be formulated in the Boolean form:

$$\bigvee_{i=0}^{k-1} r_{ij}y_i = 1, \quad \forall j = 0, \ldots, s-1. \tag{31}$$

It is also classical to represent the constraints in tabular form. For that purpose, one draws a rectangular table with k rows and s columns; the column j that corresponds to the element m_j of M has a '1' in every row i corresponding to an element p_i of P such that $m_j \leqslant p_i$. In other words, the table is directly deduced from the matrix $[r_{ij}]$ by indexing the rows and columns by means of the elements of P and M respectively and by deleting the 0's.

Theorem 16. *A subset P' is an irredundant cover of M iff the conjunction*

$$\bigwedge_{i:p_i \in P'} y_i \tag{32}$$

is a prime implicant of the switching function r defined by

$$r(y_0, y_1, \ldots, y_{k-1}) = \bigwedge_{j=0}^{s-1} \left[\bigvee_{i=0}^{k-1} (r_{ij} \wedge y_i \right]. \tag{33}$$

Proof. A subset P' covers M iff the corresponding k-tuple $(y_0, \ldots, y_{k-1})$ is a solution of each of the s Boolean equations (31), i.e. if it is a solution of the equation $r(y_0, \ldots, y_{k-1}) = 1$, i.e. iff the conjunction (32) is an implicant of r. Furthermore, it is shown that to an irredundant cover corresponds a prime implicant, and conversely. □

The implicant–implicate method described in Section 4.2.2 thus applies to the solution of the minimal covering problem while it is also possible to take into account the unate character of (33) to devise other several methods.

Example 10. The Boolean function $f(x_4, x_3, x_2, x_1\, x_0)$, depicted by Figure 10, has 6 prime implicants:

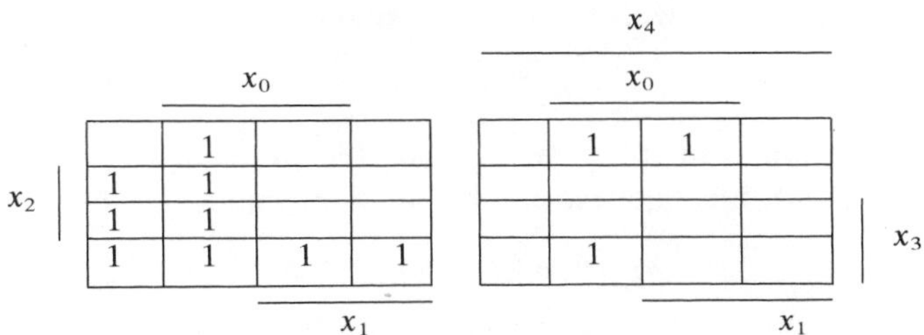

Figure 10. Karnaugh map

These prime implicants are, e.g., obtained by the implicant–implicate method of Section 4.2.2; they are:

$$p_0 = \bar{x}_1 x_3 \bar{x}_4,$$
$$p_1 = \bar{x}_1 x_2 \bar{x}_4,$$
$$p_2 = x_0 \bar{x}_1 \bar{x}_4,$$
$$p_3 = \bar{x}_2 x_3 \bar{x}_4,$$
$$p_4 = x_0 \bar{x}_1 \bar{x}_2,$$
$$p_5 = x_0 \bar{x}_2 \bar{x}_3 x_4.$$

The minterms of the Boolean function are:

$$m_0 = \bar{x}_0 \bar{x}_1 x_2 \bar{x}_3 \bar{x}_4, \qquad m_1 = \bar{x}_0 \bar{x}_1 x_2 x_3 \bar{x}_4, \qquad m_2 = \bar{x}_0 \bar{x}_1 \bar{x}_2 x_3 \bar{x}_4,$$
$$m_3 = x_0 \bar{x}_1 \bar{x}_2 \bar{x}_3 \bar{x}_4, \qquad m_4 = x_0 \bar{x}_1 \bar{x}_2 x_3 \bar{x}_4, \qquad m_5 = x_0 \bar{x}_1 x_2 x_3 \bar{x}_4,$$
$$m_6 = x_0 \bar{x}_1 x_2 \bar{x}_3 \bar{x}_4, \qquad m_7 = x_0 x_1 \bar{x}_2 x_3 \bar{x}_4, \qquad m_8 = \bar{x}_0 x_1 \bar{x}_2 x_3 \bar{x}_4,$$
$$m_9 = x_0 \bar{x}_1 \bar{x}_2 \bar{x}_3 x_4, \qquad m_{10} = x_0 \bar{x}_1 \bar{x}_2 x_3 x_4, \qquad m_{11} = x_0 x_1 \bar{x}_2 \bar{x}_3 x_4.$$

The relationship between the prime implicants and the minterms are written down in the covering table of Figure 11.

	m_0	m_1	m_2	m_3	m_4	m_5	m_6	m_7	m_8	m_9	m_{10}	m_{11}	
p_0		1	1		1	1							y_0
p_1	1	1				1	1						y_1
p_2				1	1	1	1						y_2
p_3			1		1			1	1				y_3
p_4				1	1					1	1		y_4
p_5										1		1	y_5

Figure 11. Covering table

From this covering table we deduce the switching function:

$$r(y_0, \ldots, y_5) = y_1(y_0 \vee y_1)(y_0 \vee y_3)(y_2 \vee y_4)(y_0 \vee y_2 \vee y_3 \vee y_4)$$
$$\times (y_0 \vee y_1 \vee y_2)(y_1 \vee y_2) y_3 (y_4 \vee y_5) y_4 y_5,$$

whose prime implicant with the smallest number of literals is:

$$y_1 y_3 y_4 y_5.$$

To this prime implicant of r corresponds the minimal representation of f (representing with the smallest number of implicants):

$$f = \bar{x}_1 x_2 \bar{x}_4 \vee \bar{x}_2 x_3 \bar{x}_4 \vee x_0 \bar{x}_1 \bar{x}_2 \vee x_0 \bar{x}_2 \bar{x}_3 x_4.$$

The associated optimal realization has four AND-gates and one OR-gate.

Consider an n-variable logic function f; it can be represented as a

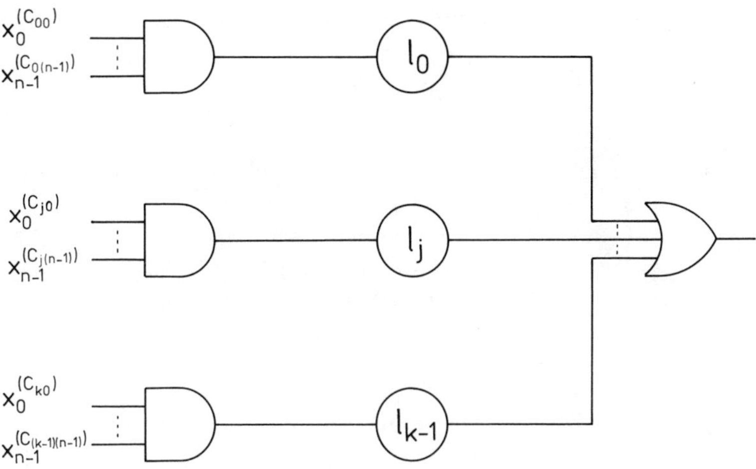

Figure 12. Two-level AND–OR network

disjunction of implicants, i.e. under the form

$$f = \bigvee_{j=0}^{k-1} p_j. \tag{34}$$

The implicants p_j are cubes:

$$p_j = \ell_j \wedge x_0^{(C_{j0})} \wedge \ldots \wedge x_{n-1}^{(C_{j(n-1)})}, \qquad C_{ji} \subseteq Z_m, \quad \forall j.$$

In Boolean functions, the implicants are the products of variables under direct and complemented forms (remember that $x_i^{(1)} = x_i$, $x_i^{(0)} = \bar{x}_i$ and $x_i^{(\{0,1\})} = 1$). With the disjunctive expression (34) can be associated the logic circuit of Figure 12 to which can be conferred some cost. Let us define two different costs:

(a) C_1 = the number of gates;

(b) C_2 = the number of gate inputs.

Note that if in the circuit of Figure 12 one has $C_{ji} = Z_m$ for some i and j, then the corresponding AND-gate input wire may be deleted. If, furthermore, this can be done for all the input wires but one, the corresponding AND-gate itself may be deleted.

For every implicant p_j denote $t(p_j)$ the number of subsets $C_{ji} \neq Z_m$. Define then the two following cost functions:

(a) $C_1(p_j) = 1, \quad \text{if } t(p_j) \geqslant 2,$
$\quad\quad = 0, \quad \text{if } t(p_j) = 0 \text{ or } 1.$

(b) $C_2(p_j) = t(p_j) + 1, \quad \text{if } t(p_j) \geqslant 2,$
$\quad\quad = 1, \quad \text{if } t(p_j) = 0 \text{ or } 1.$

The number of gates and of gate inputs in the circuit of Figure 12 is thus

$$\sum_{j=0}^{k-1} C_1(p_j) \quad \text{and} \quad \sum_{j=0}^{k-1} C_2(p_j), \text{ respectively}$$

(except in the trivial case where f itself is a cube and where the OR-gate may be deleted).

Denote by M and P the sets of minterms and of implicants of f respectively; the problem of finding a minimum cost logic realization of f reduces to the following optimal covering problem; one must find a subset P' of P such that:

(a) P' covers M;

(b) P' has minimum cost $C_1(P')$ or $C_2(P')$ according to the chosen cost.

At this point it is interesting to note that, for two implicants p_i and p_j of f, one has:

$$p_i < p_j \Rightarrow C_1(p_i) \geq C_1(p_j) \quad \text{and} \quad C_2(p_i) \geq C_2(p_j),$$

while the minterms covered by p_i are also covered by p_j. Hence, for the purpose of obtaining a minimum cost circuit realization of f we shall delete the implicants contained in others and thus only consider the prime implicants of f; for that reason in the sequel P will denote the set of prime implicants.

Note that if f is a Boolean function, then $t(p_j)$ is the number of complemented and uncomplemented variables that appear in p_j.

A *disjunctive irredundant form* is an expression of a Boolean function as a disjunction of implicants where:

1. No implicant may be deleted,

2. No letter $x_i^{(C_i)}$ may be replaced by $x_i^{(C_i')}$, $C_i \subset C_i'$, without modifying the value of the function.

For Boolean functions the statement 2 must be replaced by the following:

2′. No letter of an implicant may be deleted.

The following Theorems are an immediate consequence of the above definitions.

Theorem 17. *A disjunctive irredundant form of a Boolean function f is a disjunction of prime implicants of f.*

Theorem 18. *There exists at least one disjunctive irredundant form of a Boolean function which simultaneously minimizes the costs C_1 and C_2.*

It is possible for some Boolean functions to have one or more minterms that are covered by one, and only one, prime implicant. Prime implicants of this type are called *essential prime implicants*; any irredundant form of a Boolean function contains all its essential prime implicants.

Example 10 (continued). A disjunctive irredundant form is:

$$\bar{x}_1x_2\bar{x}_4 \vee \bar{x}_2x_3\bar{x}_4 \vee x_0\bar{x}_1\bar{x}_2 \vee x_0\bar{x}_2\bar{x}_3x_4.$$

Every prime implicant of this irredundant form is essential: p_1 is essential with respect to m_0, p_3 with respect to m_7, p_4 with respect to m_{10} and p_5 with respect to m_{11} (see Figure 11). It is the only disjunctive irredundant form for the function of Figure 10 and it represents thus a minimal cost network.

Let us summarize the results of this Section 4.2; the obtention of a minimal two-level AND–OR network may be reached by performing successively the following computations.

(1) Starting from any normal form of a Boolean function, the implicant–implicate method allows us to obtain all the prime implicants (and all the prime implicates) of this function.

(2) The optimal covering problem deduces from the set of prime implicants all the irredundant forms of the function.

(3) A cost study allows us to select the minimal irredundant form(s).

Conjunctive irredundant forms may be defined in a similar way; minimal OR–AND networks are derived from these forms. In view of Theorem 12 these irredundant forms also lead to the generation of minimal NAND–NAND and NOR–NOR networks when dealing with Boolean functions.

The problem of optimizing the two-level realization of a logic circuit constitutes a core of problems which possess several extensions of practical interest; two of these possible extensions, namely the optimization of incompletely specified functions and the optimization of multiple-output functions are considered in Sections 4.2.4 and 4.2.5, respectively.

4.2.4 Incompletely specified functions

Recall that the basic specification of a logic function is the value vector which, for switching functions, is often given in a Karnaugh map form. The value is a listing of the values of the function for the m^n possible combinations of n variables. Our design process basically consists of translating a description of logical specifications into a value vector and then finding a specific function which realizes this value vector and satisfies some criterion of minimum cost. So far, we have assumed that the function values were strictly specified for all the m^n possible input combinations. This is not always the case. Sometimes certain input combinations never occur due to various external constraints. Note that this does not mean that the circuit would not develop some output if this forbidden input occurred. Any logic circuit will respond in some way to any input. However, since the input will never occur, we don't care whether the final circuit responds with a $0, 1, \ldots,$ or $m-1$ output to this forbidden input combination.

Where such situations occur, we say that the output is *unspecified.* This is indicated on the value vector by entering an '–' as the functional value. Such conditions are commonly referred to as don't-cares and functions including don't-cares are said to be *incompletely specified.* A realization of an incompletely specified function is any circuit which produces the same outputs for all combinations for which outputs are specified.

We now give some formal definitions in order to treat the problem of optimal representation for incompletely specified discrete functions.

Definition 14. A logic function f from a proper non-empty subset Z' of Z_m^n to Z_m is an *incompletely specified logic function* (or *partial logic function*). A logic function is *compatible* with the incompletely specified function f if both take the same value at every point of Z'.

Given an incompletely specified function f, there are two logic functions, compatible with f, that play a key role; the greatest logic function f_r compatible with f and the smallest logic function f_m compatible with f. They are defined as follows:

$$\begin{aligned} &f_M(\mathbf{x}) = f_m(\mathbf{x}) = f(\mathbf{x}), && \forall \underline{x} \in Z'; \\ &f_M(\mathbf{x}) = m-1 \quad \text{and} \quad f_m(\mathbf{x}) = 0, && \forall \underline{x} \notin Z'. \end{aligned} \tag{35}$$

A disjunctive representation of an uncompletely defined function f is a disjunctive normal form of any logic function compatible with f. We may again associate a cost C_1 or C_2 and define an optimal disjunctive form as one with a minimum cost.

Theorem 19. *An optimal disjunctive form of an incompletely specified logic function f can be obtained by optimally covering the set M_m of minterms of f_m by means of the set P_M of prime implicants of f_M.*

Proof. Consider a set $\{p_0, \ldots, p_{k-1}\}$ of cubes. Their disjunction is a representation of f iff the function

$$g = \bigvee_j p_j$$

is compatible with f, i.e. iff

$$f_m \leq \bigvee_j p_j \leq f_M.$$

This last relation is clearly equivalent to the conjunction of the two following properties:

(a) Each p_i is an implicant of f_M;

(b) The minterms of f_m are covered by $\{p_0, \ldots, p_{n-1}\}$.

The proof is then completed by taking into account the implication (34). □

An example of optimizing incompletely defined functions is given at the end of the next section.

4.2.5 Multiple-output functions

So far we have considered only the implementation of a single function of a given set $\mathbf{x}$ of variables. In the design of complete systems, we frequently wish to implement a number of different functions of the same set $\mathbf{x}$ of variables. We can implement each function completely independently by the techniques already developed, but considerable savings can often be achieved by the sharing of hardware between various functions. The multiple-logic function minimization must take into account the following items.

(1) Each minimized function should have as many terms in common with those in other minimized functions as possible.

(2) Each minimized function should have a minimum number of implicants and no implicant of it can be replaced by an implicant with less variables.

The detection of the common implicants for synthesizing p functions $\{f_0, f_1, \ldots, f_{p-1}\}$ may be stated in the following way: the implicants which can be used for the synthesis of each of p functions f_i are necessarily prime implicants of the product: $\bigwedge_j f_j$. This indicates that one has not only to evaluate the prime implicants of each of the f_i but also the prime implicants of any product of q $(2 \leqslant q \leqslant p)$ functions f_i. A systematic way for producing these prime implicants is to construct an *image function*:

$$\psi = \bigwedge_{i=0}^{p-1} (z_i \vee f_i(\mathbf{x})), \tag{36}$$

in which the z_i's are auxiliary counting variables.

From (36) we deduce that the substitution by the disjunction of all its prime implicants of each $f_i \in I$ generates in (36) the prime implicants of the product $\bigwedge_{i \in I} f_i$ with $\bigwedge_{i \in J} z_j$ $(J = \{0, 1, \ldots, p-1\} \backslash I)$ as coefficient. Hence all the prime implicants of ψ which do not contain the counting variable z_k may be used for covering the function f_k.

From a practical point of view, the construction of the image function (36) is useless. Observe first that the concepts of minterm, implicant, and cube also hold for multiple output (sometimes also referred as generalized) logic functions; in this case the cubes are discrete functions of the type:

$$\ell \wedge x_0^{(C_0)} \wedge x_1^{(C_1)} \wedge \ldots \wedge x_{n-1}^{(C_{n-1})},$$

where ℓ is a p-tuple of Z_m. One may then associate a cost with every cube, implicant or prime implicant and look for optimal disjunctive representations of a given p-tuple of functions. There is, however, an important difference with the case of single-output discrete functions: we do not necessarily obtain a disjunctive form for the functions $\{f_0, \ldots, f_{p-1}\}$ which simultaneously optimizes both cost criteria c_1 and c_2.

In the case where we choose c_1 as cost function (number of gates), we may use the same method as for single-output logic functions: first find the prime implicants of the multiple-output logic function, and then search for an optimal covering of the minterms of this function by means of its prime implicants.

In the case where we choose c_2 as cost function (number of gate inputs) the problem may be solved by introducing a new ordering on the set of cubes (see, for example, Davio, Deschamps, and Thayse, 1978, ch. 8).

Example 12: *Seven-segment display circuit.* For display of decimal digits, the standard form is the seven-segment display shown in Figure 13. Each of the seven segments is a separate crystal which can be turned on or off individually. It should be apparent to the reader that each decimal digit can be formed by lighting some subset of the seven segments. To control this display, we must generate a 7-tuple Boolean function to indicate whether each segment should be on or off. If we let 0 correspond to OFF and 1 to ON, the Karnaugh map for the 7-tuple of Boolean functions will be shown as in Figure 14. Applying the methods proposed in Sections 4.2.4 and 4.2.5 we obtain the following optimal forms to which is associated the network of Figure 15.

$$f_0 = x_3 \vee x_1 \vee \bar{x}_2\bar{x}_0 \vee x_2\bar{x}_1x_0,$$

$$f_1 = \bar{x}_2 \vee \bar{x}_1\bar{x}_0 \vee x_1x_0,$$

$$f_2 = x_3 \vee x_2 \vee \bar{x}_1 \vee x_0,$$

$$f_3 = x_3 \vee x_1\bar{x}_0 \vee \bar{x}_2x_1 \vee \bar{x}_2\bar{x}_0 \vee x_2\bar{x}_1x_0,$$

$$f_4 = \bar{x}_2\bar{x}_0 \vee x_1\bar{x}_0,$$

$$f_5 = x_3 \vee x_2\bar{x}_0 \vee \bar{x}_1\bar{x}_0 \vee x_2\bar{x}_1,$$

$$f_6 = x_3 \vee x_1\bar{x}_0 \vee x_2\bar{x}_1 \vee \bar{x}_2x_1.$$

Observe that the minimization of the only function f_0 gives the normal form:

$$f_0 = x_4 \vee x_1 \quad \bar{x}_2\bar{x}_0 \vee x_2x_0.$$

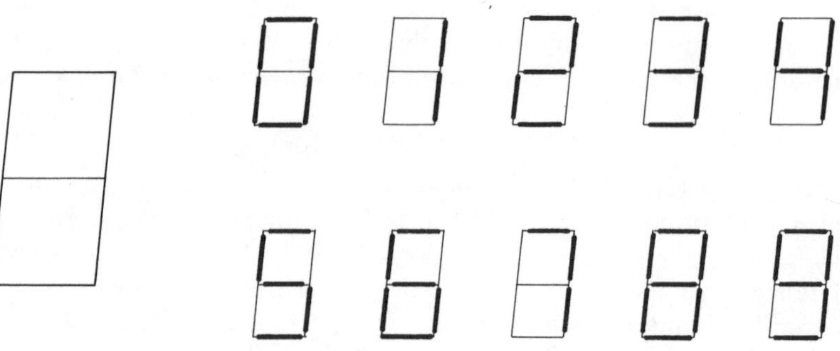

Figure 13. Seven-segment display of 10 numerals 0–9

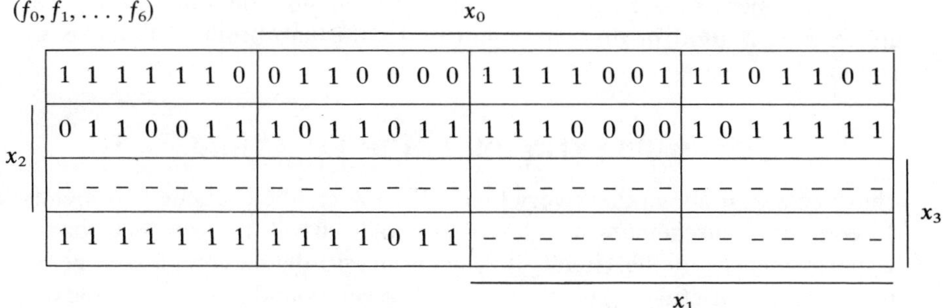

$(f_0, f_1, \ldots, f_6)$ — columns under x_0 bar: 2nd and 3rd; under x_1 bar: 3rd and 4th; rows beside x_2 bar: 2nd and 3rd; beside x_3 bar: 3rd and 4th.

1 1 1 1 1 1 0	0 1 1 0 0 0 0	1 1 1 1 0 0 1	1 1 0 1 1 0 1
0 1 1 0 0 1 1	1 0 1 1 0 1 1	1 1 1 0 0 0 0	1 0 1 1 1 1 1
- - - - - - -	- - - - - - -	- - - - - - -	- - - - - - -
1 1 1 1 1 1 1	1 1 1 1 0 1 1	- - - - - - -	- - - - - - -

Figure 14. Karnaugh map of the seven-segment display functions

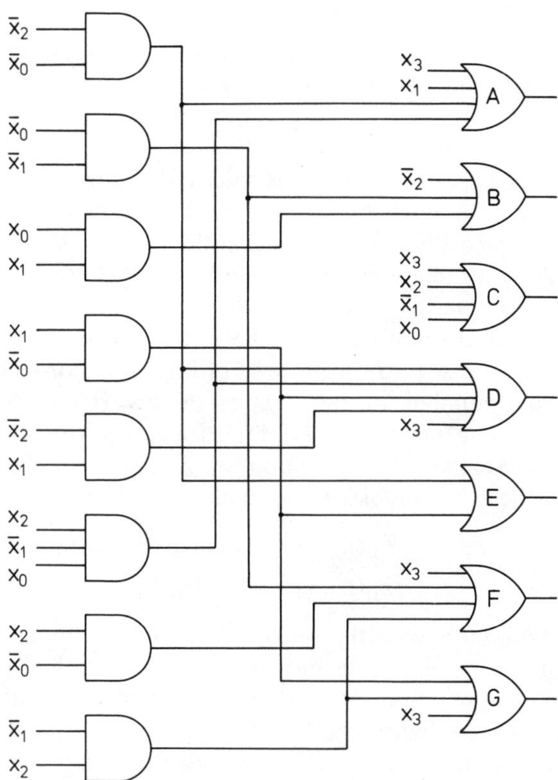

Figure 15. Seven-segment display circuit

The replacement of x_2x_0 by $x_2\bar{x}_1x_0$ does not change the value of f_0, and since $x_2\bar{x}_1x_0$ is used in the expression of f_3, this substitution allows us to drop an AND-gate.

5 COMPLEXITY OF LOGIC FUNCTIONS

It has been seen above that every logic function can be evaluated by means of a computation scheme which describes a logic circuit. In order to estimate the performances of the circuit, it is necessary to define complexity measures. It is also useful to know upper and lower bounds of the complexity measures, which hold for all logic functions. These are the two main goals of complexity theory.

5.1 Complexity measures

In this section we introduce two complexity measures for logic functions.

Consider a set Π of computation primitives. A set

$$\mathbf{f}=\{f_0, f_1, \ldots, f_{m-1}\}$$

of logic functions is Π-*realizable* if there exists at least one computation scheme A over Π that computes every function f_i, i.e.

$$\mathbf{f}\subseteq A.$$

The logic circuit associated with A is said a Π-*realization* of $\mathbf{f}$.

As it has been seen in Chapter I (Section 3.3.3), computation scheme A is characterized by a cost and a computation time. The cost corresponds to the *number of logic gates* (computation resources) of the circuit, and the computation time corresponds to the *depth* of the circuit, i.e. the maximum number of gates encountered when passing from any input to any output.

The *computation cost* $C_\Pi(\mathbf{f})$ of the set $\mathbf{f}$ of logic functions with respect to Π is the minimum number of logic gates in any Π-realization of $\mathbf{f}$. The *computation time* $D_\Pi(\mathbf{f})$ of the set $\mathbf{f}$ of logic functions with respect to Π is the minimum depth of any Π-realization of $\mathbf{f}$. The *computation complexity* $\mathscr{C}_\Pi(n, m)$ and the *time complexity* $\mathscr{D}_\Pi(n, m)$ are defined by

$$\mathscr{C}_\Pi(n, m)=\max\{C_\Pi(\mathbf{f})\},$$
$$\mathscr{D}_\Pi(n, m)=\max\{D_\Pi(\mathbf{f})\},$$

where the maximum is over the set of all possible m-tuples $\mathbf{f}$ of n-variable functions f_i. Thus, for example, the computation complexity represents the maximum number of logic gates required to realize an arbitrarily chosen m-tuple of n variable logic functions.

In order that C_Π, D_Π, $\mathscr{C}_\Pi$ and $\mathscr{D}_\Pi$ were well-defined, one must assume that Π is *functionally complete.* This means that any m-tuple $\mathbf{f}$ can actually be implemented by the computing resources of Π. Consider another set of primitives Π'. If both Π and Π' are functionally complete, then every primitive of Π is Π'-realizable, and every primitive of Π' is Π-realizable. Hence, up to a constant multiplicative factor, the complexity measures do not depend on the chosen set of computation primitives. This fact is summarized by the following Theorem 20 in which T stands for either C or D, and $\mathcal{T}$ stands for either $\mathcal{C}$ or $\mathcal{D}$. A detailed proof is given in Davio, Deschamps, and Thayse (1978) (Theorems 10.1 and 10.2).

Theorem 20

(1) *For any set of logic functions* $\mathbf{f}=\{f_0, f_1, \ldots, f_{m-1}\}$

$$K_{T1}T_{\Pi}(\mathbf{f}) \leqslant T_{\Pi'}(\mathbf{f}) \leqslant K_{T2}T_{\Pi}(\mathbf{f}),$$

where K_{T1} *and* K_{T2} *are positive rational numbers that depend on* Π *and* Π' *but not on the functions.*

(2) *For any value of the parameters* n *and* m

$$K_{T1}\mathcal{T}_{\Pi}(n, m) \leqslant \mathcal{T}_{\Pi'}(n, m) \leqslant K_{T2}\mathcal{T}_{\Pi}(n, m),$$

where K_{T1} *and* K_{T2} *are positive rational numbers that depend on* Π *and* Π' *but not on* n *and* m.

5.2 Computation cost and computation time

Let us suppose that every computation resource of Π implements, at most, q logic functions depending on, at most, p input variables (p input wires and q output wires). A logic circuit including C computing resources of Π implements, at most, $C.q$ logic functions depending on, at most, $C.p.$ variables. Hence, if $\mathbf{f}$ is a set of m different logic functions depending on n variables,

$$C_{\Pi}(\mathbf{f}) \geqslant \frac{n}{p} \quad \text{and} \quad C_{\Pi}(\mathbf{f}) \geqslant \frac{m}{q}. \tag{37}$$

Theorem 21. *If* $\mathbf{f}$ *is a set of* m *different* n *variable logic functions,*

$$C_{\Pi}(\mathbf{f}) \geqslant K_1 n \quad \textit{and} \quad C_{\Pi}(\mathbf{f}) \geqslant K_2 n, \tag{38}$$

where K_1 *and* K_2 *are positive rational numbers that depend on* Π. □

Let us consider a circuit whose depth is D and which implements m logic functions. The number of input wires of this circuit is at most equal to

$$mp^D. \tag{39}$$

This is shown in Figure 16. Hence, if $\mathbf{f}$ is a set of m different logic functions depending on n variables,

$$mp^{D_{\Pi}(\mathbf{f})} \geqslant n$$

and

$$D_{\Pi}(\mathbf{f}) \geqslant \frac{\log(n/m)}{\log p}. \tag{40}$$

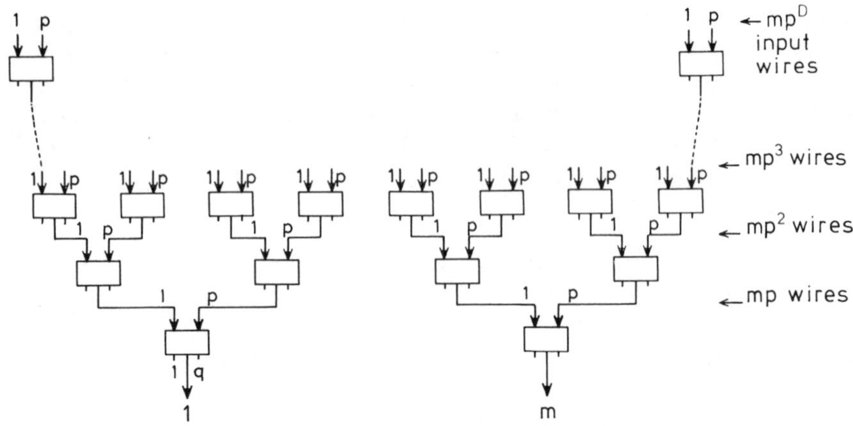

Figure 16. Tree interconnection of p-input gates

Notice that Figure 16 corresponds to the case where the m functions depend on disjoint subsets of variables. If at least one function depends on all n variables, the number of input wires corresponding to that particular function is equal to, at most,

$$p^D.$$

Hence,

$$D_\Pi(\mathbf{f}) \geqslant \frac{\log n}{\log p}. \tag{41}$$

Theorem 22. *If* $\mathbf{f}$ *is a set of m different n variable logic functions,*

$$D_\Pi(\mathbf{f}) \geqslant K \log (n/m), \tag{42}$$

where K is a positive rational number depending on Π. *If at least one of the m logic functions actually depends on all n variables*

$$D_\Pi(\mathbf{f}) \geqslant K \log n. \tag{43}$$

5.3 Computation complexity

Our main concern in this section is to trace the asymptotic dependence of $\mathscr{C}_\Pi(n, m)$. Theorem 20 allows us to choose at will the set Π. We make use of a single type of computation resource called *U-operator.* In r-ary logic, that is, when

$$S = \{0, 1, \ldots, r-1\},$$

it realizes a function U from S^{r+1} to S:

$$U(x_0, x_1, \ldots, x_{r-1}, x) = x_0 x^{\{0\}} \vee x_1 x^{\{1\}} \vee \ldots \vee x_{r-1} x^{\{r-1\}}.$$

In the binary case ($r=2$), the corresponding operator is a one control variable multiplexer (Chapter III).

The construction used in the proof of Theorem 23 is based on the next lemma.

Lemma 2. *The set of all n-variable logic functions can be realized by means of $r^{(r^n)}$ U-operators.*

Proof. Every one-variable function can be realized by means of one operator:

$$f(x)=U(f(0), f(1), \ldots, f(r-1), x).$$

One thus needs r^r operators in order to realize the r^r one-variable functions. The proof is achieved by induction on n. Suppose that one has constructed a circuit that realizes all the functions of $(n-1)$ variables. Every function which effectively depends on n variables can be written under the form

$$f(x_0, \ldots, x_{n-2}, x_{n-1})=U(f(x_0, \ldots, x_{n-2}, 0), \ldots, f(x_0, \ldots, x_{n-2}, r-1), x_{n-1}).$$

One needs one additional operator for every additional function. Hence, the total number of operators is equal to the total number of functions, that is $r^{(r^n)}$. □

Theorem 23. *The computation complexity is bounded by*

$$K_1\frac{r^s}{s}\leq \mathscr{C}_\Pi(n, m)\leq K_2\frac{r^s}{s},$$

where $s=n+\log_2 m$ and where K_1 and K_2 only depend on Π.

Proof.

(1) Lower bound. Consider a loop free circuit including N U-operators, n input variables $x_0, \ldots, x_{n-1}$, and m output variables $y_0, \ldots, y_{m-1}$. It can be drawn as shown in Figure 17. The connection network has $(n+r+N)$ source points and $(m+(r+1)N)$ destination points.

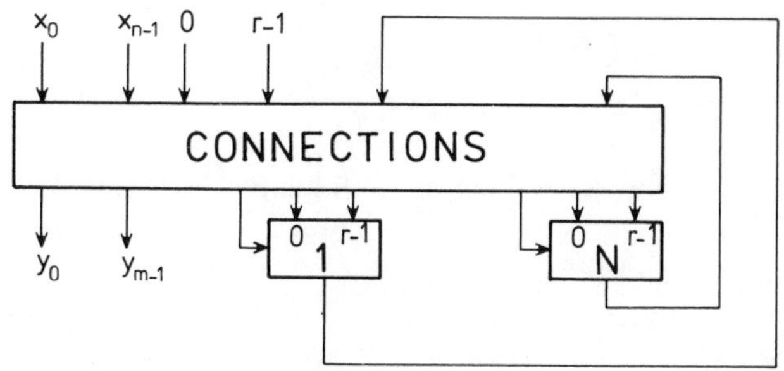

Figure 17. Interconnection of U-operators

If the circuit realizes m different functions, then

$$n+r+N \geqslant m. \tag{44}$$

If all source points are used, then

$$m+(r+1)N \geqslant n+r+N, \quad \text{i.e. } m+rN \geqslant n+r. \tag{45}$$

This occurs if the output functions depend on all n variables and if (for instance) one of the functions y is such that $y(x, \ldots, x) = x \oplus 1$, mod r: in a loop-free interconnection of U-operators, for every value of the input variables, each output wire is connected to either an input variable or to a constant value; as $y(x, \ldots, x) \neq x$, the output corresponding to y is connected to one of the constant values and all these values are used at least once. According to (44) and (45) and to the definition of $\mathscr{C}_\Pi(n, m)$ we may conclude that

$$\alpha = n+r+\mathscr{C}_\Pi(n, m) \geqslant m \tag{46}$$

and

$$m+r\mathscr{C}_\Pi(n, m) \geqslant n+r;$$

thus

$$\beta = m+(r+1)\mathscr{C}_\Pi(n, m) \geqslant \alpha. \tag{47}$$

On the other hand, the total number of different interconnections of $\mathscr{C}_\Pi(n, m)$ operators must be greater than, or equal to, the number of m-tuples of n variable functions. Hence,

$$\alpha^\beta \geqslant r^{(mr^n)} = r^{(r^s)}, \tag{48}$$

where

$$r^s = mr^h \quad \text{and thus} \quad s = n+\log_r m.$$

From (48) and (47) we deduce

$$\beta \log_r \beta \geqslant \beta \log_2 \alpha \geqslant r^s.$$

Hence,

$$\beta \geqslant \frac{r^s}{s}, \tag{49}$$

since in the contrary case

$$\beta < \frac{r^s}{s}, \qquad \log_r \beta < s - \log_r s, \qquad \beta \log_r \beta < \frac{r^s}{s}(s - \log_r s) < r^s.$$

According to (49) and (46)

$$n+r+\mathscr{C}_\Pi(n, m)+(r+1)\mathscr{C}_\Pi(n, m) \geqslant m+(r+1)\mathscr{C}_\Pi(n, m) = \beta \geqslant \frac{r^s}{s};$$

hence

$$\mathscr{C}_{\Pi}(n, m) \geqslant \left(\frac{1}{r+2}\right)\left(\frac{r^s}{s}\right)\left(1 - \frac{(n+r)s}{r^s}\right). \tag{50}$$

If s is large enough

$$\frac{(n+r)s}{r^s} \leqslant \frac{1}{r+3}$$

and thus

$$\mathscr{C}_{\Pi}(n, m) \geqslant \left(\frac{1}{r+3}\right)\left(\frac{r^s}{s}\right). \tag{51}$$

There exists only a finite number of cases under which s is smaller than a given amount, i.e. does not satisfy the above assumptions. It is possible to handle these cases by an appropriate choice of K_1.

(2) Upper bound. Any m-tuple of functions of n variables may be synthesized by the following two steps:

Step 1. Compute all the functions of k variables ($1 \leqslant k \leqslant n$). This requires

$$r^{(r^k)}$$

operators (Lemma 2).

Step 2. Compute each output function by a tree of U-operators whose input are k variable functions obtained in step 1. Indeed, a function f depending on n variables can be written under the form

$$f(\mathbf{x}) = U(f(x_0, \ldots, x_{n-2}, 0), \ldots, f(x_0, \ldots, x_{n-2}, r-1), x_{n-1}).$$

Every $(n-1)$-variable function $f(x_0, \ldots, x_{n-2}, e)$, with $e \in \{0, 1, \ldots, r-1\}$, can also be decomposed, and so on. We finally obtain a tree including

$$1 + r + r^2 + \ldots + r^{n-k-1} = \frac{r^{n-k} - 1}{r-1}$$

operators.

The total number of operators is

$$N = r^{(r^k)} + m\frac{r^{n-k} - 1}{r-1}$$

$$= r^{(r^k)} + r^{t-k} - \frac{m}{r-1}, \tag{52}$$

where

$$r^{t-k} = \frac{m}{r-1} r^{n-k}, \text{ i.e. } t = n + \log_r m - \log_r (r-1). \tag{53}$$

Notice that

$$0 \leqslant \log_r r - 1 < 1 \quad \text{and thus} \quad s - 1 < t \leqslant s. \tag{54}$$

Let us choose the value of k in order to minimize N. There are three cases.

Case 1. $1 \leq \log_r (t - \log_r t) < n.$ (55)
We choose

$$k = \lfloor \log_r (t - \log_r t) \rfloor .$$

Hence

$$\log_r (t - \log_r t) - 1 < k \leq \log_r (t - \log_r t),$$

$$\frac{t - \log_r t}{r} < r^k \leq t - \log_r t,$$

$$r^{(r^k)} \leq \frac{r^t}{t},$$

$$N < r^{(r^k)} + r^{t-k} < \frac{r^t}{t} + r^t \frac{r}{t - \log_r t} = \frac{r^t}{t}\left(1 + \frac{rt}{t - \log_r t}\right).$$

The study of

$$g(t) = \frac{t}{t - \log_r t}$$

indicates that g is maximum when $t = e$. The maximum value is

$$\frac{e}{e - \log_r e}.$$

If $r = 2$, then $t = s$ and

$$N < K_2 \frac{2^s}{s}, \quad \text{where } K_2 = 1 = \frac{re}{e - \log_2 e} = 5.26\ldots.$$

If $r \geq 3$, let us write

$$\frac{r^t}{t} = \frac{r^s}{s}\left(\frac{sr^{s-t}}{t}\right) = \frac{r^s}{s}\left(\frac{1}{t(r-1)}\right) < \frac{r^s}{s} \frac{s}{(s-1)(r-1)}.$$

According to (55)

$$t - \log_r t \geq r \text{ and thus } s \geq t \geq r \geq 3.$$

As $g(t)$, $t \geq 3$, and $s/s-1$ are decreasing functions of t and s we may write

$$N < \frac{r^s}{s} \frac{r}{(r-1)^2}(1 + rg(r)) = K\frac{r^s}{s},$$

where

$$K = \frac{r}{(r-1)^2} + \frac{r^3}{(r-1)^3}. \tag{56}$$

Case 2. $\mathrm{Log}_r (t - \log_r t) \geq n.$

Thus

$$r^n \leqslant t - \log_r t,$$

and the construction of all functions of n variables requires

$$r^{(r^n)} \leqslant \frac{r^t}{t}$$

operators. As above we may write

$$\frac{r^t}{t} = \frac{r^s}{s}\left(\frac{s}{t(r-1)}\right) = \frac{r^s}{s} \quad \text{if} \quad r = 2,$$

$$< \frac{r^s}{s}\,\frac{s}{(s-1)(r-1)} \quad \text{if} \quad r \geqslant 3.$$

as $s \geqslant t \geqslant r^n$, $s/s-1 \leqslant r^n/r^n - 1$.

Case 3. $\log_r (t - \log_r t) < 1$.
There exists only a finite number of different pairs (n, m), for given r, such that the last inequality holds. These cases are handled by an appropriate choice of K_2. □

The circuit of Figure 17 is an example of a *free universal logic module.* It consists of gates (namely U-operators), all of whose terminals are accessible and can be connected in any desired way. If N is great enough, any m-tuple of functions of n variables can be implemented by this circuit. According to Theorem 23, a necessary condition is

$$N \geqslant K_1 \frac{r^s}{s}. \tag{57}$$

Figure 18 shows a *wired universal logic module.* The gate networks consists of a series of logic gates (for instance U-operators), with fixed

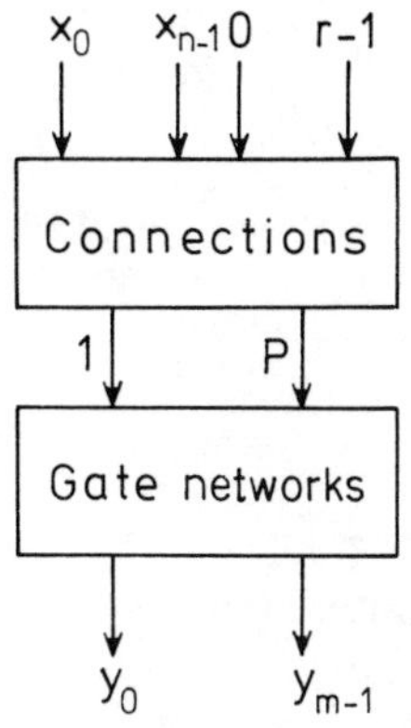

Figure 18. Wired universal logic module

interconnections. The only degree of freedom is the connection of the P inputs of the gate network to the n variables and to the r constant values. The number of different interconnections must be greater than or equal to the number of m-tuples of n variable functions:

$$(n+r)^P \geqslant r^{(mr^n)};$$

hence,

$$P \geqslant \frac{r^s}{\log_r (n+r)}.$$

If every gate has, at most, p inputs, then

$$N_p \geqslant P \quad \text{and} \quad N \geqslant \frac{r^s}{p \log_r (n+r)}.$$

There exists a constant K such that

$$N \geqslant K \frac{r^s}{\log_r n}. \tag{58}$$

Figure 19 shows a *canonical universal logic module.* The gate network consists of a series of logic gates with fixed interconnections. Furthermore, input number i of the gate network may only be connected to a subset Γ_i of the set of $(n+r+M)$ inputs of the connection network, and the number of inputs in Γ_i, denoted by γ_i, depends on the index i but not on n or m. Let us first give an example: the circuit of Figure 20 is functionally equivalent to a programmable logic array. The gate network implements the functions:

$$\bar{x}_0, \ldots, \bar{x}_{n-1},$$

$$u_j = \bigwedge_{i=0}^{n-1} t_i^j, \quad \text{where } t_i^j \in \{x_i, \bar{x}_i, 1\},$$

$$w_\ell = \bigvee_{j=0}^{k-1} v_j^\ell, \quad \text{where } v_j^\ell \in \{u_j, 0\},$$

$$y_\ell \in \{w_\ell, \bar{x}_\ell\}.$$

Hence y_ℓ is either a sum of products of variables or a product of sums of variables. If k is great enough, it is a universal module. Notice that

$$\gamma_0 = \ldots = \gamma_{n-1} = 1,$$

$$\gamma_n = \ldots = \gamma_{n+nk-1} = 3,$$

$$\gamma_{n+nk} = \ldots = \gamma_{n+nk+mk-1} = 2.$$

The number of different interconnections is greater than or equal to the number of m-tuples:

$$\gamma_1 \gamma_2 \ldots \gamma_P \geqslant r^{(mr^n)}.$$

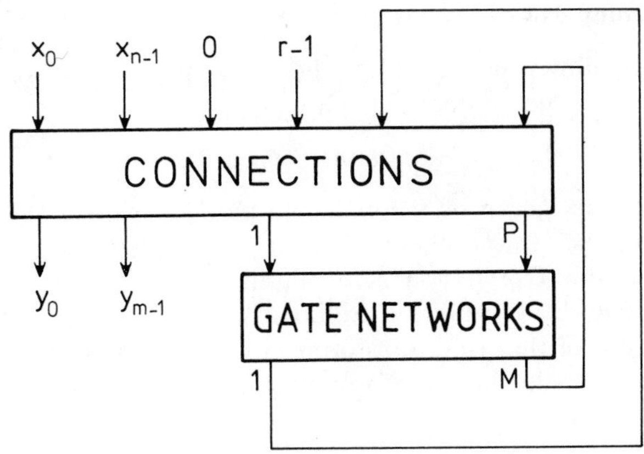

Figure 19. Canonical universal logic module

Let us denote by γ the largest γ_i. Thus,

$$\gamma^p \geq \gamma_1 \gamma_2 \dots \gamma_p \geq r^{(mr^n)},$$

and

$$P \geq \frac{r^s}{\log_r \gamma}.$$

If every gate has, at most p inputs, then

$$N^p \geq P \quad \text{and} \quad N \geq Kr^s, \tag{59}$$

where

$$K = \frac{1}{p \log_r}.$$

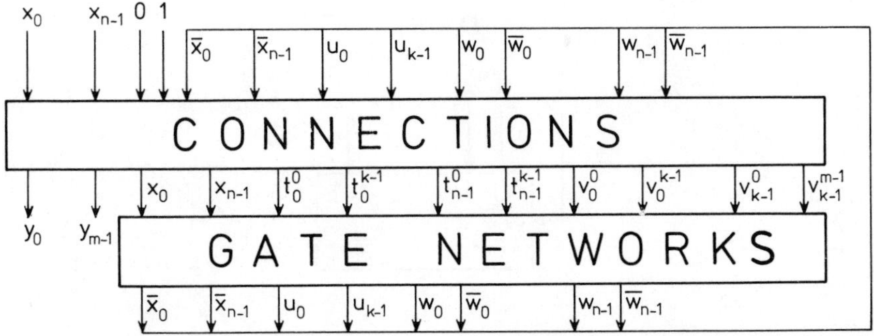

Figure 20. Example of canonical universal logic module

5.4 Time complexity

Theorem 20 allows us again to choose a single type of computation resource, namely the U-operator. Observe that

$$\mathscr{D}_{\Pi}(n, m) = \mathscr{D}_{\Pi}(n, 1)$$

since m functions f_i can be computed in parallel. Notice also that if f has computation time $D_{\Pi}(f)$, there exists a tree circuit that realizes f with computation time $D_{\Pi}(f)$: if a gate output drives k gate inputs, we may replace that gate by k copies of itself, each of them driving one gate input. A recursive use of that rule transforms any given circuit into a tree circuit, and it can eventually be transformed into a complete tree as shown by Figure 21.

Theorem 24. *The time complexity is bounded by*

$$K_1 n \leqslant \mathscr{D}_{\Pi}(n, m) \leqslant K_2 n,$$

where K_1 and K_2 only depend on Π.

Proof.

(1) Lower bound. A complete tree of U-operators, whose depth is equal to D, has

$$(r+1)^D$$

inputs. Each input is either a variable or a constant value. The number of different trees must be greater than or equal to the number of n variable functions:

$$(n+r)^{((r+1)^D)} \geqslant r^{(r^n)}.$$

Hence,

$$D \geqslant \frac{n - \log_r \log_r (n+r)}{\log_r (r+1)},$$

and there exists a constant K_1 such that

$$D \geqslant K_1 n.$$

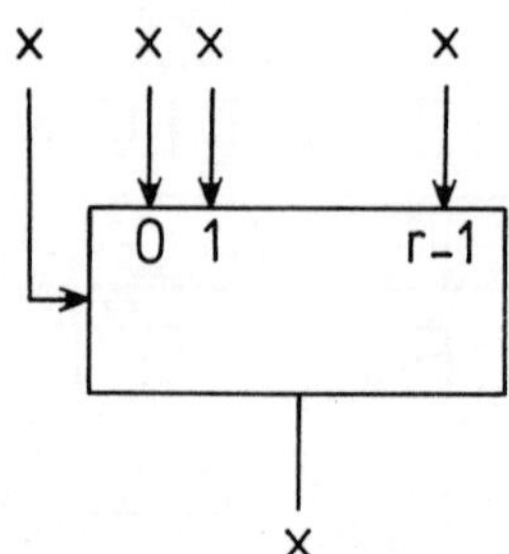

Figure 21. Single connection realized by a U-operator

(2) Upper bound. Every n variable function can be realized by a tree interconnection of $(r^n - 1)$ operators, whose depth is n: the construction is implicitly included in the proof of Lemma 2. □

BIBLIOGRAPHICAL REMARKS

The material in this chapter has been selected to meet the objective of being concise and providing the necessary mathematical background for this book.

There exist several texts dealing with lattice theory and Boolean algebra: apart from the original work by Boole (1854), let us mention: Rutherford (1965), Grätzer (1971), Birkhoff (1967), Birkhoff and Bartee (1970), Hohn (1970), Maclane and Birkhoff (1967), Sikorski (1964), and Rudeanu (1974).

Brief expositions of the foundations of lattices and of Boolean algebras are given in many books on switching theory: Kohavi (1970), Hill and Peterson (1974), Dietmeyer (1971), Davio, Deschamps, and Thayse (1978), Lee (1978), and Preparata and Yeh (1973).

The classical papers about optimization are those of Venn (1876, 1894), Veitch (1952), Quine (1952, 1953, 1955), McCluskey (1956), and Karnaugh (1953).

Most of the material of Section 5, devoted to complexity of logic functions, can be found in Davio and Quisquater (1975), Savage (1972), Muller (1956), and Davio, Deschamps, and Thayse (1978). Closer bounds for both the computation and time complexities have been derived by several authors. Let us quote Lupanov (1958), Winograd (1967), and Spira (1971a, 1971b, and 1973).

EXERCISES

E1. Show that, in any distributive lattice L

$$\forall a, b, c \in L : (a \vee b) \wedge (b \vee c) \wedge (c \vee a) = (a \wedge b) \vee (b \wedge c) \vee (c \wedge a).$$

E2. Show that, in any Boolean algebra B, the following properties hold true:

(i) $a \leqslant b \Leftrightarrow a\bar{b} = \mathbf{0} \Leftrightarrow \bar{a} \vee b = \mathbf{1}$;

(ii) (Boolean absorption):

$$\forall a, b \in B : a \vee \bar{a}b = a \vee b$$

$$a(\bar{a} \vee b) = ab;$$

(iii) $\overline{(a_0\bar{x} \vee a_1 x)} = \bar{a}_0\bar{x} \vee \bar{a}_1 x$.

E3. Analyse the circuits shown in Figure E1. Show that they both realize the modulo two sum of their inputs.

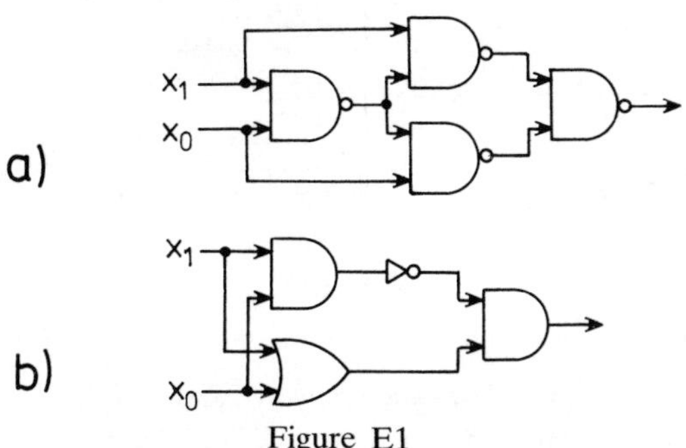

Figure E1

Show furthermore that a lamp, controlled by three two-positional switches x_0, x_1, x_2 is on iff

$$x_0 \oplus x_1 \oplus x_2 = 1.$$

under the initial condition that it is off for $x_0 = x_1 = x_2 = 0$.

E4. Analyse the circuit shown in Figure E2 and show that its output f realizes the 16 possible functions of two variables A and B when one replaces the inputs $S_3S_2S_1S_0$ by the 16 binary four-tuples.

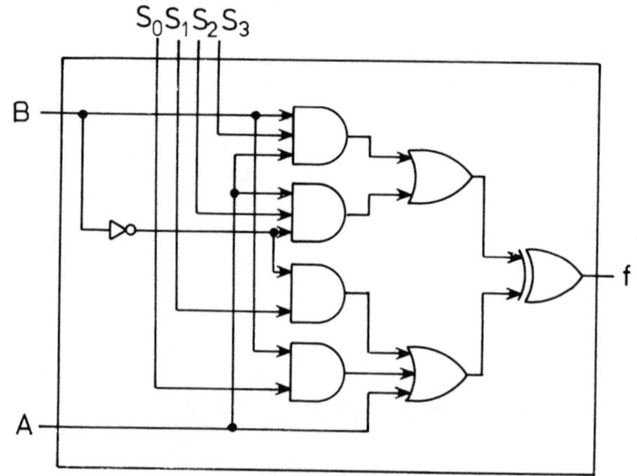

Figure E2

E5. Analyse the circuit shown in Figure E3 and show that it is a full-adder. Which of the three inputs would you select as carry input? Use this full-adder to design a radix-4 full-adder (5 inputs, 3 outputs). On the other hand design a radix-4 full-adder as a three-level circuit of NAND gates using as few gates as possible. Compare the realizations of the two obtained radix-4 full-adders.

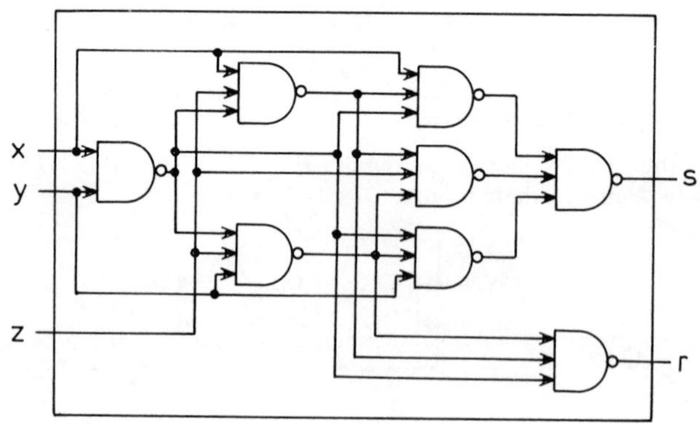

Figure E3

E6. The behaviour of a three-input two-output circuit (Figure E4) is described by:

$$\text{if } w = 0, \quad \text{then } z_0 = x_0 \quad \text{and} \quad z_1 = x_1,$$
$$\text{if } w = 1, \quad \text{then } z_0 = x_1 \quad \text{and} \quad z_1 = x_0.$$

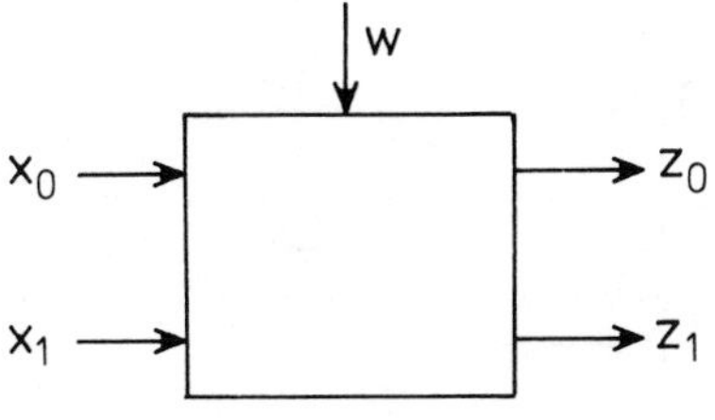

Figure E4

Write down the truth tables, the canonical disjunctive expansion, and the canonical conjunctive expansion of z_0 and z_1 in terms of x_0, x_1, w. Show that z_1 and z_0 are also given by the expressions

$$z_0 = \bar{w}x_0 \vee wx_1,$$

$$z_1 = \bar{w}x_1 \vee wx_0.$$

Design the circuit
(i) with NOR gates,
(ii) with multiplexers.

E7. A *half-adder* is a circuit having two inputs x, y and two outputs $z = xy$ and $w = x \oplus y$. Design a full-adder using half-adders and an OR-gate.

E8. Figure E5 presents a number of binary coded decimal (BCD) systems used to represent the integers $0, 1, \ldots, 9$. Design minimal two-level circuits for converting from one code to the others. Use NAND gate synthesis or NOR gate synthesis. Assume that both the input variables and their complement are available.

	8	4	2	1	Ex. – 3				2	4	2	1	2 out of 5				
0	0	0	0	0	0	0	1	1	0	0	0	0	0	0	0	1	1
1	0	0	0	1	0	1	0	0	0	0	0	1	1	1	0	0	0
2	0	0	1	0	0	1	0	1	0	0	1	0	1	0	1	0	0
3	0	0	1	1	0	1	1	0	0	0	1	1	0	1	1	0	0
4	0	1	0	0	0	1	1	1	0	1	0	0	1	0	0	1	0
5	0	1	0	1	1	0	0	0	1	0	1	1	0	1	0	1	0
6	0	1	1	0	1	0	0	1	1	1	0	0	0	0	1	1	0
7	0	1	1	1	1	0	1	0	1	1	0	1	1	0	0	0	1
8	1	0	0	0	1	0	1	1	1	1	1	0	0	1	0	0	1
9	1	0	0	1	1	1	0	0	1	1	1	1	0	0	1	0	1

Figure E5

E9. Consider the circuit displayed in Figure E6a. Obtain the truth tables and Boolean expressions of the output functions $f(x_2, x_1, x_0)$ and $g(x_2, x_1, x_0)$. Show that the design principle put at work in this circuit is general in nature. Observe that such a type of circuit makes use of four types of cells, exemplified by the cells labelled A, B, C, and D in the Figure E6. Design a cell able to exhibit these four behaviours under the control of two additional variables z_1z_0 with the conventions

$$(z_1, z_0) = (0, 0) \Rightarrow A; \qquad (z_1, z_0) = (1, 0) \Rightarrow C,$$

$$(z_1, z_0) = (0, 1) \Rightarrow B; \qquad (z_1, z_0) = (1, 1) \Rightarrow D.$$

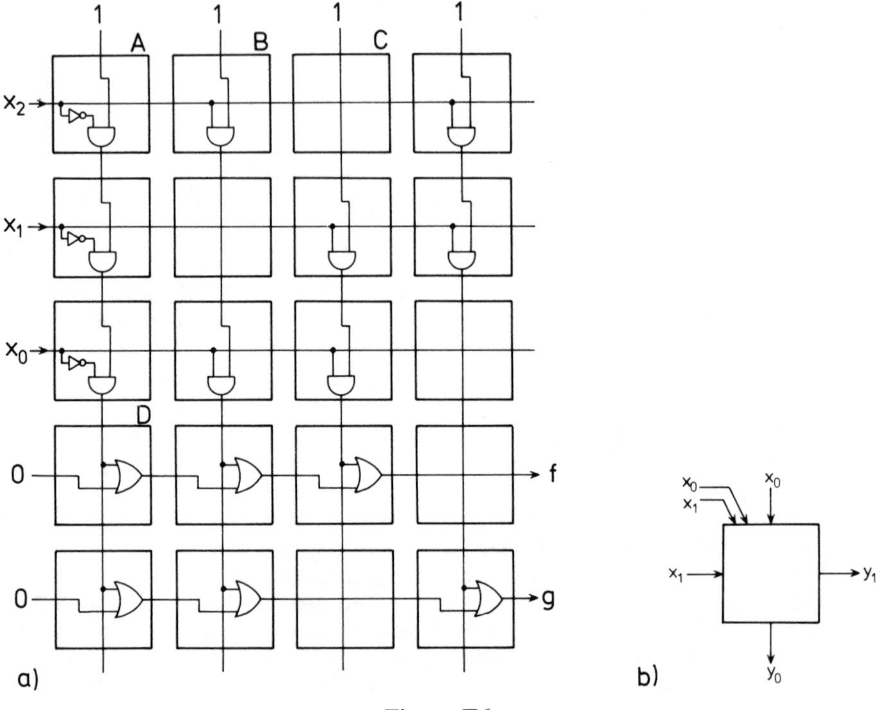

Figure E6

E10. Design a seven segment decoder having the specifications shown in Figure E7 (hexadecimal seven segment decoder).

	x_3	x_2	x_1	x_0	a	b	c	d	e	f	g	
0	0	0	0	0	0	0	0	0	0	0		0
1	0	0	0	1	1	0	0	1	1	1	1	1
2	0	0	1	0	0	0	1	0	0	1	0	2
3	0	0	1	1	0	0	0	0	1	1	0	3
4	0	1	0	0	1	0	0	1	1	0	0	4
5	0	1	0	1	0	1	0	0	1	0	0	5
6	0	1	1	0	0	1	0	0	0	0	0	6
7	0	1	1	1	0	0	0	1	1	1	1	7
8	1	0	0	0	0	0	0	0	0	0	0	8
9	1	0	0	1	0	0	0	0	1	0	0	9
10	1	0	1	0	0	0	0	1	0	0	0	A
11	1	0	1	1	1	1	0	0	0	0	0	b
12	1	1	0	0	0	1	1	0	0	0	1	C
13	1	1	0	1	1	0	0	0	0	1	0	d
14	1	1	1	0	0	1	1	0	0	0	0	E
15	1	1	1	1	0	1	1	1	0	0	0	F

Figure E7

E11. One has to determine, up to 1/10 of a turn, the angular position of a shaft. To this effect, the shaft has been provided with four concentric cams having the shapes of Figure E8. Each cam is provided with an electrical contact: a contact is closed when it faces the external circumference of the associated cam and it is open in the opposite case. Design a circuit converting the cam position in a numerical indication:

(i) in the 8421 BCD code;
(ii) in a form suitable for a 7-segment display. (Use NAND gates.)

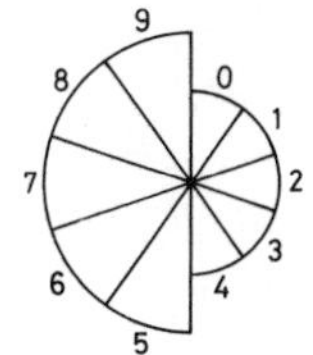

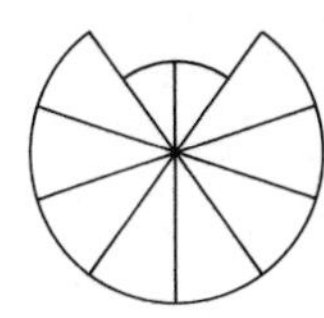
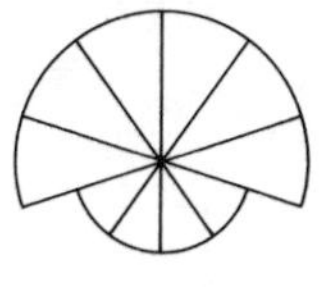
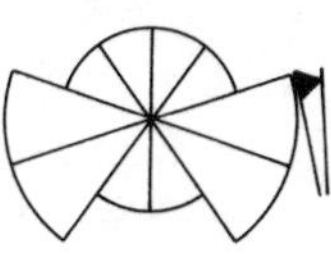

Figure E8

E12. Consider the circuits shown in Figure E9. Let $x = 2x_1 + x_0$, $y = 2y_1 + y_0$, $z = 2z_1 + z_0$, $w = 2w_1 + w_0$, $s = 2s + s_0$. Show that the equilities

(a) $x + y + r_- = 4r_+ + s$
(b) $xy + z + w = 4r + s$

determine unique integers $r_+ \in \{0, 1\}$; $r, s \in \{0, 1, 2, 3\}$. Obtain various designs of these two cells. The cell in Figure E9a is a radix-4 full-adder; the cell in Figure E9b may be used to build a radix 4 multiplier.

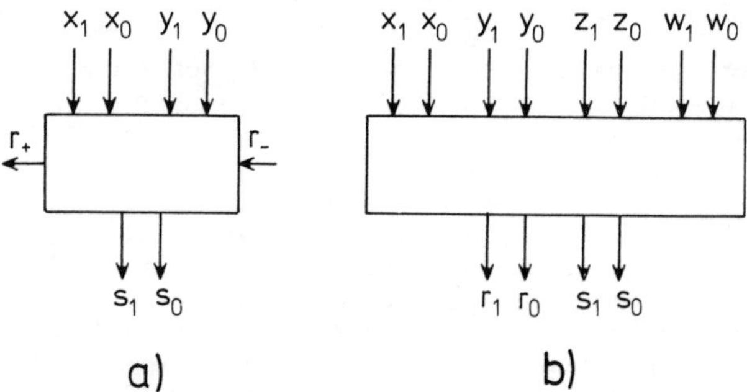

Figure E9

E13. (Marcus, 1962). Simplify the following statement: Policy no. 22 may be issued only if the applicant

1. Has been issued Policy no. 19 and is a married male, or
2. Has been issued Policy no. 19 and is married and under 25, or
3. Has been not issued Policy no. 19 and is a married female, or
4. Is a male under 25, or
5. Is married and 25 or over.

From an XYZ Insurance Company Manual.

E14. Synthesize a NAND-gate circuit for $x_1 \oplus x_0$ using the minimization techniques of Section 4. Compare your result with that of Figure E1a. This example illustrates the optimization problem for *three-level NAND synthesis.* Rigorous solutions to this problem hav been presented by Gimpel (1967) and Grasselli and Gimpel (1966). These methods are computationally complex and it is often preferable to satisfy oneself with a suboptimal solution based on the following two heuristic rules.

Rule 1. If an irredundant disjunctive normal form contains a family of term of the type

$$P\bar{x}_i \vee P\bar{x}_j \vee \ldots \vee P\bar{x}_k$$

(where P represents a product of literals), one obtains, by factoring P out, the equivalent expression:

$$P(\bar{x}_i \vee \bar{x}_j \vee \ldots \vee \bar{x}_k).$$

The resulting overall expression will be realized as a three-level circuit.

Rule 2. If, in a subset $\{p_i\}$ of prime implicants, k variables appear in each term and in such a way that a single variable is complemented in each term, then one may replace in each term the complemented literal by the sum of complemented literals which will be realized by a single NAND gate.

Use the above rules to design circuits for the following functions:

$$f = ab\bar{c} \vee ab\bar{d} \vee bc\bar{d},$$

$$g = \bar{w}xyz \vee w\bar{z} \vee w\bar{y},$$

$$h = xyz \oplus yw \oplus xw.$$

E15. Consider the canonical disjunctive expansion of a Boolean function. As we have seen, we may replace in that expansion the disjunctions '$\vee$' by modulo two additions $\oplus$. Now, for each variable, we may systematically adopt one of the three following attitudes:

(i) replace each occurrence of $\bar{x}_i$ by $1 \oplus x_i$;
(ii) replace each occurrence of x_i by $1 \oplus \bar{x}_i$; or
(iii) no operation: each term will contain either x_i or $\bar{x}_i$.

After having multiplied out and cancelled pairs of identical terms, one will obtain one among 3^n possible expansions of the given Boolean function as a modulo two sum of products. Call cost of such an expansion the number of its terms and develop a method for deriving, among the 3^n obtained expansions, an expansion with minimum cost. (Mukhopadhyay 1971, Davio, Deschamps, and Thayse 1978).

E16. *Prime implicant machine* (Mange, 1978). Consider the circuit shown in Figure E10. It has 4 inputs corresponding to the values $[f_0, f_1, f_2, f_3]$ of a binary function $f(x_1, x_0)$ of two variables. It has 9 outputs in one-to-one correspondence with the 9 possible prime implicants of a two-place function. Show that an output of this circuit takes the value 1 iff the associated product is a prime implicant of f. Describe a general algorithm for obtaining the prime implicants of a n-variable function and the corresponding combinational circuit. How many gates would have this circuit? Discuss the effect of fan-in limitations.

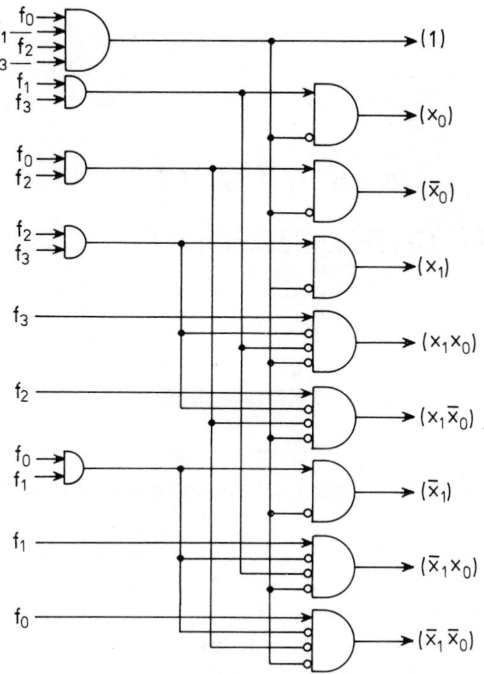
f0
f1
f2
f3
f1
f3
f0
f2
f2
f3
f3
f2
f0
f1
f1
f0
(1)
(x0)
(x̄0)
(x1)
(x1x0)
(x1x̄0)
(x̄1)
(x̄1x0)
(x̄1x̄0)

Figure E10

Chapter III
Universal combinatorial components

1 INTRODUCTION

When logic circuits were built of discrete components, or even from small unit logic blocks such as NAND or NOR gates, the cost depended directly on the number of components.

With the advent of large scale integration (LSI) the factors that determine the cost change drastically. Cost must be considered in terms of total (silicon) area as opposed to the number of devices enclosed within that area.

The main purpose of this chapter is to introduce *array logic*; it is defined as the use of memory-like structures for performing logic. Array logic is embodied in a memory-like structure wherein the variables are used as address bits and the output bits that have been selected are combined to yield a single bit which is the value of the function stored in the array for the particular combination of input bits. The array logic considered in this chapter consists of read only memory (ROM) and programmable logic array (PLA); it is generally used in logical design together with peripheral networks such as demultiplexers, address decoders, and multiplexers.

Sections 2 and 3 introduce the reader to multiplexers and demultiplexers, respectively.

Section 4 studies some logical design methods related to the use of multiplexers and demultiplexers in logical nets.

Sections 5 and 6 introduce two types of semiconductor memories, namely the ROM and the PLA, respectively.

2 MULTIPLEXERS

Multiplexing means transmitting a large number of information units over a smaller number of channels or lines. A *multiplexer* is a combinational circuit that selects data from m^p input lines and directs it to a single output line. The selction of input–output transfer paths is controlled by a set of input selection lines.

In a more precise way, a multiplexer with *p command inputs* has $p+m^p$

m-ary inputs:

$$x_0, x_1, \ldots, x_{p-1}; \qquad y_0, y_1, \ldots, y_{m^p-1},$$
(control inputs) (data inputs)

and one m-ary output z.

Let i be an integer ($i \in Z_{m^p}$) and let $(i_{p-1}, \ldots, i_1, i_0)$ its m-ary representation ($i_j \in Z_m \ \forall j$); the input–output behaviour of a multiplexer is described by the following relation:

$$z = y_i \quad \text{iff} \quad x_j = i_j \ \forall j$$

or, otherwise stated:

$$z = \bigvee_{i=0}^{m^p-1} y_i \wedge \left[\bigvee_{j=0}^{p-1} x_j^{(i_j)} \right], \qquad i = (i_{p-1}, \ldots, i_1, i_0). \tag{1}$$

The multiplexer thus consists of a switch which connects to its output z the input y_i whose index i has $(x_{p-1}, \ldots, x_1, x_0)$ as m-ary representation; it is basically a multipositional switch, as represented by Figure 1a, and it will be depicted by the gate symbol of Figure 1b.

Further on in this chapter some cost criteria for realizing multiplexers will be established; the different realization costs will always be evaluated for binary devices ($m = 2$).

Multiplexers may be realized according to different synthesis techniques. Consider first a direct synthesis of a multiplexer having p control inputs. By direct synthesis we mean, e.g., a two-level synthesis realizing the function (1); two-level synthesis illustrating the cases $p = 1$ and $p = 2$ is shown in Figures 2a and 2b respectively. The elementary building blocks are NAND-gates; we know (Chapter II, Theorem 2) that another realization is obtained by substituting for AND-gates and an OR-gate the NAND-gates of the first and second level, respectively.

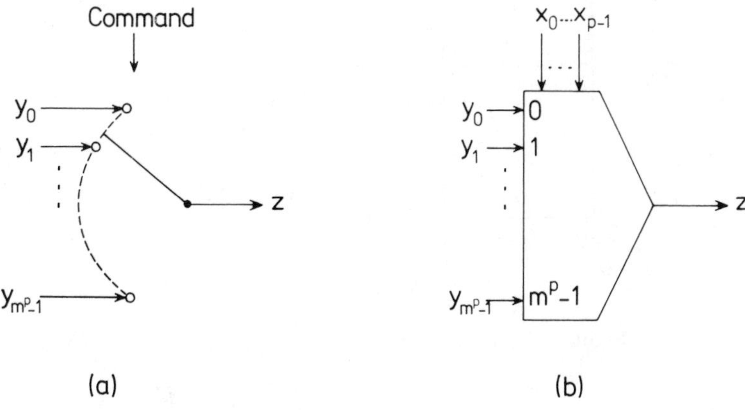

Figure 1. Multiplexer: principle and gate symbol

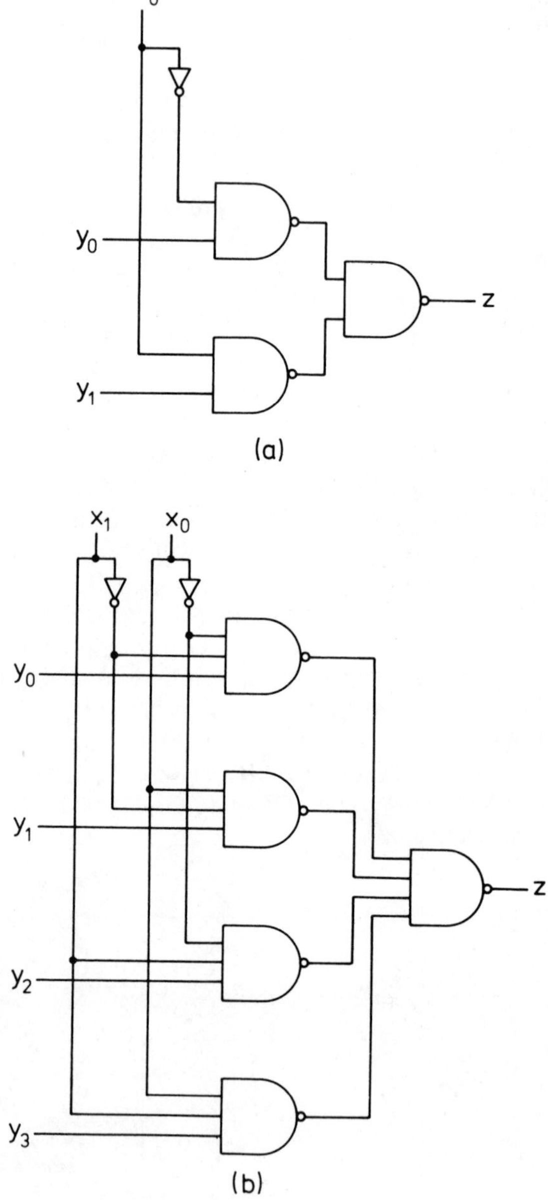

Figure 2. Two-level synthesis of a multiplexer

Let us estimate a cost for the direct realization with the number of gate inputs as the cost criterion.

The synthesis is made up of: p inverters, 2^p first level gates with $(p+1)$ inputs, 1 second level gate with 2^p inputs.

Let us describe the cost by $\Gamma_0(p)$; we have:

$$\Gamma_0(p) = p + 2^p(p+2).$$

Observe that in establishing this cost expression we assumed that there was no fan-in limitation for the gates; the main term in $\Gamma_0(p)$ is $p\ 2^p$.

A second synthesis technique for multiplexers derives from the following observation:

A multiplexer with $n = kp$ control inputs may be realized by means of multiplexers with p control inputs. This can easily be shown by using an inductive argument; any multiplexer with p control variables $(x_{2p-1}, \ldots, x_{p+1}, x_p)$ has the output function:

$$y_i = \bigvee_{\ell=0}^{m^p-1} y_{i\ell} \wedge \left[\prod_{k=p}^{2p-1} x_k^{(\ell_k)} \right]. \tag{2}$$

Substituting (2) into (1) we obtain:

$$z = \bigvee_{i,\ell=0}^{m^p-1} y_{i\ell} \wedge \left[\bigwedge_{j=0}^{p-1} \bigwedge_{k=p}^{2p-1} x_j^{(i_j)} x_k^{(\ell_k)} \right],$$

which is the output function of a multiplexer having $2p$ control variables; a multiplexer with kp control variables is then obtained by induction on the coefficient k of p.

The method is illustrated in Figure 3 by taking $n=3$, $p=1$ and $m=2$. The synthesis method produces a tree-network (i.e. a network without

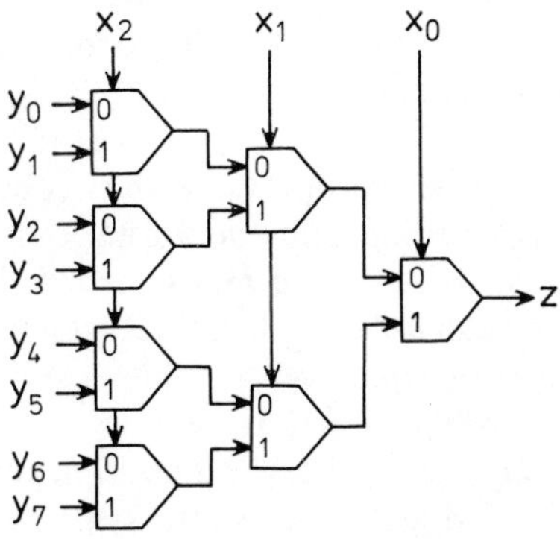

Figure 3. Inductive tree synthesis of a multiplexer

fan-out) having k levels. This network is formed by

$$\sum_{j=0}^{k-1} (2^p)^j = \frac{2^{kp}-1}{2^p-1}$$

multiplexers with p control variables; it has a cost of

$$\Gamma_1(n) = \frac{2^{kp}-1}{2^p-1}\Gamma_0(p) = \frac{2^{kp}-1}{2^p-1}(p+2^p(p+2))$$

$$\cong K_p 2^{kp} = K_p 2^n, \tag{3}$$

with

$$K_p \cong p = \frac{n}{k}.$$

The cost of a direct synthesis is approximately $n2^n$, while the cost of an inductive tree synthesis is approximately $(n2^n/k)$. Observe also that, once the parameter p (which determines the size of the multiplexer taken as elementary building block) has been chosen, the cost $\Gamma_1(n)$ of the multiplexer behaves as 2^r, i.e. exponentially as a function of the number of control variables and linearly as a function of the number of data inputs.

Let Δ_0 be the delay (which is the measure for a time criterion) of a direct realization of the elementary multiplexer with p command variables: it is assumed that the multiplexer is realized in a two-level form. Let Δ_1 be the delay of the iterative construction; we have

$$\Delta_1 = \Delta_0 \frac{n}{p}$$

and

$$\Gamma_1 \Delta_1 = \Delta_0 \frac{n}{p}\frac{n2^n}{k} = \Delta_0 n 2^n = \Delta_0 \Gamma_0. \tag{4}$$

The product $\Gamma\Delta$, which may be considered as a measure of a global cost criterion, is independent of the parameter p.

Let us now introduce another kind of cost evaluation which is more realistic in VLSI system design (e.g. in MOS bipolar technologies). Consider an n-variable multiplexer realized by means of $2^n - 1$ multiplexers with one control variable. Hence, if C_1 stands for the cost of a multiplexer with one control variable, the cost of a multiplexer with n control variables would be $C_1(2^n-1)$ under the assumption that the wiring cost is negligible. The last hypothesis does not hold if the whole multiplexer is integrated on one chip or part of one chip. For example in nMOS technology (see, e.g., Mead and Conway, 1980), the area occupied on the silicon surface by an n-command variable multiplexer is proportional to $n2^n$. It is possible to show (Deschamps, 1980) that, in the case where the multiplexer is assumed to be part of an integrated circuit and assuming a tree synthesis, the cost is $C_1^* n 2^{n-1}$ where C_1^* is the cost of a one-control variable multiplexer; the cost that is relevant in VLSI technology is proportional to the circuit area.

An interesting application for a multiplexer circuit is its use as a universal logic element, i.e. a circuit that implements any logic function of p variables. A multiplexer with p m-valued variables and m^p data inputs is a universal element for logic functions of p variables: replacing y_i in relation (1) by f_i, the ith component of the value vector of the logic function $f(x_{p-1}, \ldots, x_1, x_0)$, we obtain:

$$z = \bigvee_{i=0}^{m-1} f_i \wedge \left[\bigwedge_{j=0}^{p-1} x_j^{(i_j)} \right]$$
$$= f(x_{p-1}, \ldots, x_1, x_0) \tag{5}$$

In this function implementation, the variables are connected to the control inputs while the elements of the value vector (which are m-ary constants) are connected to the data inputs. For example, we see in Figure 4a how a

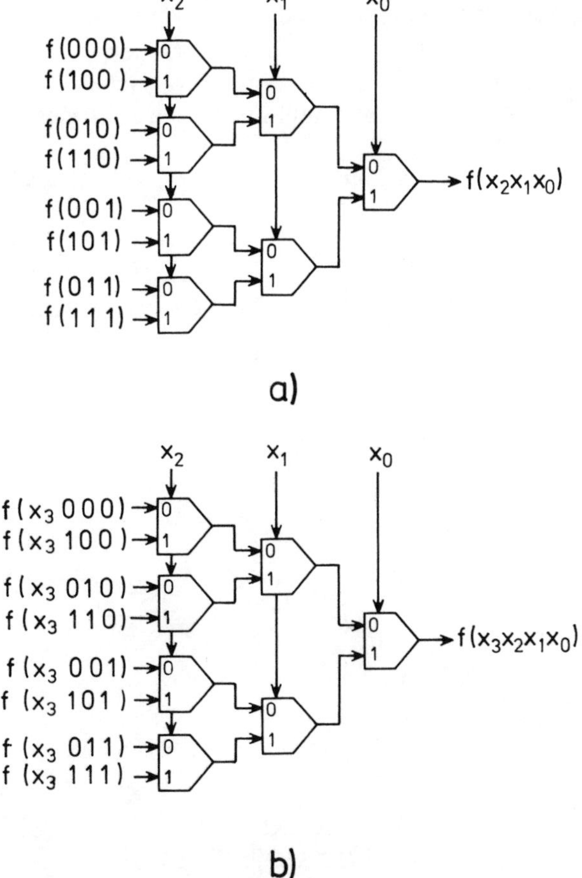

Figure 4. Universal synthesis of three and of four variables Boolean functions

binary multiplexer with three command inputs and eight data inputs can be used to implement any Boolean function of three variables.

Assume now that a function $f(x_p, x_{p-1}, \ldots, x_1, x_0)$ of $p+1$ variables is expanded according to its p variables $(x_{p-1}, \ldots, x_1, x_0)$; we have, in view of Theorem 10 in Chapter II:

$$f(x_p, \ldots, x_0) = \bigvee_{i=0}^{m^p-1} f(x_p, i_{p-1}, \ldots, i_1, i_0) \wedge \left[\bigwedge_{j=0}^{p-1} x_j^{(i_j)}\right], \tag{6}$$

which shows that any $p+1$ variable function may be implemented by means of a multiplexer with p control inputs: there are p variables $x_{p-1}, \ldots, x_1, x_0$ that are considered as control variables while the data inputs are no longer connected to constants but to the m^p one-variable functions $f(x_p, i_{p-1}, \ldots, i_0)$. These functions are members of the set $\{0, 1, x_p, \bar{x}_p\}$. For example, we see in Figure 4b how a binary multiplexer with three control inputs and eight data inputs can be used to implement any Boolean function of four variables.

Another typical application of multiplexers derives from the fact that there are many instances where one of several possible inputs to a circuit must be randomly selected and multiplexers are an ideal solution to this problem.

Assume, e.g., that given four 4-bit registers A, B, C, and D, we want to

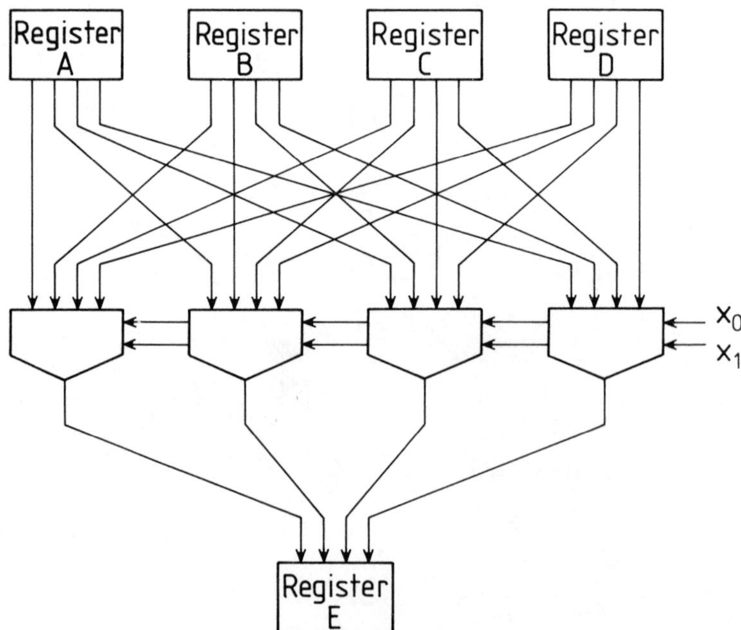

Figure 5. Use of multiplexers in register connection

design a circuit whose output puts either the contents of registers A or B or C or D in a 4-bit register E.

Since one of four possible inputs may appear on the output, this suggests that we use multiplexers with four data inputs and thus with two control inputs. A possible realization is shown in Figure 5. The output is one of the four register contents depending on the value of the control variables x_0 and x_1.

3 DEMULTIPLEXERS

Demultiplexing is a reverse operation with respect to multiplexing: it means receiving information from a small number of channels and distributing it over a large number of destinations. *Demultiplexers* reverse the multiplexer function: they take a single input line and fan it out to one of many output lines. The demultiplexing principle is illustrated in Figure 6a, while the standard demultiplexer gate symbol is depicted in Figure 6b. A demultiplexer is a combinatorial circuit that transmits data from one input to an output selected from m^p output lines. The selection of the input–output transfer path is controlled by a set of input selection lines. A demultiplexer with p control inputs has $1+m^p$ m-ary inputs:

$$\begin{array}{ll} x_0, x_1, \ldots, x_{p-1}; & y, \\ \text{(control inputs)}; & \text{(data input)}, \end{array}$$

and $m^p - 1$ m-ary outputs z_i:

$$z_i = y \wedge \left[\bigwedge_{j=0}^{p-1} x_j^{(i_j)} \right], \qquad 0 \leqslant i_j \leqslant p-1. \tag{7}$$

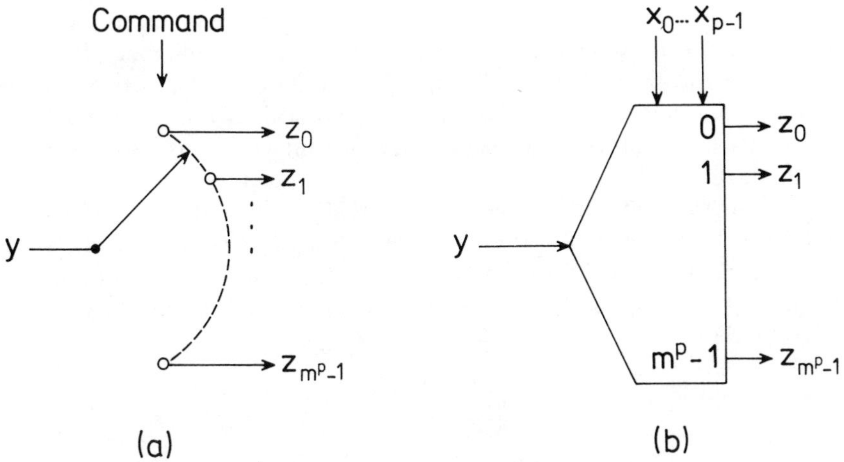

Figure 6. Demultiplexer: principle and gate symbol

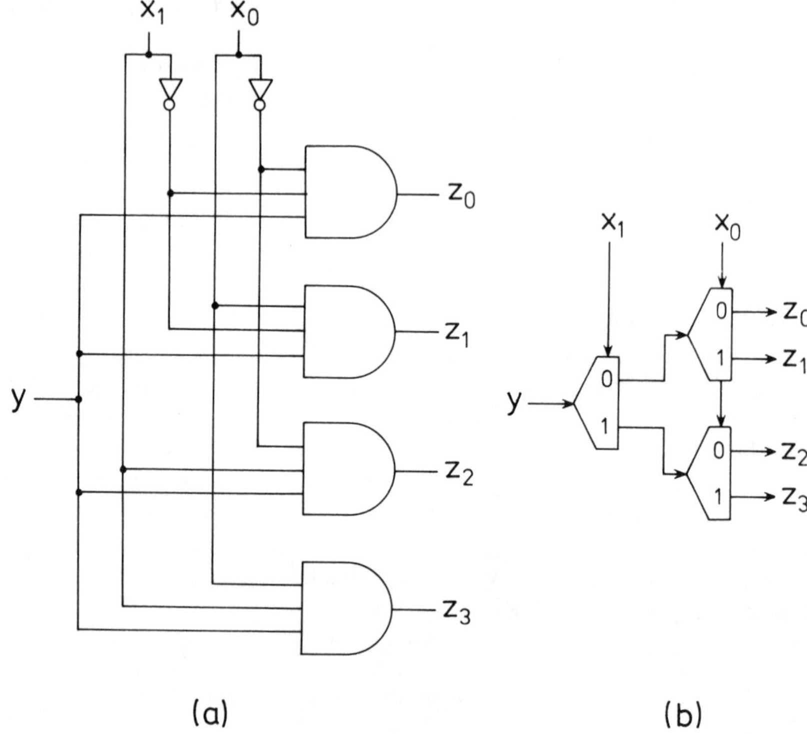

Figure 7.(a) Two-level synthesis of a demultiplexer. (b) Inductive tree synthesis of a demultiplexer

All that was said in Section 2 concerning the multiplexer syntheses and their cost remains true when dealing with demultiplexers. In particular, demultiplexers may be synthesized according to a direct synthesis and to an iterative synthesis using demultiplexers with a smaller number of control inputs. The direct synthesis is illustrated in Figure 7a for a two-variable Boolean demultiplexer while the inductive synthesis is illustrated in Figure 7b. In this last synthesis the elementary building block is a one-variable demultiplexer.

A cost study may be performed for the realization of demultiplexers in a similar way as it was done for multiplexers. It is, e.g., possible to show that the cost (cost = number of gate inputs) of a tree synthesis of a demultiplexer with $n = kp$ control variable by means of elementary demultiplexers with p control variables is

$$\Gamma \cong K'_p 2^n, \tag{8}$$

with $K'_p \cong p$.

A binary demultiplexer with n control variables can be realized by means of $2^n - 1$ demultiplexers with one command variable: if C_2 is the cost of one-variable demultiplexer and if wiring cost is negligible, the cost of an n-variable demultiplexer is equal to $C_2(2^n-1)$. Again, if the demultiplexer is part of an integrated circuit, its cost is equal to $C_2^*n(2^n-1)$ where C_2^* is the cost of a one-variable demultiplexer in the chosen technology.

A typical application of demultiplexers derives from the fact that there are many instances where one input must be connected to one of several possible outputs.

Assume, for example, that given a 4-bit register we want to be able to transfer its content to one of the four 4-bit registers A, B, C, or D. The solution of the reverse problem as shown in Figure 6 suggests the realization of Figure 8 as the solution to our problem.

In the particular case where the input value of the demultiplexer is $m-1$, the demultiplexer realizes the m^n minterms in the control variables (see relation (7)): it is an address decoder whose symbol is shown in Figure 9.

We have seen that both multiplexers and demultiplexers are expandable operators, i.e. it is possible to synthesize an operator of kp command variables by using exclusively corresponding operators of p command variables. The address decoder is not an expandable operator.

We give now a typical example of a synthesis using multiplexers or demultiplexers as building blocks.

We consider the synthesis of a switch for 2^n words of n bits; the problem

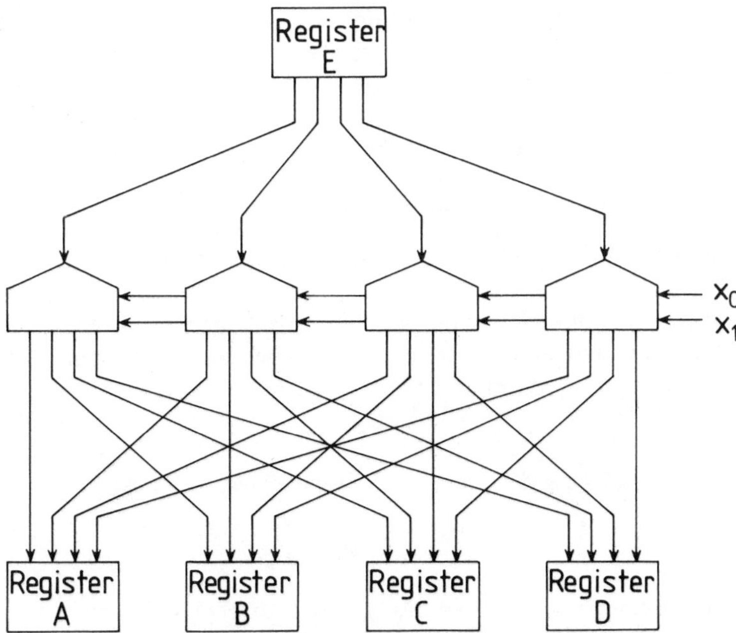

Figure 8. Use of demultiplexer in register connection

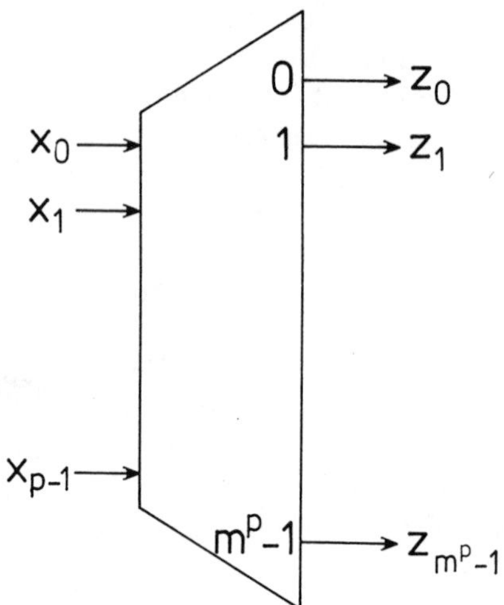

Figure 9. Address decoder

consists of designing a logical network whose output is an m-tuple $z_{m-1}, \ldots, z_1, z_0$, i.e. a word of m bits chosen between 2^n words of m bits which are the inputs to the network. This selection between these 2^n input m-tuples is realized by means of n binary control variables.

A first solution to this design problem is given by Figure 10a which details the building scheme for $n=2$ and $m=4$; this solution makes use of m multiplexers with n control variables. Adopting the number of gate inputs as cost criterion we obtain in view of (3):

$$C_a = m(K2^n). \tag{9}$$

In this solution, the 2^n minterms of the control variables are produced m times. As shown in Figure 10b, this suggests a second type of solution making use of an address decoder with n control variables and of 2^n sets of m AND-gates with 2 inputs. The cost C_b of the scheme of Figure 10b may be evaluated as follows:

$$\underset{\text{(address decoder)}}{C_b = K_p'2^n} \quad + \quad \underset{\text{(AND-gates)}}{2m2^n} \quad + \quad \underset{\text{(OR-gates)}}{m2^n} \tag{10}$$

i.e.

$$C_b = (K_p' + 3m)2^n.$$

A numerical comparison between (9) and (10) shows the interest of the second solution when m becomes sufficiently large.

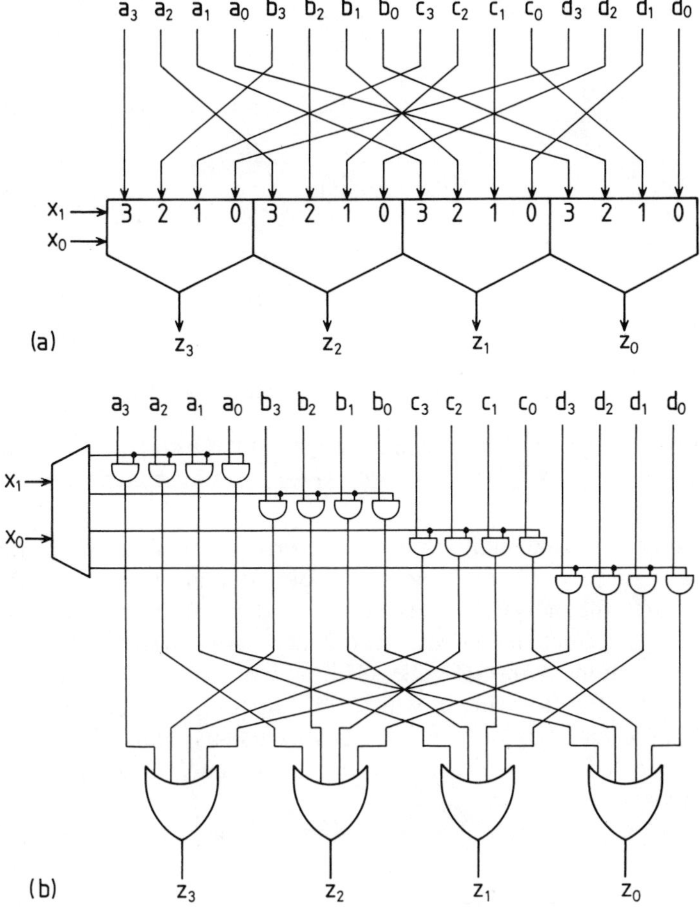

Figure 10. Synthesis of a switch for 2^n words of n bits

4 SYNTHESIS OF LOGIC FUNCTIONS WITH MULTIPLEXERS OR WITH DEMULTIPLEXERS

Remember that multiplexers are combinational devices controlled by command variables that determine which one of the several input lines is connected to the output. A multiplexer with n control variables: $\mathbf{x} = (x_{n-1}, \ldots, x_1, x_0)$ is shown in Figure 11a. We know (Section 2) that a multiplexer may be used as a universal logic element, i.e. a circuit that implements any logic function of n variables: we have to apply an adequate pattern of constants $0, 1, \ldots, m-1$ to the multiplexer inputs (see Figure 11a).

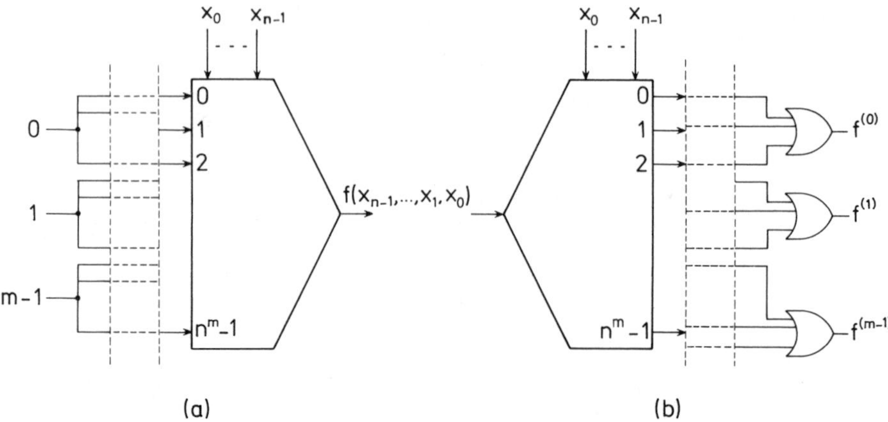

Figure 11.(a) Multiplexer with n control variables (b). Network derived from that of Figure 11a

Remember also that demultiplexers reverse the multiplexing operation: they take a single input line and fan it out to one of many output lines. Consider now the network of Figure 11b derived from the network of Figure 11a by interchanging the multiplexer with a demultiplexer and the fan-out connections with OR-gates. The network of Figure 11b is a switch which indicates the domains in which the function $f(x_{n-1}, \ldots, x_1, x_0)$ takes the constant values $0, 1, \ldots, m-1$. This switch has thus as output the m binary functions $f^j(\mathbf{x})$, $0 \leq j \leq m-1$ (remember that $f_j(\mathbf{x})$ is 1 iff $f(\mathbf{x}) = j$ and is 0 otherwise).

Multiplexers and demultiplexers are thus associated with two complementary synthesis procedures for realizing logic functions. The multiplexer synthesis produces from the set of input values, $0, 1, \ldots, m-1$ a formal expression of $f(x_{n-1}, \ldots, x_1, x_0)$, allowing us to compute the function value at any point. The demultiplexer synthesis allows us to evaluate the function value for a given vertex.

We know (see Sections 2 and 3) that n-variable multiplexers and demultiplexers are both expandable operators which can be schematized by means of p-variable multiplexers and demultiplexers respectively ($n = kp$). For both types of operators the expansion laws are the same: this means that the graph of p-variable multiplexers and of p-variable demultiplexers are identical.

Assume that the n-variable multiplexer and demultiplexer networks of Figure 11 are realized by the corresponding p-variable elements respectively. Again Figures 11a and 11b may be deduced from each other by interchanging multiplexers with demultiplexers and fan-out connections with OR-gates.

An interconnection of multiplexers and of fan-out connections may now be used for synthesizing a particular logic function (instead of synthesizing any n-variable logic function); it turns out that in this last case the synthesis may generally be greatly simplified with respect to the universal synthesis (simplification means, e.g., reduction of the number of multiplexers).

The following proposition will be stated rigorously in Chapter XII; however, it derives intuitively from the synthesis of Figures 11a and b.

From any network realizing a logic function f and made up of multiplexers and fan-out connections we derive a network realizing the m binary functions $f^{(0)}, f^{(1)}, \ldots, f^{(m-1)}$ *by interchanging multiplexers with demultiplexers and fan-out connections with OR-gates.*

Observe also that the logic function f is obtained from the binary functions $f^{(j)}$ by the relation

$$f = \bigvee_j f^{(j)} j. \tag{11}$$

With the problem of realizing a particular logic function either by means of multiplexers, or by means of demultiplexers are associated optimization problems. We may try to obtain a realization which minimizes a given cost

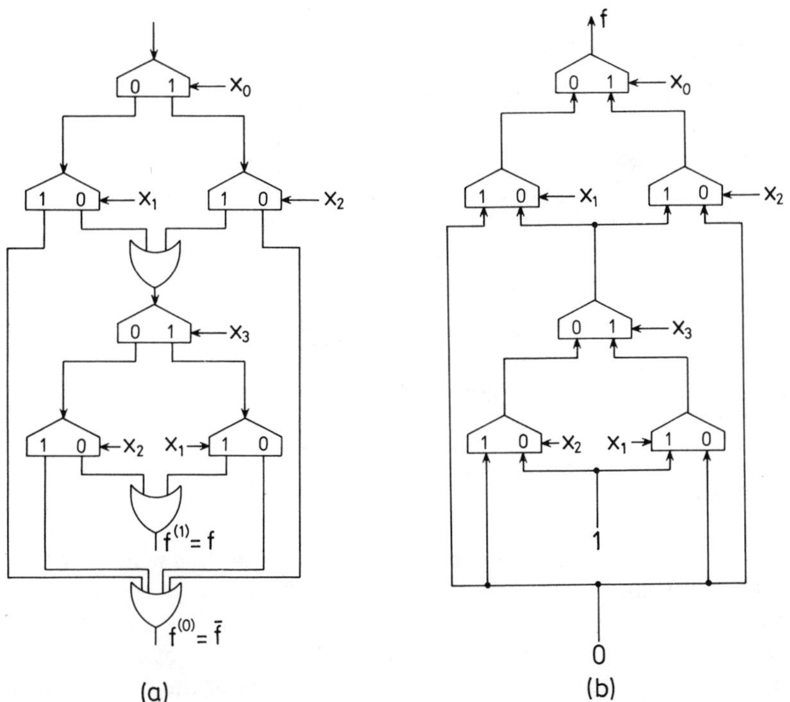

Figure 12. (a) Optimal realization of a switching function by means of demultiplexers and of OR-gates. (b) Optimal realization of a switching function by means of multiplexers.

criterion; as quoted in Section 2 the most usual criteria are, for example, the number of gates, the delay which is, e.g., reflected by the number of logical levels between input and output and the number of gate outputs.

The proposition quoted above has important consequences in an optimization process; it allows us, for example, to obtain a network scheme which simultaneously optimizes (under a given criterion) networks made up with multiplexers and networks made up with demultiplexers. This will be stated again extensively in Chapter XIII.

Example 1. Consider the switching function:

$$f = x_0x_1x_2x_3 \vee \bar{x}_0\bar{x}_1\bar{x}_2\bar{x}_3$$

and its realization by means of multiplexers or of demultiplexers with one control variable. The optimal realizations from the point of view of both the number of gates and the delay are depicted in Figures 12a and 12b. Observe that these networks may be deduced from each other by means of the above proposition.

5 READ ONLY MEMORIES

5.1 Definition and applications

A *memory array* is a logical structure organized in N words of M bits; it contains:

(1) N binary inputs $y_0, y_1, \ldots, y_{N-1}$,
(2) M m-valued outputs $z_0, z_1, \ldots, z_{M-1}$,
(3) NM elementary memory elements of 1 bit.

The contents of these elementary memories is denoted m_{ij} ($i = 0, 1, \ldots, N-1$; $j = 0, 1, \ldots, M-1$), $0 \leqslant m_{ij} \leqslant m-1$. The vector

$$(m_{i0}, m_{i1}, \ldots, m_{i(M-1)})$$

is the ith memory word. The input–output behaviour is defined by the set of M relations:

$$z_j = \bigvee_{i=0}^{N-1} y_i m_{ij}, \qquad 0 \leqslant j \leqslant M-1. \tag{12}$$

In the normal use of a memory array, one only of the y_i's is not zero; in this case the application of the value $(m-1)$ to the input y_i produces the ith memory word at the output.

The memory array is generally arranged on a rectangular matrix (see Figure 13b). In this presentation m_{ij} is the contents of the memory element at the intersection of the ith row and of the jth column. In a memory array the data content m_{ij} is fixed, read out is non-destructive, and the data are retained indefinitely (even, for example, when the power is switched off). A

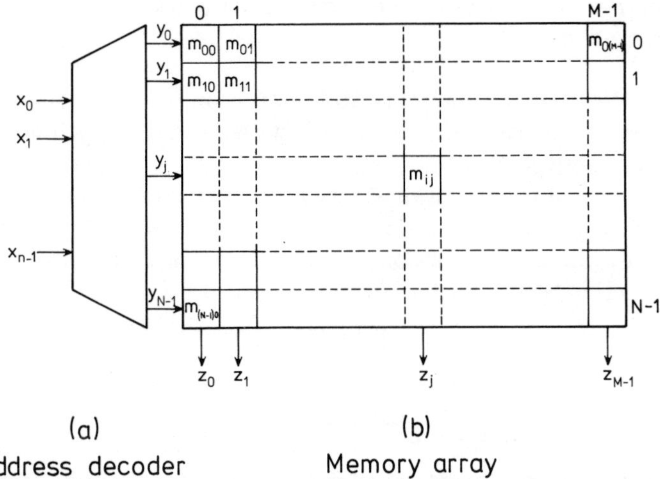

Figure 13. Read only memory

memory array is generally used together with an address decoder which realizes the condition that only one y_i at a time is different from zero (see Figure 13a).

A *read only memory* (or ROM) is a system which consists of two subsystems:

(1) an array of memory cells,

(2) an address decoder to select randomly one of the memory words.

When the input variables y_i are themselves the outputs of an address decoder, we have, in view of (7) and (12):

$$z_j = \bigvee_{i=0}^{m^n-1} \left[\bigwedge_{k=0}^{n-1} x_k^{(i_k)}\right] m_{ij},$$

$$= \bigvee_{i=0}^{m^n-1} m_i(\mathbf{x}) m_{ij}, \qquad 0 \leq j \leq M-1, \tag{13}$$

where the $m_i(\mathbf{x})$ are the minterms in the n m-valued variables $(x_{n-1}, \ldots, x_1, x_0)$.

A ROM can be considered as a logic function generator: if the memory elements m_{ij} contain the ith element of the value vector of a logic function f_j we have:

$$z_j = \bigvee_{i=0}^{m^n-1} m_i(\mathbf{x})(f_j)_i = f_j(\mathbf{x}), \tag{14}$$

and the outputs of the ROM produce the M logic functions $f_j(\mathbf{x})$, $0 \leq j \leq M-1$.

Otherwise stated, when the ROM is viewed as a logic function generator the column j of the memory array contains the value vector of the jth function to be synthesized.

Example 2. We show in Figure 14 the realization by means of a ROM of the functions f_0 and f_1 which represent the sum and carry functions of a full-adder, respectively.

A ROM could also be considered as a memory when used in microprogramming table look-up, and code conversion applications. As an illustration of a table look-up application consider the circuit of Figure 15 which gives the sinus function of the angles between 0 and 90 degrees; if:

$$x = (A2^{-5} + B2^{-4} + C2^{-3} + D2^{-2} + E2^{-1})\ 90 \text{ degrees},$$

then

$$\sin x \cong y_1 2^{-8} + y_2 2^{-7} + \ldots + y_8 2^{-1}.$$

The ROM has made possible fixed-program storage computer systems for special applications. Other applications are the decode circuitry, code conversion, and tabular implementatrion of elaborated functions (such as binary multiplication) with ROMs replacing classical logic networks.

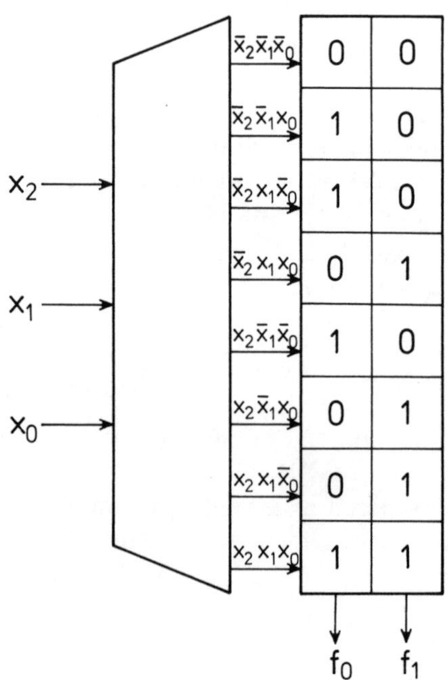

Figure 14. Realization by means of a ROM of the sum and carry functions of a full-adder

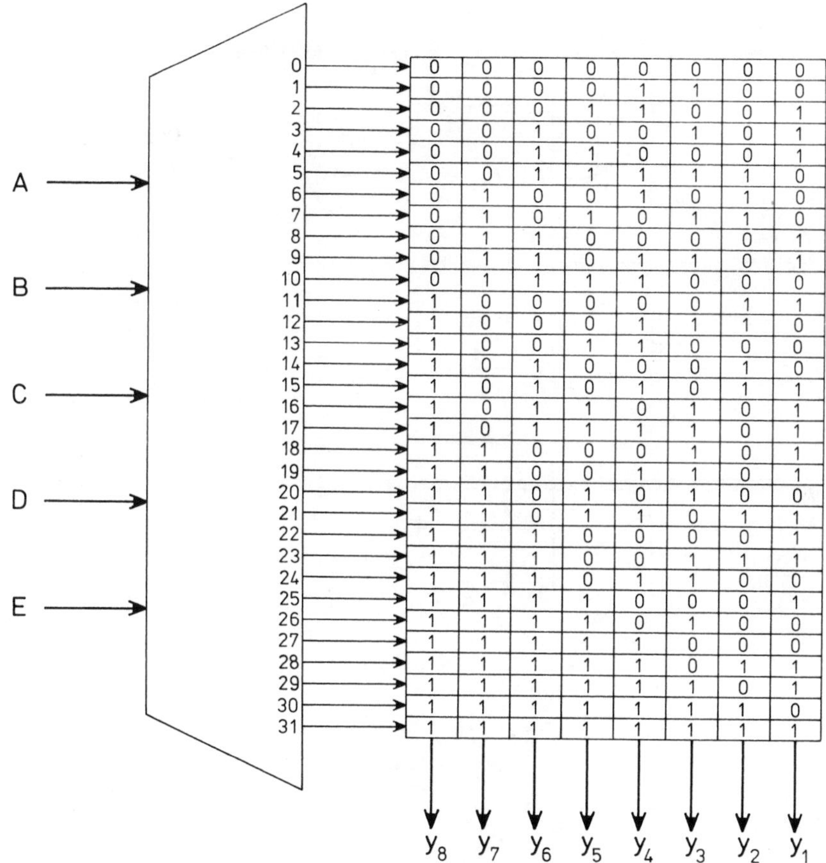

Figure 15. Sinus function of the angles between 0 and 90 degrees realized by means of a ROM

5.2 Technical realization of ROMs

Figure 16 illustrates the possibility of building the memory array of the ROM as an iterative two-dimensional array. Figure 17 contains essentially functional information: an elementary cell of memory array contains two wires and an element which possibly connects the horizontal wire to the vertical wire (see Figure 17a). We assume further on that the memory array is a two-valued array: the presence or absence of the connecting element corresponds to the two possible values in binary logic.

In the past the functions currently provided by these ROMs' connecting elements were provided by diode matrix networks (see Figure 17b). When ditigal logic first began to appear as mass-produced integrated circuits, the bipolar transistor logic was introduced for realizing the connecting element

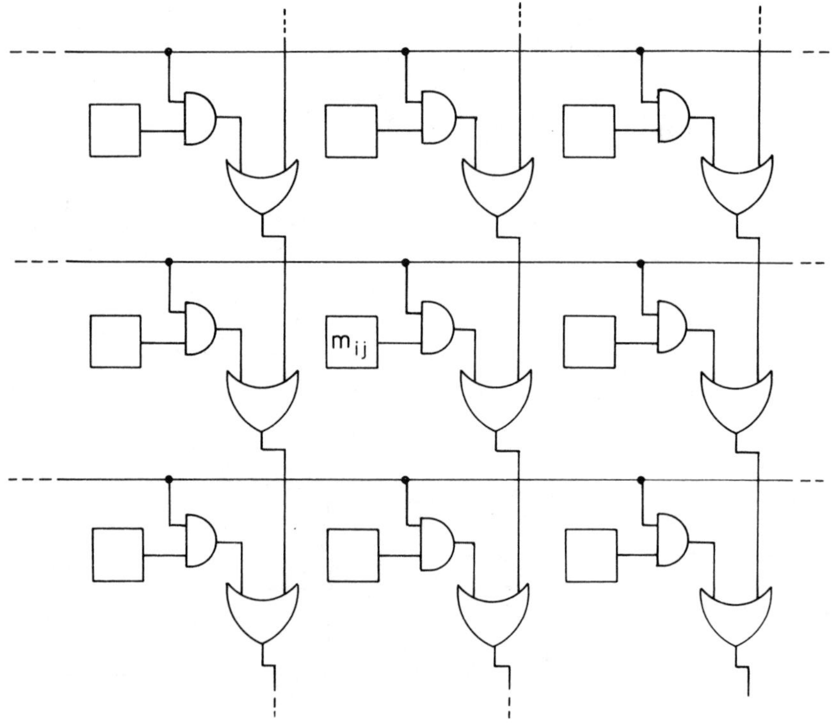

Figure 16. Iterative realization of a ROM

(see Figure 17c). More recently the MOS arrays, used for memory storage devices with their low power dissipation and high packing density, are finding a place in main ROMs, while yielding to bipolar logic only when speed is essential. When using the *n*MOS technology (see Figure 17d) the presence of a transistor corresponds to the value 0 of the memory element while the absence of a transistor corresponds to the value 1.

From a technological point of view we consider three categories of ROMs.

(1) *Read only memories* (ROMs) are intended to store non-changing information; they are thus pre-written in some permanent form. The connecting elements are generally placed by the manufacturer anywhere in the ROM but they are connected where they are needed. The user must supply the manufacturer of the ROM with the code he wishes each word to contain. Once a mask for this purpose has been done by the manufacturer, identical ROMs can be produced inexpensively.

(2) *Programmable read only memories* (PROMs) are designed to be programmed by the user instead of by the manufacturer. As supplied by the manufacturer all elementary memory elements are connected. By following

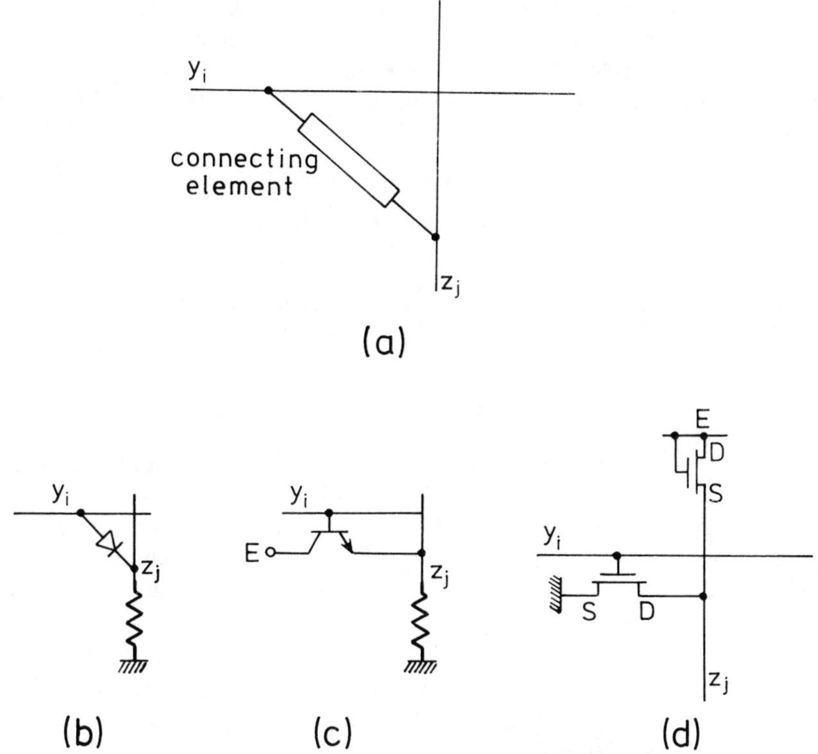

Figure 17. Realizations for a memory cell

the programming procedure, the bits at selected locations can be changed: this is done by driving high current through the connecting elements, which opens fusible links. Only bipolar PROMs are used at the present time. Also, once the PROM is programmed it cannot be reprogrammed, so programming mistakes will result in wasted units.

(3) The main problem with a large PROM is that it is a one-time programmable device. That is, once a bit pattern has been installed, the device has to be discarded once the bit pattern has to be changed. Programmable and alterable ROMs or *reprogrammable ROMs* have been designed for this reason. These devices can be programmed again and again. The best known device of this nature uses an array of MOS transistors. This ROM is made ready for programming by exposure to ultraviolet radiation which discharges the gates.

(4) *Electrically alterable ROMs* (EAROM) are reprogrammable ROMs where one memory cell may be modified at a time. A program written on a ROM may thus be transformed by modifying only an adequate set of bits.

We shall now derive some cost estimates for the realization of logic networks using ROMs.

Consider the realization of M Boolean functions by means of an address decoder and a ROM. Let C_4 be the cost per elementary memory cell of a ROM. Since the cost of an n-variable address decoder is $C_2(2^n - 1)$ (see Section 2) the cost of a ROM realization which synthesizes M Boolean function is:

$$\underset{\text{(address decoder cost)}}{C_2(2^n - 1)} + \underset{\text{(memory array cost)}}{C_4 M 2^n} \tag{15}$$

This is the cost of a one-level addressing mode network for the synthesis of M Boolean functions according to the scheme of Figure 13.

5.3 Two-level addressing mode networks

Another synthesis, grounded on a two-level addressing mode network is suggested by the network of Figure 18. In this type of network the set of variables $\mathbf{x}$ is partitioned into $\mathbf{x}_1 = (x_{n-1}, \ldots, x_{r+1}, r)$ and $\mathbf{x}_0 = (x_{r-1}, \ldots, x_1, x_0)$, the set of variables $\mathbf{x}_1$ acting on the demultiplexer networks and the set of variables $\mathbf{x}_0$ acting on the address decoder of the ROM.

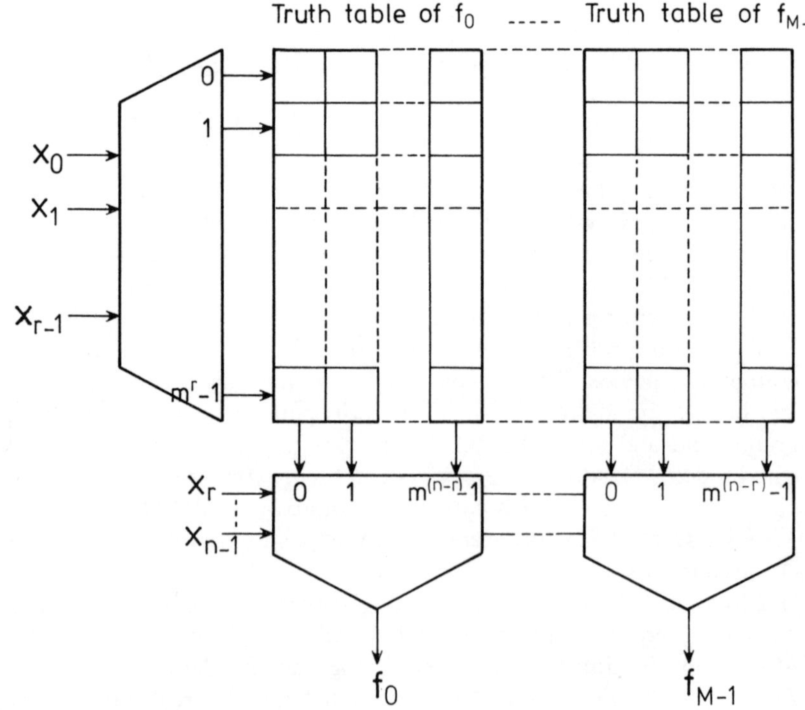

Figure 18. Two-level addressing mode networks

The two-level addressing mode networks present several advantages with respect to the one-level addressing mode networks. First of all, from a qualitative point of view the two-level mode allows us to modify the geometry of the ROM in order to adapt it to the standards of integrated networks. From a quantitative point of view, the parameter p (number of variables in the set $\mathbf{x}_0$) allows us to act on the cost of the selection circuitry (address decoder and multiplexers).

A cost function will be evaluated after the synthesis method for producing M n-variable Boolean functions has been stated in a formal way.

Consider again the set of M logic functions $\{f_0, f_1, \ldots, f_{M-1}\}$ and the partition $(\mathbf{x}_1, \mathbf{x}_0)$ of $\mathbf{x}$. The subfunctions of $\{f_0, f_1, \ldots, f_{M-1}\}$ for a fixed value of the m-valued variables $x_i \in \mathbf{x}_0$ are the functions:

$$f_j(\mathbf{e}_1, \mathbf{x}_0), \mathbf{e}_1 = (e_{n-1}, \ldots, e_{r+1}, e_r), \qquad 0 \leq e_i \leq m-1, \quad 0 \leq j \leq M-1.$$

As shown by Figure 18, the ROM produces the functions $f_j(\mathbf{e}_1, \mathbf{x}_0)$ at its outputs, which are themselves the inputs to the multiplexer networks. These last networks thus produce at their outputs z_j the functions:

$$\begin{aligned} z_j &= \bigvee_{\mathbf{e}} \bigwedge_{i=r}^{n-1} [x_i^{(e_i)}] f_j(\mathbf{e}_1, \mathbf{x}_0), \qquad && 0 \leq e_i \leq m-1 \\ &= f_j, && 0 \leq j \leq M-1. \end{aligned}$$

Let us now evaluate the cost of a two-level addressing mode network for synthesizing M Boolean functions. The cost of the network of Figure 18 is:

$C_2(2^r - 1)$	+	$C_4 M 2^n$	+	$MC_1(2^{(n-r)} - 1)$
(address decoder cost)		(memory array cost)		(cost of the M multiplexers)

$$\cong C_2 2^r + (MC_1 2^n) 2^{-r} + MC_4 2^n. \tag{16}$$

By comparing (15) with (16) we see that the cost of the memory array remains unchanged while the cost of the addressing networks becomes:

$$\Gamma = (C_2) 2^r + (MC_1 2^n) 2^{-r} = a 2^r + b 2^{-r}. \tag{17}$$

An elementary evaluation of the derivative of Γ with respect to r allows us to state:

$$\frac{d\Gamma}{dr} = 0 \Leftrightarrow r_{\text{opt}} = \tfrac{1}{2} \log_2 b, \tag{18}$$

the theoretical optimal value for Γ being:

$$\Gamma_{\text{opt}} = 2(ab)^{1/2} = 2 \cdot 2^{n/2} (MC_1 C_2)^{1/2}.$$

In view of (18) and for $C_1 = C_2$ we have:

$$r_{\text{opt}} = \tfrac{1}{2} \log_2 (M 2^n) = \frac{n}{2} + \tfrac{1}{2} \log_2 M. \tag{19}$$

Let us now slightly transform our two-level addressing mode network by replacing the set of multiplexers by an equivalent network made up with address decoders and OR- and AND-gates as suggested by the two equivalent networks of Figures 10a and 10b.

The cost of the network so obtained is evaluated by replacing in (10) 2^n by 2^{n-r}; if the address decoders are built up with elementary address decoders with p variables, we have:

$K'_p 2^r$	+	$C_4 M 2^n$	+	$(K'_p + 3M)2^{n-r}$.
(address decoder cost)		(memory array cost)		(cost of the circuit equivalent to the M multiplexers)

The cost of the addressing networks becomes for the present realization:

$$\begin{aligned} \Gamma &= K'_p 2^r + (K'_p + 3M)2^{r-n} \\ &= a2^r + b2^{-r}, \end{aligned} \tag{20}$$

with

$$a = K'_p \quad \text{and} \quad b = (K'_p + 3M)2^n.$$

In view of (18) we obtain the following values for r_{opt} and Γ_{opt} respectively:

$$\begin{aligned} r_{opt} &= \tfrac{1}{2}\log_2 \frac{(K'_p + 3M)}{K'_p} + \frac{n}{2}, \\ \Gamma_{opt} &= 2[K'_p(K'_p + 3M)2^n]^{1/2}. \end{aligned} \tag{21}$$

Comments

(a) The formulae (19, 21) have an immediate interpretation: the cost of the addressing networks are proportional to the perimeter of the memory

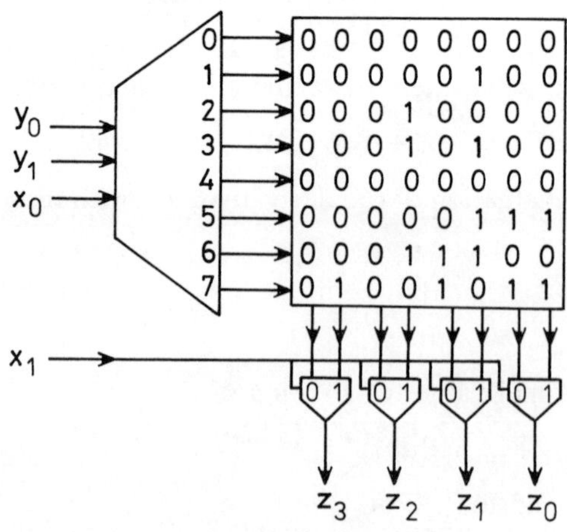

Figure 19. Synthesis of the product of two integers by means of a two-level addressing mode network

array. For a given area ($M2^n$) we know that the perimeter is minimal for a square shape.

(b) From (19) we deduce that in an approximative way the cost of the addressing networks may be neglected with respect to the cost of the memory array: the addressing network cost is asymptotically as $(M2^n)^{1/2}$ while the memory array cost is asymptotically as $M2^n$.

Example. Figure 19 illustrates the synthesis by means of a double addressing mode network of a network computing in binary numeration system the product (z_3, z_2, z_1, z_0) of two integers (x_1, x_0) and (y_1, y_0).

5.4 Optimization problems in two-level addressing mode synthesis

When the network of Figure 18 realizes M particular logic functions instead of M-output Universal-logic-module some optimization problems appear. The cost of a two-level addressing mode network depends on the following items.

5.4.1 Choice of the partition (x_1, x_0) of x

The ROM (in Figure 18) produces the $Mm^{(n-r)}$ functions:

$$f_j(\mathbf{e}_1, \mathbf{x}_0), \mathbf{e}_1 = (e_{n-1}, \ldots, e_{r+1}, e_r), \qquad 0 \leqslant e_i \leqslant m-1, \quad 0 \leqslant j \leqslant M-1,$$

each of these functions being contained in a column of the array memory. Assume that there are s different functions $f_j(\mathbf{e}_1, \mathbf{x}_0)$ which will be denoted:

$$\{h_0(\mathbf{x}_0), h_1(\mathbf{x}_0), \ldots, h_{s-1}(\mathbf{x}_0)\}$$

Denote, moreover, by

$$f_j^{(h_i)}(\mathbf{x}_1)$$

the binary function equal to 1 in the domain where $f_j = h_i$, and equal to 0 otherwise (observe that this notation is an extension of the exponentiation notation $f^{(e)}$ which means a binary function equal to 1 in the domain where $f = e$, $0 \leqslant e \leqslant m-1$, and equal to 0 otherwise). In view of these notations, the functions f_j may be expanded in the following way:

$$f_j = \bigvee_i f_j^{(h_i)}(\mathbf{x}_1) h_i(\mathbf{x}_0). \tag{22}$$

So the choice of the partition $(\mathbf{x}_1, \mathbf{x}_0)$ determines the number of functions $\{h_j(\mathbf{x}_0)\}$ and hence the number of columns of the memory array. A partition which minimizes the number of functions h_j minimizes also the number of columns of the ROM.

For a fixed choice of the partition $(\mathbf{x}_1, \mathbf{x}_0)$, the optimization process developed in Chapter XII (see also Example 3 below) allows us to synthesize an optimal multiplexer network. To each partition $(\mathbf{x}_1, \mathbf{x}_0)$ of $\mathbf{x}$ corresponds an optimal multiplexer network. The best (from an optimization point of view) multiplexer network does not necessarily correspond to the partition minimizing the number of columns of the ROM. Hence there are at this level two (non-independent) optimization problems.

5.4.2 Choice between an address-decoder and an optimal demultiplexer network

When an address-decoder is used together with a memory array (as in Figure 18) the number of rows of the memory array and the size of the address-decoder are both fixed by the number of variables of the functions to be synthesized.

The address decoder may always be replaced by a demultiplexer network; in this last case the following optimization problems may be considered.

We may first try to reduce as much as possible the size of the ROM by merging rows. We may then optimize the demultiplexer network by using the techniques developed in Chapter XII. It is worthwhile pointing out that the memory array optimization process (merging of rows) does not interact with the demultiplexer network optimization process: we may obtain as a result a network which is both optimal in the number of rows of the memory array and in the size of the demultiplexer network. Again the optimal solution depends on the chosen partition $(\mathbf{x}_1, \mathbf{x}_0)$.

Example 3. Consider the simultaneous synthesis of four Boolean functions:

$$f_0 = \bar{x}_2 \vee x_0 x_3 \vee x_1 x_3 \vee \bar{x}_0 x_1,$$
$$f_1 = x_3 \vee \bar{x}_1 \bar{x}_2 \vee x_0 \bar{x}_2 \vee x_0 x_1,$$
$$f_2 = x_0 \vee \bar{x}_2 x_3 \vee x_1 x_3 \vee x_1 x_2,$$
$$f_3 = x_1 \vee x_0 \bar{x}_3 \vee \bar{x}_2 x_3 \vee x_0 \bar{x}_2,$$

also described by the truth table of Figure 20. An optimal realization using a ROM and a one-control variable multiplexer network is shown in Figure 21.

					x_0	x_0	x_0	x_0	x_0	x_0	x_0	x_0					
	1	1	0	0	1	1	1	1	1	1	1	1	1	0	0	1	
x_2	0	0	0	0	0	0	1	1	0	1	1	1	1	0	1	1	
x_2	0	1	0	0	1	1	1	0	1	1	1	1	1	1	1	1	x_3
	1	1	1	1	1	1	1	1	1	1	1	1	1	1	1	1	x_3
									x_1	x_1	x_1	x_1	x_1	x_1	x_1	x_1	

Figure 20. Truth table

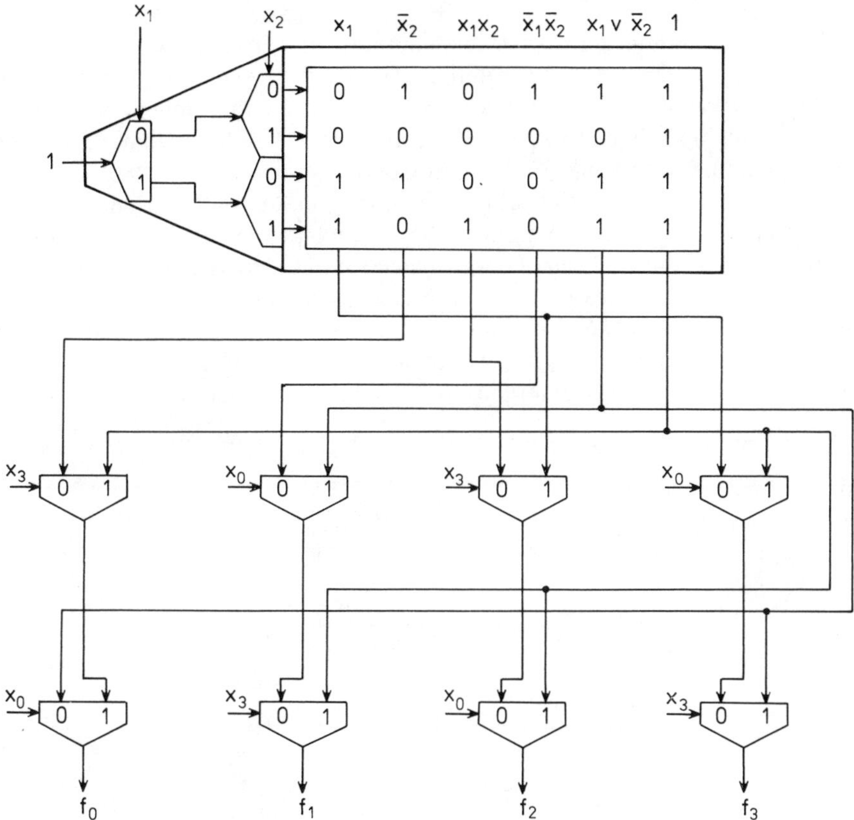

Figure 21. Optimal realization of four Boolean functions by means of a two-level addressing mode network

6 PROGRAMMABLE LOGIC ARRAYS (PLAs)

6.1 Definitions and applications

We have seen that a very general and regular way to implement a system of M Boolean functions of n variables was to use a memory array of 2^n words of M bits each. The n-inputs of the address decoder form an address into the memory array and the M-outputs are the data contained in that address. The memory array thus implements the full value vector for the M output functions. There is a major difficulty with this approach: it is often the case that most of the possible input combinations cannot occur, due to the nature of the specific problem. Otherwise stated, many combinatorial Boolean functions require only a small fraction of all 2^n minterms for a canonical sum of products' implementation. In such cases ROMs are very wasteful of area.

The *programmable logic array* (PLA) is a structure that has all the generality of a memory array for implementing logic functions. However any specific PLA structure need contain a row of circuit elements only for each of those product terms that are actually required to implement a given Boolean function. The PLA differs from a ROM only in that implicants rather than minterms of Boolean functions are stored, so that it is not necessary to have a word of storage for every input combination.

A PLA is usually implemented in the form of two arrays or matrixes (see Figure 22a). In the first part (AND matrix) input signals and their negated signals (produced, for example, by a NOT gate or by a demultiplexer gate) are selectively connected to product term lines in such a way that certain combinations of input variables (which are implicants of at least one of the *M* functions to be implemented) produce a '1' signal on one or more

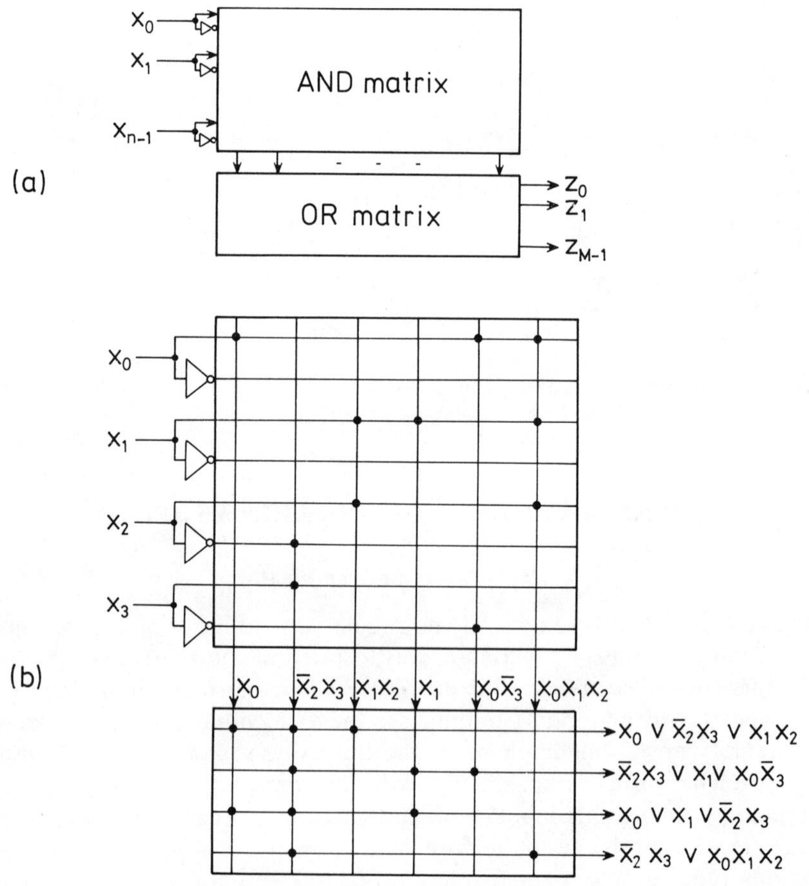

Figure 22.(a) Programmable logic array implementation. (b) Realization of four Boolean functions by means of a PLA

product term lines. These lines are input in the second matrix (OR matrix) where other selecting connections transfer the signals to the output lines.

Figure 22b shows an example of a PLA realizing the functions:

$$f_0 = x_0 \vee \bar{x}_2 x_3 \vee x_1 x_2,$$

$$f_1 = \bar{x}_2 x_3 \vee x_1 \vee x_0 \bar{x}_3,$$

$$f_2 = x_0 \vee x_1 \vee \bar{x}_2 x_3,$$

$$f_3 = \bar{x}_2 x_3 \vee x_0 x_1 x_2.$$

In this example the synthesis procedure is straightforward, that is, the selection of the connecting points is directly determined by the logical expressions expressed as a disjunction of implicants. In existing hardware, functions of these matrices are NAND (resp. NOR), so a PLA corresponds to a two-level NAND (resp. NOR) and hence to a two-level AND–OR (resp. OR–AND) circuit. NOT gates are used to generate negated input variables.

The AND-matrix contains:

(1) $2n$ binary inputs $x_0, \bar{x}_0, x_1, \bar{x}_1, \ldots, x_{n-1}, \bar{x}_{n-1}$,

(2) K binary outputs $y_0, y_1, \ldots, y_{k-1}$,

(3) rK elementary memory bits characterized by the three-valued parameters s_{ij}:

$$s_{ij} \in \{0, 1, B_2 = \{0, 1\}\}, \qquad i = 0, 1, \ldots, n-1, \qquad j = 0, 1, \ldots, K-1.$$

We use the following notation (lattice exponentiation):

$$x^{(0)} = \bar{x}, \qquad x^{(1)} = x, \qquad x^{(\{0,1\})} = 1.$$

The input–output behaviour of the AND-matrix is defined by:

$$y_j = \bigwedge_{i=0}^{N-1} x_i^{(s_{ij})}; \qquad j = 0, 1, \ldots, K-1. \tag{23a}$$

The OR-matrix is a memory array of K words of M bits; it includes K binary inputs $y_0, y_1, \ldots, y_{K-1}$ (which are the outputs of the AND-matrix) and M binary outputs $z_0, z_1, \ldots, z_{M-1}$. The KM elementary memory bits are characterized by KM parameters $r_{j\ell}$:

$$r_{j\ell} \in \{0, 1\}; \qquad j = 0, 1, \ldots, K-1; \qquad \ell = 0, 1, \ldots, M-1.$$

The behaviour of the OR-matrix is characterized by the relations:

$$z_\ell = \bigvee_{j=0}^{K-1} r_{j\ell} y_j; \qquad \ell = 0, 1, \ldots, M-1. \tag{23b}$$

By combining (23a) and (23b) we obtain the input–output behaviour of the PLA, i.e.

$$z_\ell = \bigvee_{j=0}^{K-1} r_{j\ell} \wedge \left[\bigwedge_{i=0}^{N-1} x_i^{(r_{ij})} \right]; \qquad \ell = 0, 1, \ldots, M-1. \tag{24}$$

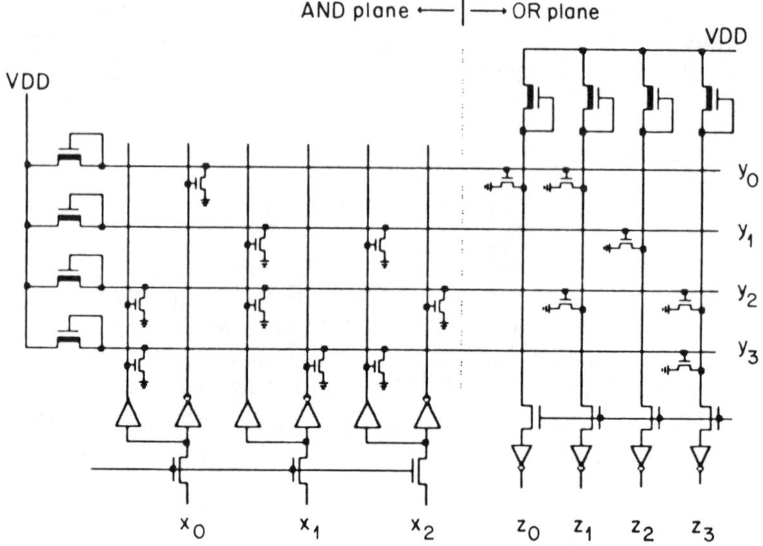

Figure 23. Programmable logic array realization in MOS technology

The circuit diagram of a specific PLA will help us to introduce the structure and function of the AND/OR matrices of the PLA for a given technology. An example of realization in nMOS technology is shown at Figure 23 (see Mead and Conway, 1980, pp. 80–82); the nMOS technology naturally leads to a NOR–NOR formulation of the input–output relations.

Example. Product terms to be realized by the AND-matrix:

$$y_0 = (\overline{x_0}) = x_0,$$

$$y_1 = \overline{(x_1 \vee x_2)} = \bar{x}_1\bar{x}_2$$

$$y_2 = \overline{(x_0 \vee x_1 \vee x_2)} = \bar{x}_0\bar{x}_1x_2,$$

$$y_3 = \overline{(x_0 \vee x_1 \vee x_2)} = \bar{x}_0x_1\bar{x}_2.$$

Outputs of the OR-matrix:

$$\bar{z}_0 = \bar{y}_0,$$

$$\bar{z}_1 = \overline{(y_0 \vee y_2)},$$

$$\bar{z}_2 = \bar{y}_1,$$

$$\bar{z}_3 = \overline{(y_2 \vee y_3)}.$$

In Figure 23 the function of the PLA's AND matrix is determined by the locations and gate connections of (pull-down) transistors connecting the horizontal lines to ground.

Each output running horizontally from the AND matrix carries the NOR combination of all input signals that lead to the gates of transistors attached to it.

The OR matrix of circuit elements is identical in form to the AND matrix, but rotated by 90 degrees. Once again, each of its outputs is the NOR of the signals leading to the gates of all transistors attached to it.

The PLA realized in nMOS technology implements the NOR–NOR normal form for Boolean functions.

PLAs may also be realized in other technologies. For example, Figure 24 represents the PLA 82S100 produced by Signetics. In the AND-matrix the connecting elements are Schottky diodes, while in the OR-matrix the connecting elements are bipolar transistors.

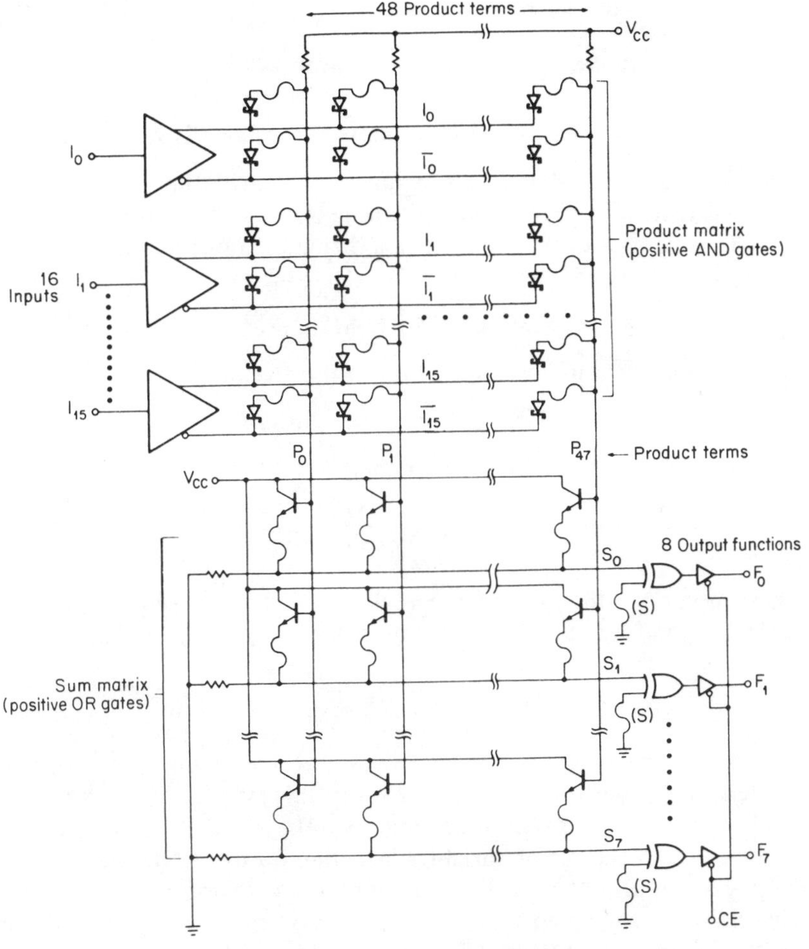

Figure 24. PLA 82S100 produced by Signetics (Reproduced with the permission of Signetics Corp. Copyright 1975)

The surface area occupied by a PLA with n input variables x_i, K product terms y_j and M output variables z_ℓ is equal to:

$$\underbrace{2nK}_{\text{AND matrix area}} + \underbrace{MK.}_{\text{OR matrix area}}$$

If C_3 is the cost per cell of the AND-plane and C_4 the cost per cell of the OR plane, the cost of the PLA of surface $(2n+M)K$ is:

$$(nC_3+MC_4)K. \tag{25}$$

Note that for the realization in nMOS technology $C_3 \cong C_4$. We know (see, for example, Kuntzmann, 1965) that the realization of n variable Boolean functions requires at most $k=2^{n-1}$ product terms and that this bound is reached for the linear function $\oplus\!\!\Sigma\, x_i$. Hence the cost of a universal PLA (i.e. a PLA able to realize any Boolean function) is

$$C_3(n+1)2^{n-1} \cong C_3 n 2^{n-1}. \tag{26}$$

6.2 Universal logic networks using PLAs

An M-tuple of n-variable Boolean functions may be implemented by the circuit of Figure 25a. The AND matrix generates the minterms while the OR plane generates every function as a disjunction of minterms. In view of (25) the cost of this implementation is equal to

$$C_3 n 2^n + C_4 M 2^n. \tag{27}$$

It is of common use in cost evaluation to use the notation:

$$S = \log_2 (M2^n) = n + \log_2 M.$$

Hence the cost (27) is written under the form:

$$C_3(S-\log_2 M)\frac{2^S}{M} + C_4 2^S. \tag{28}$$

In these notations the cost of the realization of M Boolean functions by means of an address decoder and a memory array is (see (15)):

$$C_2(2^n-1) + C_4 M 2^n = C_2\left(\frac{2^S}{M}-1\right) + C_4 2^S. \tag{29}$$

Another synthesis, grounded on a two-level addressing mode network is suggested by the network of Figure 25b. In this type of network the set of control variables is partitioned into $\mathbf{x}_1=(x_{n-1}, \ldots, x_{r+1}, x_r)$ and $\mathbf{x}_0=(x_{r-1}, \ldots, x_1, x_0)$, the set of variables $\mathbf{x}_0$ acting on the AND matrix and the set of variables $\mathbf{x}_1$ acting on the demultiplexer networks.

The circuit is shown in Figure 25; the AND matrix generates the minterms in $\mathbf{x}_0$ while the OR matrix generates the subfunctions $f_j(\mathbf{e}_1, \mathbf{x}_0)$. The multiplexers select the subfunctions under the control variables $\mathbf{x}_1$. The cost

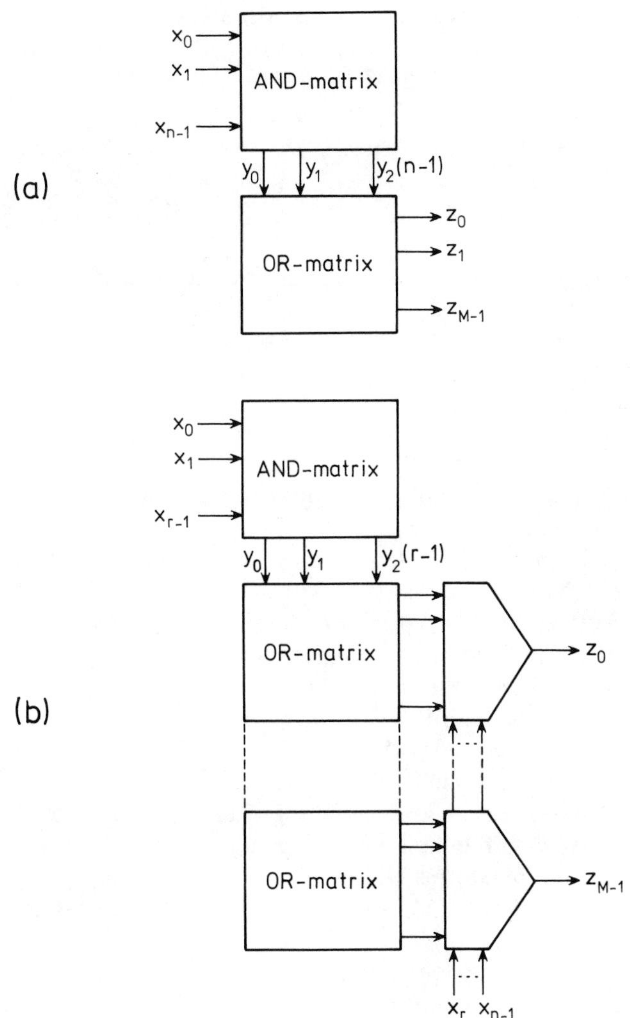

Figure 25. Implementation of an M-tuple of Boolean functions by a PLA

of the circuit is equal to:

$$C_3 r 2^2 \quad + C_4 M 2^{(n-r)} 2^r + \quad M C_1^*(n-r) 2^{n-r-1} \tag{30a}$$

(AND matrix cost) (OR matrix cost) (multiplexers cost)

(Remember that the cost of an n-variable multiplexer is $C_1^* n 2^n$ when it is assumed to be part of an integrated circuit; see also Section 6.1); using the notation $S = n + \log_2 M$, expression (30a) becomes:

$$C_3 r 2^r + C_4 2^S + \frac{C_1^*}{2}(S - \log_2 M - r) 2^{S-r} \tag{30b}$$

Let us now choose $r = \lfloor S/2 \rfloor$; hence:

$$C_3 r 2^r + \frac{C_1^*}{2}(S - \log M - r) 2^{S-r} < C_3 \frac{S}{2} 2^{S/2} + C_1^* \left(\frac{S}{2} + 1 - \log M\right) 2^{S/2}.$$

For large values of S one obtains the following estimate of the cost:

$$(C_3 + C_1^*) \frac{S}{2} 2^{S/2} + C_4 2^S. \tag{31}$$

The asymptotic value of (31) is equal to:

$$C_4 2^S.$$

In this case the lower complexity bound for a universal logic module is reached (see Theorem 23, Chapter I).

Conventional two-level PLAs, as they are defined above, are characterized by an AND matrix cascaded with an OR matrix. Several improvements with respect to this typical topology can be introduced. The main objective in the search for new configurations of PLAs is the reduction of the PLA area and hence the reduction of its cost function.

Remember that the size of a two-level AND–OR PLA for realizing M Boolean functions is $(2n + M)2^{n-1}$.

A second type of PLA, called three-level PLA, is shown in Figure 26a. In this type of PLA it is assumed that the set of input variables $\mathbf{x}$ is partitioned into $(\mathbf{x}_{r-1}, \ldots, \mathbf{x}_1, \mathbf{x}_0)$; the first matrix (generally called D matrix, Sasao, 1981) produces the maxterms with respect to each of the subsets $\mathbf{x}_i$ of this partition. This D matrix is then followed by a pair of AND–OR matrices as in the two-level PLAs.

In order to synthesize Boolean functions by means of three-level PLAs we first introduce adequate expressions of Boolean functions (see also Sasao, 1981).

Let $(\mathbf{x}_{r-1}, \ldots, \mathbf{x}_1, \mathbf{x}_0)$ be a partition of the set $\mathbf{x}$ of variables; assume that:

$$|\mathbf{x}_i| = n_i, \qquad \forall i = 0, 1, \ldots, r-1,$$

and hence $\sum_i n_i = n$. Assume moreover that $\mathbf{e}_i$ represents a vector of n_i binary components. A minterm in the variables of $\mathbf{x}_i$ will be written:

$$\mathbf{x}_i^{(\mathbf{e}_i)}.$$

An arbitrary function $f(\mathbf{x})$ is expressed in the form:

$$f(\mathbf{x}) = \bigvee_{(\mathbf{e}_i \in Z_2^{n_i}, 0 \le i \le r-1)} f(\mathbf{e}_{r-1}, \ldots, \mathbf{e}_1, \mathbf{e}_0) \mathbf{x}_{r-1}^{(\mathbf{e}_{r-1})} \ldots \mathbf{x}_1^{(\mathbf{e}_1)} \mathbf{x}_0^{(\mathbf{e}_0)},$$

which is nothing but the canonical disjunctive expansion of f.

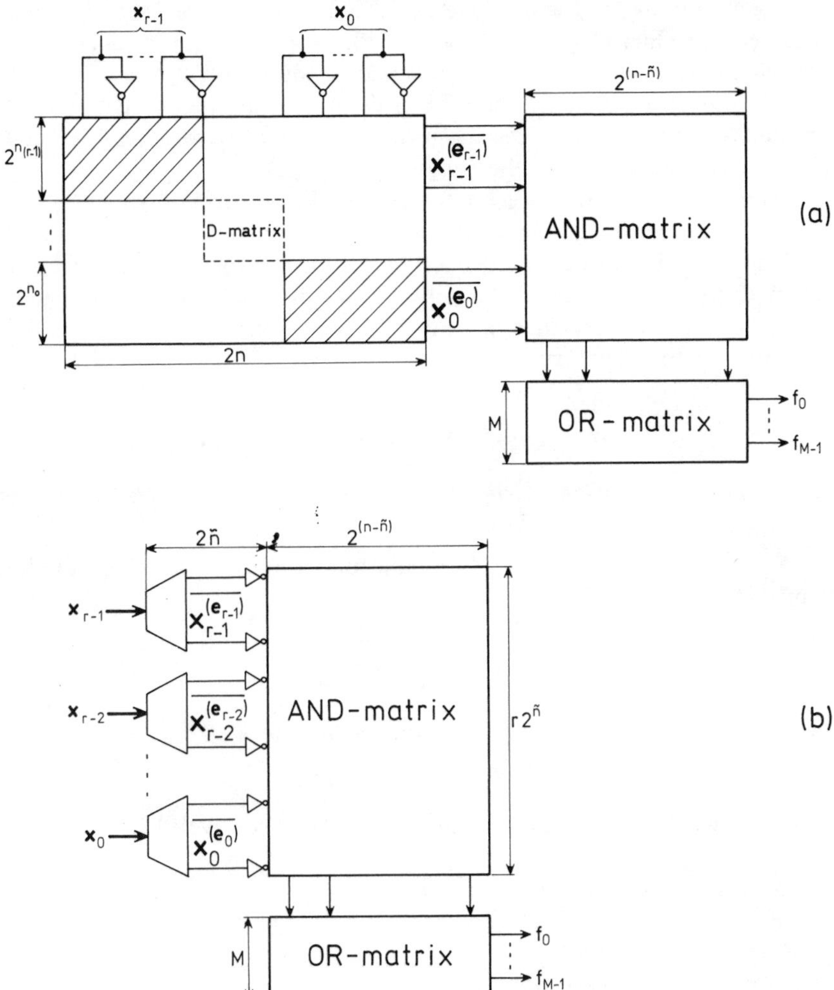

Figure 26.(a) Three-level PLA. (b) Synthesis of the D-array by means of decoders

By grouping the terms in

$$\mathbf{x}_{r-2}^{(\mathbf{e}_{r-2})} \ldots \mathbf{x}_1^{(\mathbf{e}_1)}\mathbf{x}_0^{(\mathbf{e}_0)},$$

the above expression of f is transformed into the following one:

$$f(\mathbf{x}) = \bigvee_{\mathbf{e}_{r-2},\ldots,\mathbf{e}_1,\mathbf{e}_0} \left[\bigvee_{\mathbf{e}_{r-1}} f(\mathbf{e}_{r-1}, \ldots, \mathbf{e}_1, \mathbf{e}_0)\mathbf{x}_{r-1}^{(\mathbf{e}_{r-1})}\right]\mathbf{x}_{r-2}^{(\mathbf{e}_{r-2})} \ldots \mathbf{x}_0^{(\mathbf{e}_0)}. \tag{32}$$

Without loss of generality assume that $n_{r-1} = \max\{n_i\} = \tilde{n}$. The number of terms in (32) is at most:

$$2^{(n_0+n_1+\ldots n_{r-2})} = 2^{(n-\tilde{n})}.$$

Since the functions $(\bigvee f(\mathbf{e}_{r-1}, \ldots, \mathbf{e}_1, \mathbf{e}_0)\mathbf{x}_{r-1}^{(\mathbf{e}_{r-1})})$ and $\mathbf{x}_j^{(\mathbf{e}_j)}$, $0 \leqslant j \leqslant r-2$ may be written as a product of maxterms, the synthesis of a Boolean function f by means of three-level PLAs can be reached in the following way:

(a) The D-array produces the maxterms in the variables $\mathbf{x}_j$, $0 \leqslant j \leqslant r-1$;
(b) the AND-array produces the product terms of expansion (32), and
(c) the OR-array realizes the function(s) f.

We present two different techniques for realizing the D-array.

(1) *Synthesis of the D-array by means of decoders.* Assume that the maxterms are produced by r decoders which are in one-to-one correspondence with the r subsets of $\mathbf{x}$; the network has the configuration depicted in Figure 26b if we assume that $n_i = \tilde{n}\ \forall i$; hence $n = \tilde{n}r$.

The area of this network is:

$$S = \underset{\text{(decoders)}}{r2^n \times 2\tilde{n}} \quad + \underset{\text{(AND-array)}}{r2^n \times 2^{(n-n)}} \quad + \underset{\text{(OR array)}}{M2^{(n-\tilde{n})}}$$

$$= 2n2^n + \frac{n}{\tilde{n}}2^n + M2^{(n-\tilde{n})}. \tag{33}$$

For $M=1$ (realization of one Boolean function) the value of $\tilde{n}$ which minimizes (33) is:

$$\tilde{n}_{\text{opt.}} = \frac{n-1}{2} - \log_2 n.$$

If we choose for $\tilde{n}$ the value $n/2 \cong \tilde{n}_{\text{opt.}}$, we obtain:

$$S = 2^{n+1} + (2n+1)2^{n/2}, \tag{34}$$

which is asymptotically much better than the two-level PLA area, i.e. $(2n+1)2^{n-1}$.

(2) *Synthesis of the D-array by means of an OR-array and demultiplexers.* As shown in Figure 26a a three-level PLA can be realized as an OR–AND–OR network.

The area of this network is:

$$S = \underset{\text{(D-array)}}{2n \times \left(\sum_{i=0}^{M-1} 2^{n_i}\right)} + \underset{\text{(AND-array)}}{2^{(n-\tilde{n})}\left(\sum_{i=0}^{r-1} 2^{n_i}\right)} + \underset{\text{(OR-array)}}{M2^{(n-\tilde{n})}}. \tag{35}$$

Assume n is even and consider the partition of $\mathbf{x}$ in $(n/2+1)$ subsets, $\mathbf{x}_{n/2}$ and $\mathbf{x}_i$, $0 \leqslant i \leqslant n/2-1$, with the following dimensions:

$$\text{dimension of } \mathbf{x}_{n/2} = \frac{n}{2}; \qquad \text{of } \mathbf{x}_i = 1,\ 0 \leqslant i \leqslant \frac{n}{2} - 1.$$

Hence,

$$\tilde{n} = \frac{n}{2} \quad \text{and} \quad \sum_i 2^{n_i} = \frac{n}{2} + 2^{n/2}.$$

The expression (35) becomes then:

$$S = 2^n + (3n + M)2^{n/2} + 2n^2. \tag{36}$$

Example 4. Consider the four variable Boolean function whose Karnaugh map is given in Figure 27:

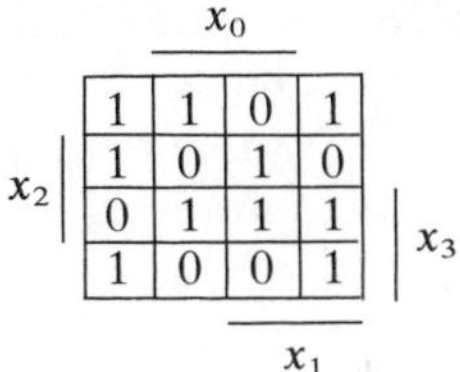

Figure 27. Karnaugh map

Two PLAs with two variable decoders are depicted in Figures 28a and 28b respectively. From these two realizations appear the fact that the way the

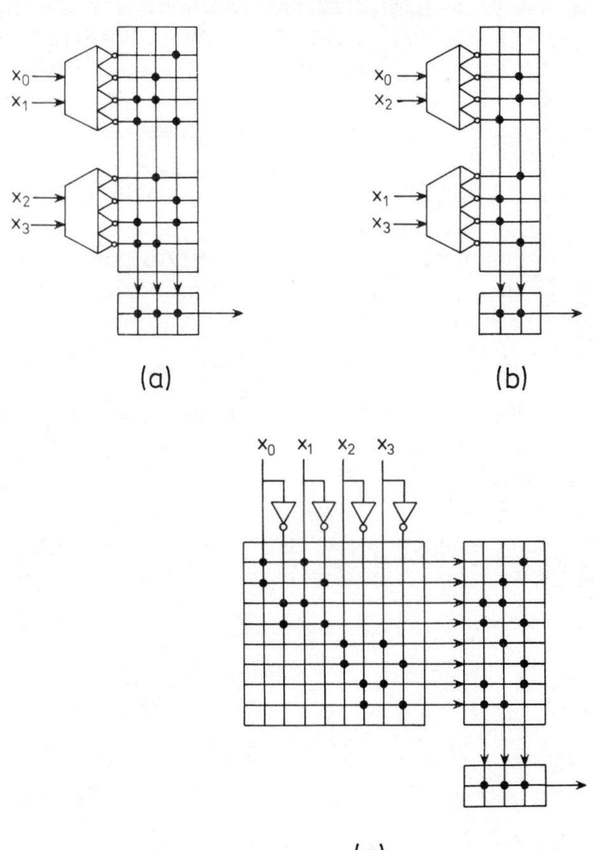

Figure 28. PLA realizations for the switching function of Figure 27

input variables are assigned to the decoders influences the size of the PLA. A three-level OR–AND–OR PLA realization is shown in Figure 28c.

6.3 Optimization problems in PLA design

One of the classical design problems in semiconductor array logic is the reduction of array area. Many techniques are discussed in the literature (see, for example, Special *IEEE Trans.* on computers on PLA, September 1979, and *IBM Journal of Research and Development*, March 1975) to compress array size. They include the use of true-complement function representation (the function f and its complement $\bar{f}$ may not require PLAs of equal complexity), maximum use of the possible don't-cares of the problem, partitioning of inputs, and the use of cascaded decoders etc.

Consider first the optimization problems that occur in the classical two-level rectangular array PLAs (see Figure 29a). A PLA being a ROM with programmable addresses, it is particularly suitable for realizing Boolean functions with many unspecified input combinations. Since the cost of a PLA is mainly determined by the number of pins (input and output connections) and the array area, the reduction of the number of input variables (if possible) is very important in pLA design: it reduces both the pin count and the chip area.

Kambayashi (1979) proposes an algorithm for obtaining a formal expression of a partially indeterminate function f with a minimum number of variables and a minimum number of literals (x_i and $\bar{x}_i$ are then considered as different literals). This algorithm is grounded on a sequential detection of redundant variables in f: if

$$f(x_k=0)=f(x_k=1),$$

x_k is said to be redundant in f. Kambayashi proves that the detection of a maximum number of redundant variables in a function f reduces to an optimal covering problem.

Example 5. Consider a function f which has to take the value 1 and 0 on the vertices satisfying $g_\alpha=1$ and $g_\beta=1$, respectively:

$$g_\alpha = x_0x_1x_2\bar{x}_3\bar{x}_4\bar{x}_5 \vee \bar{x}_0\bar{x}_1x_2\bar{x}_3\bar{x}_4\bar{x}_5 \vee \bar{x}_0\bar{x}_1x_2\bar{x}_3x_4x_5 \vee \bar{x}_0\bar{x}_1\bar{x}_2\bar{x}_3x_4x_5,$$

$$g_\beta = \bar{x}_0x_1x_2\bar{x}_3\bar{x}_4x_5 \vee \bar{x}_0\bar{x}_1x_2x_3x_4\bar{x}_5 \vee x_0x_1x_2\bar{x}_3x_4x_5.$$

We have:

$$f_{\max} = g_\alpha \vee \bar{g}_\beta \geqslant f \geqslant g_\alpha\bar{g}_\beta = f_{\min}.$$

We check that the following functions satisfy the latter inequality:

$f_0=\bar{x}_4\bar{x}_5 \vee \bar{x}_0x_4x_5$ (x_1, x_2 and x_3 are redundant variables),

$f_1=\bar{x}_3\bar{x}_5 \vee \bar{x}_1\bar{x}_3$ (x_0, x_2 and x_4 are redundant variables),

$f_2=\bar{x}_4\bar{x}_5 \vee \bar{x}_1x_5$ (x_0, x_2 and x_3 are redundant variables).

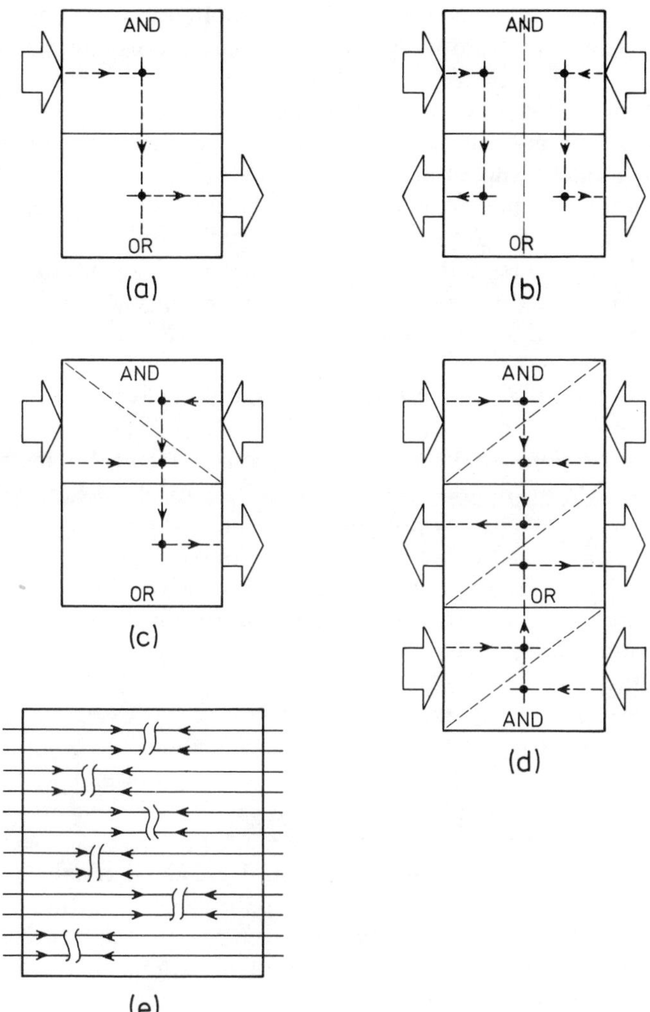

Figure 29. Possible modifications for the PLA architecture

The algorithm by Kambayashi is a systematic procedure for obtaining functions compatible with f and having a maximum number of redundant variables.

The reduction of the number of input variables reduces the number of rows of the AND matrix; it is obvious that it does not reduce the number of product terms, i.e. the number of columns of this matrix. Kambayashi gives an example where a reduction of the number of variables causes an increase in the number of product terms.

The minimization of the number of product terms may be obtained by solving a classical minimal covering problem. In this case, however, the implicants to be introduced in the covering table are not necessarily prime implicants.

Besides this Kambayashi proves that for practical problems, the reduction of the number of input variables is more important than the reduction of the number of product terms in order to optimize the PLA area.

Another way that has been proposed for decreasing the complexity of PLAs is the preprocessing of input variables: instead of introducing the literals $(x_i, \bar{x}_i)$ as inputs of the AND array, we may also introduce pairs of functions $(f(\mathbf{x}_0), \bar{f}(\mathbf{x}_0), \mathbf{x}_0 \subseteq \mathbf{x})$. These functions may, e.g., be simple logic gates realizing the AND, OR, NAND or NOR operations; they also may be more complex circuits such as, e.g., decoders.

Another method that is currently used in the design of improved PLAs is to modify the basic topology of Figure 29a.

It was shown (Wood, 1979) that an increase by a factor 2 in logic function could increase the array area by as much as a factor 4. The PLA shown in Figure 29b overcomes this by putting inputs and outputs on two sides of the arrays. This is equivalent to placing two PLAs back to back: we obtain a folded PLA. Its use is restrictive in that the logic must be capable of being separated into two parts with equal array areas allocated for each part. Many different folding techniques are possible, each putting different constraints on the logic.

Another arrangement is shown in Figure 29c. In this case only one half of the AND matrix is folded. The other inputs can connect to any point in the AND matrix. The folded sections can be used for those inputs and product terms that can be distinctly separated from each other. The through inputs are used for inputs that must connect to both product term sections. An AND matrix area saving of up to one third could be obtained with this configuration but only for the particular applications requiring a fifty-fifty fold. The geometry is too restrictive for a general masterpiece design.

Another way of handling the matrices of a PLA is in duplicating the AND-matrices and in folding the OR-matrix as shown in Figure 29d.

It becomes apparent, however, that PLA structures with predetermined folds are not suitable for a masterpiece design. A variable fold can be introduced as shown in Figure 29e. Array inputs can now face each other on either side of the AND matrix and personalized cuts can be made at any points between them.

Finally, it should be noted that at this time there exist computer programs allowing the optimization of the Boolean function implementation by PLAs described by Figure 29 (Wood, 1979; Segovia, 1980).

BIBLIOGRAPHICAL REMARKS

The material in this chapter has been selected to provide the necessary background which allows us to synthesize networks by means of multiplexers, demultiplexers, programmable logic arrays, read only-memories, etc.

There are several books which have chapters devoted to the design and use of these components: Mano (1972), Lee (1976), and Lee (1978). Moreover, since LSI and VLSI design provide strong motivation for using logic arrays for implementation of computer functions, several special issues of journals are devoted to array logic: *IBM J. Res. Develop.* **19,** No. 2, March 1975; *IEEE Transactions on Computers* **C-28,** 9, September 1979.

EXERCISES

E1. Consider the logical module shown in Figure E1. It has two data inputs a_0 and a_1, one control input x, and one output z. It realizes the Boolean function:

$$z = a_0 \oplus a_1 x. \tag{E1}$$

Show that any Boolean function of n variables may be realized by a tree consisting of $(2^n - 1)$ copies of this module. Compare this design with the multiplexer design developed in Section 2. Establish complexity bounds and devise optimization techniques.

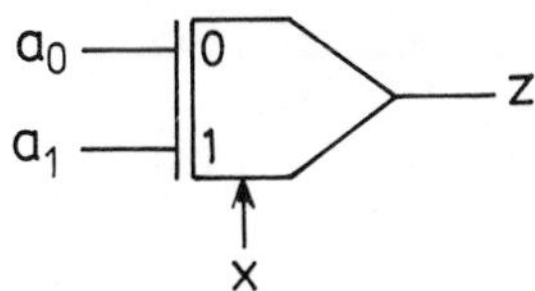

Figure E1

E2. Consider the cell shown in Figure E2a. That cell will be represented as shown in Figure E2b. Using that cell, it is possible to build rectangular arrays: a typical 2

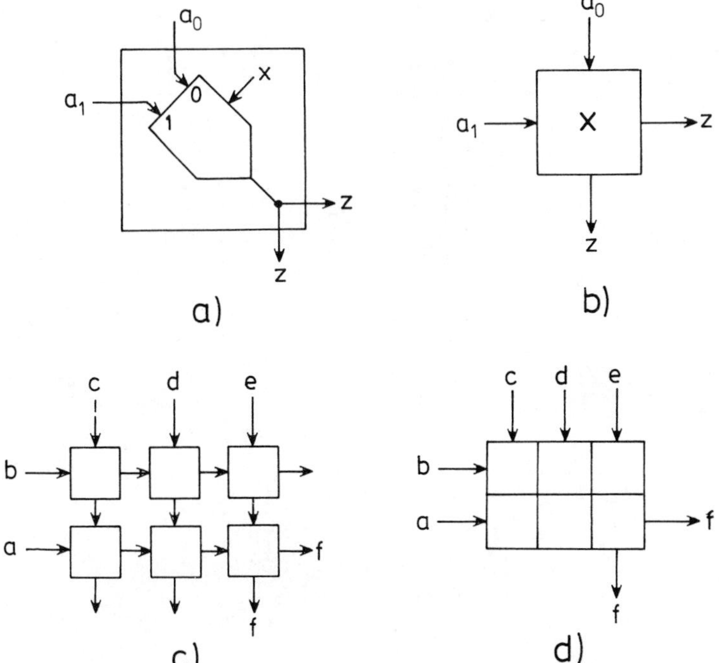

Figure E2

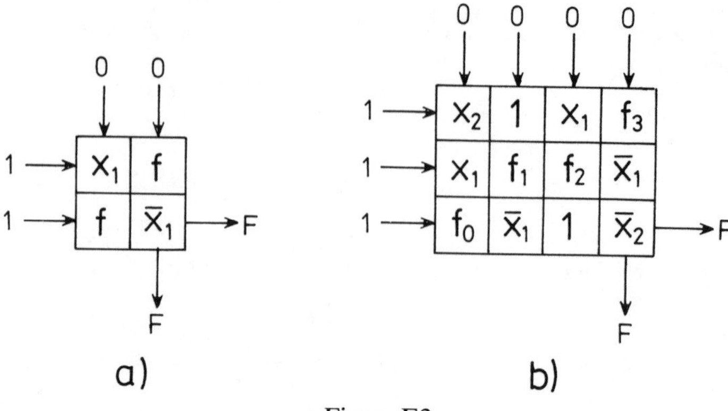

Figure E3

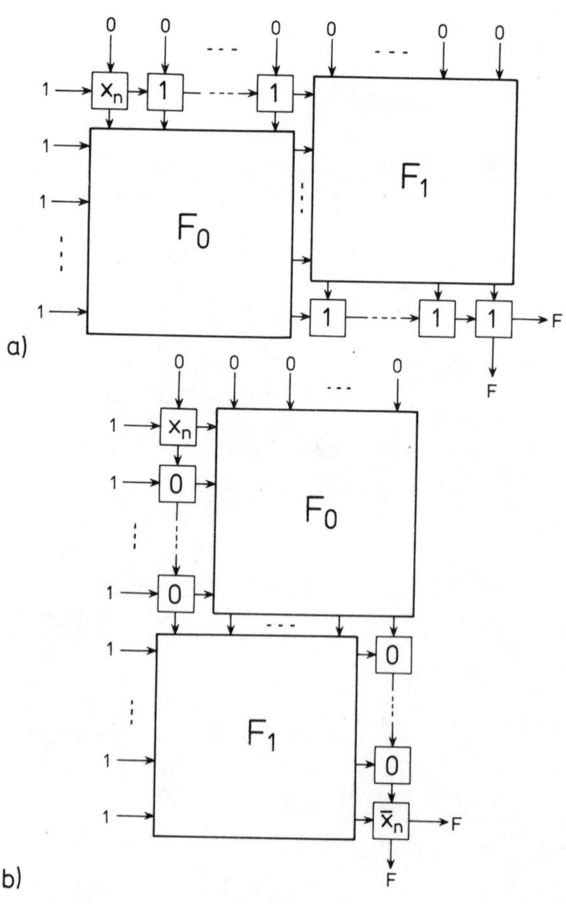

Figure E4

by 3 array is depicted in Figure E2c. The output function of that array is the function appearing at the output of its lower right cell. An array of cells will be represented as shown in Figure E2d. An array is *normal* if all its left-hand border inputs are set to 1 and if all its upper border inputs are set to 0. We say that an array *realizes* the function F if F is the output function of the array when the latter is normal.

(i) Show that the arrays of Figure E3a and Figure E3b realize arbitrary functions of two and three variables respectively.

(ii) Assume that F_0 and F_1 are arbitrary functions of n variables realized by the corresponding subnetworks of Figure E4a and Figure E4b and show that the networks of these figures realize arbitrary functions of $(n+1)$-variables.

(iii) Now start from the $(2, 2)$-array of Figure E3a and use repeatedly the construction of Figure E4a to produce a n-variable ULM. Show that the resulting cost and delay are proportional to $n \cdot 2^n$ and 2^n, respectively.

(iv) Now start from the $(2, 3)$-array of Figure E3b and use alternatively the constructions of Figure E4a and Figure E4b (starting with the former) to produce a n-variable ULM. Show that the resulting cost and delay are proportional to 2^n and $2^{n/2}$, respectively. (*Hint*: observe that, at every second step of the recursive construction, the array is 'almost square', i.e. has $(r+1)$ columns if it has r rows.)

References: Canaday (1964), Akers (1972), and Davio and Quisquater (1973a), (1973b).

E3. The majority function $M(x_2, x_1, x_0)$ is defined by:

$$M(x_2, x_1, x_0) = x_2x_1 \vee x_1x_0 \vee x_0x_2.$$

(i) Show that the majority function satisfies the following equations:

(a) $M(x_2, x_1, x_0) = x_2x_1$,

(b) $M(x_2, x_1, 1) = x_2 \vee x_1$,

(c) $M(x_2, x_2, x_0) = x_2$,

(d) $M(x_2, \bar{x}_2, x_0) = x_0$,

(e) $M(\bar{x}_2, \bar{x}_1, \bar{x}_0) = \overline{M(x_2, x_1, x_0)}$,

(f) $f(x) = M(M(x, f(1), 0), M(\bar{x}, f(0), 0), 1)$,

(g) $f(x) = M(M(x, f(0), 1), M(\bar{x}, f(1), 1), 0)$,

(h) $f(x, y, z) = M(M(x, \bar{y}, f(x, x, z)), M(\bar{x}, \bar{y}, f(x, \bar{x}, z), y)$.

(ii) Show furthermore that:

$$a \oplus b \oplus c = M(M(a, \bar{b}, c), \overline{M(a, b, c)}, b).$$

Deduce from the latter equation a synthesis of a full-adder using three majority gates and two inverters.

(iii) Consider the cascade shown in Figure E5. The **b** inputs are used as

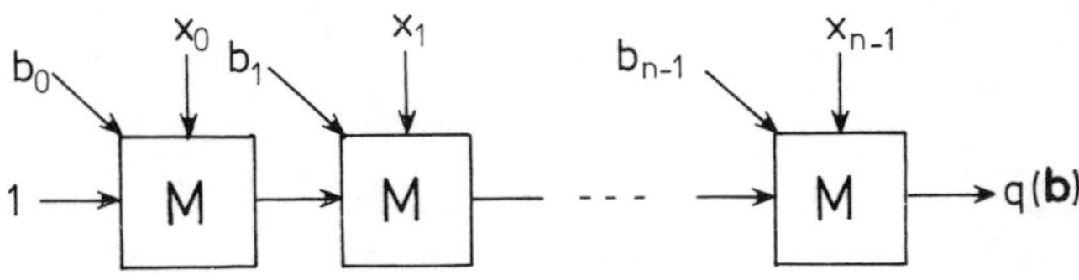

Figure E5

programming inputs. Show that the output function of this cascade is given by:

$$q(\mathbf{b}) = \bigvee_{\mathbf{e}=2^n-\mathbf{b}-1}^{2^n-1} \mathbf{x}^{(\mathbf{e})}.$$

(iv) Show that any Boolean function may be described as a modulo-2 sum of $q(\mathbf{b})$ functions. Obtain a systematic procedure for deriving such an expression. (*Hint*: Express a minterm as a sum of $q(\mathbf{b})$ functions.) Using that observation, describe an array made up of a number of cascades of the above type followed by a series of $\oplus$-gates to collect the q functions. Discuss the complexity of your design.

References. Cohn and Lindaman (1961), Amarel *et al.* (1964), Tohma (1964), Rudeanu (1965), and Kautz (1971).

E4. Consider the cascade shown in Figure E6. Show that it is able to realize an arbitrary function of three variables. Generalize this design to n variables. Discuss the cost and delay of the resulting network. Develop a procedure which is the dual of the one above.

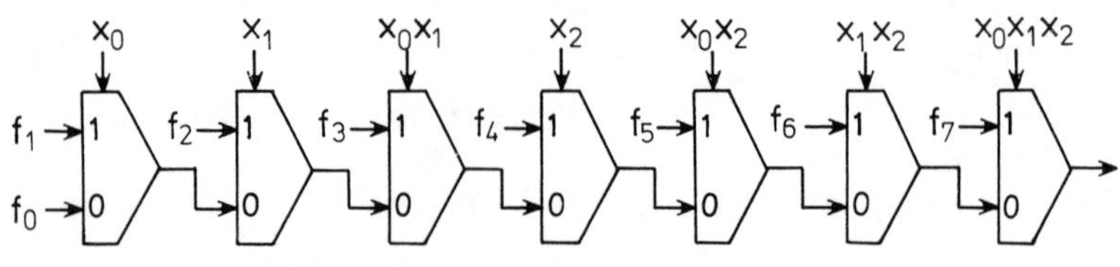

Figure E6

E5. Consider a multiplexer network such as the one displayed in Figure 12b. Such a network has the following property: each of its externally available data inputs is connected either to 'zero' or to 'one'. Such a network (with one or more outputs) will be symbolized as shown in Figure E7. Show that the two networks of Figure E8 realize the same function.

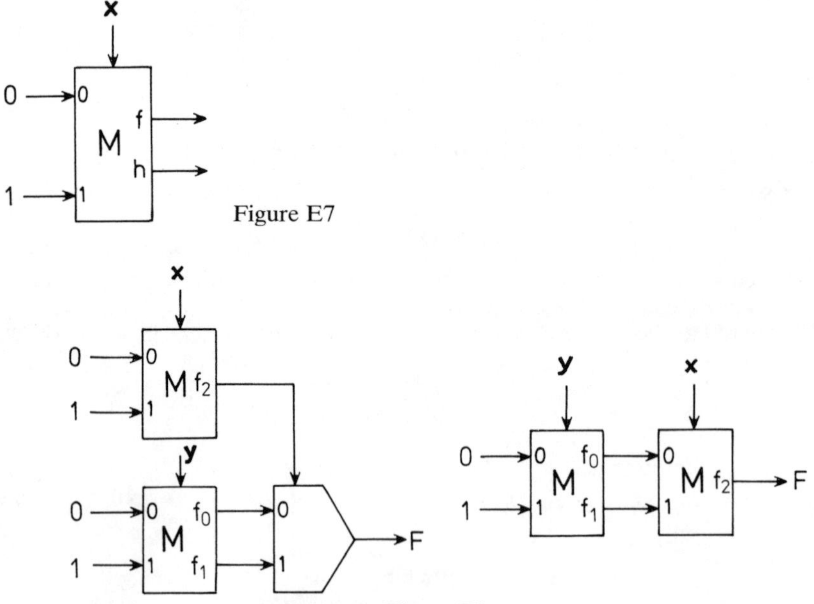

Figure E7

Figure E8

E6. Figure E9 illustrates the graphical convention in use for 16-segment alphanumeric display. The 64 displayed characters are given in correspondence with their ASCII representation (see Chapter I). Design a ROM for conversion from ASCII to the inputs of the 16-segment decoder (LITRONIX DL 1416). Organize your design so as to minimize the cost of the selection equipment (address decoder and multiplexers).

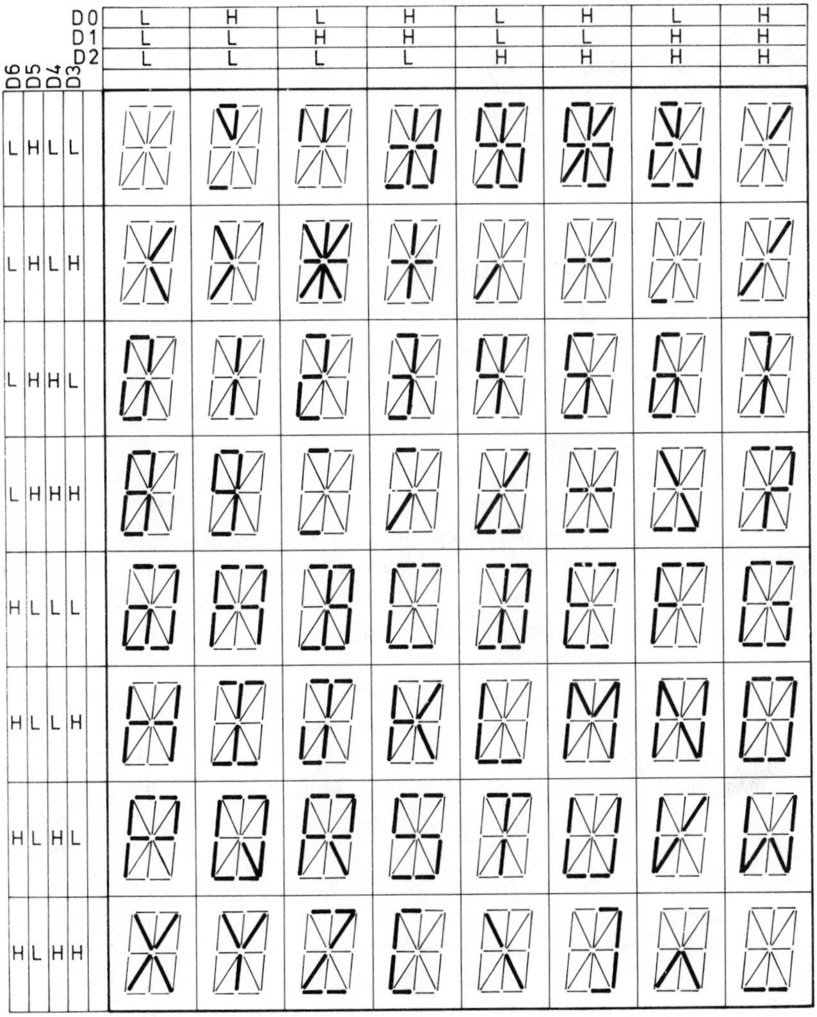

Figure E9

E7. Describe a PLA design based on a combination of a three-level PLA, as discussed in Section 6.2, and a second level of addressing realized by output multiplexers as those used in Figure 25. Discuss the cost and delay of your result.

Chapter IV
Special purpose combinational components

1 INTRODUCTION

The study of complexity theory and the study of optimization methods both point to the fact that it may be rewarding to develop special purpose algorithms and circuits for some well-chosen and frequently encountered computations. Indeed, from the point of view of complexity theory, we may hope that the cost and delay of the realization of m functions of n variables drops from $0(2^s/s)$ and $0(n)$ respectively (the universal bound) to $0(n)$ and $0(\log n)$ respectively (the case-by-case bound). On the other hand, the available optimization techniques are hardly powerful enough to cope with many practical situations, and furthermore, they suppose available as data the truth table of the function to be implemented, i.e. the results of the intended computation.

This chapter is devoted to the study of a number of special purpose algorithms: addition, amplitude comparison, radix conversion, and permutation networks. Except for the latter, the described algorithms are the theoretical basis of commercially available components, and this has been one of our choice criteria. In this chapter, all the described circuits are combinational circuits, i.e. the underlying algorithms are computation schemes. It will, however, turn out that the very same algorithms also have implementations as sequential logic circuits and, for this reason, many of these algorithms will be encountered again in later sections. In particular, arithmetic algorithms as addition, subtraction and amplitude comparison have a definite impact on the architecture of computer processing units: they indeed constitute a time-homogeneous set of primitive operators in terms of which more complex operations will be expressed. Finally, in all these case studies, we have tried to ascertain asymptotic upper bounds both for the cost and delay. A remarkable result in this respect is related to addition: it is indeed shown that addition of two n-bit summands may be realized with cost $0(n)$ and delay $0(\log n)$, these two bounds being achieved simultaneously by the same circuit.

2 ADDITION AND SUBTRACTION

2.1 Addition and positive integers

2.1.1 Principles

The radix B serial addition algorithm has been described in Section 2 of Chapter I. Given the radix B representations:

$$x_{n-1} \ldots x_1 x_0 \quad \text{and} \quad y_{n-1} \ldots y_1 y_0$$

of two positive integers X and Y ($X, Y \in Z_{B^n}$), we wish to compute the sum

$$Z = X + Y + c_{-1} \tag{1}$$

of X, Y and of an initial carry c_{-1} ($c_{-1} \in Z_2$). As it has been shown in equation (16) in Chapter I, Z has the readix B representation;

$$z_n z_{n-1} \ldots z_1 z_0; \qquad z_n \in Z_2; \qquad z_i \in Z_B. \tag{2}$$

The radix B representation of Z is obtained at the outset of an iterative process whose typical step is described by:

$$x_i + y_i + c_{i-1} = c_i B + s_i$$
$$(x_i, y_i, s_i \in Z_B; \qquad c_{i-1}, c_i \in Z_2; \qquad i = 0, 1, \ldots, n-1). \tag{3}$$

Indeed, from equations (3), one easily deduces

$$X + Y + c_{-1} = c_{n-1} B^n + \sum_{i=0}^{n-1} s_i B^i, \tag{4}$$

i.e.

$$z_n = c_{n-1} \quad \text{and} \quad s_i = z_i; \qquad (i = 0, 1, \ldots, n-1).$$

The computation primitive described by (3) is the *radix B full adder* (FA_B). As we have seen earlier, we deduce from (3):

$$c_i = \lfloor (x_i + y_i + c_{i-1})/B \rfloor, \tag{5}$$

$$s_i = [\![x_i + y_i + c_{i-1}]\!]_B. \tag{6}$$

The symbol for a FA_B is given in Figure 1a. The serial adder is then represented as in Figure 1b. It is interesting to compare equations (3) and (4). From equation (4) we indeed deduce, by analogy with (5) and (6):

$$c_{n-1} = \lfloor (X + Y + c_{-1})/B^n \rfloor, \tag{7}$$

$$S = [\![X + Y + c_{-1}]\!]_{B^n}. \tag{8}$$

As a matter of fact, equations (7) and (8) are the radix B^n transposition of equations (5) and (6), written in radix B. That simple observation immediately points to the *expandability* of the full-adder function. Stated

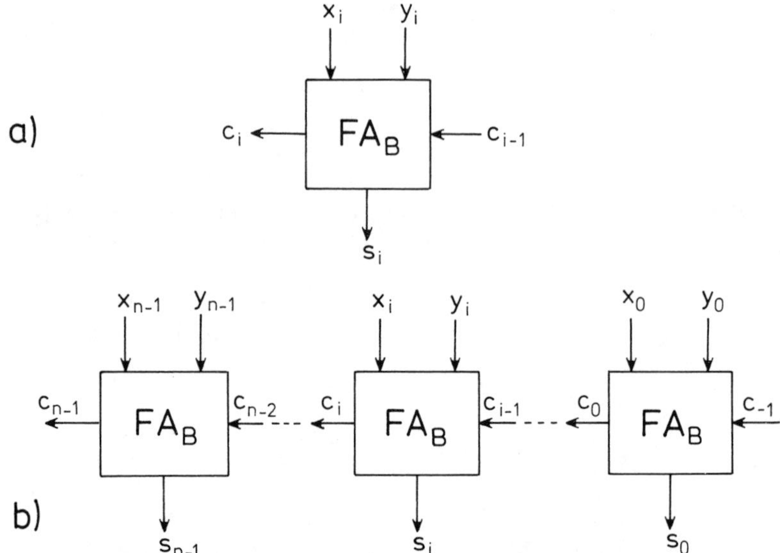

Figure 1.(a) Symbol for a radix B full-adder. (b) Serial adder

otherwise, the circuit shown in Figure 1b may be viewed as a radix B^n full-adder.

We shall describe in the following sections a number of full-adders corresponding to $B=2$, 16. We first conclude this preliminary section with some general comments.

The radix B full-adder is characterized by the following parameters:

(i) Its cost γ_B, and

(ii) a number of input–output delays:

$$\delta_B(x_i, s_i), \quad \delta_B(y_i, s_i), \quad \delta_B(c_{i-1}, s_i), \quad \delta_B(x_i, c_i), \quad \delta_B(y_i, c_i), \quad \delta_B(c_{i-1}, c_i).$$

We shall assume by symmetry that

$$\begin{aligned} \delta(x_i, s_i) &= \delta(y_i, s_i), \\ \delta(x_i, c_i) &= \delta(y_i, c_i). \end{aligned} \tag{9}$$

According to these definitions, the cost Γ_B and the delay Δ_B of the adder displayed in Figure 1b are given by

$$\Gamma_B = n\gamma_B, \tag{10}$$

$$\begin{aligned} \Delta_B = {} & \max\{\delta_B(x_i, c_i), \delta_B(c_{i-1}, c_i)\} + (n-2)\delta_B(c_{i-1}, c_i) \\ & + \max\{\delta_B(c_{i-1}, c_i), \delta_B(c_{i-1}, s_i)\}. \end{aligned} \tag{11a}$$

If all the delays have the same order of magnitude, we may write as a first approximation:

$$\Delta_B \cong n\delta_B(c_{i-1}, c_i). \tag{11b}$$

The linearity of Δ_B with respect to n is clearly the consequence of the iterative nature of the underlying algorithm. The above discussion also explains why, in practical implementations, one always tries to minimize the delay $\delta_B(c_{i-1}, c_i)$ with respect to all the other delays. The delay $\delta_B(c_{i-1}, c_i)$ is called *propagation delay*.

2.1.2 Binary full-adder

The radix 2 version of equations (3), (5), and (6) is given by:

$$x_i + y_i + c_{i-1} = 2c_i + s_i, \tag{12}$$

$$c_i = \lfloor (x_i + y_i + c_{i-1})/2 \rfloor, \tag{13}$$

$$s_i = x_i \oplus y_i \oplus c_{i-1}. \tag{14}$$

The output carry c_i and the sum bit s_i are both Boolean functions of the three binary variables x_i, y_i and c_{i-1}. These functions are also represented by the Karnaugh maps in Figure 2.

The output bit s_i and the carry bit c_i are also described by the following expressions:

$$s_i = x_i\bar{y}_i\bar{c}_{i-1} \vee \bar{x}_i\bar{y}_i c_{i-1} \vee \bar{x}_i y_i \bar{c}_{i-1} \vee x_i y_i c_{i-1}, \tag{15}$$

$$c_i = x_i y_i \vee x_i c_{i-1} \vee y_i c_{i-1}, \tag{16}$$

$$c_i = (x_i \vee y_i)(x_i \vee c_{i-1})(y_i \vee c_{i-1}), \tag{17}$$

$$c_i = x_i y_i \vee (x_i \vee y_i) c_{i-1}, \tag{18}$$

$$c_i = (x_i \vee y_i)(x_i y_i \vee c_{i-1}), \tag{19}$$

$$c_i = x_i y_i \vee (x_i \oplus y_i) c_{i-1}, \tag{20}$$

$$c_i = (x_i \vee y_i)[\overline{(x_i \oplus y_i)} \vee c_{i-1}]. \tag{21}$$

These expressions are the most frequently used computation schemes of s_i and c_i in practical implementations. Their proof is elementary and is left to the reader (Exercise 1).

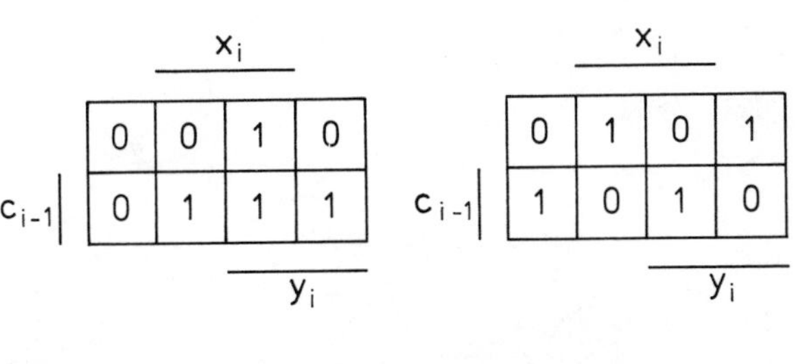

Figure 2. Karnaugh map for the binary full-adder

If we put

$$G_i = x_i y_i, \tag{22}$$

$$P_i = x_i \vee y_i, \tag{23}$$

then equations (18) and (19) may be written again as

$$c_i = G_i \vee P_i c_{i-1}, \tag{24}$$

$$c_i = P_i(G_i \vee c_{i-1}). \tag{25}$$

For obvious reasons, the quantities G_i and P_i are the *generated carry* and *propagated carry*, respectively. The comparison of (14) and of (20) shows the possibility of sharing the circuit producing $(x_i \oplus y_i)$ in a simultaneous synthesis of c_i and s_i.

As simple as it may appear, the design of optimal binary full-adders is a difficult problem. Indeed, on the one hand, one does not impose any limitation on the design itself such as gate type or two-level synthesis; on the other hand, the optimization criteria are numerous and often contradictory, for example, gate number, number of internal connections, propagation delay $\delta_B(c_{i-1}, c_i)$, input–output delay $\delta_B(x_i, s_i)$, maximum gate fan-in, maximum gate fan-out, and so on.

The paper by Liu *et al.* (1974) describes not less than 30 binary full-adders, each of which is optimal in a given environment gate type(s), fan-in, . . .) and for a specific optimality criterion. As an illustration, Figure 3 represents three possible implementations of a binary full-adder.

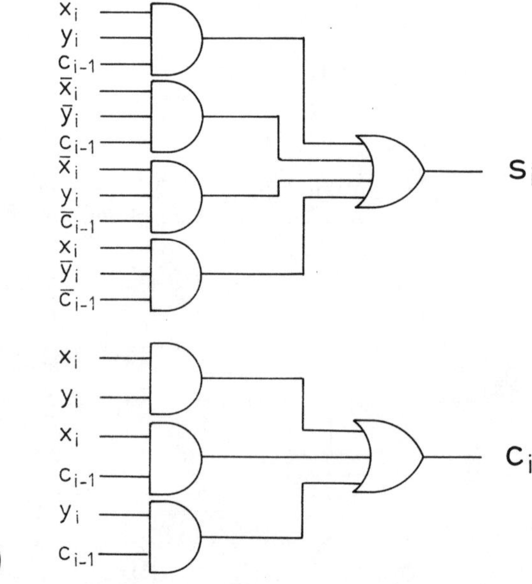

Figure 3. Implementations of binary full-adders

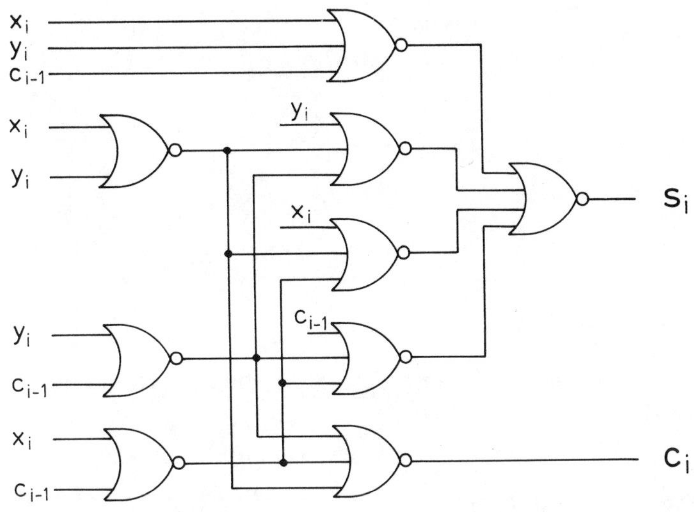

b)

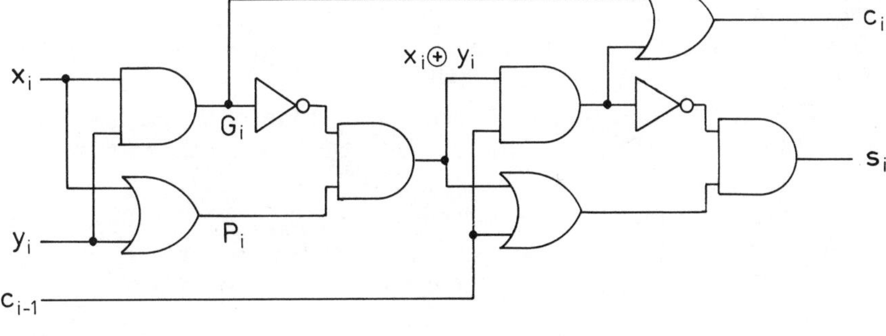

c)

(i) The circuit in Figure 3a is described by equations (15) and (16).

(ii) The circuit in Figure 3b may be justified as follows. Consider the first-level gate outputs:

$$\alpha = \bar{x}_i\bar{y}_i; \qquad \beta = \bar{x}_i\bar{c}_{i-1}; \qquad \gamma = \bar{y}_i\bar{c}_{i-1}.$$

From equation (17), one immediately deduces: $c_i = \bar{\alpha}\bar{\beta}\bar{\gamma}$. Furthermore, from the canonical disjunctive expansion of s_i:

$$s_i = (x_i \vee y_i \vee c_{i-1})(x_i \vee \bar{y}_i \vee \bar{c}_{i-1})(\bar{x}_i \vee y_i \vee \bar{c}_{i-1})(\bar{x}_i \vee \bar{y}_i \vee c_{i-1}),$$

one easily proves:

$$s_i = (x_i \vee y_i \vee c_{i-1})(x_i \vee \alpha \vee \beta)(\alpha \vee y_i \vee \gamma)(\beta \vee \gamma \vee c_{i-1}).$$

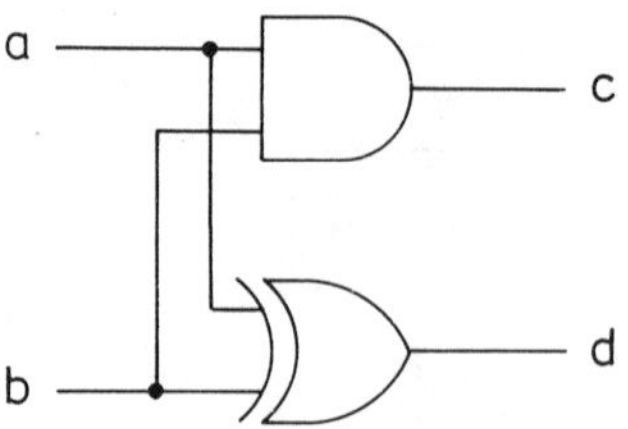

Figure 4. Half-adder

(iii) Finally, the circuit in Figure 3c is obtained from equations (14) and (20), and from:

$$x_i \oplus y_i = \bar{G}_i P_i \tag{26}$$

Observe that, in those three full-adders, the propagation delay $\delta_2(c_{i-1}, c_i)$ is equal to two gate delays. In particular, in the circuit of Figure 3c, one has destroyed the natural symmetry in x_i, y_i, c_{i-1} to minimize that delay.

Remarks.

(i) At this stage, the discussion of the binary full-adder is far from being complete. Observe, for example, that, in a binary serial adder, nothing obliges us to use a single bit to represent the *internal carries*, $c_0, c_1, \ldots, c_{n-2}$. Lai and Muroga (1979) use a two-bit coding of these carries to ensure cost minimization (see Exercise 2).

(ii) Forthcoming technologies will most probably use new full-adder design. As an example, let us mention the Manchester full-adder (Mead and Conway, 1980) which tends to become a standard in MOS design.

(iii) When one of the three quantities x_i, y_i or c_{i-1} becomes equal to zero, the FA_2 becomes a two-input two-output device called a half-adder. The equations describing a half-adder (HA) are:

$$a + b = 2c + d, \tag{27}$$

or

$$c = ab, \tag{28}$$

$$d = a \oplus b. \tag{29}$$

A circuit diagram for the half-adder is given in Figure 4.

2.1.3 Radix B^n full-adder: the carry-lookahead principle

As we have seen in the previous section, we obtain a radix B^n full-adder by cascading n FA_B's. In the carry-lookahead addition we intend to describe now, we try to lower the high overall delay (11) due to the step by step carry propagation.

We define the following two logical functions:

(i) *generated carry* G_i:

$$G_i = 1, \quad \text{iff } x_i + y_i \geqslant B, \qquad\qquad (30)$$
$$G_i = 0, \quad \text{iff } x_i + y_i < B;$$

(ii) *propagated carry* P_i:

$$P_i = 1 \quad \text{iff } x_i + y_i = B - 1,$$
$$P_i = 0 \quad \text{iff } x_i + y_i < B - 1, \qquad\qquad (31)$$
$$P_i \text{ is arbitrary iff } x_i + y_i \geqslant B.$$

These definitions extend to an arbitrary radix the concepts of generated and propagated carries introduced earlier for the binary case. They allow us to describe the output carry c_i as the Boolean function:

$$c_i = G_i \vee P_i c_{i-1}; \qquad\qquad (32)$$

the latter expression extends to an arbitrary radix the validity of (24). Consider now the equations:

$$\begin{aligned} C &= G_0 \vee P_0 c_{-1} \\ c_1 &= G_1 \vee P_1 c_0 \\ &\text{-----------------} \\ c_i &= G_i \vee P_i c_{i-1} \\ &\text{-----------------} \\ c_{n-1} &= G_{n-1} \vee P_{n-1} c_{n-2} \end{aligned} \qquad\qquad (33)$$

which describe the iterative computation of carries in the serial adder. To avoid the carry propagation effect, we have only to eliminate the intermediate carries $c_0, c_1, \ldots, c_{n-2}$. By stepwise substitutions, we obtain, from (33), the equivalent system:

$$\begin{aligned} c_0 &= G_0 \vee P_0 c_{-1} \\ c_1 &= G_1 \vee P_1 G_0 \vee P_1 P_0 c_{-1} \\ c_2 &= G_2 \vee P_2 G_1 \vee P_2 P_1 G_0 \vee P_2 P_1 P_0 c_{-1} \\ &\text{-----------------------------} \\ c_{n-1} &= G_{n-1} \vee P_{n-1} G_{n-2} \vee P_{n-1} P_{n-2} G_{n-3} \ldots \\ &\quad \vee P_{n-1} \ldots P_2 P_1 G_0 \vee P_{n-1} \ldots P_1 P_0 c_{-1} \end{aligned} \qquad\qquad (34)$$

Thanks to (34), we are now in position to compute the carries c_i directly from G_i, P_i and c_{-1}. Our new addition algorithm may then be described as follows:

Step 1. Compute the G_i's and P_i's from the digits x_i and y_i.

Step 2. (i) Compute the intermediate carries c_i by (34); (ii) compute the modulo-B sums $x_i \oplus y_i$. When $B = 2$, $x_i \oplus y_i = \overline{G_i} P_i$.

Step 3. Compute the modulo-B sum $(x_i \oplus y_i) \oplus c_{i-1}$.

A logical diagram illustrating the above principles ($B = 2$, $n = 4$) is given in Figure 5. The hexadecimal digits x_i, y_i and s_i are encoded by binary

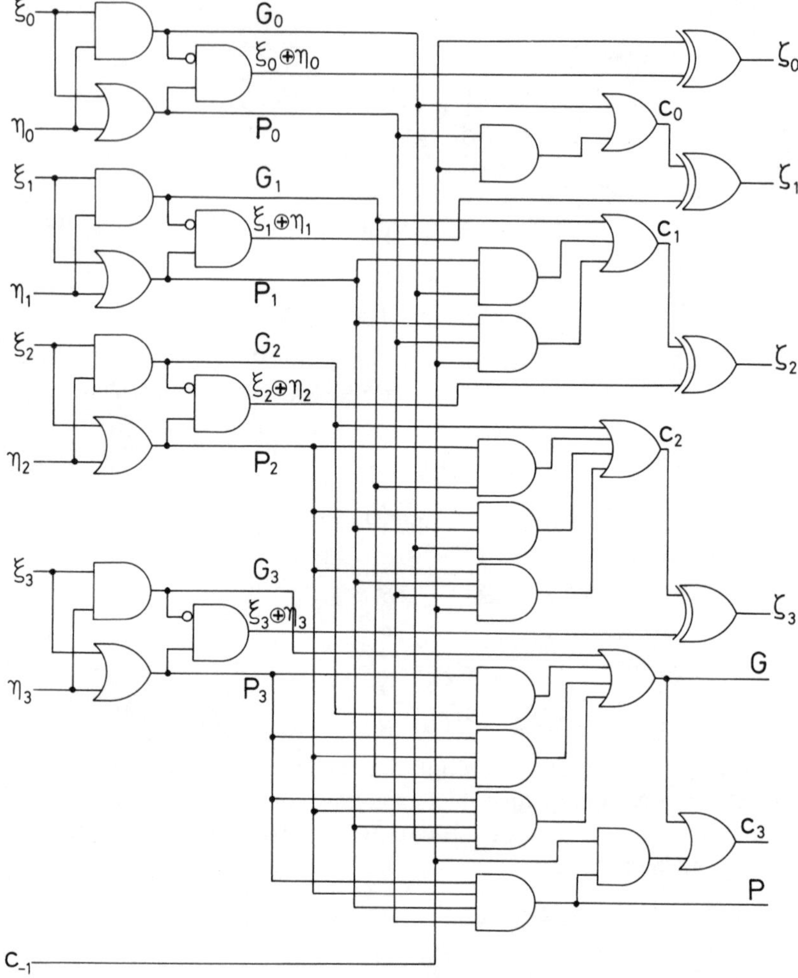

Figure 5. Carry-lookahead principle

4-tuples according to

$$x_i \simeq \xi_3\xi_2\xi_1\xi_0; \qquad y_i \simeq \eta_3\eta_2\eta_1\eta_0; \qquad s_i \simeq \zeta_3\zeta_2\zeta_1\zeta_0;$$

(the symbol $\simeq$ means 'is represented by').

Observe that in this circuit, the output carry c_3 is not designed as a two-level AND to OR circuit, as is the case for the intermediate carries c_0, c_1, c_2. Instead, c_3 is described by

$$c_3 = G \vee Pc_{-1}, \tag{35}$$

where G and P are given by

$$G = G_3 \vee P_3G_2 \vee P_3P_2G_1 \vee P_3P_2P_1G_0, \tag{36}$$

$$P = P_3P_3P_1P_0. \tag{37}$$

Clearly, G and P may be considered as hexadecimal generated and propagated carries, in the sense of definitions (30) and (31). In some commercial modules, the two quantities G and P are made available as external outputs: this is generally the case in arithmetic logic units such as the TTL 74181 to be discussed later on (24 pin case). Observe, however, that the number of external connections to implement fully the circuit in Figure 5, including the G and P outputs, the voltage source, and the ground is 18, while the standard case for hexadecimal (and decimal) adders was, at the design time, a 16 lead package. The availability of the G and P outputs makes possible an iterated use of the carry-lookahead scheme to be discussed in the next section.

The principle described above and illustrated by Figure 5 is clearly of a general nature and we shall conclude this section by appreciating its performances whenever $B = 2$, n being arbitrary. If we assume that the fan-in is at least equal to $(n+1)$, we can immediately check that the cost of the device, expressed as a number of gates is of the form:

$$\gamma_{2^n} = kn + k'n^2. \tag{38}$$

The linear term kn corresponds to the input and output modulo-two sums, while the quadratic term $k'n^2$ is due to the carry-lookahead circuitry: the latter indeed requires

$$2+3+4+\ldots+(n+1) = n(n+3)/2 \text{ gates.}$$

Similarly, if there is no fan-in limitation, the involved delays $\delta_B(x_i, s_i), \ldots, \delta_B(c_{i-1}, c_i)$ remain constant. In particular, the propagation delay $\delta_B(c_{-1}, c_{n-1})$ is equal to two gate delays. Obviously, these conclusions do not apply any more whenever n becomes large (say $n = 32$ or 64 bits). An appropriate modification is described in the next section.

2.1.4 Block carry-lookahead. Carry-lookahead generator

We define an *augmented radix B full-adder* (AFA(B)) as a device having three inputs x_i, y_i, c_{i-1} and four outputs s_i, c_i, G_i, P_i $(x_i, y_i, s_i \in Z_B;$ $c_{i-1}, c_i, P_i, G_i \in Z_2)$. The generated and propagated carries are defined by (30) and (31). The circuit in Figure 5 may thus be viewed as an AFA (16).

Let us now consider two-integers having $n = t^s$ radix B digits:

$$X \simeq x_{t^s-1} \ldots x_1x_0; \qquad Y \simeq y_{t^s-1} \ldots y_1y_0. \tag{39}$$

We may as well consider that these integers have t radix $B^{n/t}$ digits

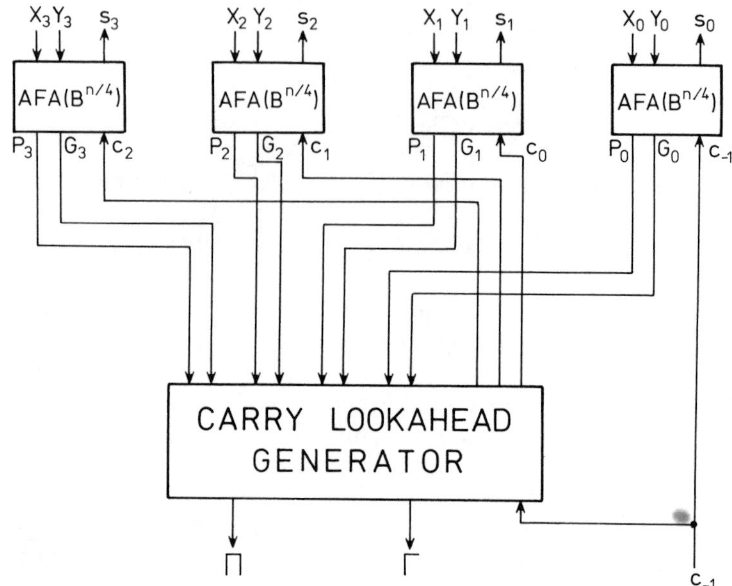

Figure 6. Basic step of carry-lookahead generation

$(n/t = t^{s-1})$. Thus, we may write

$$X \simeq X_{t-1} \ldots X_1 X_0; \qquad Y \simeq Y_{t-1} \ldots Y_1 Y_0, \tag{40}$$

with

$$0 \leqslant X, Y \leqslant B^{n/t} - 1. \tag{41}$$

The basic step of block carry-lookahead generation is now illustrated by Figure 6 ($t = 4$). We assume available t AFA($B^{n/t}$) and show how they may be completed to yield an AFA(B^n). For $t = 4$, the carries (at the radix $B^{n/t}$ level) are given by the usual equations:

$$c_0 = G_0 \vee P_0 c_{-1}, \tag{42}$$

$$c_1 = G_1 \vee P_1 G_0 \vee P_1 P_0 c_{-1}, \tag{43}$$

$$c_2 = G_2 \vee P_2 G_1 \vee P_2 P_1 G_0 \vee P_2 P_1 P_0 c_{-1}, \tag{44}$$

$$c_3 = \Gamma_0 \vee \Pi_0 c_{-1}, \tag{45}$$

with

$$\Gamma_0 = G_3 \vee P_3 G_2 \vee P_3 P_2 G_1 \vee P_3 P_2 P_1 G_0, \tag{46}$$

$$\Pi_0 = P_3 P_2 P_1 P_0. \tag{47}$$

The equations (42)–(44), (46) and (47) giving c_0, c_1, c_2, Γ_0 and Π_0 describe a new component, the *carry-lookahead generator* illustrated in Figure 7 (commercial component 74182). The most interesting point about this device is that its binary description only depends on t (here $t = 4$) and not on

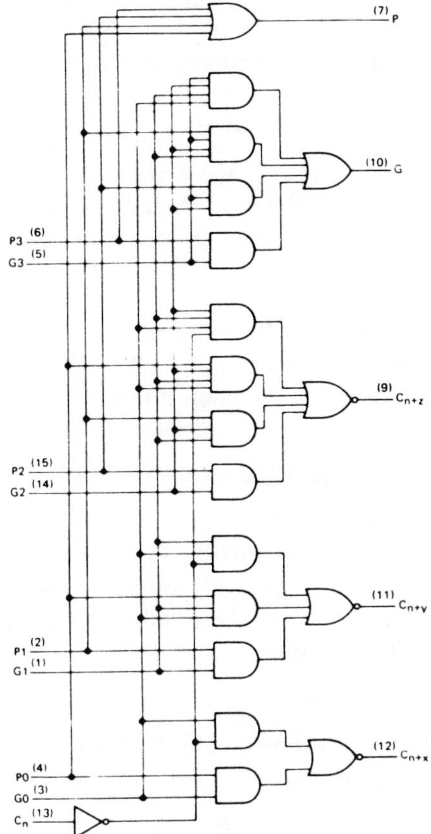

Figure 7. Carry-lookahead generator

the radix B. It may thus be used as well for decimal applications as for binary applications.

It should now be clear that Figure 5 only describes the first step of a recursive process: indeed, the AFA($B^{n/t}$) may in turn be decomposed in AFA(B^{n/t^2}), using the same carry-lookahead generator. After s steps, one will be left with n AFA(B) and a tree of carry-lookahead generators containing

$$1+t+t^2+\ldots+t^{s-1}=(t^s-1)/(t-1)$$

carry-lookahead generators. It is thus not difficult to obtain estimates of the performances achieved by a carry-lookahead addition. Assuming that the number of digits of the two summands is $n=t^s$, one easily shows:

(i) that the addition cost Γ exhibits a linear increase with n; thus

$$\Gamma \cong kn; \tag{48}$$

in particular, the cost of the carry-lookahead generators is, in first approximation, of the form t^2n.

(ii) that the addition delay Δ is proportional to the logarithm of n; more precisely:

$$\Delta \simeq 4 \log_t n \tag{49}$$

Both the asymptotic behaviours of Γ and Δ are thus optimal from the point of view of complexity theory. This interesting property is shared by the so-called conditional sum adder which, however, exhibits a higher numerical cost value.

Remarks. We shall not consider here the problem of multisummand addition, which will be briefly discussed in Chapter IX in the context of multiplication. BCD adders are considered in the exercises.

2.1.5 Subtraction of positive integers

Consider two integers X and Y given by their radix B representations $x_{n-1} \ldots x_1 x_0$ and $y_{n-1} \ldots y_1 y_0$. In this section, we try to obtain the radix B representation of the difference $X-Y$ and, so far, we only can succeed if $X-Y$ is a positive quantity. The serial subtraction algorithm, studied in (Chapter I, Exercise E6), is based on the iteration

$$x_i - y_i - b_{i-1} = -Bb_i - d_i, \tag{50}$$

where b_{i-1} and b_i are internal borrows and where d_i is the local difference digit. Eliminating the internal borrows from equations (50), one obtains:

$$X - Y = -B^n b_{n-1} + D, \tag{51}$$

i.e.

(i) if $b_{n-1} = 0$: $X - Y = D$,
(ii) if b_{n-1}: $X - Y = -B^n + D < 0$.

Observe that, in the latter case, $\|X - Y\| = B^n - D$.

Examine then what happens if, instead of subtracting Y from X, we add $B^n - Y$ to X. We thus compute the quantity $D^* = X + B^n - Y$, and D^* is related to D by

$$D^* = B^n - B^n b_{n-1} + D. \tag{52}$$

As above, two cases are possible:

(i) $b_{n-1} = 0$. In this case:

$$D^* = B^n + D \geqslant B^n \tag{53}$$

and

$$D^* = D (\text{mod } B^n) \tag{54}$$

(ii) $b_{n-1} = 1$. In this case:

$$D^* = D \tag{55}$$

and

$$\|X-Y\| = B^n - D^*. \tag{56}$$

The above discussion shows that it is possible to replace the subtraction $X-Y$ by the addition $X+(B^n-Y)$. The result is accepted if the output carry c_{n-1} appearing in this addition is equal to one (see (54)). The last n result digits then provide the correct answer (see (55)). The situation $X-Y<0$ is the *overflow* situation: an overflow situation clearly corresponds to $c_{n-1}=0$ in the addition $X+(B^n-Y)$. In that situation, the absolute value $\|X-Y\|$ may be restored by computing B^n-D^*.

Numerical example. $B=10$, $n=3$.

(a) $X=576$; $\quad Y=352$
$10^3-Y=648$
$D^*=576+648=1224$
$D^*\geq 1000 \Rightarrow X-Y\geq 0$
$X-Y=[\![D^*]\!]_{1000}=224$

(b) $X=352$; $\quad Y=576$
$10^3-Y=424$
$D^*=35+424=776$
$D^*<1000 \Rightarrow X-Y<0$
$\|X-Y\|=1000-D^*=224.$

The above procedure rests upon the preliminary computation of B^n-Y. That computation is based on the familiar identity:

$$B^n-1=\sum_{i=0}^{n-1}(B-1)B^i. \tag{57}$$

From the latter identity, we immediately deduce:

$$B^n-Y=\sum_{i=0}^{n-1}(B-1-y_i)B^i+1. \tag{58}$$

Hence, to obtain the representation of B^n-Y, we first replace each digit y_i by $(B-1-y_i)$ and add 1 to the obtained number.

The above subtraction method is called *radix-complement subtraction* (not to be confused with the radix-complement representation, to be studied in the next section). The number having the digits $(B-1-y_i)$ is also called *diminished radix complement* of Y.

In the binary case, equation (58) becomes

$$2^n-Y=\sum_{i=0}^{n-1}\bar{y}_i+1. \tag{59}$$

Hence:

$$D^*=\sum_{i=0}^{n-1}x_i2^i+\sum_{i=0}^{n-1}\bar{y}_i2^i+1. \tag{60}$$

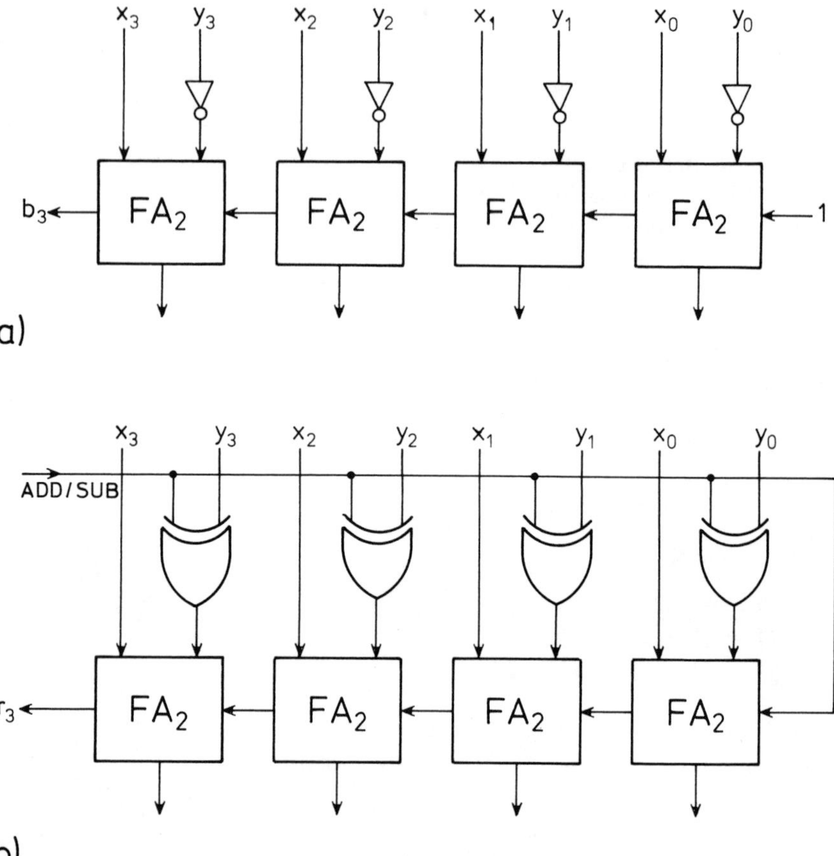

Figure 8.(a) Radix complement subtractor. (b) Programmable adder subtractor

The result is accepted if the output carry c_{n-1} is equal to one. This output carry is not part of the result. A subtractor based on this principle is displayed in Figure 8a: observe that the carry input is used to introduce the unit appearing in the expression of D^*.

Radix-complement subtraction has a double advantage over the direct serial subtraction discussed earlier. First it yields a slightly simpler circuit. Moreover, it is easily combined with the serial adder design to yield a programmable adder/subtractor. This point is illustrated in Figure 8b. The overflow situation is $r_3 = 1$ for addition and $r_3 = 0$ for subtraction. Note finally that, as subtraction has been reduced to addition, the fast addition techniques such as the carry-lookahead addition are immediately transposed to the subtraction.

2.2 Addition and subtraction of rational integers

2.2.1 Sign magntitude representation

In Section 2.2 we extend to signed integers the addition and subtraction methods discussed earlier for positive integers. We first require a representation of signed integers and introduce here such a representation called *sign magnitude representation.*

Any rational integer X such that

$$-(B^n-1) \leqslant X \leqslant B^n-1 \tag{61}$$

may be represented by a $(n+1)$-tuple $x_s, x_{n-1} \ldots x_1 x_0$, where

$$\begin{aligned} x_s &= 0, \quad \text{iff } X \geqslant 0, \\ x_s &= 1, \quad \text{iff } X \leqslant 0, \end{aligned} \tag{62}$$

and where $x_{n-1} \ldots x_1 x_0$ is the radix B representation of the absolute value $\|X\|$ of X. This is the sign-magnitude representation of X. x_s is the *sign bit.* Observe that zero has the two representations $0,00\ldots0$ and $1,00\ldots0$. Figure 9 illustrates that representation for $B=2$ $n=3$. Similarly, in BCD:

+182 is represented by 0 0001 1000 0010,
−036 is represented by 1 0000 0011 0110.

The sign change is obviously a trivial operation as it amounts to the complementation of the sign bit; the subtraction is thus easily reduced to addition. The discussion of the addition yields a deceiving conclusion: sign magnitude addition is a rather complex operation, and, in particular, it requires amplitude comparisons when the summands have opposite signs.

X	x_3	x_2	x_1	x_0
7	0	1	1	1
6	0	1	1	0
5	0	1	0	1
4	0	1	0	0
3	0	0	1	1
2	0	0	1	0
1	0	0	0	1
0	0	0	0	0
0	1	0	0	0
−1	1	0	0	1
−2	1	0	1	0
−3	1	0	1	1
−4	1	1	0	0
−5	1	1	0	1
−6	1	1	1	0
−7	1	1	1	1

Figure 9. Sign magnitude representation

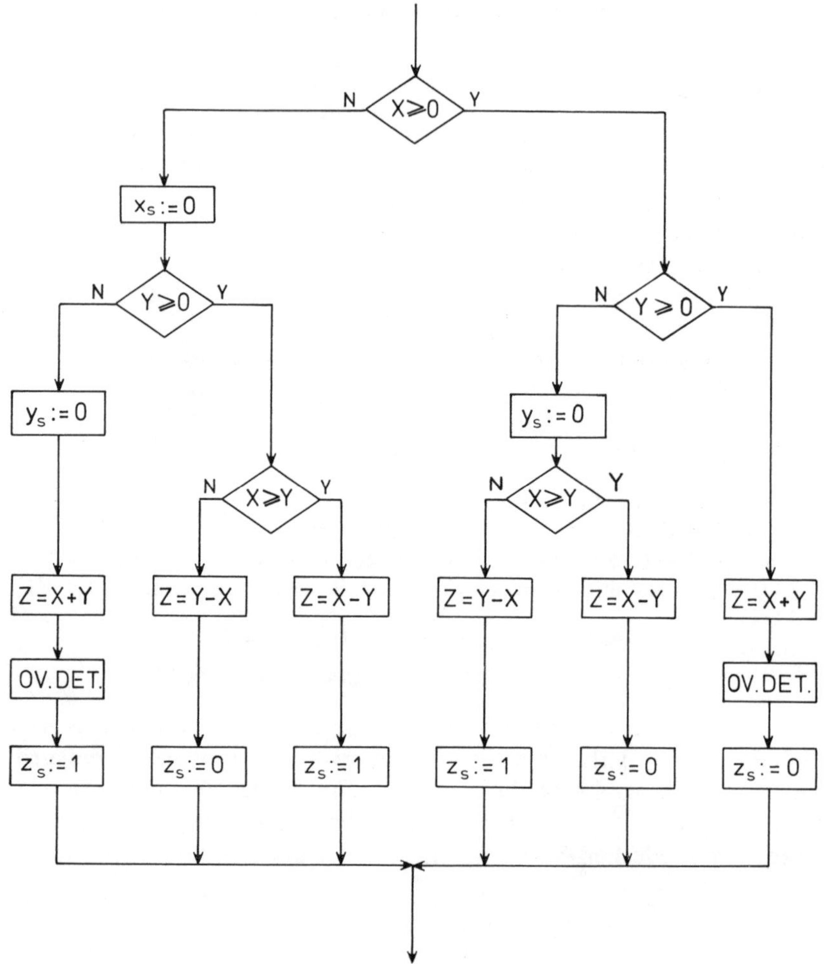

Figure 10. Flow chart of sign magnitude addition

This fact is illustrated by Figure 10 which represents a typical flowchart of sign-magnitude addition. The operands are X and Y, the result is Z; the respective sign bits of these quantities are x_s, y_s and z_s. Further comments on this algorithm will appear in Chapter IX.

2.2.2 Radix-complement representation

The study of the radix-complement subtraction, carried out in Section 2.1.5 suggests the 'simulation' of the negative integer $-\|Y\|$ by the positive integer $B^n - \|Y\|$. We may formalize that idea as follows:

Let $\tilde{Z}$ represent an interval of B^n consecutive integers. We assume that $\tilde{Z}$

contains both negative and positive integers; in particular, $0 \in \tilde{Z}$. The mapping

$$S: \tilde{Z} \to Z_{B^n}: X \to [\![X]\!]_{B^n} \tag{63}$$

is a bijection. The positive integer $S(X)$ is the *simulation* of X. The radix B representation of $S(X)$ is the *radix-complement representation* of X. Observe that:

$$X \geqslant 0 \Rightarrow S(X) = X, \tag{64}$$

$$X < 0 \Rightarrow S(X) = B^n + X = B^n - \|X\|. \tag{65}$$

These equations will allow us to obtain relations between X and its radix-complement representation $x_{n-1} \ldots x_1 x_0$. By definition, we have:

$$S(X) = \sum_{i=0}^{n-1} x_i B^i. \tag{66}$$

Hence:

$$\text{(i) if} \quad X \geqslant 0: X = x_{n-1} B^{n-1} + \sum_{i=0}^{n-2} x_i B^i, \tag{67}$$

$$\text{(ii) if} \quad X < 0: X = -(B - x_{n-1}) B^{n-1} + \sum_{i=0}^{n-2} x_i B^i. \tag{68}$$

These two relations show that the most significant digit x_{n-1} is the only digit to influence the sign. On the other hand, we clearly expect to be able to deduce the sign of X from its representation $x_{n-1} \ldots x_1 x_0$. Combining these two observations, we reach the convention

$$\begin{aligned} &x_{n-1} \in \{0, 1, \ldots, K-1\} \Rightarrow X \geqslant 0, \\ &x_{n-1} \in \{K, K+1, \ldots, B-1\} \Rightarrow X < 0, \\ &K \in Z_B. \end{aligned} \tag{69}$$

Such a convention completes the definition of the representation domain $\tilde{Z}$. An elementary computation shows that

$$\tilde{Z} = [-(B-K)B^{n-1}, KB^{n-1} - 1]. \tag{70}$$

If the radix B is even, we usually choose

$$K = (B - K) = B/2. \tag{71}$$

In this case

$$\tilde{Z} = [-B^n/2, (B^n/2) - 1]. \tag{72}$$

This choice provides $\tilde{Z}$ with the maximum possible symmetry around zero.

Example. $B = 10$, $K = 5$, $n = 3$. Here $\tilde{Z} = [-500, 499]$. The relation between X and $S(X)$ runs as follows:

X	-500	$-499 \ldots -100 \ldots$	-1	0	1	$2 \ldots 499$
$S(X)$	500	$501 \ldots \quad 900 \ldots$	999	0	1	$2 \ldots 499$

X	x_3	x_2	x_1	x_0
7	0	1	1	1
6	0	1	1	0
5	0	1	0	1
4	0	1	0	0
3	0	0	1	1
2	0	0	1	0
1	0	0	0	1
0	0	0	0	0
−1	1	1	1	1
−2	1	1	1	0
−3	1	1	0	1
−4	1	1	0	0
−5	1	0	1	1
−6	1	0	1	0
−7	1	0	0	1
−8	1	0	0	0

Figure 11. 2's complement representation

In radix 2 (2'*s complement representation*), the above convention becomes:

$$x_{n-1}=0 \Rightarrow X \geqslant 0, \qquad x_{n-1}=1 \Rightarrow X<0, \tag{73}$$

x_{n-1} is again called *sign bit.* The representation domain Z is now given by:

$$\tilde{Z}=[-2^{n-1}, 2^{n-1}-1] \tag{74}$$

and both formulae (67) and (68) are condensed in the single expression

$$X=-x_{n-1}2^{n-1}+\sum_{i=0}^{n-2} x_i 2^i. \tag{75}$$

A representation of 2's complement is illustrated in Figure 11 for $n=4$. This table should be compared with the one in Figure 9. Observe in particular that 0 only has a single representation.

2.2.3 Radix-complement addition

Let $X, Y \in \tilde{Z}$. The situation corresponding to $X+Y \notin \tilde{Z}$ is the *overflow* situation. When there is no overflow, one has:

$$S(X)+S(Y)=[\![X]\!]_{B^n}+[\![Y]\!]_{B^n}, \tag{76}$$

$$S(X+Y)=[\![X+Y]\!]_{B^n}, \tag{77}$$

hence

$$S(X-Y)=[\![S(X)+S(Y)]\!]_{B^n}. \tag{78}$$

The restriction about overflow is required by the fact that, if $X+Y\notin\tilde{Z}$, $S(X+Y)$ is undefined. As a result of (78), we conclude that the simulation $S(X+Y)$ is obtained by adding the simulations $S(X)$ and $S(Y)$, neglecting eventual carries from position $(n-1)$ to position n. It remains for an overflow detection procedure to be derived. We shall restrict our discussion to the case of even radices and of symmetric representation domain $\tilde{Z}$ (equations (71) and (72)).

To any integer $X\in\tilde{Z}$, we associate a binary variable x^* defined by

$$\begin{aligned} x^*=0 &\Leftrightarrow X\geq 0 \Leftrightarrow x_{n-1}\in\{0,1,\ldots,(B/2)-1\},\\ x^*=1 &\Leftrightarrow X<0 \Leftrightarrow x_{n-1}\in\{B/2,\ldots,B-1\}. \end{aligned} \tag{79}$$

(when $B=2$, $x^*=x_{n-1}$). The binary variables y^* and z^* are similarly associated with Y and $[\![S(X)+S(Y)]\!]_{B^n}$. Now, the situations

$$x^*=y^*=z^*=0;\qquad x^*=y^*=z^*=1 \tag{80}$$

obviously identify overflow conditions as the sum of two integers having the same sign should have the same sign. We show below that there are no other overflow conditions than those covered by (80). First we observe that an overflow is only possible if X and Y have the same sign. There are thus only two cases to discuss:

(i) $X\geq 0$ and $Y\geq 0$ $(x^*=y^*=0)$.
In this case:

$$0\leq X<B^n/2;\qquad 0\leq Y<B^n/2,$$

and the overflow corresponds to

$$B^n/2\leq X+Y<B^n,$$

i.e. as $S(X)=X$ and $S(Y)=Y$:

$$B^n/2\leq S(X)+S(Y)<B^n.$$

The latter condition is only fulfilled if the digit z_{n-1} of the sum $S(X)+S(Y)$ is at least equal to $B/2$, i.e. if $z^*=1$.

(ii) $X<0$ and $Y<0$ $(x^*=y^*=1)$
In this case

$$-(B^n/2)\leq X<0;\qquad -(B^n/2)\leq Y<0$$

and the overflow condition corresponds to

$$-B^n\leq X+Y<-(B^n/2),$$

i.e. as $S(X)=B^n+X$ and $S(Y)=B^n+Y$,

$$B^n\leq S(X)+S(Y)<\frac{3B^n}{2},$$

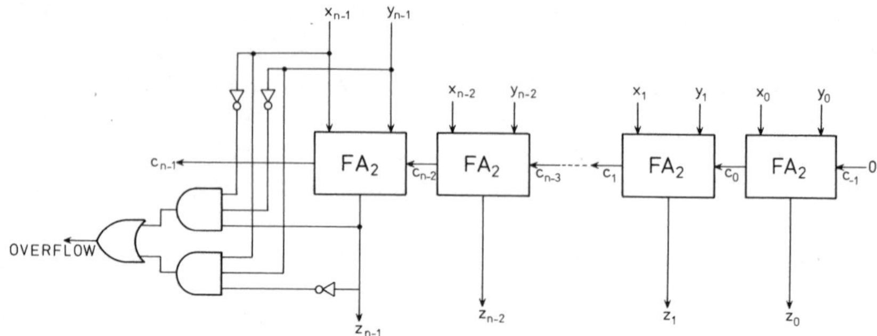

Figure 12. 2's complement adder

or finally

$$0 \leqslant [\![S(X)+S(Y)]\!]_{B^n} < \frac{B^n}{2}.$$

The later condition in turn implies $z_{n-1} < B/2$, i.e. $z^* = 0$.

Summarizing, a necessary and sufficient overflow condition is:

$$x^* z^* \bar{z}^* \vee \bar{x}^* \bar{y}^* z^* = 1. \tag{81}$$

The preceding discussion is illustrated for $B = 2$ by Figure 12 which presents a 2's complement adder together with its overflow detection circuitry.

2.2.4 Sign change and subtraction

Let us first observe that

$$X \neq 0 \Rightarrow S(-X) = B^n - S(X). \tag{82}$$

Indeed,

(i) if $X > 0$, then $S(X) = X$ and $S(-X) = B^n - X = B^n - S(X)$,

(ii) if $X < 0$, then $S(X) = B^n + X$ and $S(-X) = -X = B^n - S(X)$.

Making use again of (66), we obtain

$$S(-X) = \sum_{i=0}^{n-1} (B-1-x_i)B^i + 1. \tag{83}$$

This also solves the subtraction problem, as, to subtract a number, one simply has to add its opposite. As an illustration let us design an adder–subtractor in 2's complement notation. Let t be the control variable ($t = 0$: addition; $t = 1$: subtraction): the circuit will be the same as in Figure 8b: it will be completed by an overflow detection circuit producing the function

$$d = \bar{t}(x_{n-1}y_{n-1}\bar{z}_{n-1} \vee \bar{x}_{n-1}\bar{y}_{n-1}z_{n-1}) \vee t(x_{n-1}\bar{y}_{n-1}\bar{z}_{n-1} \vee \bar{x}_{n-1}y_{n-1}z_{n-1}). \tag{84}$$

3 AMPLITUDE COMPARISON

3.1 Problem statement

Consider two positive integers X and Y given by their radix B representations $x_{n-1} \ldots x_1 x_0$ and $y_{n-1} \ldots y_1 y_0$. We wish to generate three binary functions F, G and H defined by:

$$\begin{aligned} F &= 1, \quad \text{if } X > Y, \\ G &= 1, \quad \text{iff } X = Y, \\ H &= 1, \quad \text{iff } X < Y. \end{aligned} \tag{85}$$

We may at once observe that the amplitude comparison problem is related to the subtraction problem. For example, F may be viewed as the output borrowed in the subtraction $Y - X$.

For each index i, we define, as a counterpart to (85), the local comparison functions

$$\begin{aligned} \phi_i &= 1, \quad \text{iff } x_i > y_i, \\ \gamma_i &= 1, \quad \text{iff } x_i = y_i, \\ \eta_i &= 1, \quad \text{iff } x_i < y_i. \end{aligned} \tag{86}$$

3.2 THE RECURRENCE EQUATIONS

3.2.1 The lexicographic recurrence

The amplitude comparison techniques are generally grounded on the properties of the lexicographic order that have been studied in Chapter I. The obvious method is that which we use when we consult a dictionary, comparing first the leftmost literals, then the second literal, and so on. That method may be formalized as follows: define the three binary functions

$$\begin{aligned} f_i &= 1, \quad \text{iff } x_{n-1} \ldots x_{i+1} x_i > y_{n-1} \ldots y_{i+1} y_i, \\ g_i &= 1, \quad \text{iff } x_{n-1} \ldots x_{i+1} x_i = y_{n-1} \ldots y_{i+1} y_i \\ h_i &= 1, \quad \text{iff } x_{n-1} \ldots x_{i+1} x_i < y_{n-1} \ldots y_{i+1} y_i. \end{aligned} \tag{87}$$

These functions are related by the following recurrence equations:

$$f_i = f_{i+1} \vee g_{i+1} \phi_i, \tag{88}$$

$$g_i = g_{i+1} \gamma_i, \tag{89}$$

$$h_i = h_{i+1} \vee g_{i+1} \eta_i. \tag{90}$$

These equations, initialized by $(f_n, g_n, h_n) = (0, 1, 0)$ may be used for an iterative amplitude comparison scheme. From a practical point of view, a second set of recurrence equations, described below, is generally used instead of (88)–(90).

3.2.2 The inverse lexicographic recurrence

It is also possible to perform amplitude comparison starting with the least significant bits. Let us define the three binary functions:

$$\begin{aligned} F_i &= 1, \quad \text{iff } x_i \ldots x_1 x_0 > y_i \ldots y_1 y_0, \\ G_i &= 1, \quad \text{iff } x_i \ldots x_1 x_0 = y_i \ldots y_1 y_0, \\ H_i &= 1, \quad \text{iff } x_i \ldots x_1 x_0 < y_i \ldots y_1 y_0. \end{aligned} \tag{91}$$

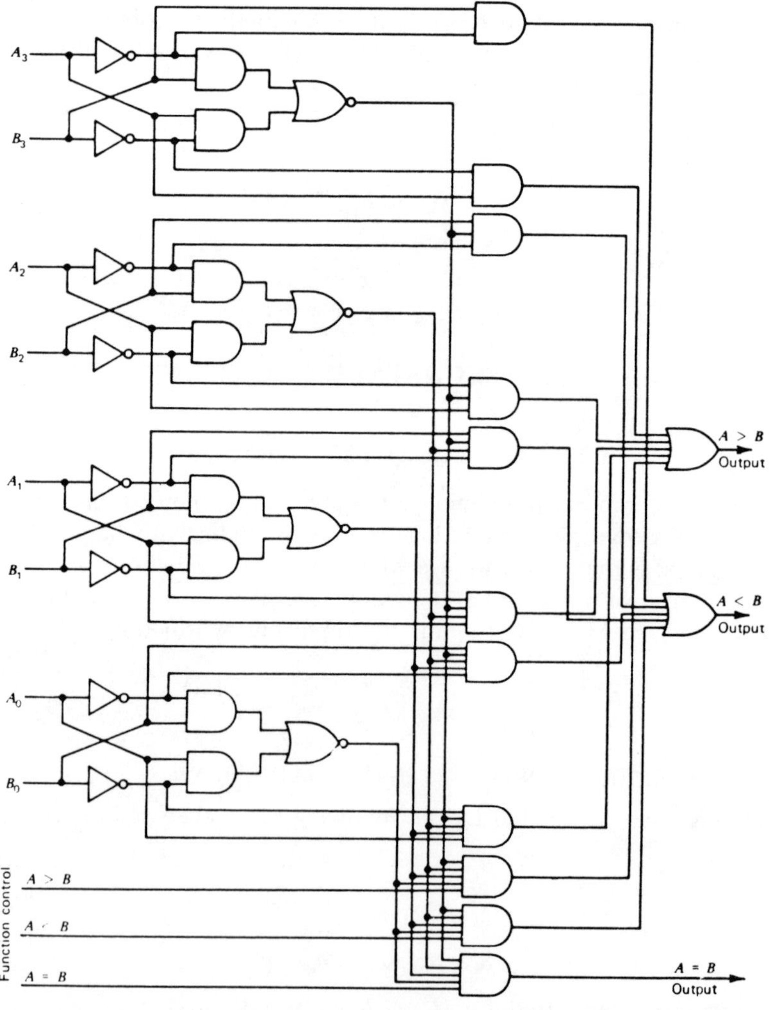

Figure 13. Module to compare blocks of bits

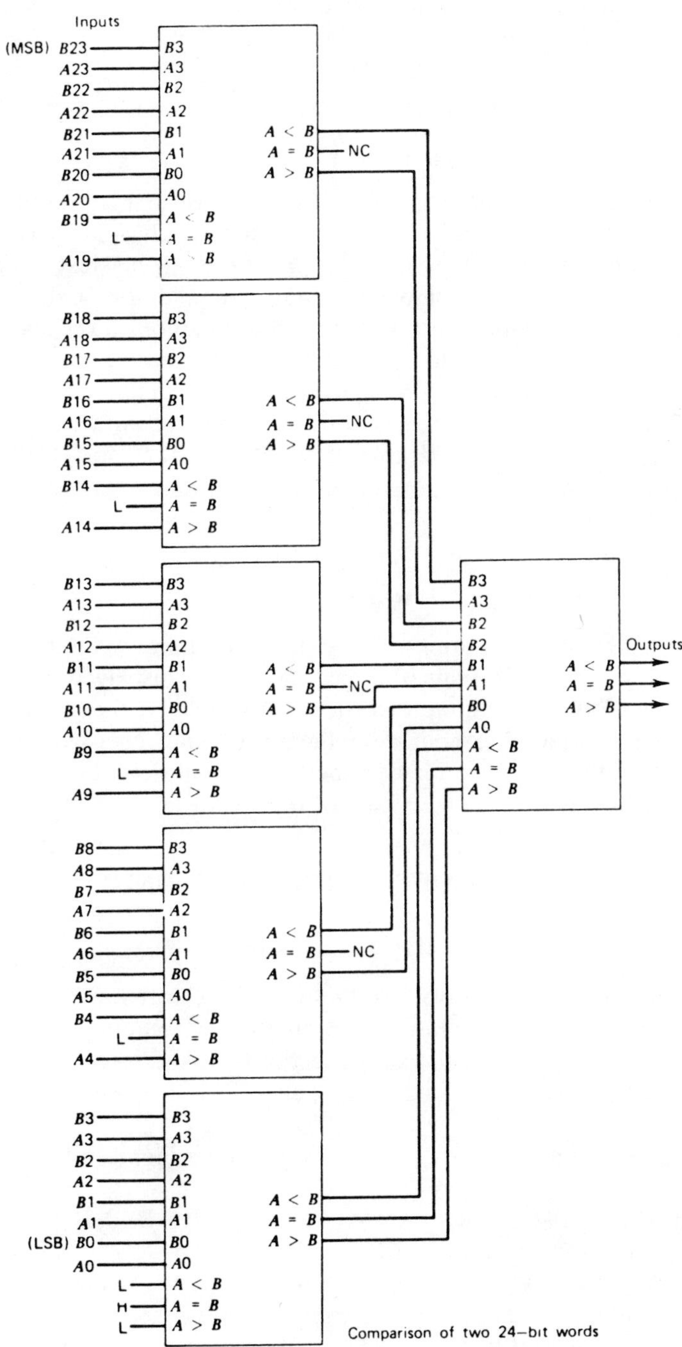

Figure 14. TTL module 7485

The recurrence equations are now:

$$F_{i+1} = \phi_{i+1} \vee \gamma_{i+1} F_i, \tag{92}$$

$$G_{i+1} = \gamma_{i+1} G_i, \tag{93}$$

$$H_{i+1} = \eta_{i+1} \vee \gamma_{i+1} H_i. \tag{94}$$

These equations call for two remarks. Let us first observe that, in (92, . . . , 94), the functions F_i, G_i and H_i are completely separated. This was not the case for the functions f_i, g_i and h_i in (88)–(90). Next we note the affinity of the amplitude comparison with the addition and subtraction operations. If we compare for instance, the equations (32) and (92) we conclude that F_i may be considered as a carry and that ϕ_{i+1} and γ_{i+1} play the roles of a generated and of a propagated carry, respectively. The theory of amplitude comparison circuits may thus be developed in a way parallel to that used for addition. For example, if we observe that, for $B = 2$:

$$\phi_i = x_i \bar{y}_i, \tag{95}$$

$$\gamma_i = \overline{(x_i \oplus y_i)}, \tag{96}$$

$$\eta_i = \bar{x}_i y_i, \tag{97}$$

and if we perform in equations (92)–(94) substitutions similar to those of Section 2.1.3, we shall obtain the classical circuit displayed in Figure 14 (TTL module 7485). As in the carry-lookahead theory, the process may be used recursively and the module in Figure 13 may be used to compare blocks of bits. One obtains, for example, the circuit described in Figure 14 which is analogous to the upper part of the Figure 6.

4 ARITHMETIC AND LOGIC UNITS

4.1 Description

We present in this section a brief description of the arithmetic and logic unit 74181: such a component illustrates a number of computation techniques that we have studied in the preceding sections. Furthermore, it will play an essential role in the general purpose sequentialized system to be studied later on.

The logical diagram of the circuit is given in Figure 15. The circuit terminals are:

(i) Nine binary data inputs: $A = A_3A_2A_1A_0$, $B = B_3B_2B_1B_0$, $\bar{C}_n$.

(ii) Five binary data outputs: $\bar{C}_{n+4}$, $F = F_3F_2F_1F_0$.

(iii) Three auxiliary outputs G, P and $A \equiv B$. A glance at the circuit allows one to identify the G and P outputs with hexadecimal generated and propagated carries that will be used in conjunction with a carry-lookahead generator (Figure 7). The output $A \equiv B$ will allow one to detect the identity of the input vectors A and B.

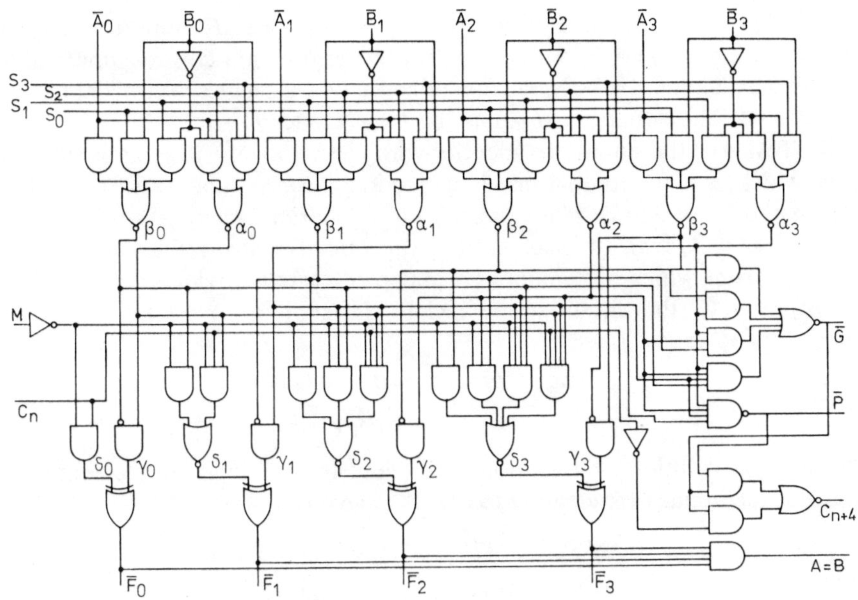

Figure 15. Logical diagram of the arithmetic and logic unit 74181

(iv) Five binary control inputs: the mode control input M and the function control inputs $S = S_3S_2S_1S_0$. The circuit will thus exhibit 32 distinct input–output behaviours.

The analysis of the circuit is straightforward, thanks to the internal symmetries of this circuit. The essential analysis formulae are gathered in Figure 16. Let us first examine the formulae giving the quantities δ_i.

$$\alpha_i = \overline{(S_3A_iB_i \vee S_2A_iB_i)}$$

$$\beta_i = \overline{(A_i \vee S_1B_i \vee S_0B_i)}$$

$$\gamma_i = \alpha_i \oplus \beta_i$$

$$\delta_0 = \overline{\bar{M}\bar{C}_n}$$

$$\delta_1 = \overline{\bar{M}(\beta_0 \vee \alpha_0\bar{C}_n)}$$

$$\delta_2 = \overline{\bar{M}(\beta_1 \vee \alpha_1\beta_0 \vee \alpha_1\alpha_0\bar{C}_n)}$$

$$\delta_3 = \overline{\bar{M}(\beta_2 \vee \alpha_2\beta_1 \vee \alpha_2\alpha_1\beta_0 \vee \alpha_2\alpha_1\alpha_0\bar{C}_n)}$$

$$F_i = \gamma_i \oplus \delta_i$$

$$C_{n+4} = \overline{\beta_3 \vee \alpha_3\beta_2 \vee \alpha_3\alpha_2\beta_1 \vee \alpha_3\alpha_2\alpha_1\beta_0 \vee \alpha_3\alpha_2\alpha_1\alpha_0C_n)}$$

Figure 16. Analysis formulae

(i) If $M=1$, $\delta_i=1$ and $F_i=\bar{\gamma}_i=\overline{\alpha_i \oplus \beta_i}$. In this case, F_i only depends on the two inputs A_i and B_i. Under control of the function control inputs S, F_i realizes any of the 16 possible functions of two variables (see Chapter II, Exercise 4). The situation $M=1$ is the logic mode.

(ii) If $M=0$, the equations identify with carry-lookahead equations (in a form which is dual to that used in Section 2.3): $\bar{\beta}_i$ plays the role of a propagated carry and $\bar{\alpha}_i$ plays the role of a generated carry. As one always has $P_i \oplus G_i = A_i \oplus B_i$, it is clear that the circuit realizes the sum of two numbers from which the generated and propagated carries are produced. More precisely, the circuit outputs yield the sum $X+Y$ of two numbers whose bits x_i and y_i satisfy

$$\begin{aligned} x_i y_i &= \bar{\alpha}_i = A_i(S_3 B_i \vee S_2 \bar{B}_i), \\ x_i \vee y_i &= \bar{\beta}_i = A_i \vee S_1 \bar{B}_i \vee S_0 B_i. \end{aligned} \tag{98}$$

The general solution of equations (98) may be obtained by conventional Boolean techniques (Rudeanu, 1974). It is given by

$$\begin{aligned} x_i &= A_i(S_3 B_i \vee S_2 \bar{B}_i) \vee \lambda (A_i \vee S_1 \bar{B}_i \vee S_0 B_i), \\ y_i &= A_i(S_3 B_i \vee S_2 \bar{B}_i) \vee \bar{\lambda} (A_i \vee S_1 \bar{B}_i \vee S_0 B_i), \end{aligned} \tag{99}$$

where λ is an arbitrary Boolean function. An appropriate choice of λ allows one to run along all the pairs $\{X, Y\}$ of integers having the appropriate sum $X+Y$. For example, if $(S_3 S_2 S_1 S_0)=(0\ 1\ 1\ 0)$, one obtains from (99):

$$\begin{aligned} x_i &= A_i \bar{B}_i \vee \lambda (A_i \vee \bar{B}_i), \\ y_i &= A_i \bar{B}_i \vee \bar{\lambda} (A_i \vee \bar{B}_i), \end{aligned}$$

S_3	S_2	S_1	S_0	$M=1$ F_i	$M=0$ $16C_{n+4}+\Sigma_i F_i 2^i$	
0	0	0	0	$\bar{A}_i$	$C_n+\Sigma_i A_i 2^i$	
0	0	0	1	$\bar{A}_i \bar{B}_i$	$C_n+\Sigma_i (A_i \vee B_i) 2^i$	
0	0	1	0	$\bar{A}_i B_i$	$C_n+\Sigma_i (A_i \vee \bar{B}_i) 2^i$	
0	0	1	1	0	C_n+15	
0	1	0	0	$\bar{A}_i \vee \bar{B}_i$	$C_n+\Sigma_i A_i 2^i$	$+\Sigma_i (A_i \bar{B}_i) 2^i$
0	1	0	1	$\bar{B}_i$	$C_n+\Sigma_i (A_i \vee B_i) 2^i$	$+\Sigma_i (A_i \bar{B}_i) 2^i$
0	1	1	0	$A_i \oplus B_i$	$C_n+\Sigma_i A_i 2^i$	$+\Sigma_i \bar{B}_i 2^i$
0	1	1	1	$A_i \bar{B}_i$	C_n+15	$+\Sigma_i (A_i \bar{B}_i) 2^i$
1	0	0	0	$\bar{A}_i \vee B_i$	$C_n+\Sigma_i A_i 2^i$	$+\Sigma_i (A_i B_i) 2^i$
1	0	0	1	$A_i \oplus B_i$	$C_n+\Sigma_i A_i 2^i$	$+\Sigma_i B_i 2^i$
1	0	1	0	B_i	$C_n+\Sigma_i (A_i \vee \bar{B}_i) 2^i$	$+\Sigma_i (A_i B_i) 2^i$
1	0	1	1	$A_i B_i$	C_n+15	$+\Sigma_i (A_i B_i) 2^i$
1	1	0	0	1	$C_n+2\Sigma_i A_i 2^i$	
1	1	0	1	$A_i \vee \bar{B}_i$	$C_n+\Sigma_i (A_i \vee B_i) 2^i$	$+\Sigma_i A_i 2^i$
1	1	1	0	$A_i \vee B_i$	$C_n+\Sigma_i (A_i \vee \bar{B}_i) 2^i$	$+\Sigma_i A_i 2^i$
1	1	1	1	A_i	C_n+15	$+\Sigma_i A_i 2^i$

Figure 17. Input–output behaviour of the 74181

and the choice $\lambda = B_i$ yields in particular $x_i = A_i$ and $y_i = \bar{B}_i$. The other cases are handled in a similar way. The situation $M = 0$ is the *arithemetic mode* (see Figure 17).

4.2 Applications

In circuit diagrams, we shall represent the arithmetic and logic unit (ALU) by the symbol in Figure 18a. As an application, we show how it is possible to use this ALU to build a 16-bit two's complement adder–subtractor.

Observe first that, in the mode $(M, S_3S_2S_1S_0) = (0, 1001)$, the circuit is nothing other than an hexadecimal adder with input carry C_n and output carry C_{n+4}. Cascading four of these units, we shall thus obtain the required adder. Let us now consider the mode $(M, S_3S_2S_1S_0) = (0, 0110)$. Reference to (61), which also applies to $S(X) + S(-Y)$ in two's complement arithmetic, shows that, in the considered mode, the circuit actually behaves as a subtractor if we impose $C_n = 1$. The overall circuit, answering the control bit t, is shown in Figure 18b. This circuit should be completed with an overflow detection network behaving as indicated by (84). Observe also that carry propagation could be avoided by using the 74181 module in conjunction with its companion module 74182 (Figure 7).

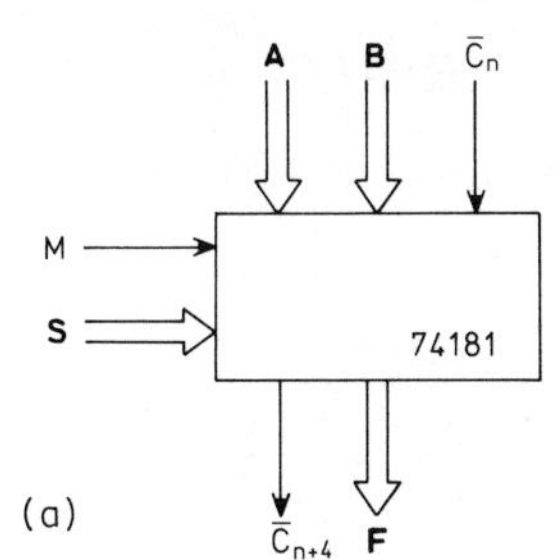

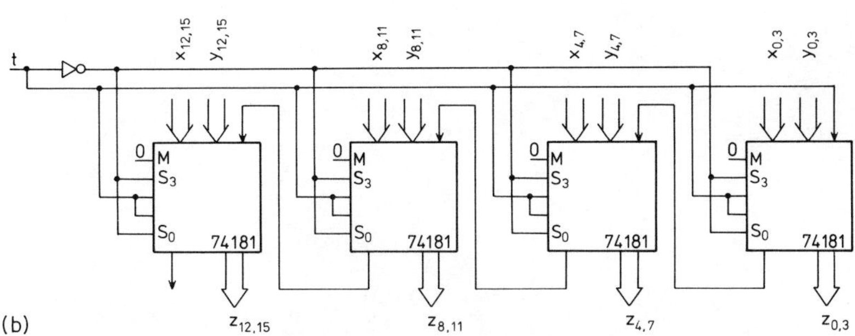

Figure 18.(a) Arithmetic and logic unit. (b) Control bit t circuit

5 RADIX CONVERSION

5.1 Basic conversion algorithms

Let A be an integer. Consider two radices b and β (for instance, $b = 2$ and $\beta = 10$). We call b the *object radix* and β the *image radix*. The integer A has, with respect to these radices, the representations $a_{n-1} \ldots a_1 a_0$ and $\alpha_{p-1} \ldots \alpha_1 \alpha_0$ satisfying:

$$A = \sum_{i=0}^{n-1} a_i b^i, \tag{100}$$

$$A = \sum_{j=0}^{p-1} \alpha_j \beta^i, \tag{101}$$

The problem is the following one: given the radix b representation $a_{n-1} \ldots a_1 a_0$ of A, obtain its radix β representation $\alpha_{p-1} \ldots \alpha_1 \alpha_0$. This problem may be considered as solved if we are able to perform, in the radix β system, the computations implied by formula (100). In practice, (100) will be used in its *Hörner scheme* form:

$$A = ((\ldots((0 \cdot b + a_{n-1})b + a_{n-2})b + \ldots)b + a_1)b + a_0. \tag{102}$$

The computation suggested by (102) appears as an iteration at each step of which a partial result R, initially 0, is multiplied by the object radix b and is added to one of the object digits a_i. This is denoted by

$$R := bR + a_i. \tag{103}$$

In order to perform in the radix β system the computations implied by

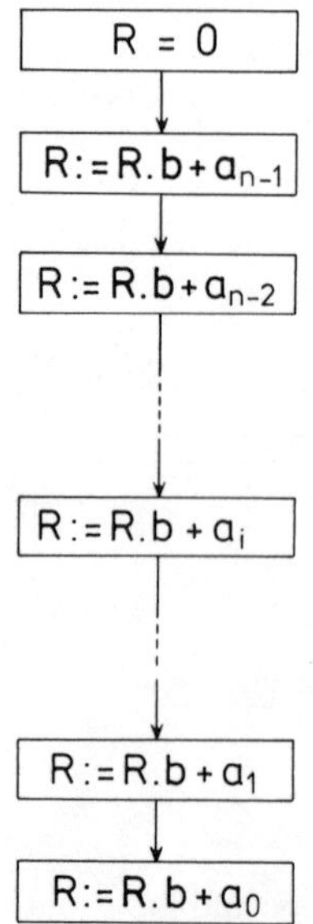

Figure 19. Computation organization

(103), one should thus

(i) know the radix β representations of the object digits a_i:
(ii) be able to multiply by b in the radix β system; and
(iii) be able to add in the same system.

If these conditions are met, the computation may be organized as in Figure 19. It only remains to indicate how to realize the typical step (103) of the conversion process.

Assume given the radix β representation $\rho_{m-1}\ldots\rho_1\rho_0$ of R. By definition, there exists a series of quotients q_i such that:

$$\begin{aligned} R &= q_0\beta + \rho_0 \\ q_0 &= q_1\beta + \rho_1 \\ &\cdots \\ q_{i-1} &= q_i\beta + \rho_i \\ &\cdots \\ q_{n-2} &= 0 + \rho_{m-1} \end{aligned} \tag{104}$$

These equations allow us to obtain the radix β representation $\rho^*_{s-1}\ldots\rho^*_m\rho^*_{m-1}\ldots\rho^*_1\rho^*_0$ of $bR + a_i$. From (104), we first deduce:

$$\begin{aligned} bR &= q_0b\beta + \rho_0b \\ bq_0 &= q_1b\beta + \rho_1b \\ &\cdots \\ bq_{i-1} &= q_ib\beta + \rho_ib \\ &\cdots \\ bq_{m-2} &= 0 + \rho_{m-1}b \end{aligned} \tag{105}$$

From the first equation (105), we obtain

$$bR + a_i = q_0b\beta + \rho_0b + a_i, \tag{106}$$

and, if we wish to obtain the digit ρ^*_0 of $bR + a_i$, we are bound to put

$$\rho_0b + a_i = q^*_0\beta + \rho^*_0 \tag{107}$$

in such a way that

$$bR + a_i = (q_0b + q^*_0)\beta + \rho^*_0. \tag{108}$$

The computation may be pursued along the same lines. From the second equation (105) one indeed deduces

$$bq_0 + q^*_0 = q_1b\beta + \rho_1b + q^*_0 \tag{109}$$

and the situation is thus comparable to that encountered in (106). Finally, we can write

$$\begin{aligned} \rho_0b + a_i &= q^*_0\beta + \rho^*_0 \\ \rho_1b + q^*_0 &= q^*_1\beta + \rho^*_1 \\ &\cdots \\ \rho_jb + q^*_{j-1} &= q^*_j\beta + \rho^*_j \\ &\cdots \\ \rho_{m-1}b + q^*_{m-2} &= q^*_{m-1}\beta + \rho^*_{m-1} \\ q^*_{m-1} &= q^*_m\beta + \rho^*_m \\ &\cdots \\ q^*_{s-2} &= 0\cdot\beta + \rho^*_{s-1} \end{aligned} \tag{110}$$

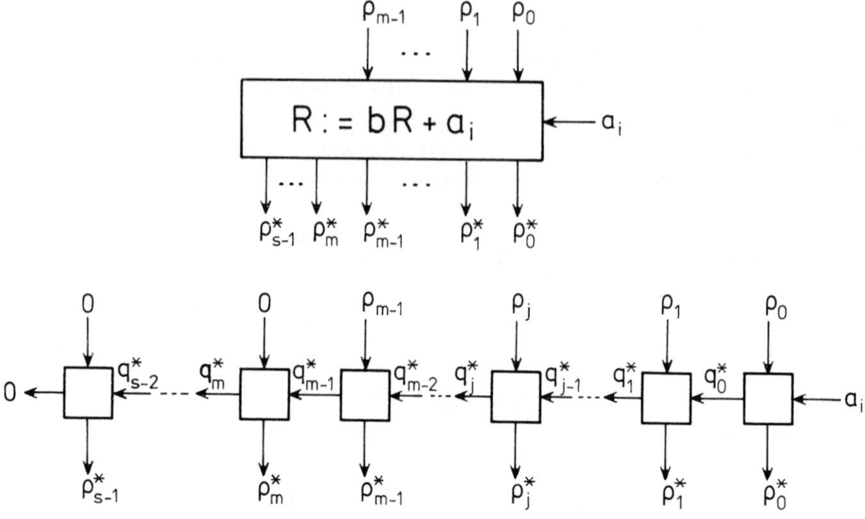

Figure 20. Basic step of the Hörner conversion scheme

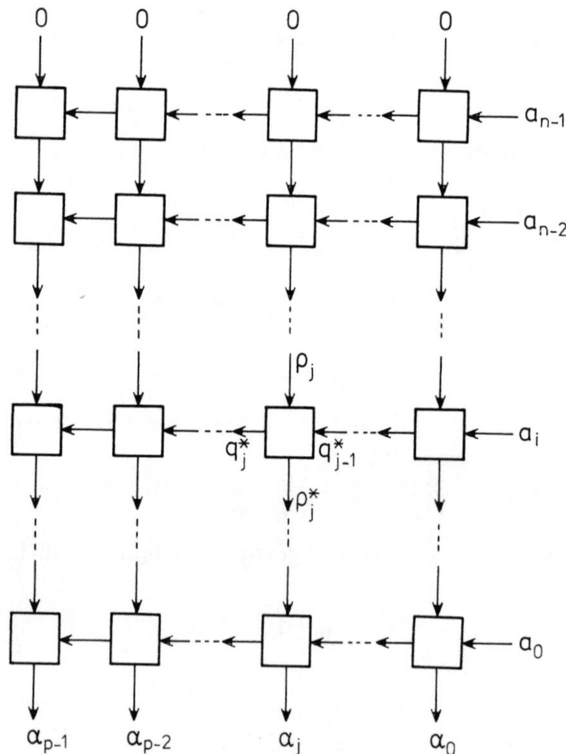

Figure 21. Two-dimensional array which implements the Hörner conversion scheme

and progressive substitutions of type (108) indeed prove that $[\rho^*_{s-1} \ldots \rho^*_1 \rho^*_0]$ is the radix β representation of $bR + a_i$. The equations (110) give the solution of the conversion problem. They indeed show, as illustrated by Figure 20, that the basic step (103) of the Hörner conversion scheme itself splits up into a series of partial steps typically described by:

$$\rho_j b + q^*_{j-1} = q^*_j \beta + \rho^*_j. \tag{111}$$

Globally, the circuit appears as a two-dimensional array, illustrated by Figure 21, in which each cell has the behaviour (111). As a matter of fact, the two members of (111) are the representations of an integer of $Z_{B\beta}$ in the two mixed radix number system with base vectors $[\beta, b]$ and $[b, \beta]$, respectively. This implies in particular that $q^*_j \in Z_B$.

5.2 Circuit application: BCD to binary conversion

(see also Lanning (1975)).

The conversion from BCD to binary may be viewed as a binary coded conversion from object radix $b = 10$ to some image radix $\beta = 2^s$. For cost and delay reasons discussed in the next subsection, the choice $s = 2$, i.e. $\beta = 4$, may be considered as an adequate trade-off. The behaviour of the basic cell is described by:

$$\rho_j 10 + q^*_{j-1} = q^*_j 4 + \rho^*_j, \tag{112}$$

where $\rho_j, \rho^*_j \in Z_4$ and $q^*_{j-1}, q^*_j \in Z_{10}$. The 4-valued quantities will be encoded by two bits: $\rho_j \simeq (x_1, x_0)$, $\rho^*_j \simeq (z_1, z_0)$. Similarly, the decimal digits are encoded by 4 bits: $q^*_{j-1} \simeq (y_3, y_2, y_1, y_0)$, $q^*_j = (w_3, w_2, w_1, w_0)$. Our basic

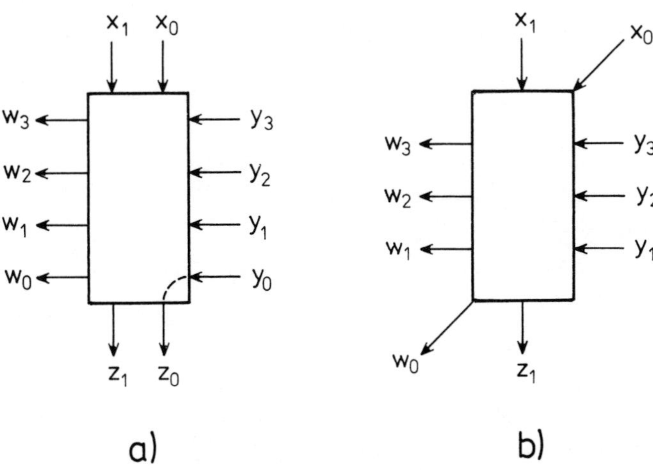

Figure 22. Typical cells for BCD to binary conversion

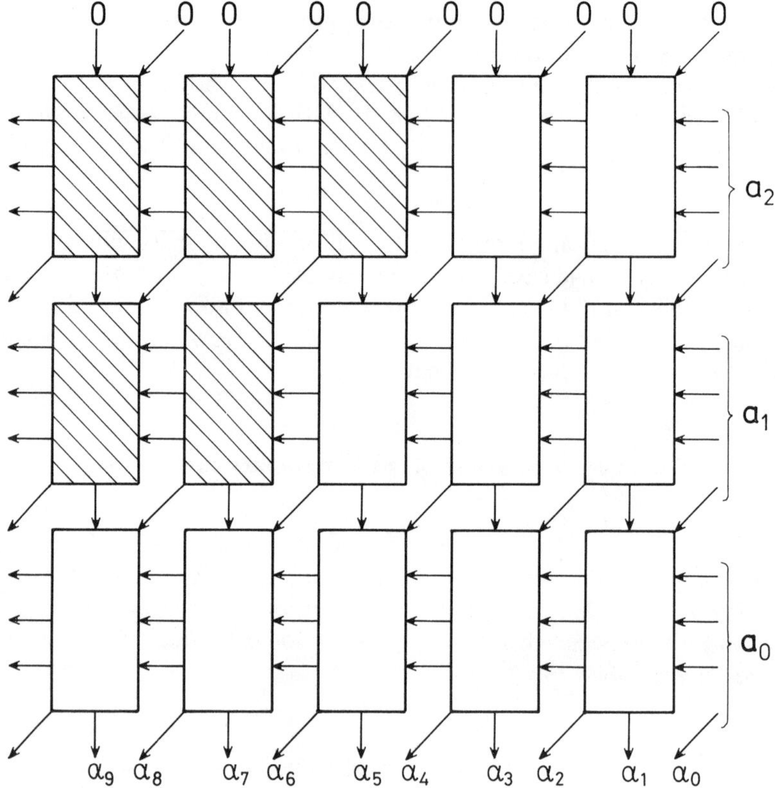

Figure 23. Typical array for BCD to binary conversion

conversion cell, displayed in Figure 22a, is thus described by:

$$(2x_1+x_0)10+(8y_3+4y_2+2y_1+y_0)=(8w_3+4w_2+2w_1+w_0)4+(2z_1+z_0). \tag{113}$$

It is at once clear that $y_0=z_0$ since these bits represent, in their respective members, the parity of the represented number. The other input and output variables are now related by:

$$(2x_1+x_0)5+4y_3+2y_2+y_1=(8w_3+4w_2+2w_1+w_0)2+z_1. \tag{114}$$

The cell in Figure 22a may thus be replaced by the 5-input 5-output cell displayed in Figure 22b. Figure 23 represents a typical conversion array for a number having three decimal digits. The hatched cells may be deleted from that array as all the inputs (and outputs) are necessarily equal to zero.

The cell used in that array and defined by (114) is practically implemented as a ROM (TTL 74184 module).

5.3 Discussion

It remains to discuss the cost and delay of the obtained networks. Let n_b represent the maximum number of digits in the object radix and β represent the maximum number of digits in the image radix. Once chooses

$$n_\beta = \lceil n_n \cdot \log_\beta b \rceil \qquad (115)$$

so as to allow the conversion of any input number. We may neglect the eventual network truncations without impairing the asymptotic behaviour. In this case, the number of cells is given by:

$$\Gamma = n_b n_\beta = n_b \lceil n_b \log_\beta b \rceil, \qquad (116)$$

and the delay Δ, expressed as number of cell delays is given by:

$$\Delta = n_b + n_\beta = n_b + \lceil n_b \log_\beta b \rceil. \qquad (117)$$

(The situations $b = 10$, $\beta = 2$ and $b = 2$, $\beta = 10$ are an exception to (117). In these cases, one shows that $\Delta = n_2$. This is due to a particular degeneracy of the conversion cell in these cases.)

If the conversion cell is realized as a ROM, it will have $\lceil \log_2 b \rceil + \lceil \log_2 \beta \rceil$ input and output wires so that the cost, expressed in bits, is given by:

$$\Gamma = n_b \lceil n_b \log_\beta b \rceil (\lceil \log_2 b \rceil + \lceil \log_2 \beta \rceil) 2^{(\lceil \log_2 b \rceil + \lceil \log_2 \beta \rceil)}.$$

As a first approximation, we may accept, neglecting the $\lceil \quad \rceil$:

$$\Gamma \cong n_b^2 b\beta \log_2 b (1 + \log_\beta b). \qquad (118)$$

Let $b = 10$ and assume that β runs along the sequence $2, 4, \ldots, 2^k, \ldots$. Γ is clearly an increasing function of β. Under the same conditions, the delay Δ runs along the sequence

$$\Delta = n_2, n_{10} + \frac{n_2}{2}, \ldots, n_{10} + \frac{n_2}{k}, \ldots$$

and, if we adopt the approximate rule $n_{10} \cong n_2/3$:

$$\Delta = n_2, \frac{5n_2}{6}, \ldots, \frac{n_2}{3} + \frac{n_2}{k}.$$

Clearly, Δ is a decreasing function of k, and the technological choice $\beta = 4$ appears as a compromise between cost and delay. The cost and delay of the obtained array are quadratic and linear functions of n_b, respectively.

6 PERMUTATION NETWORKS

6.1 Connection networks

It will frequently happen, in systems to be studied in later chapters, that information items, such as binary coded alphanumeric quantities, have to be

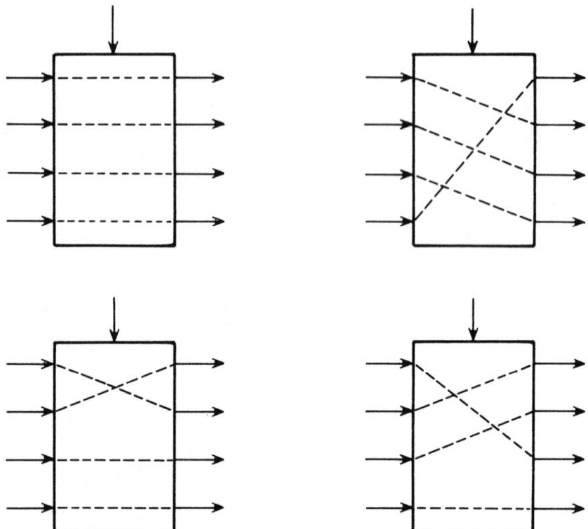

Figure 24. States of a permutation network

transferred from one point to another point of the system. Furthermore, the origins and destinations of the transmitted information vary from time to time. A system endowed with facilities of this type will contain some form of switching device, analogous to a telephone exchange. We may thus expect to observe, in the hardware implementation of such a system, a number of multiplexers, demultiplexers or functionally equivalent devices, playing the required routing role. This has already been exemplified in Chapter III.

If one wishes to obtain quantitative estimates on the performances of such a system, one would have to be able to ascertain the cost and delay of intervening connection networks. In the present section, we briefly present a typical example of such a computation by studying permutation networks. A *N-input N-output permutation network* is defined as a network able to establish, in all the possible ways, N simultaneous connections each involving a single input and a single output; in algebraic terms, a permutation network should be able to realize the $N!$ bijections from its input set onto its output set. Figure 24 illustrates some states of a permutation network for $N=4$.

6.2 A lower complexity bound

It is easy to give a lower bound on the cost Γ of a N by N permutation network. Indeed, such a network has $N!$ possible states. Hence, the number P of its control variables should satisfy:

$$P \geqslant \log_2 (N!). \tag{119}$$

From Stirling's formula: $N! \simeq N^N e^{-N}\sqrt{2\pi N}$, one deduces as a first approximation $\log_2(N!) \cong N \log_2 N$, and this yields:

$$P \geqslant N \log_2 N. \tag{120}$$

Finally, as we know (see Chapter II) that the cost increase of a network is never less than linear with respect to the number of its variables, we obtain

$$\Gamma \geqslant kN \log_2 N. \tag{121}$$

6.3 Waksman's construction and the looping algorithm

An elementary design of a permutation network is obtained by using N multiplexers having N data inputs each. That elementary construction, together with the results of Chapter III on the cost of a multiplexer, is enough to show that the cost Γ of a N by N permutation network has an upper bound of the form:

$$\Gamma \leqslant K'N^2. \tag{122}$$

Observe, by the way, that the number of control inputs to that elementary circuit would be of the form $\lceil N \log_2 N \rceil$.

Our purpose in the present section is however to exhibit a permutation network having a cost of the form $KN \log_2 N$, i.e. asymptotically optimal in the sense of (121). The construction to be described is due to Waksman (1968); it uses as the building block a 2 by 2 permutation network also known as a (2,2)-cross-bar switch. A possible implementation of the (2,2)-crossbar switch is shown in Figure 25a. In what follows, we shall represent this building block by the symbol shown in Figure 25b. The (2,2)-crossbar switch is also described by the Boolean equations:

$$\begin{aligned} y_0 &= \bar{w}x_0 \vee wx_1, \\ y_1 &= \bar{w}x_1 \vee wx_0. \end{aligned} \tag{123}$$

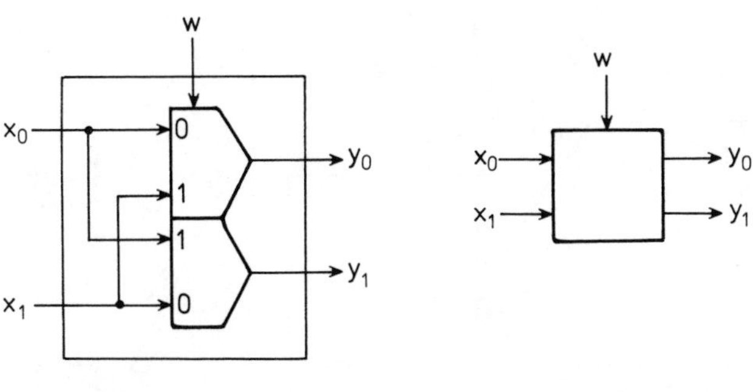

Figure 25.(a) Implementation of a (2–2) crossbar switch. (b) Symbol of a (2–2) crossbar switch

The recurrence construction, with the (2,2)-crossbar switch as the initial step, shows how it is possible to build a permutation network acting on $N = 2^n$ points from

(i) two permutation networks acting on $N/2 = 2^{n-1}$ points, and

(ii) $2^n - 1$ copies of the (2,2)-crossbar switch.

This construction is illustrated by Figure 26. In this circuit diagram, the boxes $\mathcal{N}_0$ and $\mathcal{N}_1$ are the permutation networks and the boxes numbered $0, 1, \ldots, 2^{n-1}-1$ are the (2,2-crossbar cells. The interstage connection pattern ties, e.g. output i of the input crossbar cell number j, to input j of $\mathcal{N}_i$. This connection pattern is known as a *perfect shuffle*.

We now have to show that the network in Figure 26 is able to realize an arbitrary bijection ψ from its inputs x_i to its outputs y_j. Such a bijection is defined from a permutation π on Z_{2^n} by

$$\psi\colon x_i \to y_j; \qquad j = \pi(i); \qquad i, j \in Z_{2^n}. \tag{124}$$

To achieve the proof we show that, given π, it is possible to choose the states of the $(2^n - 1)$ crossbar cells and to define the permutations π_0 and π_1 (to be realized by $\mathcal{N}_0$ and $\mathcal{N}_1$) in such a way that the overall network actually realizes π. The algorithm yielding that result is the *looping algorithm*.

Step 1. From input 0, there is a single path (over $\mathcal{N}_0$) allowing one to reach output $i = \pi(0)$. The observation of this path fixes the image $\pi_0(0)$ in $\mathcal{N}_0$ and the state of the crossbar cell to which i belongs. Let j be the second output of this crossbar cell $(j = i \pm 1)$.

Step 2. From output j, there is a single path over $\mathcal{N}_1$ allowing input $k = \pi^{-1}(j)$. The observation of this path fixes one pair of points to be related

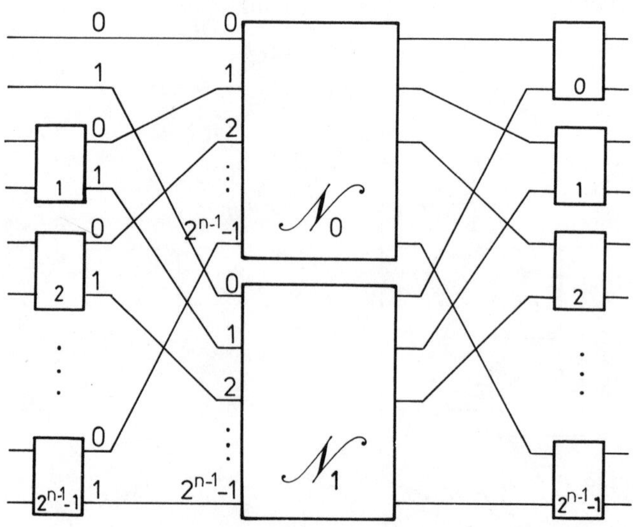

Figure 26. Perfect shuffle

by π_1. Furthermore:

(i) If $\pi^{-1}(j) \neq 1$, it fixes the state of the input cell to which k belongs. If ℓ is the second input to this cell, the algorithm is restarted from step with 1ℓ as origin.

(ii) If $\pi^{-1}(j) = 1$, the algorithm is restarted from step 1 with an arbitrary unused origin.

This to and from movement is continued until all the points are connected. If at some step one reaches an input cell whose second terminal is already connected, one restarts the algorithm from step 1 with an arbitrary origin.

Let us illustrate the looping algorithm for $N=8$ and for the permutation defined by:

i	0	1	2	3	4	5	6	7
$\pi(i)$	7	4	1	0	2	6	3	5

The computation, illustrated by Figure 27, runs as follows:

$$0 \to a \to d' \to 7$$
$$5 \leftarrow \gamma \leftarrow \delta' \leftarrow 6$$
$$4 \to c \to b' \to 2$$
$$6 \leftarrow \delta \leftarrow \beta' \leftarrow 3$$
$$7 \to d \to c' \to 5$$
$$1 \leftarrow \alpha \leftarrow \gamma' \leftarrow 4$$
$$(*)2 \to b \to a' \to 1$$
$$3 \leftarrow \beta \leftarrow \alpha' \leftarrow 0$$

In that computation, the choice of 2 (marked with an asterisk) is arbitrary.

It remains for an estimate of the cost of the circuit to be given. Let $P(2^k)$ represent the number of (2,2)-crossbar cells in a 2^k by 2^k permutation

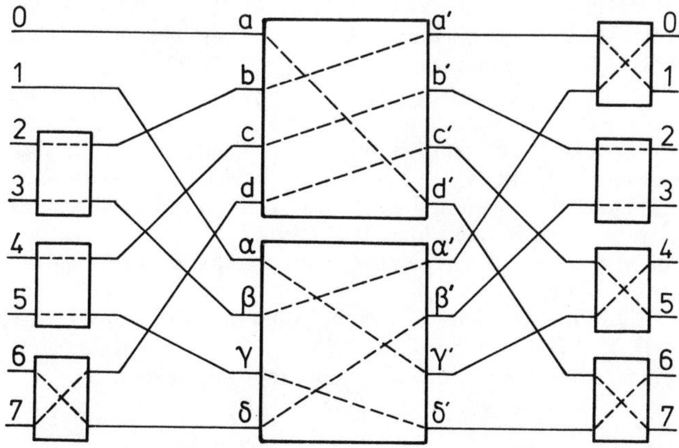

Figure 27. Illustration of the looping algorithm

network. Then, we have

$$\begin{aligned} P(2) &= 1, \\ P(2^n) &= 2P(2^{n-1}) + 2^n - 1. \end{aligned} \tag{125}$$

The recurrence (125) is easily solved:

$$P(2^n) = (n-1)2^n + 1. \tag{126}$$

For a large n, we may thus write, as $N = 2^n$,

$$P(N) \cong N \log_2 N. \tag{127}$$

BIBLIOGRAPHIC REMARKS

Two types of sources may be consulted about the material presented in this chapter: on the one hand, some recent textbooks such as Greenfield (1977) are devoted to the description of digital integrated modules: on the other hand, there exists a considerable amount of literature on computer arithmetic: since the early book by Flores (1963) appeared, a number of more recent studies have been published; in this respect, the book by Hwang (1979) deserves a special mention for its completeness. Note also that many textbooks on computer organization include comprehensive chapters on computer arithmetic; see, e.g., Hamacher, Vranesic, and Zaky (1978).

The basic papers about fast addition, including carry-lookahead techniques and conditional sum adders, remain those of Slansky (1960a, 1960b) and McSorley (1961). The complexity aspects of that question have been discussed by Spira (1973) and Winograd (1965).

There is less bibliography about radix-conversion. The first paper on that subject seems that of Couleur (1958). Since then, some papers by Lanning (1975), Nicoud (1969), and Benedek (1977) have appeared.

In contrast, there is a considerable literature about permutation networks. The reader may refer to the recent paper by Feng (1981) for an up to date bibliography (75 references).

EXERCISES

E1. Radix B binary coded full adder. Let the radix B satisfy: $2^{n-1} < B \leqslant 2^n$, so that a radix B digit has a n-bit binary representation. The radix B full adder may be described by the equation

$$X + Y + c_- = c_+ B + S,$$

where X, Y and S are radix-B digits and where c_- and c_+ are binary. We assume that the digits X, Y and S have the binary representations $x_{n-1} \ldots x_1 x_0$, $y_{n-1} \ldots y_1 y_0$ and $s_{n-1} \ldots s_1 s_0$. Figure E1 illustrates a possible realization of radix B full-adder. It combines

(i) a binary adder which generates the binary representation of the sum $Z = X + Y + c_-$;

(ii) an amplitude comparator which generates the output carry: $c_+ = 1$ if $Z \geqslant B$, $c_+ = 0$ otherwise; and

(iii) a binary subtractor which generates the output digit S:

$$\begin{aligned} S &= Z \quad \text{if} \quad c_+ = 0, \\ S &= Z - B \quad \text{if} \quad c_+ = 1. \end{aligned}$$

Assume $B = 10$.

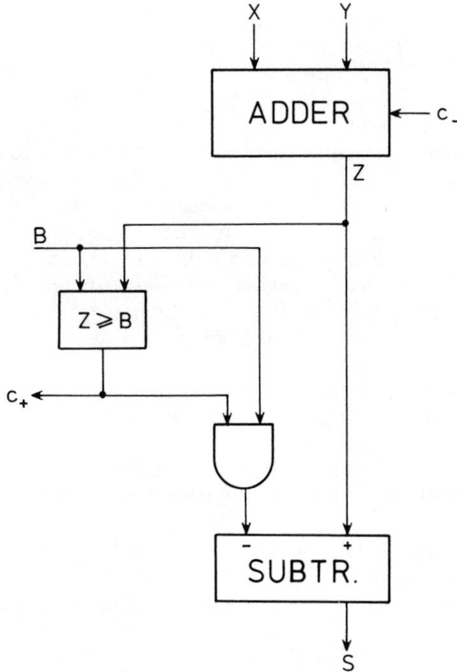

Figure E1

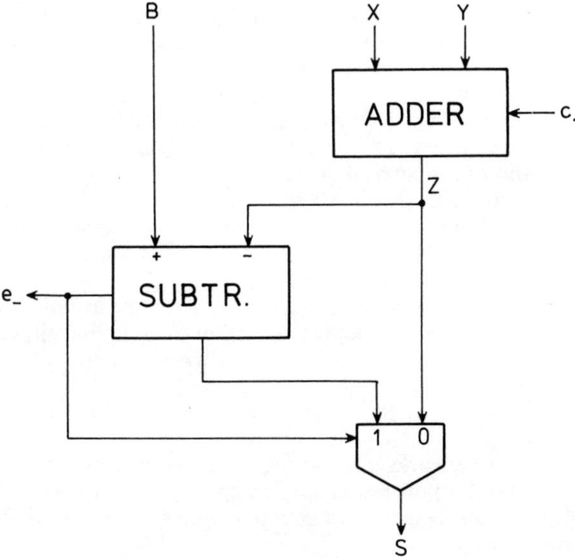

Figure E2

(i) Develop the comparator $Z \geqslant 10$ as a two-level AND to OR circuit and obtain a first BCD full-adder scheme.

(ii) Express directly the 5 output bits c_+, S as functions of the 5 bits Z. Realize these functions as a two-level AND to OR circuit. Compare this second design to the first one.

(iii) Using the same principles, design a BCD augmented full adder (with decimal generated and propagated carries) able to cooperate with TTL 74182 modules. Design a fast 16-decimal-digit BCD adder.

We now return to the general situation where B is arbitrary. The bits of the representation of B are now inputs to the system. Show that the circuit in Figure E2 may be used as a substitute for that in Figure E1. Give a detailed design of this adder for $n = 5$. In general, what are the best time and cost performances that may be expected from these designs? Express your result in terms of n.

E2. BCD arithmetic is sometimes performed in the excess-3 code. (see Chapter II, Figure E4). Design a BCD-adder in the excess-3 code. Compare this design with those obtained in E1.

E3. Redundant signed digit representations have been introduced in Chapter I, Exercise E4. For $B = 10$, describe a carryless adder performing according to that scheme.

E4. Consider the integer A having the binary representation $a_{N-1} \ldots a_1 a_0$. One wishes to compute the binary representation of $[\![A]\!]_B$, where $2^{n-1} < B \leqslant 2^n$; we assume that $N \gg n$. The classical method for computing A is described by

$$[\![A]\!]_B = \left[\!\!\left[\sum_{i=0}^{N-1} a_i [\![2^i]\!]_B \right]\!\!\right]_B,$$

and, in a computer environment, the residues $[\![2^i]\!]_B$ are stored in some memory location. These residues $[\![2^i]\!]_B$ may, however, be computed recursively. Indeed:

$$[\![2^{i+1}]\!]_B = [\![2[\![2^i]\!]_B]\!]_B$$

so that the principles outlined in E1 apply to the computation of $[\![2^{i+1}]\!]_B$. Starting from the above equations, devise a computation scheme for obtaining $[\![A]\!]_B$. Your circuit should have cost $0(Nn)$ and delay $0(N \log n)$. What happens if B is of the form $2^r - 1$?

E5. Residue arithmetic is particularly simple when performed with respect to moduli of the form $2^n - 1$. Design an adder modulo $2^n - 1$ (see, for example, Bioul, Davio, and Quisquater, 1975).

E6. Design a binary to BCD $(1, 2, 4, 8)$ converter. Describe the basic cell and the complete network. Compare your result with those obtained in Section 5.2. Design a binary to BCD excess-3 converter and a BCD excess-3 to binary converter.

E7. Extend the Waksman technique to the case $N = r^n$; use as building blocks r by r crossbar switches. Does the looping algorithm work in that situation? What is the cost of the obtained network? (See, for example, Neiman 1969; Davio, Deschamps, and Liénard, 1977.)

E8. Discuss the time required to perform the looping algorithm as described in Section 6.3.

E9. Some permutation networks are not required to perform the $N!$ bijections of their inputs to their outputs but merely an appropriate subset of this set. As an example, a rotator on N points (numbered from 0 to $N-1$) is able to realize the N permutations

$$\rho_k : x_i \to y_j; \qquad j = i \oplus k \bmod N; \qquad k = 0, 1, \ldots, N-1.$$

Show that it is possible to design rotators with cost $0(N \log N)$ and delay

0(log N). Discuss the possibility of designing large rotators from smaller ones. (See, for example, Davis, 1974; Gajsky and Tulpule, 1978; Davio and Ronse, 1980.)

E10. A masking network on N points is a logical circuit with N outputs $(y_0, y_1, \ldots, y_{N-1})$, two enabling inputs E_0 and E_1 and p control inputs $(z_0, z_1, \ldots, z_{p-1})$. We assume $p \geq \lceil \log_2 N \rceil$ and $z = \sum_i z_i 2^i < N$. The input–output behaviour of the masking network is then described by

$$y_j = E_0 \quad \text{if} \quad j < z$$
$$y_j = E_1 \quad \text{if} \quad j \geq z.$$

In particular $y_{N-1} = E_1$. The symbol used in the sequel to represent a masking network is illustrated in Figure E3a for $N = 8$. Give a direct design of such a masking network by means of N multiplexers having p control inputs. Give estimates of the performances of your design.

Now let $N = n_1 n_0$. Consider the construction illustrated in Figure E3b for $n_1 = 8$ and $n_0 = 3$. Show that this construction is general and actually yields a mask on N ponts. Use this construction to design a mask on 16 points. Show finally that, if the construction is used repeatedly, one obtains, for $N = 2^r$, a cost 0(N) and delay 0(log N).

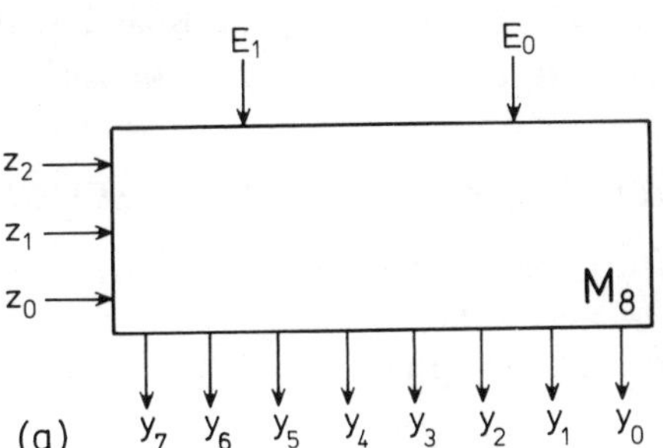

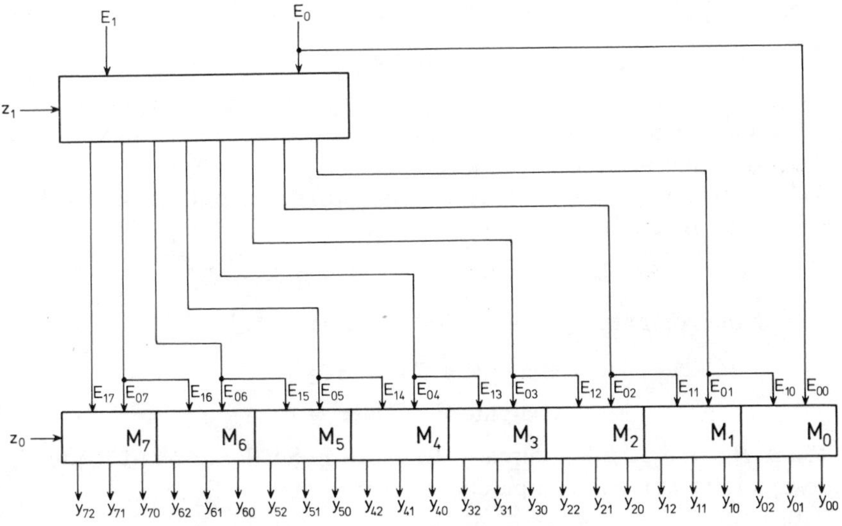

Figure E3

Chapter V
Iterations and finite automata

This chapter is devoted to the study of a first particular class of algorithms, namely *iterative computation schemes*. Their study is justified by their frequent occurrence, and by their close relation with automata and thus with finite state machines (Chapter VI).

The first section gives the main definitions. Section 2 is a short introduction to automata theory. Section 3 gives some information concerning the implementation of automata. Finally, Section 4 briefly evokes three classical optimization problems: state minimization, decomposition, and encoding.

1 ITERATIVE COMPUTATION SCHEMES

In the course of this first section we aim at specifying what is intended by *iterative computation* or *iteration*.

Let us suppose that we have to perform a computation whose data

$$\mathbf{x}^0, \mathbf{x}^1, \ldots, \mathbf{x}^{N-1}$$

belong to a set $S^n(\mathbf{x}^i = x^i_{n-1}, \ldots, x^i_1, x^i_0)$. The computation generates results

$$\mathbf{z}^0, \mathbf{z}^1, \ldots, \mathbf{z}^{N-1}$$

belonging to a set $S^m(\mathbf{z}^i = z^i_{m-1}, \ldots, z^i_1, z^i_0)$. Furthermore, the computation uses an initial condition $\mathbf{y}^0$ and generates intermediate conditions $\mathbf{y}^1, \ldots, \mathbf{y}^N$, belonging to the set $S^p(y^i = y^i_{p-1}, \ldots, y^i_1, y^i_0)$.

The computation is decomposed into partial computations, which are described by functions M and N.

$$M: S^n \times S^p \to S^p \quad \text{and} \quad N: S^n \times S^p \to S^m.$$

Step i of the computation runs as follows ($i = 0, 1, \ldots, N-1$):

$$\text{compute } \mathbf{y}^{i+1} = M(\mathbf{x}^i, \mathbf{y}^i), \tag{1}$$

$$\text{compute } \mathbf{z}^i = N(\mathbf{x}^i, \mathbf{y}^i). \tag{2}$$

Radix B serial addition, already described in Section 3.2 of Chapter I, is an example of iterative computation:

$S = \{0, 1, \ldots, B-1\}$;

Data: $\mathbf{x}^0 = (a_0, b_0)$, $\mathbf{x}^1 = (a_1, b_1), \ldots, \mathbf{x}^{N-1} = (a_{N-1}, b_{N-1})$;

Results: $s_0, s_1, \ldots, s_{N-1}$;
Conditions: $c_0=0,\ c_1, c_2, \ldots, c_N$.
Functions M and N are defined by

$$M(a_i, b_i, c_i) = \left\lceil \frac{a_i+b_i+c_i}{B} \right\rceil,$$

$$N(a_i, b_i, c_i) = [\![a_i+b_i+c_i]\!]_B.$$

The computation yields the sum $s=a+b$, where

$$a=\sum_{i=0}^{N-1} a_i 2^i; \qquad b=\sum_{i=0}^{N-1} b_i 2^i; \qquad s=\sum_{i=0}^{N-1} s_i 2^i;$$

c_N is the overflow condition.

As was done in Section 3.2 of Chapter I in the particular case of radix B serial addition, every iterative computation, whose equations are (1) and (2), can be described by a computation scheme: the set of variables is

$$\{\mathbf{x}^0, \mathbf{x}^1, \ldots, \mathbf{x}^{N-1}\},$$

the only computation primitive $\mathbf{\Pi}: S^{n+p} \to S^{m+p}$ is defined by

$$\mathbf{\Pi}(\mathbf{x}, \mathbf{y}) = (N(\mathbf{x}, \mathbf{y}), M(\mathbf{x}, \mathbf{y})).$$

The computation scheme is written as follows ($s=p+n+1$):

$$\text{step } 0 \left\{ \begin{array}{l} N_0 = y_0^0 \\ \text{-----------} \\ N_{p-1} = y_{p-1}^0 \\ N_p = x_0^0 \\ \text{-----------} \\ N_{p+n-1} = x_{n-1}^0 \\ N_{p+n} = \mathbf{\Pi}(N_{p+n-1}, \ldots, N_0) \end{array} \right.$$

$$\text{step } 1 \left\{ \begin{array}{l} N_s = \mathrm{pr}_0\,(N_{p+n}) \\ \text{-----------} \\ N_{s+p-1} = \mathrm{pr}_{p-1}\,(N_{p+n}) \\ N_{s+p} = x_0^1, \\ \text{-----------} \\ N_{s+p+n-1} = x_{n-1}^1 \\ N_{s+p+n} = \mathbf{\Pi}(N_{s+p+n-1}, \ldots, N_s) \end{array} \right.$$

$$\text{step } i \left\{ \begin{array}{l} N_{is} = \mathrm{pr}_0\,(N_{(i-1)s+p+n}) \\ \text{-----------} \\ N_{is+p-1} = \mathrm{pr}_{p-1}\,(N_{(i-1)s+p+n}) \\ N_{is+p} = x_0^i \\ \text{-----------} \\ N_{is+p+n-1} = x_{n-1}^i \\ N_{is+p+n} = \mathbf{\Pi}(N_{is+p+n-1}, \ldots, N_{is}) \end{array} \right.$$

$$N_{Ns-1} = \mathbf{\Pi}(N_{(N-1)s+p+n-1}, \ldots, N_{(N-1)s})$$

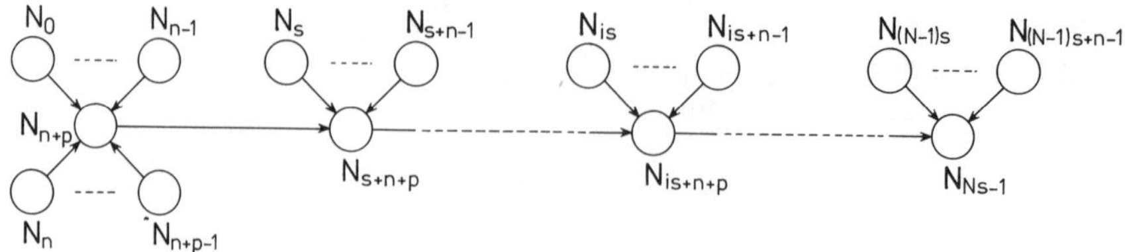

Figure 1. Precedence graph

It includes N steps which are identical, up to a shift of the indices by s. Some intermediate results, which are computed during step $(i-1)$, are used during step i:

$$N_{is} = \mathrm{pr}_0\,(N_{(i-1)s+p+n}), \ldots, N_{is+p-1} = \mathrm{pr}_{p-1}\,(N_{(i-1)s+p+n}).$$

The precedence graph is shown in Figure 1.

This computation scheme can be implemented by an iterative circuit (Figure 2a), or by a circuit containing a memory element and a feedback loop (Figure 2b). The first implementation is a parallel-in, parallel-out system, while the second implementation is a serial-in, serial-out system; the second implementation thus requires a synchronization signal.

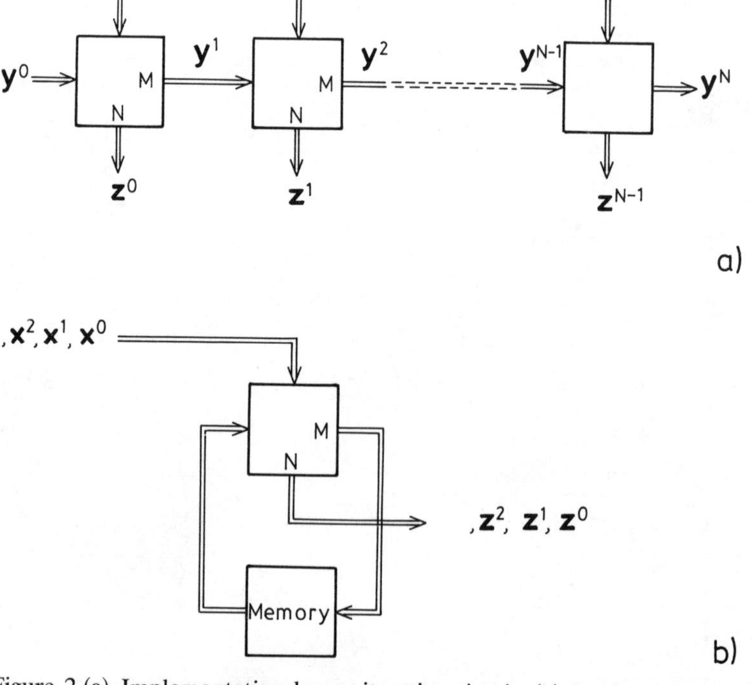

Figure 2.(a) Implementation by an iterative circuit. (b) Implementation by a sequential circuit

2 MOORE AND MEALY AUTOMATA

A *Mealy automaton* is an algebraic structure

$$A = \langle \Sigma, \Omega, Q, \mathbf{M}, \mathbf{N} \rangle,$$

where

(1) $\Sigma = \{\sigma_1, \sigma_2, \ldots, \sigma_N\}$ is the *input alphabet*; its elements are the *input letters.*

(2) $\Omega = \{\omega_1, \omega_2, \ldots, \omega_p\}$ is the *output alphabet* whose elements are the *output letters.*

(3) $Q = \{q_1, q_2, \ldots, q_M\}$ is the set of *internal states* of the automaton.

(4) $\mathbf{M} = \{M_{\sigma_1}, M_{\sigma_2}, \ldots, M_{\sigma_N}\}$ is a set of functions $M_{\sigma_i} : Q \to Q$; the image of q_j by M_{σ_i} is denoted by $q_j M_{\sigma_i}$. They are the *transition functions.* Hence, with every element (σ_i, q_j) of $\Sigma \times Q$ we may associate the element $q_j M_{\sigma_i}$ of Q, and define a mapping from $\Sigma \times Q$ into Q.

(5) $\mathbf{N} = \{N_{\sigma_1}, N_{\sigma_2}, \ldots, N_{\sigma_N}\}$ is a set of functions $N_{\sigma_i} : Q \to \Omega$; the image of q_j by N_{σ_i} is denoted by $q_j N_{\sigma_i}$. They are the *output functions.* With every element (σ_i, q_j) of $\Sigma \times Q$ we may associate the element $q_j N_{\sigma_i}$ of Ω, and define a mapping from $\Sigma \times Q$ into Ω.

The iterative computation of Section 1 defines a Mealy automaton for which

$$\Sigma = S^n, \qquad \Omega = S^m, \qquad Q = S^p,$$

$$\forall \mathbf{x} \in S^n = \Sigma \quad \text{and} \quad \mathbf{y} \in S^p = Q: \qquad M_{\mathbf{x}}(\mathbf{y}) = M(\mathbf{x}, \mathbf{y}), N_{\mathbf{x}}(\mathbf{y}) = N(\mathbf{x}, \mathbf{y}).$$

Conversely, it is possible to associate an iterative computation with every Mealy automaton: it suffices to encode the elements of Σ, Ω and Q by means of elements of S. Integers n, p, and m must be chosen in such a way that

$$|S|^n \geqslant |\Sigma| = N, \qquad |S|^m \geqslant |\Omega| = M, \qquad |S|^p \geqslant |Q| = P.$$

Very often, an automaton is described by a *flow table.* It is a matrix including P rows and N columns. The rows correspond to the elements of Q, and the columns to the elements of Σ. The entry appearing at the intersection of row j and column i is the element $(q_j M_{\sigma_i}, q_j N_{\sigma_i})$ of $Q \times \Omega$.

Another classical description method is the *flow graph.* It contains P nodes, which correspond to the elements of Q. There is an arc from node j to node k, bearing the indication (σ_i, ω_r), if

$$q_j M_{\sigma_i} = q \quad \text{and} \quad q_j N_{\sigma_i} = \omega_r.$$

The previous definitions hold in the case of *complete automata.* We may slightly generalize the definitions and consider the case where M_{σ_i} and N_{σ_i} are *partial functions,* i.e. *functional relations*: a relation from E_1 to E_2, that is a subset of $E_1 \times E_2$, is functional if for every $e_1 \in E_1$, there is at most one $e_2 \in E_2$ such that (e_1, e_2) belongs to the relation. It may thus arise that some $e_1 \in E_1$ has no image.

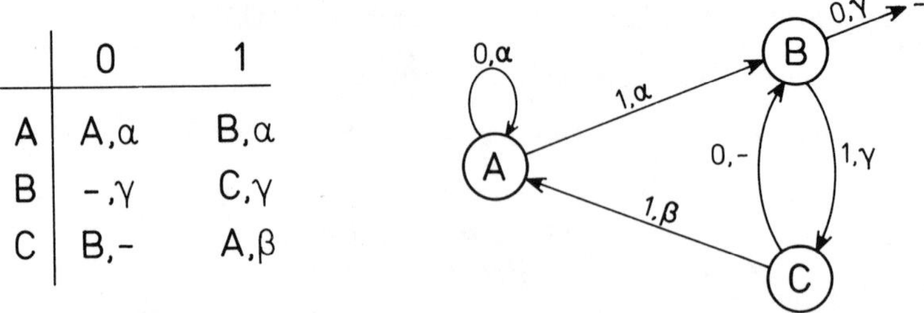

	0	1
A	A,α	B,α
B	-,γ	C,γ
C	B,-	A,β

Figure 3. Flow table and flow graph of an incomplete automaton

Let us turn back to automata. If state q_j has no image by M_{σ_i} we simply write

$$q_jM_{\sigma_i} = -;$$

similarly, if q_j has no image by N_{σ_i} we write

$$q_jN_{\sigma_i} = -.$$

Thanks to those conventions, *incomplete automata* can still be described by flow tables and flow graphs. Figure 3 shows the flow table and the flow graph of an incomplete automaton:

$$\Sigma = \{0, 1\}, \qquad \Omega = \{\alpha, \beta, \gamma\}, \qquad Q = \{A, B, C\},$$

$$M_0 = \{(A, A), (C, B)\}, \qquad M_1 = \{(A, B), (B, C), (C, A)\},$$

$$N_0 = \{(A, \alpha), (B, \gamma), \qquad N_1 = \{(A, \alpha), (B, \gamma), (C, \beta)\}.$$

A *Moore automaton* is an algebraic structure

$$A = \langle \Sigma, \Omega, Q, \mathbf{M}, N \rangle$$

where Σ, Ω, Q and $\mathbf{M}$ are defined as above for a Mealy automaton, and where N is a function from Q to Ω; the image of q_i by N is denoted by q_iN.

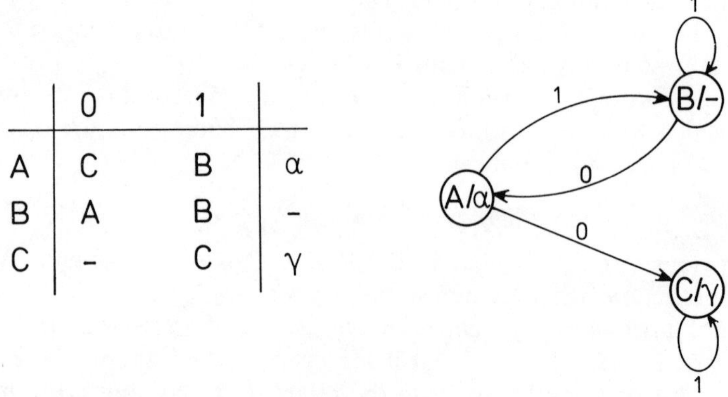

	0	1	
A	C	B	α
B	A	B	-
C	-	C	γ

Figure 4. Flow table and flow graph of a Moore automaton

It is the output function. Hence, the output letter is fixed by the internal state and does not depend on the input letter. Figure 4 shows the flow table and the flow graph of a Moore automaton. Notice that the representation conventions are not exactly the same as for Mealy automata.

$$\Sigma=\{0,1\},\qquad \Omega=\{\alpha,\beta,\gamma\},\qquad Q=\{A,B,C\},$$
$$M_0=\{(A,C),(B,A)\},\qquad M_1=\{(A,B),(B,B),(C,C)\},$$
$$N=\{(A,\alpha),(C,\gamma)\}.$$

In spite of appearances, the Moore model allows the same input–output behaviours to be described as the Mealy model (see, for instance, Hartmanis and Stearns, 1966).

Example 1. A base two serial adder can be described by a Mealy automaton:

$$\Sigma=\{(0,0),(0,1),(1,0),(1,1)\},$$
$$\Omega=\{0,1\},\ Q=\{0,1\}.$$

Its flow table and its flow graph are given in Figure 5.

Example 2. Among the thirty-two binary 5-tuples $x_4x_3x_2x_1x_0$, ten include two 1's and three 0's. They are sometimes used in order to encode the integers 0, 1, . . . , 9 (two-out-of-five encoding). The redundancy of the representation can serve to detect transmission or computation errors. Hence, it can be used to design a circuit which sequentially examines the five bits $x_4x_3x_2x_1x_0$, and decides whether the 5-tuple belongs to the two-out-of-five code, or not.

The input variables are successively examined in increasing order of their indices. When x_i is examined, it is sufficient to know whether the sum $x_0+\ldots+x_{i-1}$ is equal to

$$0, 1, 2, 3 \text{ or more},$$

in order to get the same information about the sum $x_0+\ldots+x_i$. Let us formalize the iterative computation:

$$\Sigma=\{0,1\};\qquad Q=\{q_0,q_1,q_2,q_3\};\qquad \Omega=\{0,1\}.$$

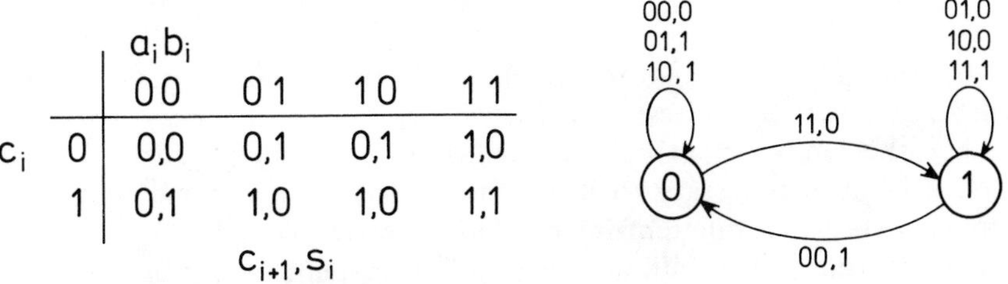

Figure 5. Flow table and flow graph for Example 1

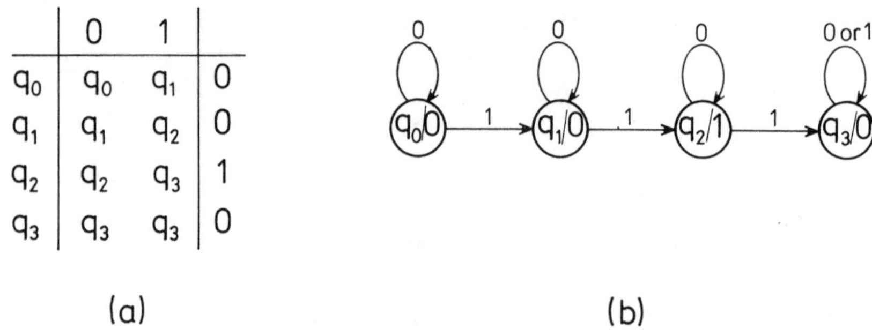

	0	1	
q_0	q_0	q_1	0
q_1	q_1	q_2	0
q_2	q_2	q_3	1
q_3	q_3	q_3	0

Figure 6. Flow table and flow graph for a computation step

Conditions q_0, q_1, q_2, q_3 correspond to the four pieces of information: the sum of already examined variables is equal to $0, 1, 2, 3$ or more. The computation result is 1 if the sum if equal to 2, and 0 in the contrary case. Step i of the computation is described by the flow table of Figure 6a. The corresponding flow graph is shown in Figure 6b. The output value generated at step i depends only on the fact that the sum is equal to two, or not; hence it depends only on the element of Q which is generated. The corresponding automaton is a Moore automaton.

Remark. When the automaton of Example 2 lies in state q_1, and when input letter 1 occurs, we implicitly assumed that it undergoes a transition from state q_1 to state q_2, and generates output letter 0. Similarly, when it is in state q_2 and when input letter 1 occurs, we supposed that it goes to state q_3 and generates output letter 0. The output letter, which is generated at step i, thus depends on the final state which is reached during the step, and not on the initial state. In terms of iterative computation, step i runs as follows:

$$\mathbf{y}^{i+1} = M(\mathbf{x}^i, \mathbf{y}^i), \tag{3}$$

$$\mathbf{z}^i = N(\mathbf{y}^{i+1}) = N(M(\mathbf{x}^i, \mathbf{y}^i)). \tag{4}$$

Nevertheless, in some cases the Moore model describes an iterative computation whose basic relations are

$$\mathbf{y}^{i+1} = M(\mathbf{x}^i, \mathbf{y}^i), \tag{5}$$

$$\mathbf{z}^i = N(\mathbf{y}^i). \tag{6}$$

To equations (1, 2), (3, 4) and (5, 6), correspond the elementary circuit structures shown in Figures 7a, b, and c.

When the input–output behaviour of a system is described by a Moore automaton, it is necessary to be precise about the exact meaning of the output function. To different interpretations correspond

(1) a one cell spatial shifting in the case of implementation by an iterative circuit, and

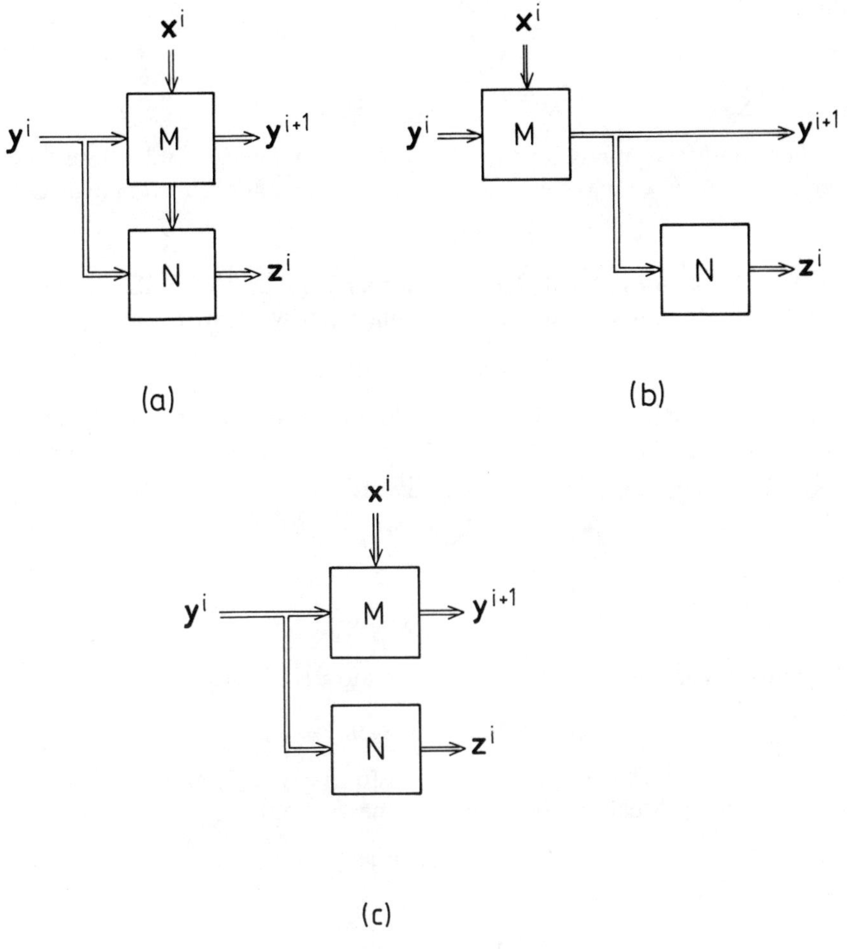

Figure 7. Elementary circuit structures

(2) a one period time shifting in the case of implementation with memory element and feedback loop.
The existence of a time shifting must be taken into account if several automata are working in cooperation. This arises in circuits based on the model of Chapter X: the operational automaton is a Moore automaton of the second type which may be implemented by the circuit of Figure 7c.

The transition and output functions can be generalized to all words of Σ^*. Let us consider complete Mealy automata. Let

$$w = \sigma_{i1}\sigma_{i2} \ldots \sigma_{ir}$$

be a word of the input dictionary. Functions M_w and N_w are defined as follows:

$$M_w = M_{\sigma_{i1}} M_{\sigma_{i2}} \ldots M_{\sigma_{ir}}, \tag{7}$$

$$N_w = M_{\sigma_{i1}} M_{\sigma_{i2}} \ldots M_{\sigma_{ir-1}} N_{\sigma_{ir}}. \tag{8}$$

Furthermore, if λ stands for the *empty word* (imaginary word whose length is equal to zero), we may define M_λ as being the identity function over Q:

$$\forall q \in Q: \qquad qM_\lambda = q.$$

Function N_λ is not defined. The output word, generated by the automaton starting from initial state q_0 and receiving input word w, is

$$q_0 N_{\sigma_{i1}}, q_0 N_{\sigma_{i1}\sigma_{i2}}, \ldots, q_0 N_{\sigma_{i1}\sigma_{i2}\ldots\sigma_{ir}}. \tag{9}$$

Let us now consider a Moore automaton. Function M_w is defined as above, by (7). There are two definitions of N_w. If the Moore automaton performs transformations (3) and (4), then

$$N_w = M_{\sigma_{i1}} M_{\sigma_{i2}} \ldots M_{\sigma_{ir}} N = M_w N. \tag{10}$$

In this case we also define N_λ:

$$N_\lambda = M_\lambda N = N.$$

The output word generated by the automaton is given by (9):

$$q_0 M_{\sigma_{i1}} N, q_0 M_{\sigma_{i1i2}} N, \ldots, q_0 M_{\sigma_{i1}\sigma_{i2}\ldots\sigma_{ir}} N. \tag{11}$$

If the Moore automaton performs transformations (5) and (6), then N_w is defined as for a Mealy automaton with the convention:

$$N_{\sigma_1} = N_{\sigma_2} = \ldots = N_{\sigma_N} = N.$$

Hence, according to (8):

$$N_w = M_{\sigma_{i1}} M_{\sigma_{i2}} \ldots M_{\sigma_{ir-1}} N. \tag{12}$$

The output word is, according to (9) and (12):

$$q_0 N, q_0 M_{\sigma_{i1}} N, \ldots, q_0 M_{\sigma_{i1}\ldots\sigma_{ir-1}} N. \tag{13}$$

These definitions hold for incomplete automata. The functions become functional relations, and the composition of functional relations still yields functional relations.

3 IMPLEMENTATION PRINCIPLES

Some information concerning the implementation of automata has already been given in the last section. Let us now detail the general principles somewhat.

We look for an implementation of automaton $A = \langle \Sigma, \Omega, Q, \mathbf{M}, \mathbf{N} \rangle$ by a digital circuit. If we use binary combinational components, our first task is the binary encoding of sets Σ, Ω and Q. Let

$$f_\Sigma : \Sigma \to B_2^n, \qquad f_\Omega : \Omega \to B_2^m, \qquad f_Q : Q \to B_2^p$$

define three one-to-one mappings. Hence

$$|\Sigma| \leqslant |B_2^n| = 2^n, \qquad |\Omega| \leqslant |B_2^m| = 2^m, \qquad |Q| \leqslant |B_2^p| = 2^p,$$

and

$$n \geqslant \log_2 |\Sigma|, \qquad m \geqslant \log_2 |\Omega|, \qquad p \geqslant \log_2 |Q|.$$

Once the encoding is chosen, we associate with automaton A an iterative computation. The data belong to B_2^n, the results to B_2^m, and the initial and intermediary conditions to B_2^p. Initial condition $\mathbf{y}^0$ is the image by f_Q of the initial state of A. Data $\mathbf{x}^0, \mathbf{x}^1, \ldots$ are the images by f_Σ of the input letters successively received by A. Functions M and N satisfy the two conditions

$$M(f_\Sigma(\sigma), f_Q(q)) = f_Q(qM_\sigma), \tag{14}$$

$$N(f_\Sigma(\sigma), f_Q(q)) = f_\Omega(qN_\sigma), \tag{15}$$

$\forall \sigma \in \Sigma$ and $q \in Q$. These conditions are unambiguous since f_Σ, f_Ω, and f_Q are one-to-one-mappings.

The next step of the implementation consists in associating a digital circuit with the iterative computation. Functions M and N are Boolean functions:

$$M : B_2^{n+p} \to B_2^p, \qquad N : B_2^{n+p} \to B_2^m.$$

They are realized by an interconnection of elementary combinatorial circuits (gates, programmable array, etc.). The complete circuit may belong to either one of two types.

(1) *Iterative circuit.* As far as the number of input letters applied to A is constant, it may be implemented by an iterative circuit. This fact is illustrated by Figure 2a.

(2) *Synchronized sequential machine.* Figure 2b shows another realization method, based on the use of only one combinational circuit implementing M and N. Data are sequentially introduced, and the results are sequentially generated at the same rate. The intermediate conditions, which are generated at some step, are used at the next step. This accounts for the presence of a memory element; it is implemented by means of flip-flops. The rate at which the data are introduced is fixed by an external synchronizing signal (or chocking signal). Flip-flops and synchronized sequential machines will be studied in Chapter VI.

4 OPTIMIZATION PROBLEMS

4.1 State minimization

Given automaton $A = \langle \Sigma, \Omega, Q, \mathbf{M}, \mathbf{N} \rangle$, it can arise that some internal states may be merged, either because they are actually equivalent, or because (in the case of an incomplete automaton) they are compatible; the meaning of the last term will be defined precisely later. From a practical point of view, the reduction of the number of states may lead to a cost lowering.

(1) The number of points of B_2^{n+p} where M and N are defined is equal to the number of elements in $\Sigma \times Q$. A reduction of that number may lead to a cost reduction of the circuits implementing M and N. As an example, if functions M and N are implemented by a read only memory, a reduction of $|\Sigma \times Q|$ corresponds to a reduction of the number of rows which are actually used; saved rows can be assigned to other functions.

(2) The elements of Q are encoded by binary p-tuples. A division of $|Q|$ by a factor two yields a one unit decreasing of m. This fact corresponds to a reduction of the number of interconnections in the case of implementation by an iterative circuit, and, in addition, to a reduction of the number of flip-flops in the case of implementation by a synchronized sequential machine.

Let us now turn to the definition of **compatibility** between states. States q_1 and q_2 are compatible if the next implication holds for every word w:

$$\textit{if } q_1N_w \text{ and } q_2N_w \text{ are both specified, } \textit{then } q_1N_w = q_2N_w.$$

It is a reflexive and symmetric relation. If A is an incomplete automaton, the relation is generally not transitive.

A *compatibility class* is a subset of pairwise compatible states.

A *cover* of Q is a set $\{C_1, C_2, \ldots, C_q\}$ of compatibility classes whose union is Q:

$$C_1 \cup C_2 \cup \ldots \cup C_q = Q.$$

A cover $\mathscr{C} = \{C_1, C_2, \ldots, C_q\}$ is *closed* if, for every index $i \in \{1, 2, \ldots, q\}$ and every input letter $\sigma \in \Sigma$, there exists at least one index $j \in \{1, 2, \ldots, q\}$ such that

$$C_iM_\sigma \subseteq C_j. \tag{16}$$

If $\mathscr{C}$ is a closed cover, we may define the *quotient automaton* $A/\mathscr{C}$ by

$$A/\mathscr{C} = \langle \Sigma, \Omega, \mathscr{C}, \mathbf{M}', \mathbf{N}' \rangle,$$

$C_iM'_\sigma = C_j$, the class C_j being chosen from among those verifying (16),

$C_iN'_\sigma = C_iN_\sigma$;

notice that C_iN_σ is either the empty set or a singleton since C_i is a compatibility class.

The next example comes from Hill and Peterson (1968 p. 330). The automaton of Figure 8 has the closed cover

$$\{C_1 = \{R, 1\}, C_2 = \{R, 2\}\}.$$

The quotient automaton is shown in Figure 9.

Let q be a particular state of A, and C_i a class which contains q. If

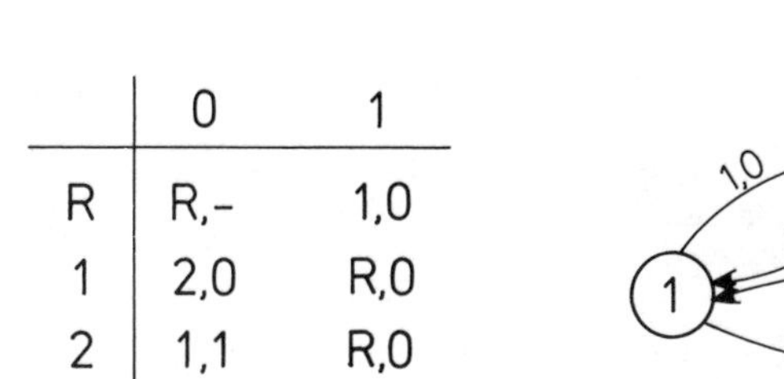

	0	1
R	R,-	1,0
1	2,0	R,0
2	1,1	R,0

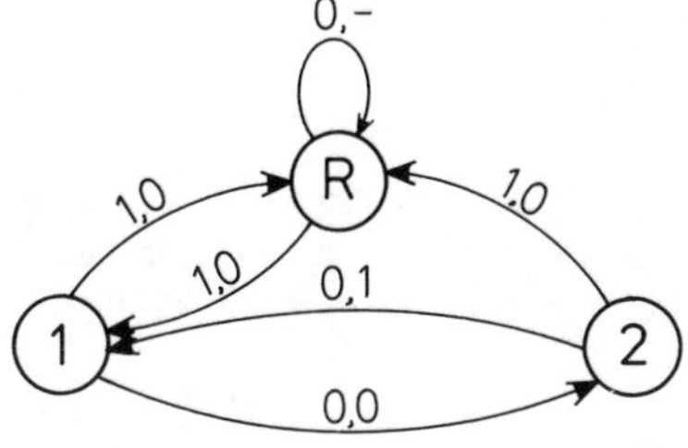

Figure 8. Automaton

automata A and $A/\mathscr{C}$ are in initial states q and C_i, respectively, and both receive input word w, it can easily be proven that the output word generated by $A/\mathscr{C}$ is identical to the output word generated by A in every place where the output letter of A is defined. In other words, '$A/\mathscr{C}$ does at least as much as does A'. We say that $A/\mathscr{C}$ *covers* A. Up to a very particular case (Zahnd, 1969; Davio and De Lit, 1970), every automaton covering A can be deduced from a quotient automaton of A.

The optimization process runs as follows:

(1) Research of the compatibility classes (see, for instance, Paull and Unger, 1959).

(2) Solution of the covering–closure problem (Meisel, 1967; Zahnd and Mange, 1975).

The optimization process is particularly simple if A is a complete automaton. The compatibility relation is an equivalence relation:

$$q_1 \equiv q_2 \quad \text{if} \quad q_1 N_w = q_2 N_w, \forall w.$$

The set of equivalence classes is a closed cover: if states q_1 and q_2 are equivalent then, for every input letter $\sigma \in \Sigma$, states $q_1 M_\sigma$ and $q_2 M_\sigma$ are also equivalent. The corresponding quotient automaton has the minimum number of states among the automata which cover A. The optimization problem thus simply consists in looking for the equivalence classes.

The computation of the equivalence classes can be performed in a recurrent way. Let us introduce a new definition: two states q_1 and q_2 are

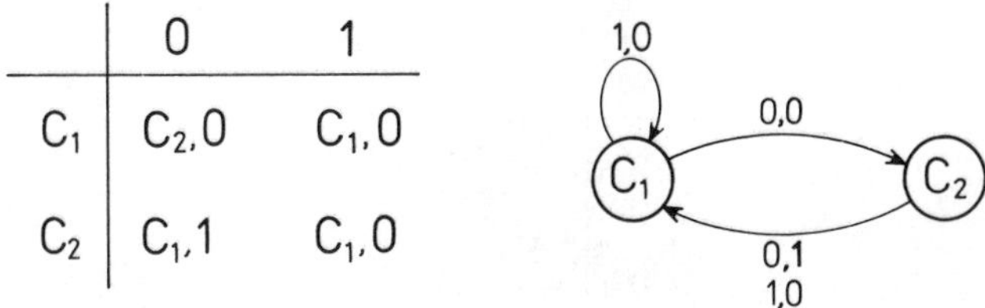

	0	1
C_1	C_2,0	C_1,0
C_2	C_1,1	C_1,0

Figure 9. Quotient automaton

ℓ-equivalent ($q_1 \overset{\ell}{\equiv} q_2$), where ℓ is a positive integer, if

$$q_1 N_w = q_2 N_w$$

for every word w whose length is smaller than or equal to ℓ. An algorithm searching for the equivalence classes can be based on the following remarks:

(1) Two states, q_1 and q_2, are 1-equivalent if

$$q_1 N_\sigma = q_2 N_\sigma, \quad \forall \sigma \in \Sigma.$$

Hence, the equivalence classes of this particular equivalence relation can be deduced directly from the flow table.

(2) Two states q_1 and q_2 are ℓ-equivalent if they are $(\ell-1)$-equivalent, and if, for every $\sigma \in \Sigma$, states $q_1 M_\sigma$ and $q_2 M_\sigma$ are also $(\ell-1)$-equivalent.

(3) If the two equivalence relations $\overset{\ell-1}{\equiv}$ and $\overset{\ell}{\equiv}$ are identical, then all equivalence relations $\overset{i}{\equiv}$, $i \geqslant \ell - 1$, are identical. It is the relation $\equiv$.

Example (Hill and Peterson, 1974, p. 250). Binary messages are encoded in such a way that they do contain neither two successive 1's, nor four successive 0's. Let us design a system which detects the erroneous messages. For that purpose, we define an automaton whose input and output alphabets are {0, 1}. It generates an output letter 1 every time it has just received two consecutive 1's, or four consecutive 0's. The flow table is shown in Figure 10; it can be easily understood by noticing that state s_{ijk} is reached when the three last input letters were i, j, and k. Let us search for the partitions associated with $\overset{1}{\equiv}$, $\overset{2}{\equiv}$, etc.

$$\overset{1}{\equiv} \rightarrow (s_{000})(s_{001}, s_{010}, s_{011})(s_{100}, s_{101}, s_{110}, s_{111});$$

$$\overset{2}{\equiv} \rightarrow (s_{000})(s_{001})(s_{010}, s_{011})(s_{100}, s_{101}, s_{110}, s_{111});$$

$$\overset{3}{\equiv} \rightarrow (s_{000})(s_{001})(s_{010}, s_{011})(s_{100}, s_{101}, s_{110}, s_{111}).$$

Hence $\overset{2}{\equiv} = \overset{3}{\equiv} = \ldots = \equiv$. The quotient automaton is described in Figure 11.

	0	1
s_{000}	s_{000}, 1	s_{100}, 0
s_{001}	s_{000}, 0	s_{100}, 0
s_{010}	s_{001}, 0	s_{101}, 0
s_{011}	s_{001}, 0	s_{101}, 0
s_{100}	s_{010}, 0	s_{110}, 1
s_{101}	s_{010}, 0	s_{110}, 1
s_{110}	s_{011}, 0	s_{111}, 1
s_{111}	s_{011}, 0	s_{111}, 1

Figure 10. Flow table

		0	1
(s_{000})	$\rightarrow q_1$	$q_1/1$	$q_4/0$
(s_{001})	$\rightarrow q_2$	$q_1/0$	$q_4/0$
(s_{010}, s_{011})	$\rightarrow q_3$	$q_2/0$	$q_4/0$
$(s_{100}, s_{101}, s_{110}, s_{111})$	$\rightarrow q_4$	$q_3/0$	$q_4/1$

(a)

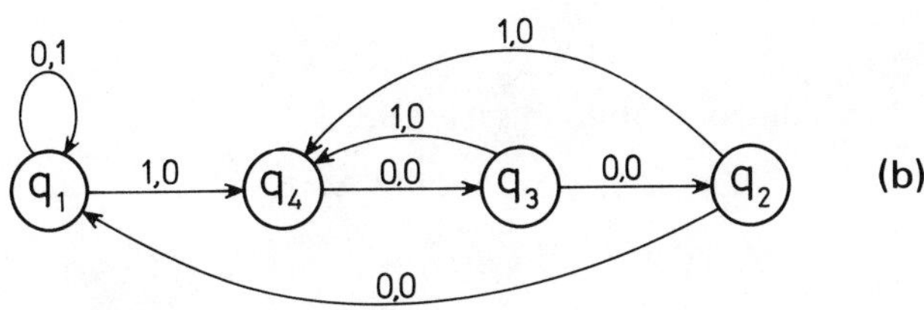

Figure 11. (a) Flow table of the quotient automation.
(b) Flow graph of the quotient automaton

Remark. If A is an incomplete automaton, one may think of minimizing A by choosing in all possible ways the unspecified values, and minimizing all corresponding complete automata. Nevertheless, this method does not necessarily yield the optimal solution. As an example, let us consider the automaton of Figure 8 which is covered by the two state automaton of Figure 9. If the only unspecified entry of Figure 8 is replaced by either a 1 or a 0, the corresponding three state complete automata cannot be further simplified.

4.2 Decomposition

The decomposition of a complex system into simpler subsystems is a currently used technique, resulting from a series of observations. Among other points are the following.

(1) The system organization is made clearer: this is an important requirement for getting well-documented and easily maintainable systems.

(2) Every subsystem can be separately optimized, whereas it can be impossible directly to optimize the whole system.

Hence, the question naturally arises whether a given automaton can be decomposed into a series of smaller automata. Let us first recall the classical definition of cascade connection of two automata. Let us consider two Mealy automata A and B:

$$A = \langle \Sigma, \Omega^A, Q^A, \mathbf{M}^A, \mathbf{N}^A \rangle,$$
$$B = \langle \Sigma \times \Omega^A, \Omega, Q^B, \mathbf{M}^B, \mathbf{N}^B \rangle.$$

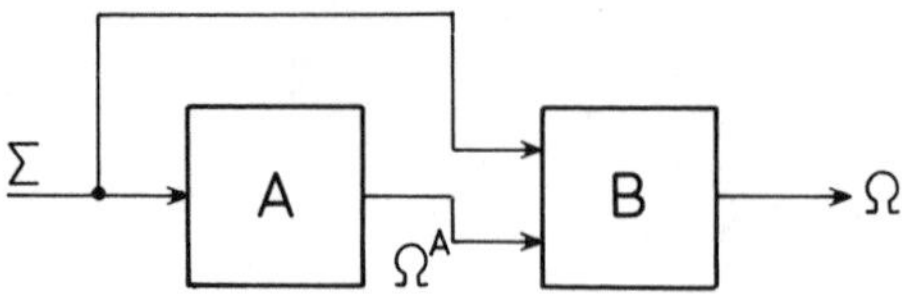

Figure 12. Automaton C

Automaton C is defined as follows (see also Figure 12):

$$C = \langle \Sigma, \Omega, Q^A \times Q^B, \mathbf{M}, \mathbf{N} \rangle, \qquad (q^A, q^B)M_\sigma = (q^A M_\sigma^A, q^B M^B_{(\sigma, q^A N_\sigma^A)}), \tag{17}$$

$$(q^A, q^B)N_\sigma = q_B N^B_{(\sigma, q^A N_\sigma^A)}. \tag{18}$$

Notice that $(q^A, q^B)M_\sigma$ or $(q^A, q^B)N_\sigma$ can be undefined.

The two possible implementations (iterative or sequential circuit) are shown in Figure 13.

The research of cascade decompositions of a given automaton is related to the detection of admissible decompositions of the set of states. A *decomposition* of the set of states of automaton $\langle \Sigma, \Omega, Q, \mathbf{M}, \mathbf{N} \rangle$ is a set $\{S_1, S_2, \ldots, S_k\}$ of subsets of Q, which covers Q, i.e.

$$\forall i \in \{1, 2, \ldots, k\} : S_i \subseteq Q; S_1 \cup S_2 \cup \ldots \cup S_k = Q.$$

It is an *admissible decomposition* if for every index $i \in \{1, 2, \ldots, k\}$, and every input letter $\sigma \in \Sigma$, there exists at least one index $j \in \{1, 2, \ldots, k\}$ such that

$$S_i M_\sigma \subseteq S_j. \tag{19}$$

In the particular case where the elements of the decomposition are non-empty pairwise subsets of Q, it is a *partition*. An admissible partition is called partition with the substitution property.

Let us turn back to automaton C, where is the cascade (or series) interconnection of automata A and B. We may define a partition of the set $Q^A \times Q^B$ of states of C: if

$$Q^A = \{q_1^A, q_2^A, \ldots, q_t^A\},$$

the t partition blocks are

$$\{(q_1^A, q^B) \mid q^B \in Q^B\}, \ldots, \{(q_t^A, q^B) \mid (q^B \in Q^B\}.$$

It is obviously a partition with the substitution property. The *head automaton* (i.e. automaton A) determines to which partition block the present state of C belongs; the *tail automaton* (i.e. automaton B) gives the additional information allowing which element of the partition block is the present state of C to be determined.

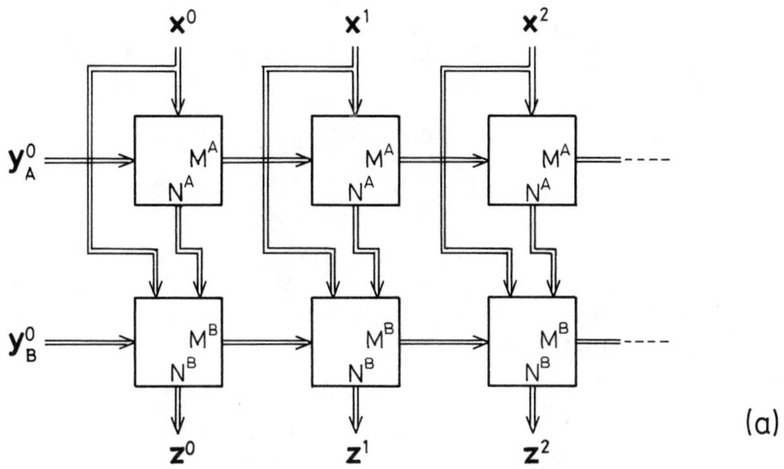

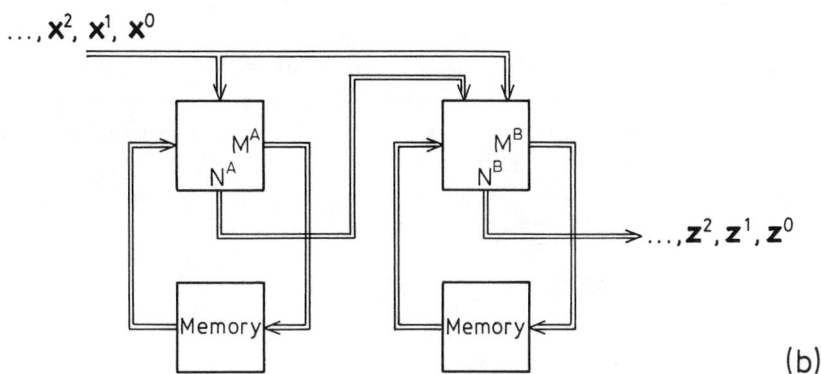

Figure 13. Two possible implementations of the automaton C

This is a general principle. From the knowledge of a partition with the substitution property of the states of an automaton, or, more generally, from the knowledge of an admissible decomposition, we may deduce a head automaton which is the quotient automaton corresponding to the partition or to the decomposition. It thus includes a number of states equal to the number of blocks of the partition or decomposition. The tail automaton defines a particular state inside a given block. It includes as many states as the number of elements in a block of maximum size.

Let us illustrate the general principle by an example. The automaton of Figure 14 is a parity bit generator: every four time units, it generates the parity bit of the word formed by the three last input bits. Otherwise, its output is 0. Let us recall that the parity bit of a binary word w is equal to 1 if w includes an odd number of 1's, and equal to 0 if it includes an even

	$x=0$	$x=1$
A	B, 0	C, 0
B	D, 0	E, 0
C	E, 0	D, 0
D	F, 0	G, 0
E	G, 0	F, 0
F	A, 0	A, 0
G	A, 1	A, 1

Figure 14. Automaton representing a parity bit generator

number of 1's. As an example, here are an input sequence and the corresponding output sequence:

$$\text{input} = 0\ 0\ 1\ 0\ 1\ 1\ 0\ 1\ 0\ 1\ 1\ 0\ 1\ 1\ 1\ 1\ldots,$$
$$\text{output} = 0\ 0\ 0\ 1\ 0\ 0\ 0\ 0\ 0\ 0\ 0\ 0\ 0\ 0\ 0\ 1\ldots.$$

The following partition has the substitution property:

$$(A), (B, C), (D, E), (F, G).$$

The corresponding quotient automaton is shown in Figure 15. Notice it is an *autonomous system*: its working does not depend on the input signals. The tail automaton includes two states q_1 and q_2. Figure 16 shows an encoding of the states of the initial automaton, by means of the states of the head and tail automata. It remains to define the transition and output functions of the tail automaton. As an example, let us compute q_1M_{0,p_1} and q_1N_{0,p_1}:

$$(p_1, q_1) \to A;$$
$$AM_0^A \to B, \qquad AN_0^A \to 0;$$
$$B \leftarrow (p_2, q_1);$$
$$\Rightarrow q_1M_{0,p_1} = q_1, \quad q_1N_{0,p_1} = 0.$$

The flow table of the tail automaton is shown in Figure 17.

We conclude this section with some indications concerning the parallel

			$x=0$	$x=1$
(A)	→	p_1	p_2, p_2	p_2, p_2
(B, C)	→	p_2	p_3, p_3	p_3, p_3
(D, E)	→	p_3	p_4, p_4	p_4, p_4
(F, G)	→	p_4	p_1, p_1	p_1, p_1

Figure 15. Corresponding quotient automaton

	q_1	q_2
p_1	A	A
p_2	B	C
p_3	D	E
p_4	F	G

Figure 16. Encoding of the states of the automaton of Figure 14

	$x=0$				$x=1$			
	p_1	p_2	p_3	p_4	p_1	p_2	p_3	p_4
q_1	$q_1, 0$	$q_1, 0$	$q_1, 0$	$-, 0$	$q_2, 0$	$q_2, 0$	$q_2, 0$	$-, 0$
q_2	$q_1, 0$	$q_2, 0$	$q_2, 0$	$-, 1$	$q_2, 0$	$q_1, 0$	$q_1, 0$	$-, 1$

Figure 17. Flow table of the tail automaton

interconnection of automata. Let us consider two Mealy automata:

$$A = \langle \Sigma, \Omega^A, Q^A, \mathbf{M}^A, \mathbf{N}^A \rangle,$$

and

$$B = \langle \Sigma, \Omega^B, Q^B, \mathbf{M}^B, \mathbf{N}^B \rangle.$$

Automaton C is defined as follows:

$$C = \langle \Sigma, \Omega^A \times \Omega^B, Q^A \times Q^B, \mathbf{M}, \mathbf{N} \rangle,$$
$$(q^A, q^B)M_\sigma = (q^A M_\sigma^A, q^B M_\sigma^B),$$
$$(q^A, q^B)N_\sigma = (q^A N_\sigma^A, q^B N_\sigma^B).$$

Notice again that $(q^A, q^B)M_\sigma$ or $(q^A, q^B)N_\sigma$ can be undefined.

It is a particular case of cascade connection in which function $M_{\sigma,\omega}^B$ does not actually depend on ω. Automata A and B play equivalent roles, and both may be considered as the head automaton. Hence, the decomposition defines two partitions, of the states of C, with the substitution property. They furthermore have the following characteristic: the intersection of a block of the first partition and of a block of the second partition is a singleton.

More generally, the parallel decompositions of an utomaton are related to the existence of two admissible decompositions which have the next property: the intersection of a block of the first decomposition and of a block of the second decomposition is either the empty set or a singleton.

Remark. We briefly recalled classical decomposition methods, based on admissible decompositions of the set of states. According to our experience, those methods, when applied to practical systems, lead to very long and tedious computations, except in rather trivial situations. In some cases, an attentive examination of the problem can directly yield a useful decomposition. The head automaton of the parity bit generator (Figure 15) is an autonomous system, namely a four state counter. As a matter of fact, the existence of a four state counter as part of the whole circuit is quite natural since the working of the system can be decomposed into four cycles: examination of the first bit, of the second bit, and of the third bit, and computation of the parity bit.

4.3 Encoding

It has been seen in Section 3 that the first step of the implementation of an automation is the binary encoding of sets Σ, Ω and Q. The encoding of Σ

	y_0 0	1
y_2y_1 00	A	A
01	B	C
10	D	E
11	F	G

Figure 18. Encoding of the states of the automaton of Figure 14

and Ω often depends on external conditions: the input signals come from other circuits, and the output signals feed other circuits. On the other hand, the designer may arbitrarily choose the encoding of set Q.

The choice of f_Q (see Section 3) determines functions M and N according to equations (14) and (15). There is thus an optimization problem: the complexity of functions M and N, with respect to some given criterion, depends on the chosen encoding.

Let us first illustrate by an example the relation between encoding and decomposition. The automaton of Figure 14 can be decomposed into the cascade connection of two automata. Every state of the initial automaton is represented by a pair (p_i, q_j), according to Figure 16, where p_i stands for a state of the head automaton, and q_j for a state of the tail automaton. This observation suggests the encoding of the states of the automaton by means of three variables, y_0, y_1, and y_2, as shown in Figure 18. Functions M and N are given in Figure 19. The main remark is that functions M_2 and M_1 do not depend on y_0: it is a direct consequence of the fact that state variables y_2 and y_1 describe the blocks of a partition with the substitution property. After, this partial degeneracy is considered as a simplifying factor.

If an automaton C is decomposed into the parallel connection of automata A and B, its states an be encoded by associating h variables $y_0, \ldots, y_{h-1}$ to the states of A, and $(p-h)$ variables $y_h, \ldots, y_{p-1}$ to the states of B. Functions $M_0, \ldots, M_{h-1}$ then only depend on $y_0, \ldots, y_{h-1}$, and functions $M_h, \ldots, M_{p-1}$ only depend on $y_h, \ldots, y_{p-1}$. This type of partial degeneracy is well related to the existence of partitions with the substitution property.

The question naturally arises whether there are other types of partial

	x 0	1
$y_2y_1y_0$ 000	010, 0	011, 0
001	010, 0	011, 0
010	100, 0	101, 0
011	101, 0	100, 0
100	110, 0	111, 0
101	111, 0	110, 0
110	00–, 0	00–, 0
111	00–, 1	00–, 1

$M = M_2M_1M_0, N$

Figure 19. Coded flow table for the functions M and N

	x_1x_2 00	01	11	10
1	1, 1	2, 1	3, 1	4, 1
2	3, 1	4, 1	1, 1	2, 1
3	2, 0	1, 0	4, 0	3, 0
4	4, 0	3, 0	2, 0	1, 0

Figure 20. Automaton with two partition pairs

degeneracy which can be introduced by a suitable encoding. The theory of *partition pairs* (Hartmanis and Stearns, 1966) gives an affimative answer. Let us briefly evoke the basic ideas. Let us consider automaton $\langle \Sigma, \Omega, Q, \mathbf{M}, \mathbf{N} \rangle$, and partitions

$$\pi_1 = \{S_1, S_2, \ldots, S_k\}, \qquad \pi = \{T_1, T_2, \ldots, T_\ell\}$$

of the set of states. The two partitions form a pair (π_1, π_2) if, for every index $i \in \{1, 2, \ldots, k\}$, and every input letter $\sigma \in \Sigma$, there exists at least one index $j \in \{1, 2, \ldots,\}$ such that

$$S_i M_\sigma \subseteq T_j.$$

Let us show by an elementary example the usefulness of this new concept. The automaton of Figure 20 has two partition pairs:

$$(\pi_1, \pi_2) \quad \text{and} \quad (\pi_2, \pi_1),$$

where

$$\pi_1 = \{(1, 2), (3, 4)\} \quad \text{and} \quad \pi_2 = \{(1, 3), (2, 4)\};$$

furthermore, the intersection of a block of π_1 and of a block of π_2 is a singleton. The state encoding of this automaton can be performed by means of two state variables y_1 and y_2: y_1 distinguishes the particular block of π_1, and y_2 the particular block of π_2. As (π_1, π_2) and (π_2, π_1) are partition pairs, function M_1 only depends on x_1, x_2, and y_2 and function M_2 only depends on x_1, x_2, and y_1 (Figure 21).

		x_1x_2 00	01	11	10
y_1y_2	00	00, 1	01, 1	10, 1	11, 1
	01	10, 1	11, 1	00, 1	01, 1
	10	01, 0	00, 0	11, 0	10, 0
	11	11, 0	10, 0	01, 0	00, 0

$M = M_1M_2, N$

Figure 21. Encoding of the automaton of Figure 20

We complete this section by a remark similar to that concluding the preceding section: methods based on partitions with the substitution property and partition pairs most often lead to very long computations. In many cases, the choice of a good encoding results from an attentive examination of the problem.

BIBLIOGRAPHICAL REMARKS

The basic papers about automata are those of Moore (1956), Kleene (1956), Mealy (1956), and Rabin and Scott (1959). Many classical textbooks, such as Harrison (1965), Ginzburg (1968), Kohavi (1970), and Zahnd (1980), give fairly extensive developments to that theory. Basic to our presentation, however, are the books by Hennie (1961, 1968) which stress the identity between the concepts of iteration and of finite automaton.

EXERCISES

E1. It is easy to check that a finite, fixed length iteration is described by the PASCAL statement

$$\text{for } i := 1 \text{ to } n \text{ do } P,$$

where P represents the computation performed (by a combinational switching circuit) during a typical iteration step. How does the iteration terminate in the case of a sequential implementation? Discuss similarly the 'while' statement

$$\text{while } i \leqslant n \text{ do } P.$$

E2. Many of the algorithms discussed in Chapter IV may be viewed as iterations. Discuss their sequential implementations.

E3. (Unger, 1977.) Consider a finite automaton $A = \langle \Sigma, \Omega, Q, \mathbf{M}, \mathbf{N} \rangle$, with the transition and output functions

$$M_\sigma : Q \to Q \quad \text{and} \quad N_\sigma : Q \to \Omega; \qquad \sigma \in \Sigma.$$

The transition function is easily extended from the input alphabet Σ to the input dictionary Σ^*. Let indeed $w = \sigma_1\sigma_2 \ldots \sigma_n$; $w \in \Sigma^*$ and define

$$M_w = M_{\sigma 1} M_{\sigma 2} \ldots M_{\sigma n},$$

where the multiplicative notation stands for the composition of mappings. The finite set

$$\mathcal{M} = \{M_w \mid w \in \textstyle\sum^*\}$$

is a semigroup under the consumption and it is called semigroup of the automaton A. Consider now the four mappings

$$P_1 : \Sigma \to \mathcal{M},$$
$$P_2 : \mathcal{M} \times \mathcal{M} \to \mathcal{M},$$
$$P_3 : Q \times \mathcal{M} \to Q,$$
$$P_4 : \Sigma \times Q \to \Omega.$$

The mapping P_1 associates, with each input literal σ the (encoded) name of the corresponding mapping M_σ. The mapping P_2 is first applied to pairs of consecutive input literals, say σ, τ, and computes the name of the mapping $M_\sigma M_\tau$ corresponding to the input word $\sigma\tau$ from the names M_σ and M_τ. In fact P_2 is the multiplication table of the automaton semigroup. The operation P_2 will next be repeated to obtain the names of the mappings M_w corresponding to groups of $4, 8, \ldots, 2^m$ consecutive output literals. The mapping P_3 may be seen as a table of the transition function extended to the input dictionary. It allows one to compute, from a state q and from the name M_w of a mapping in $\mathcal{M}$ the next state the system would have reached if it had been started in q and had received the input word w. Thus, thanks to the results available from P_2, it is possible to compute the final state, the states reached after the input words of length 2^{m-1}, $2^{m-2}, \ldots$, and finally, all the intermediate states the iterative sytem would have passed through while receiving an input word of length 2^m. Finally, P_4 is nothing but the output function of the automaton. Show that the above construction replaces an iterative circuit of length n (delay $0(n)$) by a tree circuit with delay $= 0(\log n)$. Derive conditions on $\mathcal{M}$ cardinality such that the cost of the obtained circuit remains a linear function of n. Apply the method to the serial binary addition to obtain the carry-lookahead adder.

E4. Consider the literals $a, b, \ldots, z$ of the English alphabet A and represent them by the integers $0, 1, \ldots, 25$ (thus $a \to 0$, $b \to 1, \ldots, z \to 25$). The well known de Vigenere encipherment scheme is then described by

$$E : x \to x \oplus k \pmod{26}.$$

Thus, the encipherment scheme replaces an integer $x \in \{0, 1, \ldots, 25\}$ representing a literal of the clear text message by the new integer $x \oplus k$ representing a literal of the ciphertext. Develop iterative schemes for performing the encoding and decoding functions (assume the literals represented in the ASCII code). Design the corresponding circuits. Do you think it possible to apply Unger's technique in this case?

E5. (Huffman, 1959; Kurmit, 1974.) An automaton $A = \langle \Sigma, \Omega, Q, \mathbf{M}, \mathbf{N} \rangle$ is a lossless automaton of order r if, given an initial state q and an output word $\tilde{w}$, it is possible to recover the corresponding input word after a delay at most equal to r. Show that, if A is lossless of order r, the knowledge of an initial state q and of an output word of length $(r+k)$ allows one to recover the k first literals of the input word. Discuss the decompositions of lossless automata of order r.

Chapter VI
Sequential networks

1 INTRODUCTION

In this chapter we first investigate one of the most common information storage devices that are found in digital networks and systems, i.e. *flip-flops.* Flip-flops are commonly used to construct registers and other digital networks that require a small amount of flexible information storage capability. Our main concern is with external properties of flip-flops and how they are used in the design of digital networks and system.

A *synchronous sequential network* is a sequential network in which the contents of the basic information storage elements can change only during the occurrence of a *clock pulse.* Between clock pulses logical operations are performed on the input and on stored information, but there is no change in the information contained in the information storage elements.

We assume that the information storage devices of the synchronous sequential networks are flip-flops. Starting with this information we first show how the operation of a given sequential network can be analysed. We then go on to consider the problem of designing a sequential network to carry out specific information processing tasks.

2 MEMORIZATION AND SYNCHRONIZATION FLIP-FLOPS

In the basic logic elements discussed in Chapter II, the output signals depend only on the instantaneous value of the various input signals. In such combinational elements, the outputs do not depend on the history of previous states, because they contain no memory function.

Elements whose operation is conditioned by their history (or internal states) are called *sequential elements.* The most important representative of this class is the flip-flop: it is a bistable element with one or more inputs and two complementary outputs. A flip-flop remains in its last state until a specific input signal causes it to change state. Because of its ability to store bits of information, the flip-flop is a basic building block in digital circuitry.

There are many forms of flip-flops, each of which has its specific features.

2.1 The synchronous set-reset flip-flops

Consider the logic network of Figure 1a; it may be described by the equations:

$$P = \bar{S}\bar{Q}, \qquad Q = \bar{R}\bar{P}. \tag{1}$$

From these equations we deduce:

$$(S, R) = (1, 0) \Rightarrow (P, Q) = (0, 1),$$
$$(S, R) = (0, 1) \Rightarrow (P, Q) = (1, 0),$$
$$(S, R) = (1, 1) \Rightarrow (P, Q) = (0, 0).$$

We call *total state* of the logic network the set of values (S, R, Q, P). The above analysis allows us to conclude that the total states

$$\{(S, R, Q, P)\} = \{(1, 0, 1, 0), (0, 1, 0, 1), (1, 1, 0, 0)\}$$

are *stable states.* This means that the system, once in this state, remains in it indefinitely so long as the input variables S and R do not change.

The input state $(S, R) = (0, 0)$ requires more attention. From the equations (1) we deduce that for this input state $Q = \bar{P}$ so that the total states

$$\{S, R, Q, P)\} = \{(0, 0, 1, 0), (0, 0, 0, 1)\}$$

are both stable states.

Let us now consider the logic network shown in Figure 1 where the inherent delay Δt of each NOR logic element is indicated.

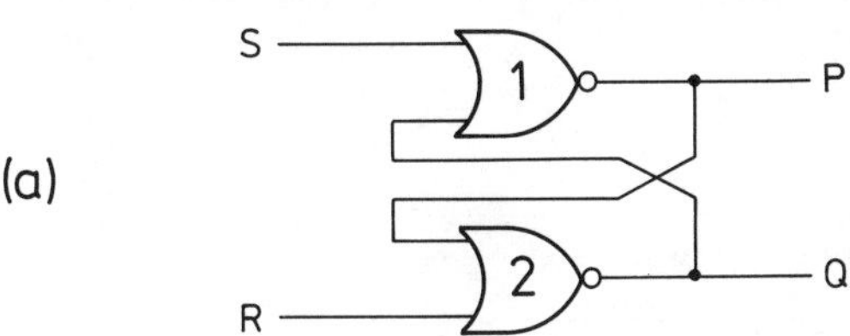

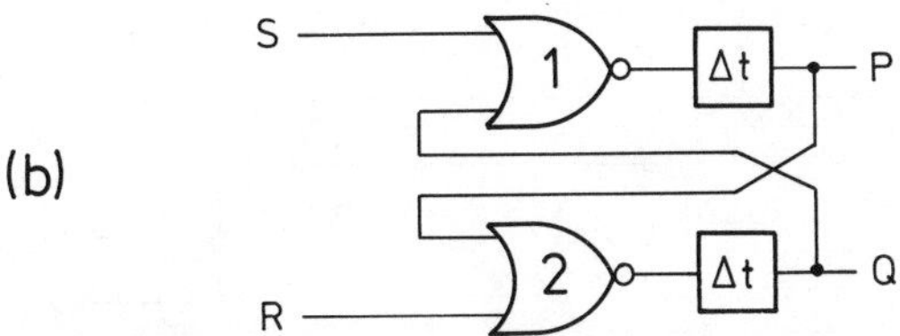

Figure 1.(a) Synchronous set–reset flip-flop. (b) Synchronous set-reset flip-flop with delay elements

Let us assume that the network is in an initial total stable state $S=0$, $R=0$, $Q=0$, $P=1$. At time t_1 assume that a 1 is applied to the input S and the R input remains 0. When this input occurs the output of NOR element 1 goes to 0 and after Δt P goes to 0. This in turn drives the output of NOR element 2 to 1 and after another Δt time Q goes to 1. Now if S goes to 0 and if R remains 0, the network will hold the condition $Q=1$, $P=0$. When this happens we say that the network is in the state 1.

Next consider what happens when an input condition of $S=0$, $R=1$ occurs while this flip-flop is in the state 1. When this happens the output of logic element 2 becomes 0 and the output of logic element 1 becomes 1. We say that the flip-flop is in the state 0.

A complete timing diagram illustrating the dynamic behaviour of this circuit is shown in Figure 2. Note that after the change of state occurs the inputs can both be set to 0 and the outputs hold a constant value. The only requirement is that the inputs remain applied for a time greater than $2\,\Delta t$.

More generally, we shall say that a sequential network operates in *fundamental mode* if it must reach a stable state before the next input change may occur and that at most one input variable may change at a time. In the present case the fundamental mode operation involves an input signal remaining constant for at least $2\,\Delta t$ units of time.

The circuit of Figure 1 is called a *set–reset flip-flop*, or in short SR flip-flop, since a 1 on the S input sets the circuit to the 1 state while a 1 applied to the R input resets the circuit to the 0 state. This renders obvious the memory function of the flip-flop. It should finally be noted that the input condition $S=1$, $R=1$ should never be allowed to occur. The reason for this restriction can be understood if we see that $S=R=1$ implies $P=Q=0$,

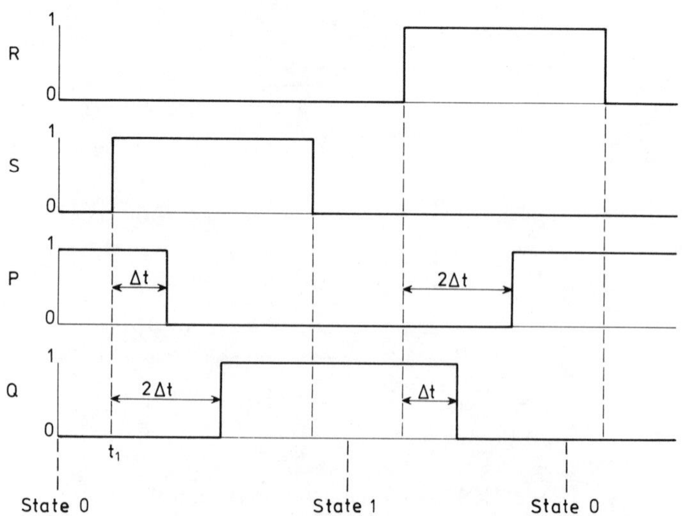

Figure 2. Timing diagram for the SR flip-flop

which is inconsistent with our definition of a flip-flop that requests $P = \bar{Q}$. Another reason for forbidding the situation $S = R = 1$ will be discussed later on.

Before continuing the analysis of the behaviour of the SR flip-flop we introduce a general model for the description of a sequential circuit.

A switching network is called a sequential switching network if its outputs depend on both the present inputs to the network and on the past history of inputs into the network; rather than using the past history of the inputs and thus the time as an explicit variable, one generally uses the concept of *state* of the sequential network and makes the outputs of the sequential network at any time t depend on the inputs and the state at time t. The general model for sequential circuits, shown in Figure 3, consists of two parts, a combinational logic section and a memory section. The memory stores information about past events required for proper functioning of the circuit. This information is requested in the form of p binary outputs, $y_0, y_1, \ldots, y_{p-1}$, known as state variables. Each of the 2^p combinations of state variables defines a state of the memory; in general, each state corresponds to a particular combination of past events.

The combinational logic section receives as inputs the state variables and the n circuit inputs $x_0, x_1, \ldots, x_{n-1}$. It generates m outputs $z_0, z_1, \ldots, z_{m-1}$ and p excitation variables $Y_0, Y_1, \ldots, Y_{p-1}$, which specify the next state to be assumed by the memory. The combinational logic network may be defined by Boolean equations of the standard forms, i.e.:

$$\begin{aligned} z_i &= f_i(x_0, \ldots, x_{n-1}, y_0, \ldots, y_{p-1}), & i &= 0, 1, \ldots, m-1, \\ Y_j &= Y_j(x_0, \ldots, x_{n-1}, y_0, \ldots, y_{p-1}), & j &= 0, 1, \ldots, p-1. \end{aligned} \tag{2}$$

These are time-independent equations, i.e. they are valid at any time and all the inputs and outputs are stable as for combinational circuits.

The combinational logic section is formed by an interconnection of ideal logic gates, i.e. gates without any time delay.

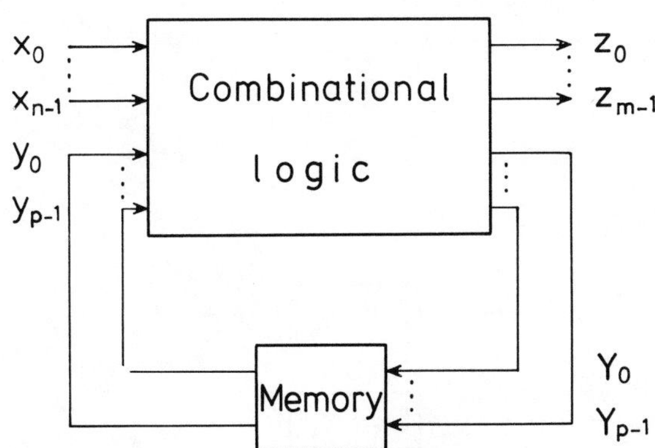

Figure 3. Sequential network model

We have not yet specified the relationship between the inputs and outputs of the memory portion of Figure 3. This relationship will depend on whether the memory is realized in terms of an array of memory elements or flip-flops or is merely a set of feedback loops with time delay.

We first assume that feedback loops contain only time delays as represented by Figure 4. Furthermore, we consider circuits for which all inputs are *levels*, i.e. signals which may remain at either value 0 or 1 for arbitrary periods and may change at arbitrary times: such circuits will be called *level mode sequential circuits.* Moreover, if we place a restriction on the inputs that only one variable may change at a time and that no further changes may occur until all the outputs and state variables have stabilized, we will say that the level mode circuit is operating in the fundamental mode.

The effect of a pure delay is for its output to reflect the value of its input Δt time units prior; if the delays in Figure 4 are *pure delays,* the state and excitation variables are related by the following relation:

$$y_j(t+\Delta t) = Y_j(t), \qquad j = 0, 1, \ldots, p-1. \tag{5}$$

In practice the delays present in feedback loops are not pure delays but *inertial delays:* an inertial delay has a similar effect to a pure delay except that an input must persist for at least Δt time units before being reflected at the output of the delay. Thus pulses of width less than Δt are ignored.

If τ is the width of an input pulse, the presence of an inertial delay in the feedback loop instead of a pure delay is reflected by the replacement of relations (3) by relations (4):

$$\begin{aligned} \text{for} \quad \tau < \Delta t&: \; y_j(t) = \text{constant}, \\ \tau > \Delta t&: \; y_j(t+\Delta t) = Y_j(t). \end{aligned} \tag{4}$$

Relations (2) and (4) completely describe the behaviour of a level mode sequential circuit operating in fundamental mode.

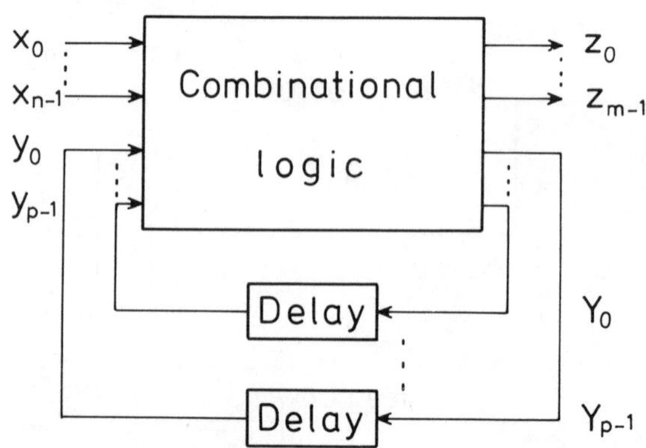

Figure 4. Feedback loop containing delays

The SR flip-flop may now be described in the basic form for the level-mode circuit (see Figure 5a; relations (5) and their corresponding excitation table given in Figure 5b):

$$P = \bar{S}\bar{q},$$
$$Q = \bar{R}\bar{p}. \tag{5}$$

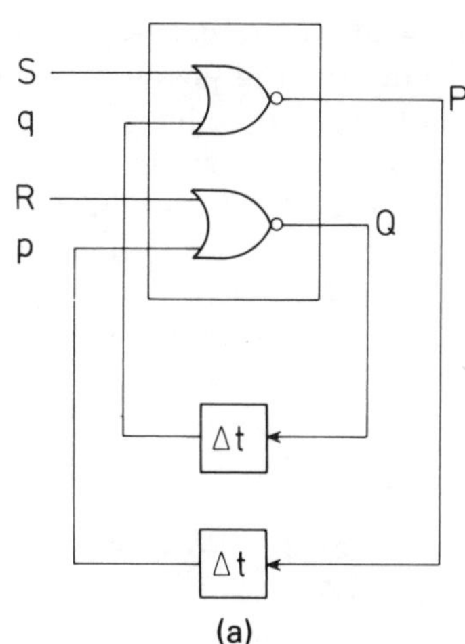

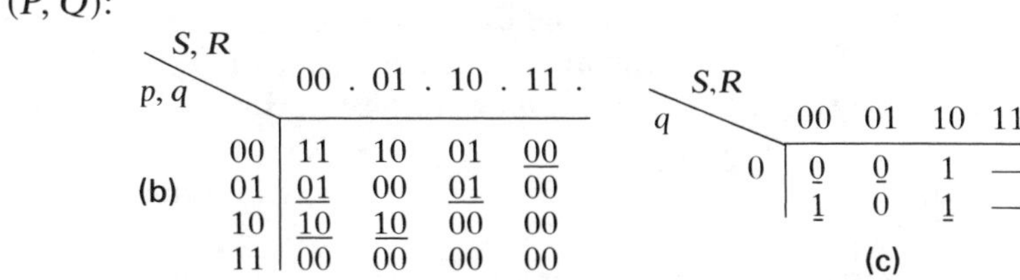

(P, Q):

(b)

p, q \ S, R	00	01	10	11
00	11	10	01	$\underline{00}$
01	$\underline{01}$	00	$\underline{01}$	00
10	$\underline{10}$	$\underline{10}$	00	00
11	00	00	00	00

q \ S,R	00	01	10	11
0	$\underline{0}$	$\underline{0}$	1	—
	$\underline{1}$	0	$\underline{1}$	—

(c)

Figure 5. (a) Basic form for the SR flip-flop. (b) Excitation table. (c) Modified excitation table

The *excitation table* is a map representing the excitation equation in function of the input variables (which generally determine the columns of the table) and of the state variables (which generally determine the rows of the table).

The stable states are underlined in the excitation tables (see Figure 5b). The dynamical analysis of the SR flip-flop may be continued with the aid of Figure 5b.

Assume that the flip-flop is in the stable state

$$(S, R, q, p) = (1, 1, 0, 0);$$

when the input variable S changes from 1 to 0, the excitation variable Q changes and the system goes into the stable state

$$(S, R, q, p) = (0, 1, 0, 1).$$

If the input variables S, R change both from $(1, 1)$ to $(0, 0)$ excitation variables P and Q may both change and the system reaches either the stable state

$$(S, R, q, p) = (0, 0, 0, 1),$$

or the stable state

$$(S, R, q, p) = (0, 0, 1, 0),$$

according to the relative value of the delays present in the two feedback loops of the circuit of Figure 5a. The possibility of reaching two different stable states consecutively to an input change is known as a critical race. Clearly, in order that a sequential network will operate correctly, critical races must be avoided. In the S-R flip-flop the critical race is avoided by forbidding the input condition $S=1$, $R=1$ to occur. Assuming, moreover, that the delay element in the feedback loop associated with the pair of variables (p, P) is omitted (otherwise stated the relation $P=p(t+\Delta t)$ becomes $P=p$), the equations (5) are successively written in the following forms:

$$\begin{aligned} Q &= \bar{R}\bar{p} = \bar{R}\bar{P} \\ &= \bar{R}(S \vee q) = \bar{R}S \vee \bar{R}q \vee RS \\ &= S \vee \bar{R}q. \end{aligned} \tag{6}$$

To this last relation is associated the excitation table of Figure 5c. Further on we shall assume that the input–output behaviour of the SR flip-flop is described either by the equation (6) or, equivalently, by the excitation table of Figure 5c.

Before continuing the study of flip-flops we briefly introduce an essential characteristic for sequential networks.

Sequential circuits may be either *synchronous* or *asynchronous*. Synchronization is usually achieved by a timing device called a *clock* which produces a train of equally spaced pulses. The clock for a particular circuit may have a number of outputs, on which pulses appear at certain intervals and with fixed relation between the pulses on the various outputs. This process ensures an orderly execution of the various operations and logical decisions to be made by the circuit. Asynchronous circuits, on the other hand, are usually faster because they are almost free-running and do not depend on the frequency of a clock, which in most cases is well below the speed of operation of a free-running gate. The orderly execution of operations in asynchronous circuits is controlled by a number of completion and initiation signals, so that the completion signal of one operation initiates the execution of the next consecutive operation, and so on.

The SR flip-flop considered in this section is an asynchronous sequential network. We show in the next sections how any type of flip-flop (and in particular the SR flip-flop) can be synchronized by a clock. In particular the so obtained synchronized flip-flops will have two asynchronous inputs generally called *Preset* and *Clear* and a synchronous input generally denoted *CK* (clock input).

2.2 The D flip-flop

As it will be seen in Chapter VII, the design of a sequential circuit will be greatly simplified if state changes can take place only at periodic points in time. This can be assumed if we synchronize all state changes with pulses from an electronic clock. A pulse is simply a signal which normally remains at one level (usually 0) and goes to the other level only for a short duration. A typical clock waveform is illustrated in Figure 6b.

A sequential circuit synchronized with this clock will change state only on the occurrence of a clock pulse and will change state no more than once for each clock pulse. Under these conditions, the circuit is said to be operating in the clocked mode.

Figure 6a illustrates how a conventional (non-clocked) flip-flop may be converted to a clocked flip-flop.

The basic SR flip-flop of Figure 1 is first transformed to a clocked SR flip-flop by adding two AND-gates which are the inputs for the clock signal: the outputs of the two AND-gates remain at 0 as long as the clock pulse (abbreviated *CK*) is 0.

A second modification with respect to the SR flip-flop consists of the introduction of an inverter which reduces the number of inputs from two to one; this flip-flop receives the designation of D flip-flop from its ability to transfer data into a flip-flop. The *D* input goes directly to the *S* input and its

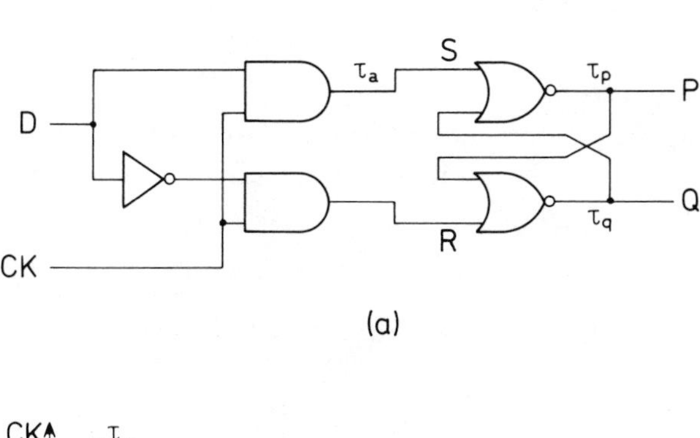

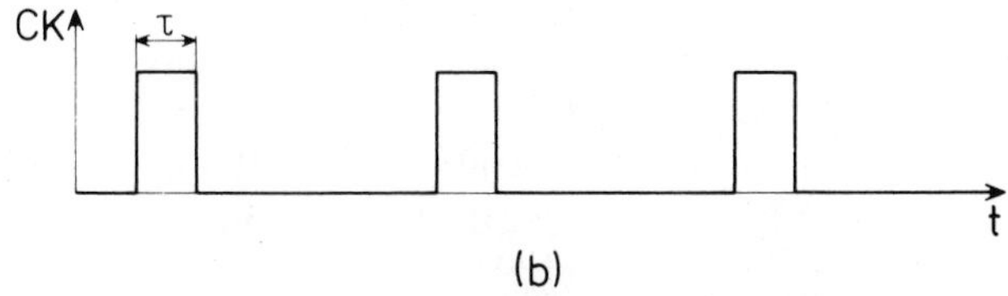

Figure 6.(a) Clocked D flip-flop. (b) Clock wave form

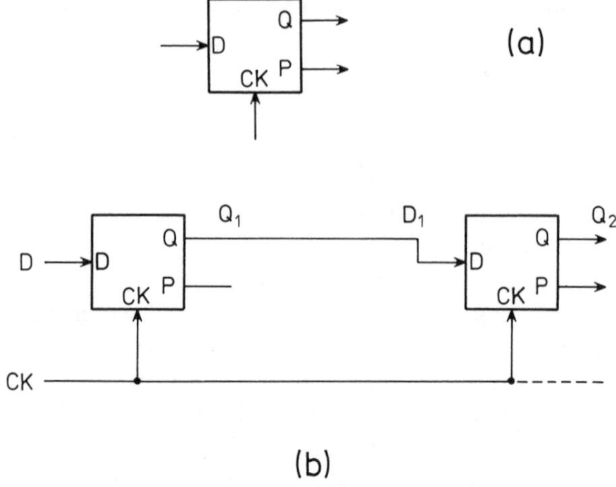

Figure 7.(a) Symbol for the clocked D flip-flop. (b) Typical interconnection for clocked D flip flop

complement is applied to the R input: the relation $SR=0$ which is requested for a SR flip-flop to operate correctly is thus automatically encountered by the scheme of Figure 6a.

As long as the clock input is 1 the SR flip-flop is connected to the D input which necessitates the internal state of the network to be D, i.e.

$$Q=D. \tag{7a}$$

The outputs of the two AND gates remain at 0 as long as the clock signal is 0: the SR flip-flop is deconnected from the D input and the state of the flip-flop cannot change, i.e.

$$Q=q. \tag{7b}$$

In view of (7a,b) the input–output behaviour of the D flip-flop of Figure 6a is described either by the excitation equation (7c) or equivalently by the excitation table of Figure 8:

$$Q=q\cdot\overline{CK}\vee D\cdot CK\vee qD. \tag{7c}$$

In order that the clocked D flip-flop should operate correctly the impulse width τ of the signal CK must satisfy a constraint of the form $\tau_{\min}<\tau<\tau_{\max}$.

If we assume that inertial delays τ_a, τ_p and τ_q are present in the network

Q \ CK, D q	00	01	11	10
0	**0**	**0**	1	**0**
1	**1**	**1**	**1**	0

Figure 8. Excitation table

(see Figure 6a), the impulse width τ must be large enough so that the input data can be memorized, i.e.

$$\tau > \max(\tau_a, \tau_p, \tau_q).$$

The symbol for a clocked D flip-flop is shown in Figure 7a; a typical interconnection scheme for these flip-flops is shown in Figure 7b. In order that this interconnection scheme should operate correctly, it is necessary that the clock signal CK has the value 0 when a change of the signal Q_1 reaches the input D_1 of the second flip-flop. We verify that this requirement is reflected by the following constraint on the impulse width

$$\tau < \tau_a + \min(\tau_p, \tau_q).$$

The above elementary analysis allows us to conclude that a correct synchronous memorization process happens if the following conditions are satisfied:

(1) D does not change when $CK = 1$;

(2) $\tau_{\min} < \tau < \tau_{\max}$.

Below we shall use a notation of the form

$$Q := D$$

in order to reflect that during the next 1 value of the clock signal, the output Q is replaced by the value of the input D as long as the two above conditions are satisfied.

If $\tau_a < \tau_q$ and if the impulse width τ has a value $\tau_a < \tau < \tau_q$, the above condition 2 cannot be satisfied and the network of Figure 7b does not operate correctly.

The master-slave flip-flop introduced in next section eliminates the requirement that a condition of the form (2) should be satisfied.

2.3 The master–slave flip-flop and the two-phase clock scheme

The need to limit the maximum clock pulse width can be avoided by using a more sophisticated flip-flop called a *master–slave flip-flop*. One form of a clocked master–slave is depicted in Figure 9a together with a timing diagram in Figure 9b.

Assume that initially the flip-flop output is $Q_2 = 0$ as shown in Figure 9b. In the absence of positive set or reset inputs or of a clock pulse, the output Q_1 of the first flip-flop—the master flip-flop—is the same as Q_2. Thus initially we have $Q_1 = 0$. The set input goes to 1 before the change from 0 to 1 of the clock pulse 2 (see Figure 9b). Therefore, the value of S_1 goes to 0 one gate delay (for the sake of simplicity assume that all the gates have identical inherent delay elements) following the edge of pulse 2. After one more gate delay unit, the output Q_1 of the master element goes to 1. Notice that the output Q_2 of the second flip-flop, the slave flip-flop, remains 0 during the 1 pulse of the clock.

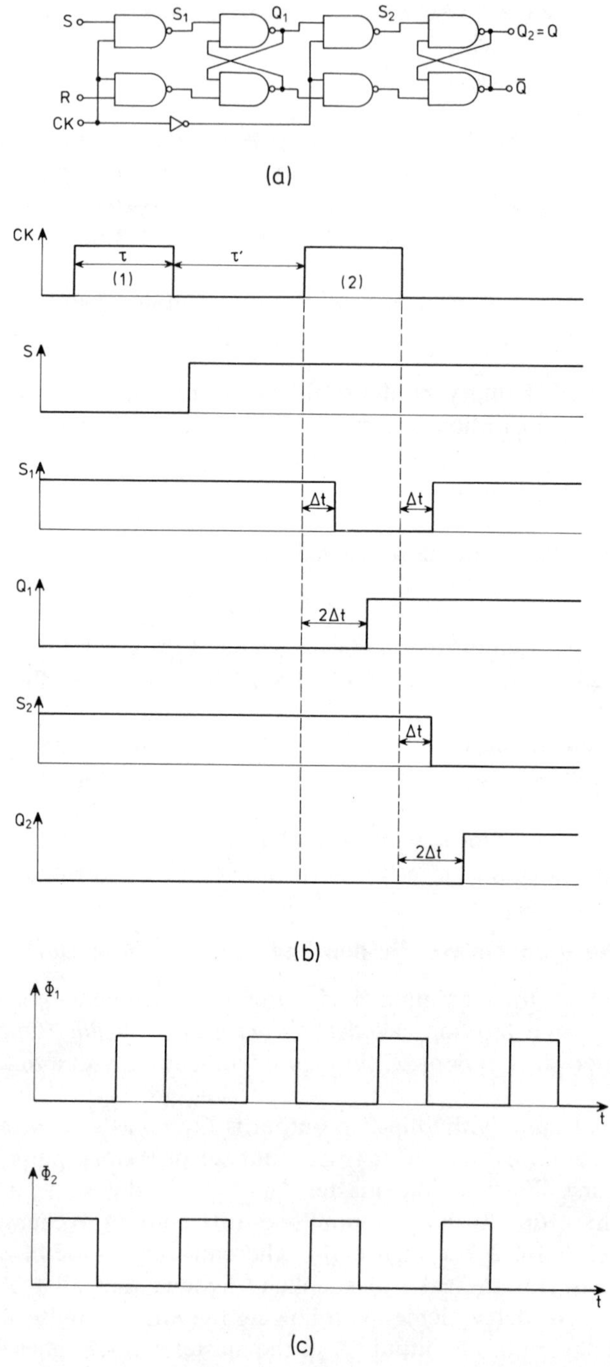

Figure 9.(a) Master–slave flip-flop. (b) Time diagram for the master–slave flip-flop. (c) Two-phase non-overlapping clock signals

When the clock returns to 0, S_2 goes to 0 and after another gate delay, Q_2 goes to 1. The symmetry of the network is apparent, so that the dual two-step process would occur following a 1 signal on the reset input.

Now consider a clocked sequential network containing only master–slave memory elements. Assume that the clock inputs to all the flip-flops are synchronized. Therefore, at each raise of a positive clock signal, some of the master flip-flops will change state, but the outputs of the slave flip-flops will remain at the previous values. After the clock pulse returns to 0, no further set or reset signals can be propagated into the master flip-flops. At this same time, some of the slave flip-flops' outputs will change state. Clearly, none of these new state values will have an effect on any of the master flip-flops until the next clock pulse. As a conclusion: no more than one state change of the master–slave flip-flop can take place for each clock pulse regardless of the clock pulse width (see Figure 9b).

The above transient analysis allows us to state the conditions under which a master–slave flip-flop should operate correctly.

(1) The inputs S and R do not change when $CK = 1$.

(2) The 1-input duration τ and the 0-input duration τ' must satisfy:

$\tau \geqslant K$ (correct memorization of the master flip-flop),

$\tau' \geqslant K'$ (correct memorization of the slave flip-flop).

In most integrated circuit technologies, gate delays exhibit a substantial statistical variance. For these circuits it is much easier to satisfy a condition of the form: $\tau \geqslant K$, $\tau' \geqslant K'$, (master–slave synchronization condition) than a condition of the form: $\tau_{min} < \tau < \tau_{max}$ (flip-flop synchronization condition).

The design methodology which uses a two-place non-overlapping clock scheme naturally derives from the master–slave flip-flop behaviour, if we assume that the master and slave flip-flops are synchronized by means of two different clock systems: instead of synchronizing the slave flip-flop by the clock $\overline{CK}$ as made in Figure 9a it is connected to a clock signal $(CK)'$. This strategy is briefly introduced below.

In Chapter VII a particular form of clocking scheme will often be made use of to control the movement of data through MOS circuits and other analog structures. By clocking scheme we mean a strategy for defining the times during which data is allowed to move into and through successive processing stages in a system, and for defining the intervening times during which the stages are isolated from one another.

Many alternative clocking schemes are possible, and a variety are in current use in different integrated systems. The clocking scheme used in an integrated system is closely coupled with the basic circuit or subsystem structuring and has major architectural implications. One of the most widely used clocking schemes is the *two-phase, non-overlapping clock signals.* This scheme is used consistently throughout the following chapters and is well matched to the type of basic structures possible in MOS technology, for example.

The two clock signals are often denoted ϕ_1 and ϕ_2; they are plotted as a function of time in Figure 9c. Note that both signals are non-symmetric and have non-overlapping 1-pulses. As usual, the 1-pulses are somewhat shorter than the 0-pulses.

2.4 Edge-triggered flip-flops

The state changes in a sequential circuit composed of master–slave flip-flops are synchronized by the values (0 or 1) of the clock signal. Another

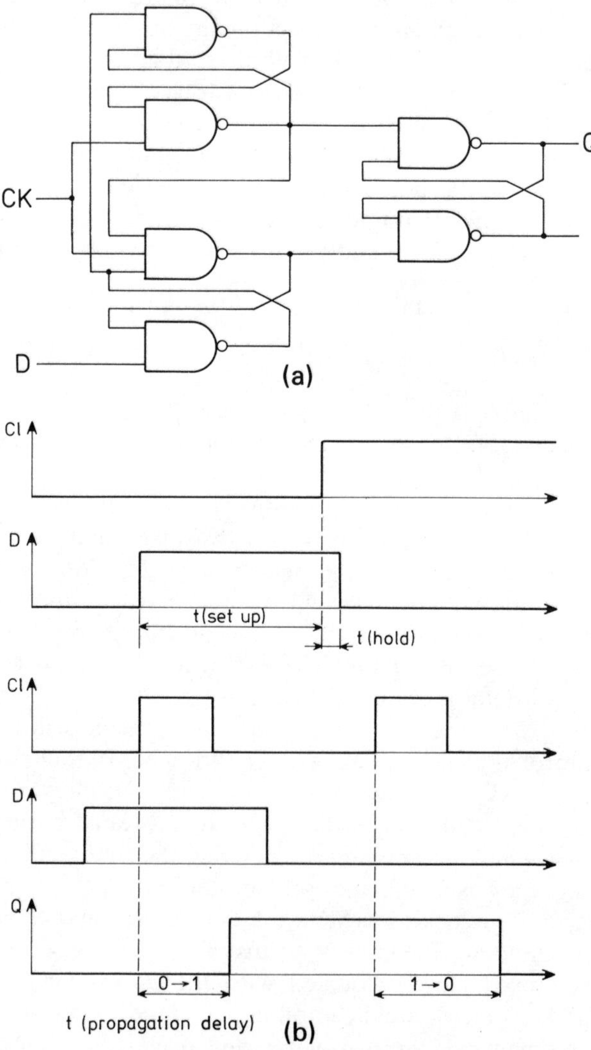

Figure 10.(a) Edge triggered flip-flop. (b) Signal constraints

commonly used type of flip-flop synchronizes the state change with the variation of the value of the clock signal: such devices are generally called *edge-triggered flip-flops.*

The flip-flop of Figure 10a (the 5474/7474 TTL D flip-flop) has the property that its output Q never changes in response to a change on the data input D. The output can only change in response to a change on the clock signal CK.

The circuit of Figure 10a can have its output change only when the clocking input changes from 0 to 1. This is sometimes called leading-edge triggering.

It is also possible to design D flip-flops which can have the output change only when the clocking input changes from 1 to 0. This is called *trailing-edge triggering.*

In order to ensure proper operation of a flip-flop it is usually necessary to place some constraints on the time interval between input signal changes. Since the state of the flip-flop depends on the input signal just prior to the active edge of the clocking signal, if the input signal changes just before the clocking edge, the flip-flop cannot respond properly. The input signal must be stable long enough for the internal circuitry to settle down before being clocked. The length of time that the input data must be stable before the flip-flop is clocked is defined as the set-up time. For the TTL D flip-flop type 5474/7474 the typical set-up time is 15 nanoseconds.

It is also necessary to have the data input stable for some time after the active clocking edge. This is called the hold time. A typical value for the 5474/7474 flip-flop is 2 nanoseconds.

Another important time interval which is not a restriction on the inputs but a measure of the circuit performance is the propagation delay. This is defined as the time interval from the occurrence of the signal which causes a change to the appearance of that change at the flip-flop output. The propagation delays differ for signals changing to 1 and signals changing to 0. Typical of the 5474/7474 are the values of 16 and of 22 nonoseconds for these two signal changes respectively.

These signal constraints are schematized in Figure 10b.

2.5 MOS flip-flops

It is possible to construct flip-flops by interconnecting NAND gates, made up of MOSFET's, according to the logic diagrams described above. They are called *static* memory elements. Nevertheless there exist realizations which are characteristic of MOS technology.

Figure 11a shows a *dynamic* register cell: it includes a transmission gate (T_1) and an inverter (T_2, T_3). The stray capacitance C, present at the gate T_2, is able to temporarily store one bit of data. It serves the same function as a D flip-flop. When the pass transistor T_1 is turned on, i.e. when $CK = 1$, the stray capacitance C will charge to the logic level of the input line. When T_1

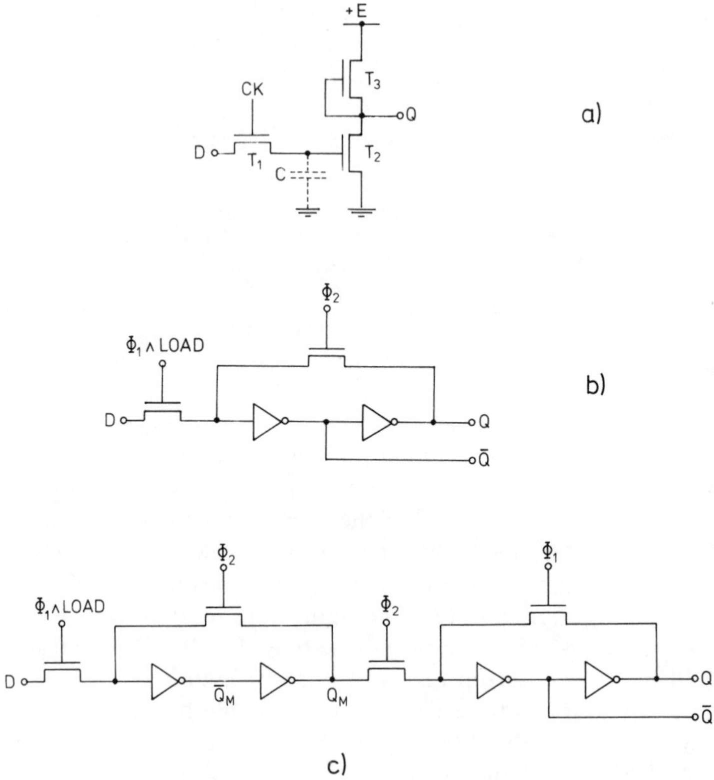

Figure 11.(a) Dynamic cell register. (b) Selectively loadable D flip-flop. (c) Master–slave D flip-flop

turns off, i.e. when $CK=0$, the data bit is stored by C. Nevertheless, the charge on C will eventually leak off; this is due to leakage through the gate insulator of T_2 and through the reversed biased drain-substrate junction of T_1. Typically, C is able to hold a charge for some milliseconds. This type of D flip-flop can be used in cases where there is a continuous flow of new data, with a clock signal whose period is sufficiently short. Typical applications are shift registers (Chapter VII), finite state machines (Section 3) and pipelined processing structures (Chapter VIII).

Figure 11b shows a selectively loadable D flip-flop, controlled by clock signals ϕ_1 and ϕ_2, and by a LOAD signal. When $\phi_2=1$, and thus $\phi_1=0$, it consists of a closed loop including two inverters, and may be in two stable states. When $\phi_1=\text{LOAD}=1$, and thus $\phi_2=0$, it stores a new input data bit. Let us notice that during the time intervals where $\phi_1 \wedge \text{LOAD}=\phi_2=0$, its contents are held because of the stray capacitances present at the inverter gates. Hence, it is sometimes called a *quasistatic memory element.*

A master–slave D flip-flop is shown in Figure 11c.

Many others MOS memory elements may be found in Penney and Lau (1972), Taub and Schilling (1977), Mead and Conway (1980).

3 SYNTHESIS OF SYNCHRONOUS SEQUENTIAL NETWORKS

3.1 Introduction

In chapter V, Section 3, we gave the implementation principles of automata. These are based on the coding of Σ (input alphabet), Ω (output alphabet), and Q (set of internal states), and on the synthesis of either an iterative circuit or a *synchronous sequential machine finite state machine.* In this section we consider the second implementation method.

The input letters are encoded by means of n *input variables* $\mathbf{x} = x_0, x_1, \ldots, x_{n-1}$; the output letters are encoded by m *output variables* $\mathbf{z} = z_0, z_1, \ldots, z_{m-1}$; the internal states are encoded by p *internal state variables* $\mathbf{y} = y_0, y_1, \ldots, y_{p-1}$. Functions M and N define $p+m$ Boolean functions of $n+p$ variables, according to equations (14) and (15) in Chapter V. To the transition function M correspond p *excitation functions*

$$Y_k(\mathbf{x}, \mathbf{y}), \qquad k = 0, 1, \ldots, p-1; \tag{8}$$

they give the next internal state variable values in function of the present input and internal state variable values. To N correspond m *output functions*

$$z_\ell(\mathbf{x}, \mathbf{y}), \qquad \ell = 0, 1, \ldots, m-1. \tag{9}$$

The general circuit architecture was shown in Figure 2 in Chapter V. It includes a combinational circuit implementing M and N, that is, the $p+m$ Boolean functions Y_k and z_ℓ, and a memory circuit able to store the present

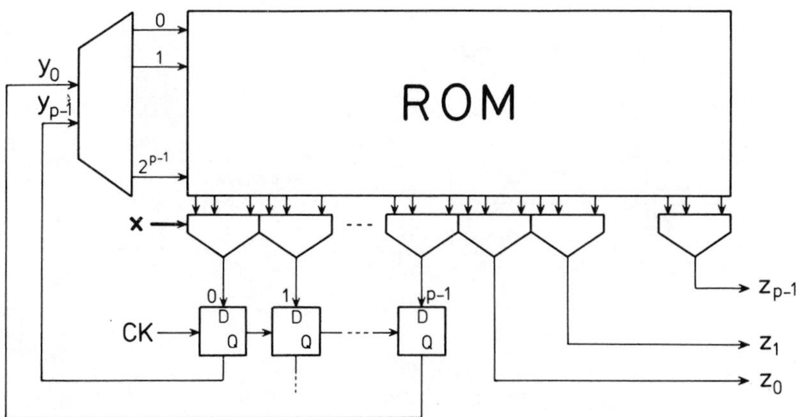

Figure 12. Circuit implementation with a read only memory

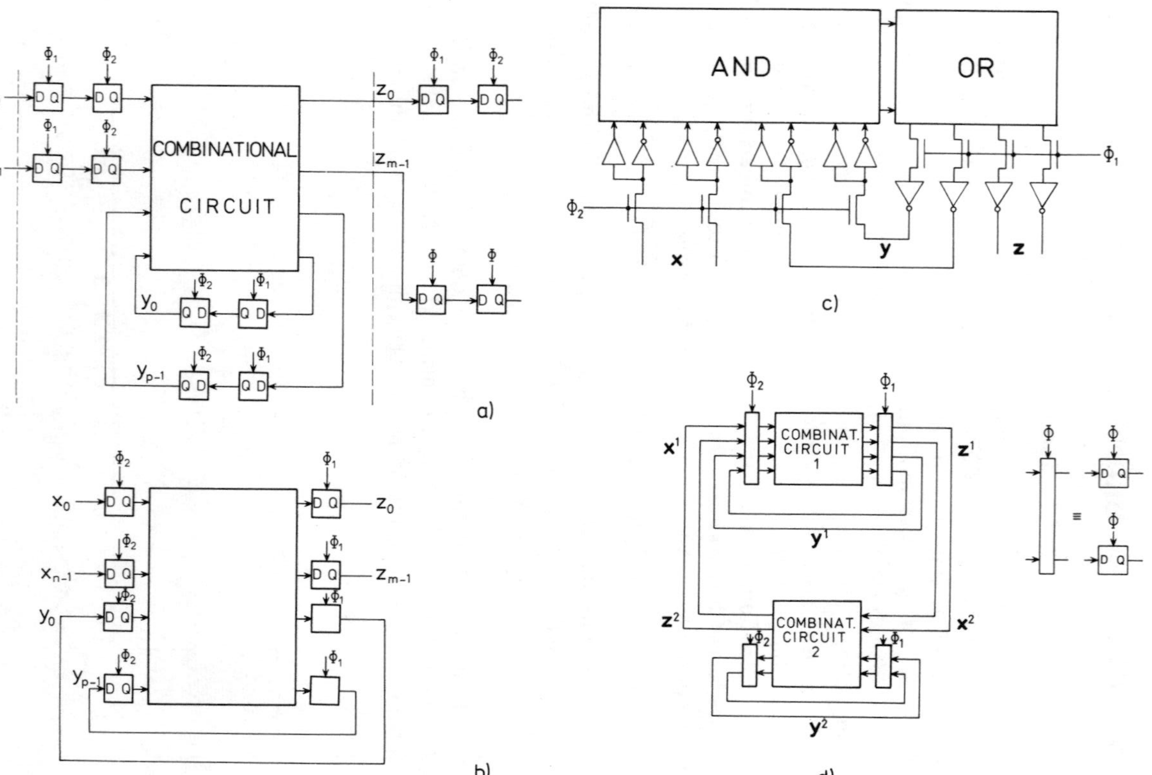

Figure 13.(a) Implementation based on a two-phase clocking scheme. (b) Basic two-phase configuration. (c) A programmable logic array implementation. (d) Input and output register cells

value of the internal state variables $\mathbf{y}$ during the computation of function Y_k and z_ℓ.

A direct implementation method consists in using p D-type flip-flops, each of which stores one internal state variable. Figure 12 shows an example of implementation in which the combinational circuit is made up of a read only memory, with a two-level addressing mode. The state variables thus only change at some particular instants: $0, T, 2T, \ldots, kT, \ldots,$ where T stands for the clock period.

As a matter of fact, the input variables must keep a constant value for a sufficiently long time interval, in order that the combinational circuit properly computes the next-state and output functions. If several machines are interconnected and concurrently work, it is necessary to provide buffering of the input variables. Figure 13a shows an implementation based on a two-phase clocking scheme. If output variables y are input variables of other machines, the whole circuit can be differently partitioned; the basic two-phase configurations is shown in Figure 13b. A programmable logic array implementation, with dynamic register cells, is described in Figure 13c. If the whole circuit consists of two interconnected automata, it is possible to save time and material by clocking them on opposite phases; the input and output register cells are then shared by both automata (Figure 13d).

3.2 Solution of the input–output equations

In the introductory Section 3.1, we implemented a finite state machine by means of D flip-flops (Figure 12). However, it is also possible to use other types of flip-flops. Let us define a flip-flop as a finite state machine, including two input variables A and B, one internal state variable Q, and two output variables Q and $\bar{Q}$. Its excitation equation is:

$$Q = F(A, B, q).$$

A machine, whose excitation and output equations are given by (2), can be realized by the circuit shown in Figure 14. Assume that we have p flip-flops satisfying the excitation equations:

$$Q_i = F_i(A_i, B_i, y_i), \qquad i = 0, 1, \ldots, p-1.$$

Assume, moreover, that there exist $2p$ Boolean functions $A_i(\mathbf{x}, \mathbf{y})$, $B_i(\mathbf{x}, \mathbf{y})$ of the $n+p$ variables $\mathbf{x}, \mathbf{y}$, satisfying the following equalities:

$$F_i(A_i(\mathbf{x}, \mathbf{y}), B_i(\mathbf{x}, \mathbf{y}), y_i) = Y_i(\mathbf{x}, \mathbf{y}), \qquad i = 0, 1, \ldots, p-1, \tag{10}$$

where F_i is the excitation equation of flip-flop number i. The equation (10), that we rewrite in the form:

$$F(A(\mathbf{x}, \mathbf{y}), B(\mathbf{x}, \mathbf{y}), y) = Y(\mathbf{x}, \mathbf{y}), \tag{11}$$

is called the input function of the flip-flop. The possibility of synthesizing a sequential synchronous system by means of flip-flops of this type reduces to

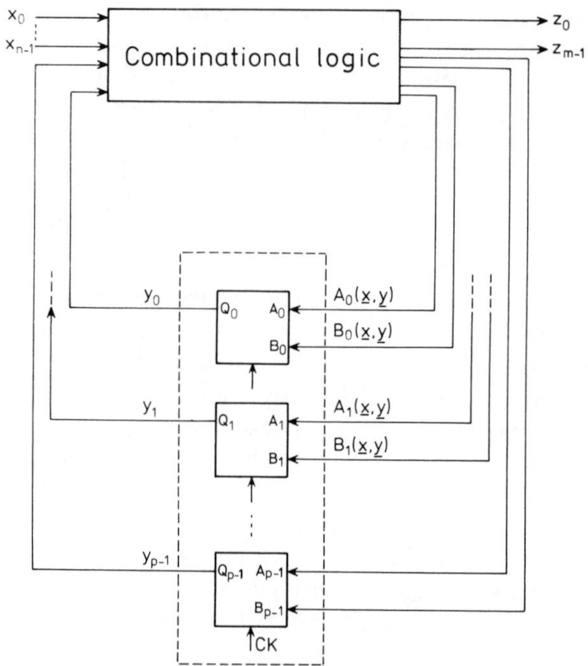

Figure 14. Sequential machine realization by means of clocked flip-flops

the possibility of solving the equation (11) whatever the function $Y(\mathbf{x}, \mathbf{y})$ might be.

We show in this section that, for the various types of flip-flops considered, it is not necessary to state precisely the function Y and that it is thus possible to give a general solution for (11) whatever the value of Y might be.

Remember first (see relation (7a)) that a D flip-flop is characterized by the input function:

$$Y(\mathbf{x}, \mathbf{y}) = D. \tag{12}$$

A Toggle flip-flop (or T flip-flop) has the input function:

$$Y(\mathbf{x}, \mathbf{y}) = T \oplus \mathrm{y} \tag{13}$$

Thus if $T = 0$ the flip-flop will remain in its present state, while if $T = 1$ the next-state of the flip-flop will be the complement of its present state.

The synthesis of synchronous systems by means of D flip-flops has been considered in the introductory Section 3.1.

A T flip-flop is obtained from a D flip-flop by considering the following input function (see also Figure 15a):

$$T \oplus \mathrm{y} = D. \tag{14}$$

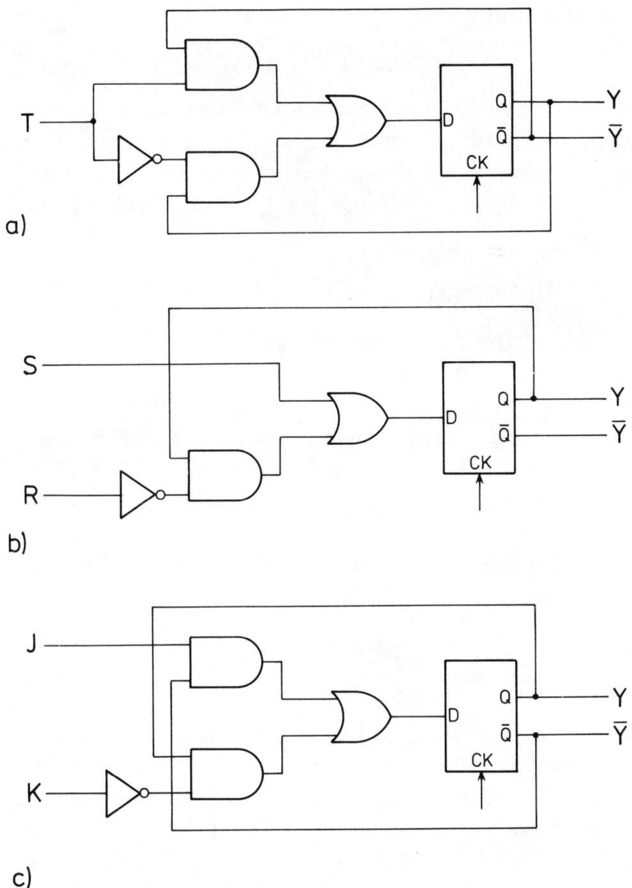

Figure 15. Synthesis of flip-flops

S-R and J-K flip-flops are characterized by the following excitation equations respectively:

$$Y = S \vee \bar{R}y, \tag{15}$$

$$Y = J\bar{y} \vee \bar{K}y. \tag{16}$$

The construction of SR and JK flip-flops by means of a D flip-flop is obtained from the following input functions respectively (see also Figures 15b and 15c).

$$S \vee \bar{R}y = D, \tag{17}$$

$$J\bar{y} \vee \bar{K}y = D. \tag{18}$$

The closed JK flip-flop is a two state machine which has two inputs J and K;

it has as input function:

$$Y(\mathbf{x}, \mathbf{y}) = \bar{J}\bar{y} \vee \bar{K}y. \tag{19}$$

The operation of the JK flip-flop is exactly the same as that of the SR flip-flop except when both inputs J and K are 1; then its next-state is defined and is equal to the complement of its present state. By identifying the input function (19) for $y = 0$ and $y = 1$ we obtain respectively:

$$\begin{aligned} &y = 0\colon J = Y(0); && K = \text{indeterminate}, \\ &y = 1\colon J = \text{indeterminate}; && K = \bar{Y}(1). \end{aligned} \tag{20}$$

The identification of the excitation function of a D flip-flop:

$$Y = D,$$

with the conditions (20) allows us to synthesize a D flip-flop by means of a JK flip-flop (see Figure 16):

$$J = D; \qquad K = \bar{D}. \tag{21}$$

The clocked SR flip-flop is described by its input function together with a constraint in the inputs:

$$\begin{aligned} Y(\mathbf{x}, \mathbf{y}) &= S \vee \bar{R}y, \\ SR &= 0. \end{aligned} \tag{22}$$

Besides the relations (22), the input–output behaviour of the SR flip-flop is described by the excitation table of Figure 7. From this table we deduce that S must take the value 1 if $q = 0$ and $Q = 1$; S may take the value 1 if $Q = 1$. Similarly R must take the value 1 if $q = 1$ and $Q = 0$; R may take the value 1 if $Q = 0$. This allows us to express S and r in terms of q and Q, i.e. adopting the notation (22) in terms of y and Y:

$$\begin{aligned} S &= \bar{y}Y \vee \mu_S Y, \\ R &= y\bar{Y} \vee \mu_R \bar{Y}, \end{aligned} \tag{23}$$

with μ_R, μ_S arbitrary Boolean functions.

Note that the relations (23) could also be obtained by solving (22) considered as system of two Boolean equations in the two unknowns, S and R (Hammer and Rudeanu, 1968); this system reduces to the unique

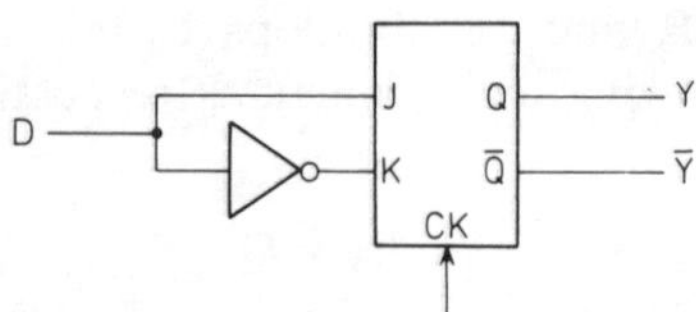

Figure 16. Synthesis of a D flip-flop

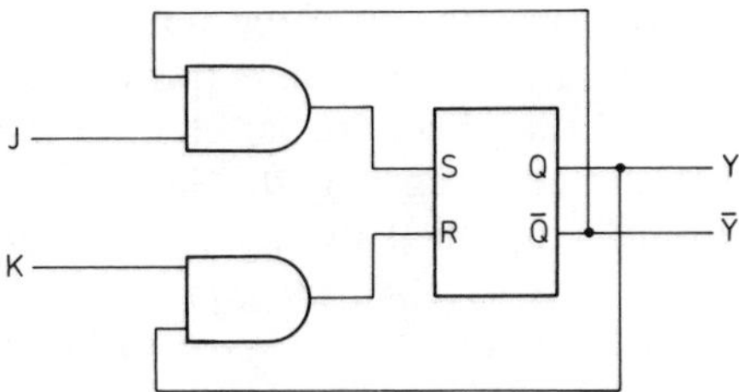

Figure 17. Synthesis of a JK flip-flop

equation:

$$[(S \vee \bar{R}y) \oplus Y] \vee RS = 0. \tag{24}$$

From (23) and (16) we obtain the realization of a JK flip-flop by means of an SR flip-flop (see also Figure 17):

$$\begin{aligned} S &= (\bar{y} \vee \mu_S)Y \\ &= (\bar{y} \vee \mu_S)(J\bar{y} \vee \bar{K}y) \\ &= J\bar{y} \text{ (taking } \mu_S = 0), \end{aligned} \tag{25a}$$

$$\begin{aligned} R &= (y \vee \mu_R)\bar{Y} \\ &= (y \vee \mu_R)(\bar{J}\bar{y} \vee Ky) \\ &= Ky \text{ (taking } \mu_R = 0). \end{aligned} \tag{25b}$$

3.3 Design procedure

The design procedure for synthesizing synchronous sequential networks to be described in this section can be used with any flip-flop. It is necessary that the flip-flop characteristic table or function will be known. The excitation table or equations of the network to be synthesized are used as explained in the preceding section.

The design procedure is summarized by a list of consecutive recommended steps as follows.

Step 1. The word description of the circuit behaviour is stated. This word specification usually assumes that the designer is familiar with digital logic terminology. It is necessary that the designer uses his intuition and experience in order to arrive at the correct interpretation of the circuit specifications because word descriptions may be incomplete. However, once such specification has been set down and the finite automaton model obtained, it is possible to make use of a formal procedure to design the network.

Example: *parity generator.* A parity bit is a scheme for detecting errors during transmission of binary information. A parity bit is an extra bit included with a binary message to make the number of 1's either odd or even. The message, including the parity bit, is transmitted and then checked at the receiving end for errors. An error is detected if the checked parity does not correspond to the one transmitted. The circuit that generates the parity bit in the transmitter is called the *parity generator*; the circuit that checks the parity in a receiver is called the *parity checker.*

As an example, consider a three-bit message to be transmitted with an even parity bit. The parity bit generator, placed in an initial state, receives three bit words which are the inputs to the circuit. For even parity, the P-bit is generated so as to make the total number of 1's even (including P). The three-bit message and the parity bit are transmitted to their destination, where they applied to a parity checker circuit. An error occurs during transmission if the parity of the four bits received is odd, since the binary information transmitted was originally even.

For the present example, the generation of a state table describing the input–output behaviour of the parity bit generator to be synthesized does not present any difficulty. We need seven internal states: the initial state 0 and the states 1_i, 2_i, and 3_i, $i = 0, 1$ (see Figure 18). The state j_i is memorized after the system received j input signals; if the number of 1 received is even $i = 0$, if it is odd $i = 1$. The corresponding next-state table is given in Figure 18.

Step 2. *Reduction of the number of states.* The number of states of sequential circuits may be reduced by classical state reduction methods quite similar to those used in the generation of minimal irredundant forms (Chapter II). The problem of state reduction may be cited as follows: given a finite automaton A is it possible to derive an automaton B having a fewer number of states than A and an identical input–output behaviour. This problem has been considered in Chapter V, Section 4.1; we verify that the table of Figure 18 is minimal.

	Input	
Present state	0	1
0	1_0; 0	1_1; 0
1_0	2_0; 0	2_1; 0
1_1	2_1; 0	2_0; 0
2_0	3_0; 0	3_1; 0
2_1	3_1; 0	3_0; 0
3_0	0; 0	0; 0
3_1	0; 0	0; 1

Figure 18. (Next state; output) table

Step 3. *Binary assignment of the internal states and obtention of the excitation equations.* It is now necessary to assign binary values to each state to replace the letter symbol used thus far in the stable state. Remember that the binary values of the states is immaterial as long as their sequence maintains the proper input–output relations. For this reason, any binary number assignment is satisfactory as long as each state is assigned a unique number. The complexity of the obtained combinational circuit depends on the chosen binary state assignment. Two examples of possible binary assignments for our working example are given below. The first assignment has been chosen arbitrarily; the second assignment has been chosen in order to render some of the excitation variables Y_j independent of some next state variables y_j. We verify that this last assignment leads to simpler excitation equations, and thus to a less complex combinatorial circuit.

Various systematic procedures have been suggested that lead to a particular binary assignment from the many available. The most common criterion is that the chosen assignments should result in a simple combinational circuit for the flip-flop inputs. Some techniques for dealing with the state assignment problem have been considered in Chapter V, Section 4.1. However, to date, there are no state assignment procedures that guarantee a minimal cost combinatorial circuit.

A. *First binary assignment.* The state q is coded with the three binary variables q: (y_2, y_1, y_0);

$$\begin{aligned}\text{states: } & 0 \to (0,0,0),\ 1_0 \to (0,0,1),\ 1_1 \to (0,1,0),\\ & 2_0 \to (0,1,1),\ 2_1 \to (1,0,0),\ 3_0 \to (1,0,1),\\ & 3_1 \to (1,1,0).\end{aligned}$$

From this assignment and from the table of Figure 18 we deduce the following excitation table.

$(y_2 y_1 y_0;\ z)$

			$x = 0$					$x = 1$				
y_2	y_1	y_0										
0	0	0	0	0	1	;	0	0	1	0	;	0
0	0	1	0	1	1	;	0	1	0	0	;	0
0	1	0	1	0	0	;	0	0	1	1	;	0
0	1	1	1	0	1	;	0	1	1	0	;	0
1	0	0	1	1	0	;	0	1	0	1	;	0
1	0	1	0	0	0	;	0	0	0	0	;	0
1	1	0	0	0	0	;	1	0	0	0	;	1
1	1	1	—	—	—		—	—	—	—		—

Figure 19. Excitation table

From Figure 19 we deduce the following excitation equations:

$$
\begin{aligned}
Y_2 &= y_1 y_0 \vee \bar{x}\bar{y}_2 y_1 \vee x\bar{y}_2 y_0 \vee y_2\bar{y}_1\bar{y}_0,\\
Y_2 &= \bar{x}\bar{y}_2\bar{y}_1 y_0 \vee \bar{x} y_2\bar{y}_1\bar{y}_0 \vee x\bar{y}_2\bar{y}_0 \vee x\bar{y}_2\bar{y}_1,\\
Y_0 &= \bar{x}\bar{y}_2\bar{y}_1 \vee \bar{x}\bar{y}_2 y_0 \vee x y_2\bar{y}_1\bar{y}_0 \vee x\bar{y}_2 y_1\bar{y}_0,\\
z &= y_2 y_1.
\end{aligned}
\tag{26}
$$

B. *Second binary assignment* The second type of binary assignment is oriented from the possibility of reducing the problem at hand to two parts:

(a) A counter with four states (0, 1, 2, 3) indicating the length of the word received.

(b) A parity detector with two states (0, 1) indicating the parity of the number of 1's received in the word under test; this parity detector is reset to the 0-state at the 4th clock pulse.

We shall choose accordingly the state variables y_2 and y_1 for the counter and the state variable y_0 for the parity detector. The new assignment will be:

$$
\begin{aligned}
\text{states:}\quad & 0 \to (0,0,0),\ 1_0 \to (0,1,0),\ 1_2 \to (0,1,1),\\
& 2_0 \to (1,0,0),\ 2_1 \to (1,0,1),\ 2_0 \to (1,1,0),\\
& 3_1 \to (1,1,1).
\end{aligned}
$$

From this assignment and from the table of Figure 18 we deduce the excitation table in Figure 20. From Figure 20 we deduce the following excitation equations:

$$
\begin{aligned}
Y_2 &= \bar{y}_2 y_1 \vee y_2\bar{y}_1,\\
Y_1 &= \bar{y}_1,\\
Y_0 &= x\bar{y}_1\bar{y}_0 \vee \bar{x}\bar{y}_2 y_0 \vee \bar{x}\bar{y}_1 y_0 \vee x\bar{y}_2\bar{y}_0,\\
z &= y_2 y_1 y_0.
\end{aligned}
\tag{27}
$$

Observe that Y_1 and Y_2 are independent of x and of y_0; moreover, Y_0 may

$(Y_2 Y_1 Y_0; z)$

			x									
y_2	y_1	y_0	0					1				
0	0	0	0	1	0	;	0	0	1	1	;	0
0	0	1	—	—	—	;	—	—	—	—	;	—
0	1	0	1	0	0	;	0	1	0	1	;	0
0	1	1	1	0	1	;	0	1	0	0	;	0
1	0	0	1	1	0	;	0	1	1	1	;	0
1	0	1	1	1	1	;	0	1	1	0	;	0
1	1	0	0	0	0	;	0	0	0	0	;	0
1	1	1	0	0	0	;	1	0	0	0	;	1

Figure 20. Excitation table

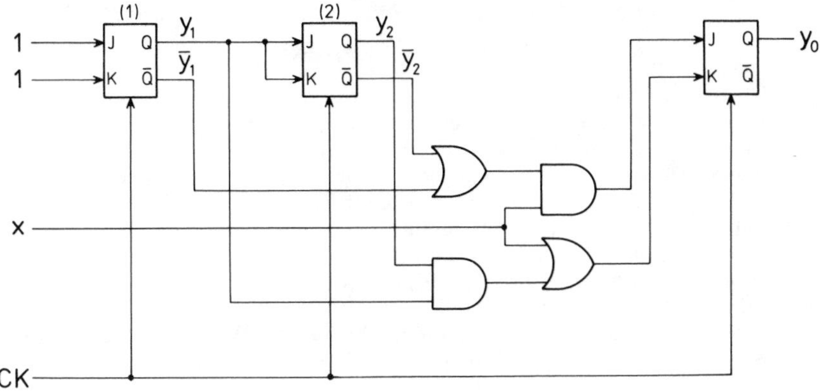

Figure 21. Synthesis of a parity bit generator

be written in the form;

$$Y_0 = (x\bar{y}_0 \vee \bar{x}y_0)(\bar{y}_2 \vee \bar{y}_1).$$

Step 4. Synthesis of the excitation equations by means of flip-flops. Our next objective is to implement by means of *J-K* flip-flops the input–output behaviour as described by the equations (27). From the method described in Section 3.2 we deduce the following expressions for the functions J_i, K_i, $0 \leqslant i \leqslant 2$, of the three *J–K* flip-flops respectively:

$$\begin{aligned} J_0 &= x(\bar{y}_2 \vee \bar{y}_1), & K_0 &= x \vee y_1 y_2, \\ J_1 &= 1, & K_1 &= y_1, \\ J_2 &= 1, & K_2 &= y_1. \end{aligned} \tag{28}$$

Equations (28) lead to the implementation of Figure 21.

BIBLIOGRAPHICAL REMARKS

Most of the material presented in this chapter may be found, at least partly, in classical books dealing with switching theory and logical design; see, e.g., Hill and Peterson (1974), Marcus (1967), McCluskey (1965), Kohavi (1970), and Mange (1978).

EXERCISES

E1. Perform the complete analysis of the clocked D flip-flop in Figure 6. Start with the computation of the total stable states and examine in turn all the possible transitions from these stable states on, assuming that the gate delays are all equal to τ. Show that the maximum delay between an input change and the complete system stabilization is 3τ. Design a D flip-flop completing its evolutions within maximum time 2τ (Maley & Earle, 1963)

E2. A P-K synchronous flip-flop is described by the excitation equation

$$Q^{+}=P\bar{K}\vee Q(P\vee\bar{K}).$$

Solve its input equation.

E3. Obtain from a constructor catalogue the scheme of a 54/7473 flip-flop and complete its dynamic analysis.

E4. A mode 16 counter may be described by the iteration while not stop do $C:=C+1$ (mod 16). Design such a counter by the above techniques. Use gates for the involved combinational circuit and D flip-flops.

E5. Consider the excitation table in Figure E1. Give various designs of the corresponding synchronous circuits. Modify
(i) the flip-flop type: D, RS, JK, T;
(ii) the combinational circuit type: NAND gates, NOR gates, ROM, and PLA.
It is interesting to use JK flip-flops in conjunction with a ROM.

y_1y_0 \ x_1x_0	00	01	10	11
00	00	01	10	11
01	01	10	11	00
10	10	11	00	01
11	11	00	01	00

Figure E1

E6. Design an autonomous sequential system which, placed in an appropriate starting state, generates the periodic sequence

$$01011011 \quad 01011011 \quad 0101\ldots.$$

Chapter VII
Synchronous sequential components

1 REGISTERS AND MEMORIES

1.1 Registers: generalities

The term *register* designates a synchronous system, made up of an ordered set of flip-flops, capable of storing a binary vector, and of submitting the vector to functional transformations, according to external control signals. This rather general definition is fundamental: a binary vector most often represents encoded alphanumeric quantities; hence, registers are basic building blocks of systems which process digital information.

The external behaviour of a register R, whose contents is denoted $[R]$, can be described in a transfer language:

$$[R] := f([R_1], \ldots, [R_p]).$$

We call *value transfer* or *load operation* a transfer

$$[R] := D,$$

which replaces the contents of register R by a constant numerical value D.

Various types of registers will be studied in the following sections.

1.2 Memories

As registers, *memories* are circuits intended to store information. Most often they are structured, and contain N words of M bits. They thus appear (at a functional level) as the juxtaposition of N registers. Some additional control circuits are provided in order to select a particular word.

Let us first state the main criteria which are used in order to classify memories.

(1) *Type of access.* In a *random access memory*, the access time to a given location does not depend on the chosen particular location. In a *sequentially accessed memory*, words are accessed in sequence, according to a previously fixed order; hence, the access time depends on the particular chosen location.

(2) *Writing possibility.* In a *writable memory* it is possible both to read

data and to write data, i.e. to modify the memory contents by recording new information. In a *read only memory*, it is only possible to read data. The memory contents are established by the manufacturer or user, and generally cannot be altered by the user.

(3) *Volatility*. In a *volatile memory*, the stored data are lost once the power supply is switched off. A *non-volatible memory* is a permanent storage device.

(4) *Static and dynamic memories. Static memories* are based on the use of bistable multivibrators (flip-flops). *Dynamic memories* use the storing capability of a capacitor, namely the input capacitance of a MOS inverter. As a matter of fact, dynamic memories must be periodically *refreshed* in order to compensate for leakage currents; hence they are volatile. We may extend the definition of static memory and designate by this term a system which does not require refreshing cycles.

We now describe the types of memories which are most frequently encountered in practice.

(1) *Random access memories* (RAM). This term always refers to writable random access memories. There exist both static and dynamic RAMs. Most often they are volatile.

(2) *Sequentially accessed memories.* Magnetic bubble memories and magnetic tapes are examples of writable, non-volatile, sequentially accessed memories. Punched paper tape is an example of read only, non-volatile, sequentially accessed memory. *Charge coupled devices* (see Section 2.1) are used to realize writable, volatile, sequentially accessed memories.

(3) *Semirandom access memories.* Magnetic disks and drums contain several independent rotating tracks. If each track has its own read–write head, they can be accessed randomly, although access within each track is sequential. They are writable and non-volatile. Such systems are called *semirandom* or *direct access* memories.

(4) *Read only memories* (ROM). This term always refers to random access, non-volatile, read only memories. *Programmable read only memories* (PROM) contain fusible links and can be programmed by the user. The basic cell of an *optically erasable programmable read only memory* (EPROM) is a MOSFET including both a control gate and a floating gate; it can be programmed by the user, and erased by exposing the cell to ultraviolet radiation. *Electrically alterable read only memories* use the metal–nitride–oxide–semiconductor (MNOS) structure; they can be programmed and altered by the user, without the necessity of erasing the entire array. EPROMs and EAROMs are sometimes called *read mostly memories* instead of read only memories.

Figure 1 shows a writable memory storing one n bit word, i.e. a n bit register. It includes n D flip-flops. The loading is performed under the control of input W (Write).

RAMs store several words under the control of address variables. Figures 2 and 3 show two examples of organization. The memory of Figure 2 stores

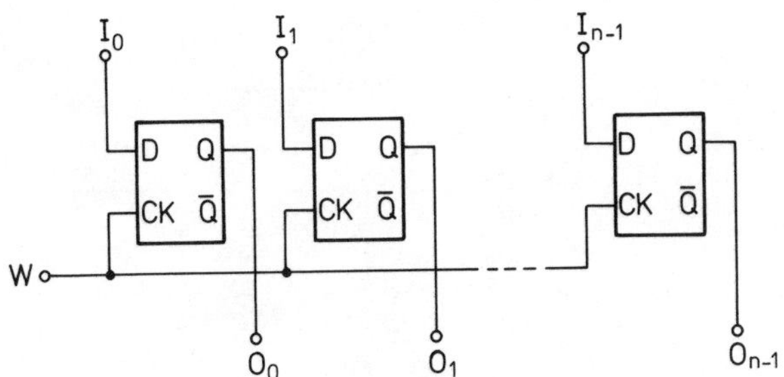

Figure 1. n bit register

2^n m-bit words; it includes an address decoder, $m \cdot 2^n$ flip-flops, m *input/output* (I/O) circuits. These I/O circuits are controlled by a *read/write* (R/W) variable: it indicates whether the present operation is a reading or a writing. The memory of Figure 3 stores 2^{n+m} one-bit words. It includes 2^{n+m} flip-flops, two address decoders, and one I/O circuit.

RAMs are widely used in practice. They are a basic building block of many digital systems and thus deserve special attention. Technological improvements lead to the diminution of cost, power consumption, area, access time, and to the achievement of ever increasing memory sizes. These technological matters being beyond the scope of this book, we quote two

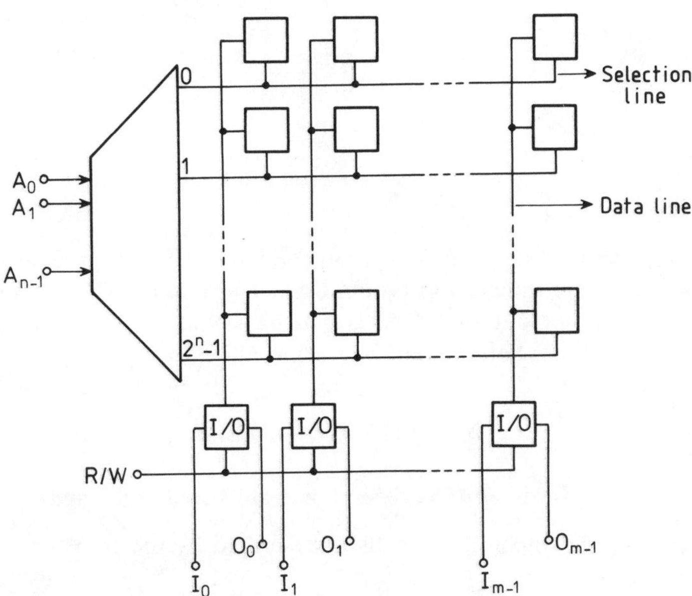

Figure 2. 2^n m-bit word RAM

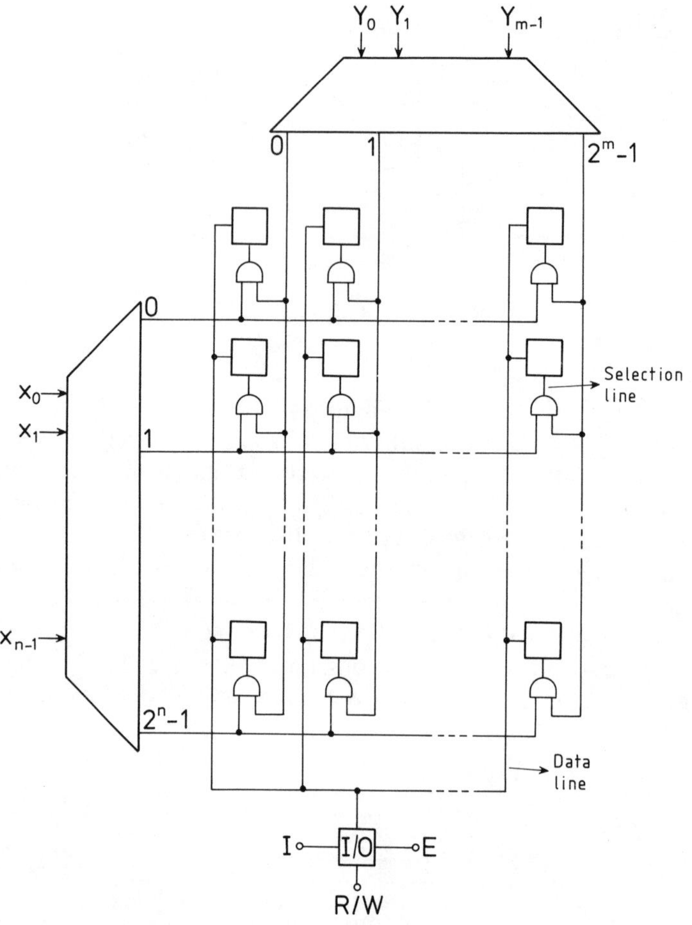

Figure 3. 2^{n+m} one-bit word RAM

recent numbers of the *IEEE Trans.* on solid-state circuits: the special issue on very large scale integration (volume SC-15, number 5, October 1980).

Let us also point out that RAMs can be asynchronous or synchronous with respect to the whole system operation.

2 SHIFT REGISTERS

2.1 Basic shift registers: left shift and right shift

A *left shift register* including n cells is described by the transfer:

$$[R] := 2[R] + x_0 \pmod{2^n}. \tag{1}$$

Let us denote by y_i the output of cell i. Equation (1) is equivalent to the n

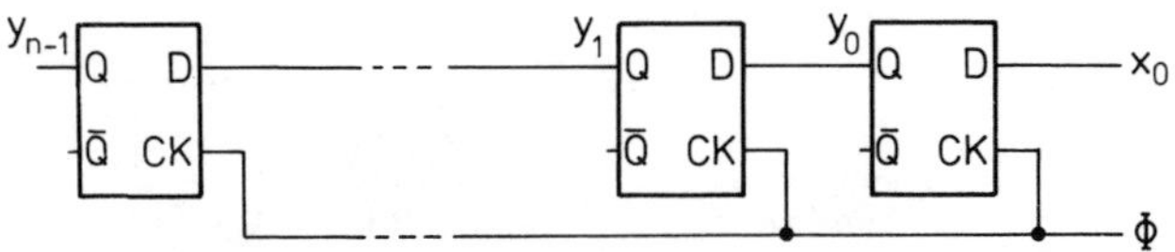

Figure 4. Static shift register

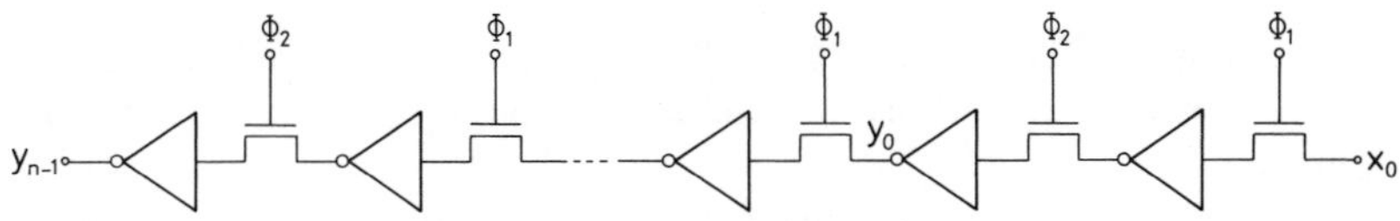

Figure 5. Dynamic shift register

excitation equations:

$$\begin{aligned} y_0 &:= x_0, \\ y_i &:= y_{i-1}, \quad (i = 1, 2, \ldots, n-1). \end{aligned} \tag{2}$$

The synthesis of a left shift register by means of D flip-flops is shown in Figure 4. It is an example of *static shift register.* Figure 5 shows a two-phase *dynamic shift register.* Another type of MOS dynamic shift register is based on the principle of the *charge coupled device* (CCD). It is an extension of the MOSFET, consisting of a source and a drain separated by a long row of gates on which are applied clock signals. Charge packets injected by the source, and successively trapped under every gate, eventually reach the drain. These devices have a low power distribution and a high packing density. They are also used for the processing of sampled analog signals: filters, delay lines, image sensors, time division multiplexers (Séquin and Tompsett, 1975).

Similarly, a *right shift register* is defined by

$$[R] := \lfloor [R]/2 \rfloor + x_{n-1} 2^{n-1}. \tag{3}$$

2.2 Variants

Variants of shift registers are numerous. They concern:

(1) the shifting possibilities: unidirectional or bidirectional shift;
(2) the loading possibilities: serial or parallel loading, and
(3) the reading possibilities: serial or parallel reading.

The ability to load and read serially is already included in the basic circuits of Section 2.1. The parallel reading ability mainly depends on the number of cells and on the number of pins available on the package. The possibility of loading in parallel also implies a sufficient number of pins, along with additional logic.

S_1	S_0	function	transfer
0	0	inhibit	$[R]:=[R]$
0	1	right-shift	$[R]:=\lfloor[R]/2\rfloor+x_{n-1}2^{n-1}$
1	0	left-shift	$[R]:=2[R]+x_0(\text{mod } 2^n)$
1	1	load	$[R]:=(A, B, \ldots, H)$

Figure 6. Definition of the 74198 circuit

As an example, TTL circuit 74198 has four possible behaviours, depending on the values of two control variables S_0 and S_1, according to the table of Figure 6.

The logic diagram of the circuit is shown in Figure 7. It suggests several comments.

(1) All operations, loading included, are synchronous. Nevertheless, there exist registers in which the loading operations are asynchronous: this is easily performed if flip-flops, with asynchronous set and reset inputs, are used.

(2) Disabling of the register is realized by inhibiting the clock signal. It is obviously possible to design a circuit in which the clock is never inhibited. Suitable signals must be applied on inputs S and R of all flip-flops, in any one of the four possible working modes.

(3) Formally speaking, the behaviour of the registers is described by the next equation:

$$[R]:=\bar{S}_1\bar{S}_0[R]\vee\bar{S}_1S_0(\lfloor[R]/2\rfloor+x_{n-1}2^{n-1})$$
$$\vee S_1S_0[\![2[R]+x_0]\!]_{2^n}\wedge S_1S_0 \quad \text{(DATA)}$$

(4) Shift registers are easily *expendable devices*: it is possible to build a $k\cdot n$-bit shift register by cascading k copies of an n-bit shift register. Two copies of the register shown in Figure 7, with indices 0 and 1, respectively,

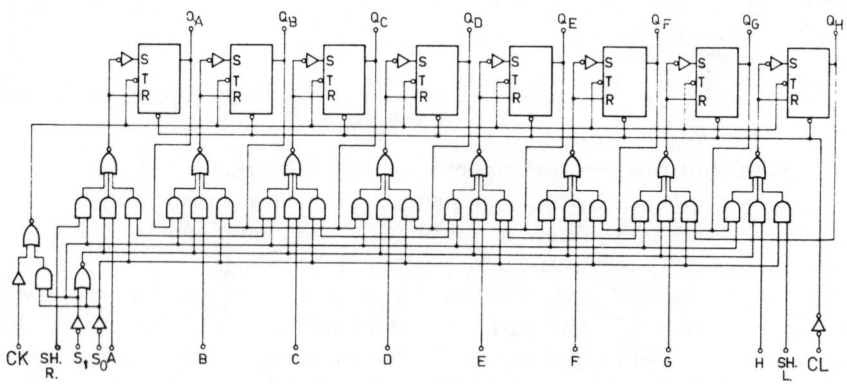

Figure 7. Block diagram of the 74198 circuit

are connected according to the following rules:

$$\text{CLOCK}_0 = \text{CLOCK}_1; \quad S1_0 = S1_1; \quad S0_0 = S0_1;$$
$$\text{CLEAR}_0 = \text{CLEAR}_1; \quad (\text{SHIFT RIGHT})_1 = QH_0;$$
$$(\text{SHIFT LEFT})_0 = QA_1.$$

2.3 Applications

Practical applications of shift registers are numerous. Let us first observe that these registers are particularly convenient for storing data and results, in the sequential implementation of Mealy automata. This fact is illustrated by Figure 8, which shows the logic diagram of a *serial accumulator*. The circuit includes two shift registers, one D flip-flop, and one full-adder. At time $t = 0$, numbers a and b are loaded into the registers (digits between parentheses). At time $t = n + 1$, the upper register contains the sum $s = a + b$ (digits between square brackets).

Figure 9 shows another application of shift registers: the *ring counter*. The initial condition is

$$(Q_0, Q_1, \ldots, Q_{n-1}) = (1, 0, \ldots, 0).$$

Every output generates a '1' every n periods of the clock signal. Hence, it divides by n the clock frequency. This component is used in counting, scaling, and addressing circuits. For instance, its outputs may serve to distinguish time slots in a time multiplexing system.

Two ring counters can be interconnected, as shown in Figure 10; they form a divide by n^2 circuit. Notice that the clock input of the second counter is connected to an output of the first counter. Hence, it is no longer a strictly synchronous circuit, and precautions have to be taken with respect to its dynamic behaviour.

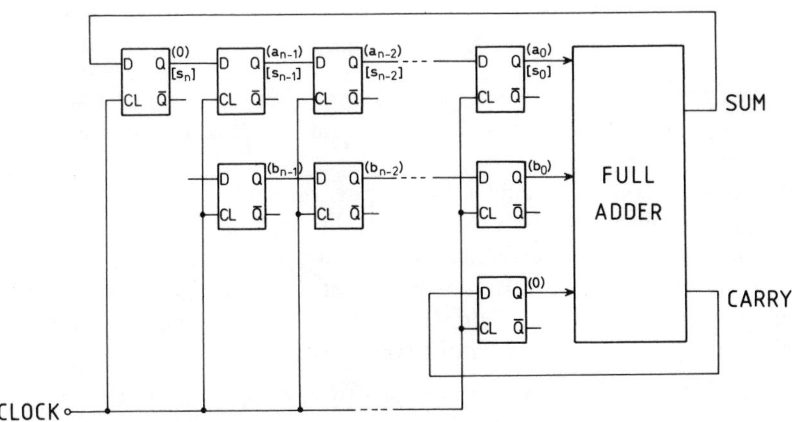

Figure 8. Serial accumulator

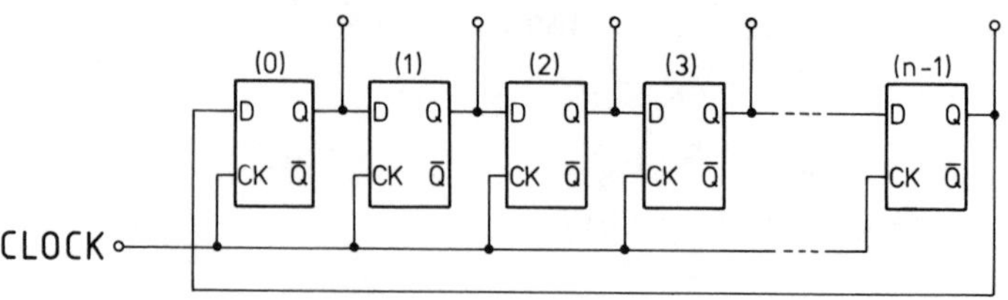

Figure 9. Ring counter

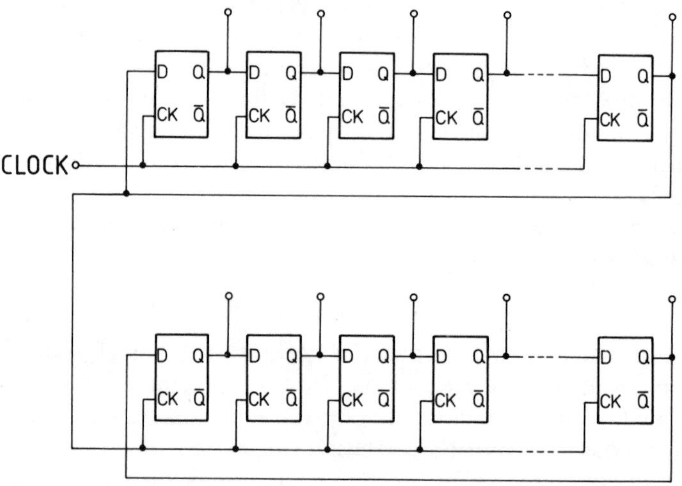

Figure 10. Cascade connection of two ring counters

3 COUNTERS

3.1 Definitions

Counters are used in a wide variety of computing applications, in scientific instruments, industrial process controls, communication equipments, and in many other areas.

An n-stage counter includes one input, on which is applied a series of pulses, and n outputs. At every moment, the n outputs represent in encoded form the number of pulses which have been applied on the input wire since initial time $t=0$ when the counter was reset. Of course, there is a maximum number m of pulses that the circuit can count. After m pulses, the system is reset; hence, the successive output states represent

$$0, 1, 2, \ldots, m, 0, 1, 2, \ldots.$$

n output wires allow the encoding of, at most, 2^n different numbers, and thus

$$m \leqslant 2^n - 1.$$

A divide by d circuit includes one input and one output. If a sequence of pulses is applied on its input, then its output generates one pulse every d input pulses. Hence, it can serve as a *frequency divider.* The circuit necessarily contains at least d different internal states. There is a narrow link between counters and dividers, as proven by the two following remarks.

(1) A counter which successively generates numbers $0, 1, \ldots, m$, can be realized by a divide by $(m+1)$ circuit, if all internal state variables of the divider are available; a combinational circuit can transform every divider state into the corresponding counter state.

(2) Conversely, a divide by d circuit can be realized by means of a counter which successively goes over states $0, 1, \ldots, d-1, 0, 1, \ldots$. It remains to detect one particular counter state by means of a combinational circuit.

Counters and dividers thus appear as one input devices. Furthermore, they are driven by pulses applied on their input; in other words, according to the input signal level, they lie in two different working phases: memorization phase (input 0) or evolution phase (input 1); in the last case, there is only one internal state transition corresponding to the counting of one input pulse. The similarity of this input signal with the clock signal of a synchronous system leads to the next definition: a *synchronous counter* is an autonomous synchronous system, that is a system which includes only one input letter, and is entirely defined by a mapping M from Q to Q, where Q is the set of internal states of the counter or divider. The only input letter is normally implied. The graph describing the relation may contain several connected components which are called *generalized cycles*: as an example, see Figure 11.

Previous observations lead to a classification of counters based on characteristics of their transition mapping M. In a *permutation counter,* M is a permutation; the generalized cycles thus are cycles. The transition mapping of a *circular counter* is a circular permutation. If the transition graph of a counter only contains one generalized cycle, it is a *connected counter.*

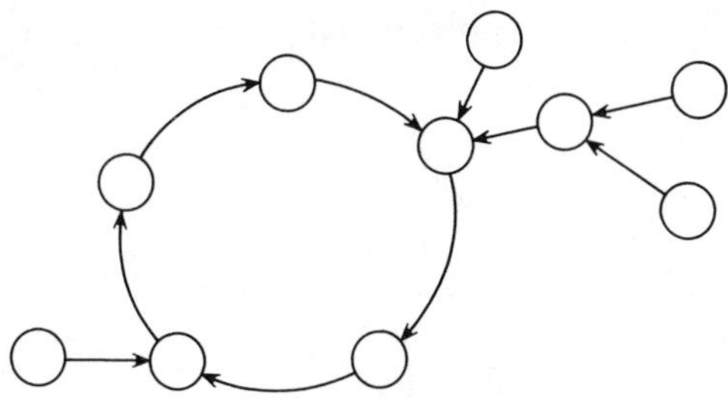

Figure 11. Generalized cycle

The useful effect of a counter or divider relies on the existence, in its transition graph, of a cycle whose length is imposed: $L = m+1$ for a counter and $L = d$ for a divider. This cycle is called the *principal cycle* of the counter or divider, while the states which do not belong to the principal cycle are called *peripheral states.* An n stage counter is *irredundant* if the length L of its principal cycle satisfies

$$2^{n-1} < L \leqslant 2^n.$$

3.2 Circular counters and variants

Circular counters, whose transition graph reduces to its principal cycle whose length is 2^n, play an important role: indeed, their implementation is particularly simple; furthermore, they can easily be modified in order to obtain any desirable cycle length. This last point will be emphasized in the study of synchronous dividers.

The simplest circular counter successively counts up numbers $0, 1, \ldots, 2^n - 1$, expressed in the binary numeration system. This counter C is defined by the functional transfer

$$[C] := [C] + 1 \quad (\text{mod } 2^n). \tag{5}$$

Let us describe two implementations.

(1) If $a_{n-1} \ldots a_1 a_0$ and $b_{n-1} \ldots b_1 b_0$ stand for $[C]$ and $[C]+1$, in binary numeration, then

$$b_0 \neq a_0,$$
$$b_i \neq a_i \quad \text{iff} \quad a_{i-1} = a_{i-2} = \ldots = a_0 = 1.$$

Hence, transfer (5) is equivalent to the next excitation equations:

$$\begin{aligned} y_0 &:= y_0 \oplus 1, \\ y_i &:= y_i \oplus \bigwedge_{j=0}^{i-1} y_j, \qquad i = 1, 2, \ldots, n-1. \end{aligned} \tag{6}$$

These equations suggest the use of T flip-flops and lead to the logic diagram of Figure 12.

(2) Another method consists in associating D flip-flops, which store the present value $[C]$, and a combinational circuit, which computes the next

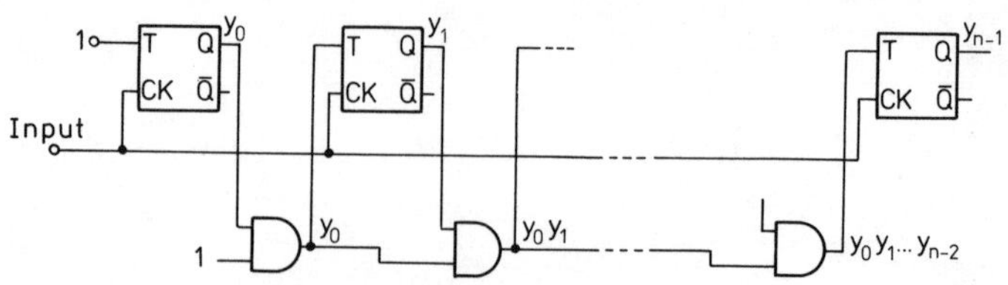

Figure 12. Circular counter

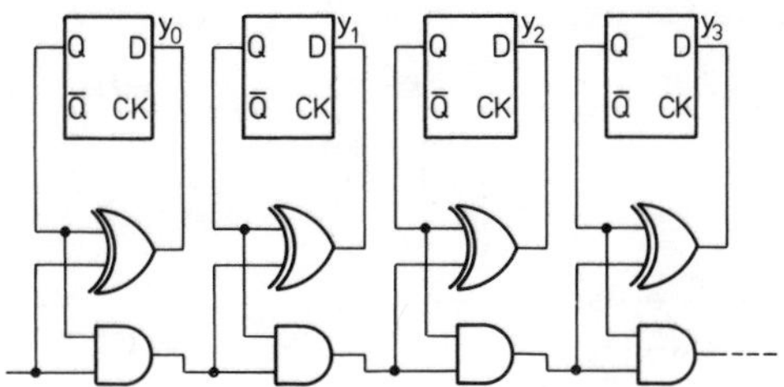

Figure 13. Circular counter

value $[C]+1$. The combinational circuit is made up of the cascade connection of half adders. The resulting logic diagram is shown in Figure 13. Notice that the interconnection of a D flip-flop and of an exclusive OR gate yields a T flip-flop. This proves the equivalence of the two obtained circuits.

As shift registers, counters have several variants. Let us quote three of them.

(1) A *down counter C* is defined by the transfer

$$[C]:=[C]-1 \quad (\text{mod } 2^n), \tag{7}$$

or by the excitation equations

$$\begin{aligned} y_0 &:= y_0 \oplus 1, \\ y_i &:= y_i \oplus \bigwedge_{j=0}^{i-1} \bar{y}_j, \qquad i=1,2,\ldots,n-1. \end{aligned} \tag{8}$$

(2) An *up–down counter* is able to perform both the forward and the backward counting, according to the value of a control variable x. It realizes the next transfer:

$$[C]:=\bar{x}([C]+1)\vee x([C]-1) \quad (\text{mod } 2^n). \tag{9}$$

The basic cell of an up–down counter is shown in Figure 14.

(3) *A programmable counter* is able, under the control of variale L, either to increment its contents, or to replace its contents by an external datum D.

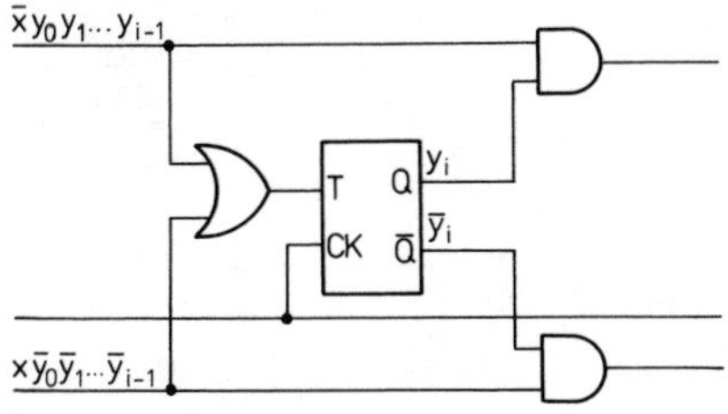

Figure 14. Basic cell of an up–down counter

$y_i x_i$ \ LP_i	00	01	11	10
00	0	1	0	0
01	0	1	1	1
11	1	0	1	1
10	1	0	0	0

(a)

$y_i x_i$ \ LP_i	00	01	11	10
00	0	1	0	0
01	0	1	1	1
11	—	—	—	—
10	—	—	—	—

(b)

$y_i x_i$ \ LP_i	00	01	11	10
00	—	—	—	—
01	—	—	—	—
11	0	1	0	0
10	0	1	1	1

(c)

Figure 15. Synthesis of the basic cell by means of JK flip-flops

It realizes the transfer

$$[C] := \bar{L}([C]+1) \vee L \cdot D \quad (\text{mod } 2^n). \tag{10}$$

Let us use the next notation:

$$P_i = \bigwedge_{j=0}^{i-1} y_j.$$

The excitation equation of cell number i is

$$y_i := \bar{L}(y_i \oplus P_i) \vee L \cdot D_i, \tag{11}$$

where D_i is bit number i of D. Figure 15 describes the computations relative to the synthesis of the basic cell by means of J–K flip-flops. Figure 16 shows the corresponding logic diagram.

Figure 17 shows the logic diagram of TTL circuit 74191 which has the three working modes described above. This counter includes four master–slave flip-flops whose outputs are triggered on a positive ($0 \rightarrow 1$) transition of the clock signal. Hence, level changes at the input should be made only

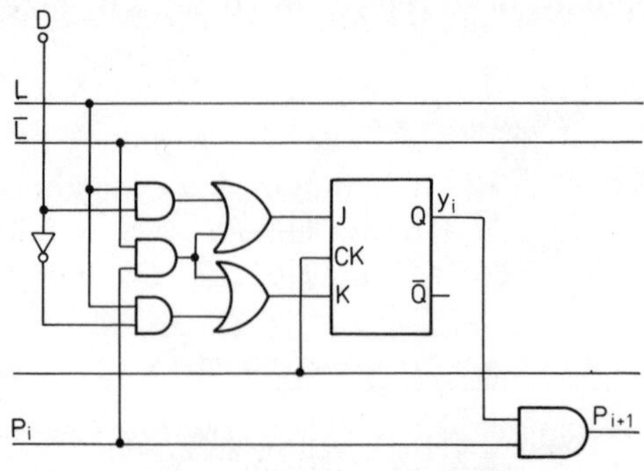

Figure 16. Basic cell of a programmable counter

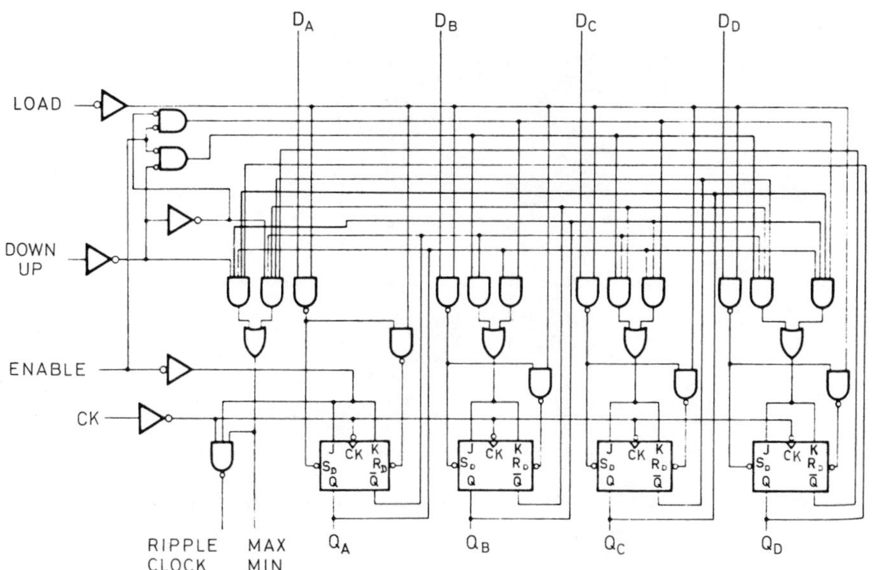

Figure 17. Synchronous binary up–down counter

ENABLE	DOWN/UP	OPERATION
0	0	$[C]:=[C]+1 \pmod{2^n}$
0	1	$[C]:=[C]-1 \pmod{2^n}$
1	—	$[C]:=[C]$

Figure 18. Synchronous operations of circuit 74191

when the clock signal is high (1), and new data are stored by the master when the clock signal is low (0). The counter performs three synchronous operations (LOAD = 1) and one asynchronous load operation (LOAD = 0). The three synchronous operations are controlled by inputs ENABLE and DOWN/UP, according to the table of Figure 18.

Furthermore, the circuit has two additional outputs MAX/MIN and RIPPLE CLOCK:

$$\text{MAX/MIN} = \overline{(\text{DOWN/UP})}\,Q_A Q_B Q_C Q_D \vee (\text{DOWN/UP})\,\bar{Q}_A \bar{Q}_B \bar{Q}_C \bar{Q}_D,$$

$$\text{RIPPLE CLOCK} = \overline{\overline{\text{CLOCK}} \cdot \overline{\text{ENABLE}} \cdot (\text{MAX/MIN})}.$$

The MAX/MIN output produces a pulse when the counter overflows or underflows. The RIPPLE CLOCK output is a synchronous version of the MAX/MIN output. The counters can be cascaded by feeding the RIPPLE CLOCK output to the ENABLE input of the succeeding counter.

3.3 Decimal counters

In some applications, among others when counting long series of pulses, it can be better directly to perform a decimal counting rather than a binary counting, in order to avoid binary to decimal decoding.

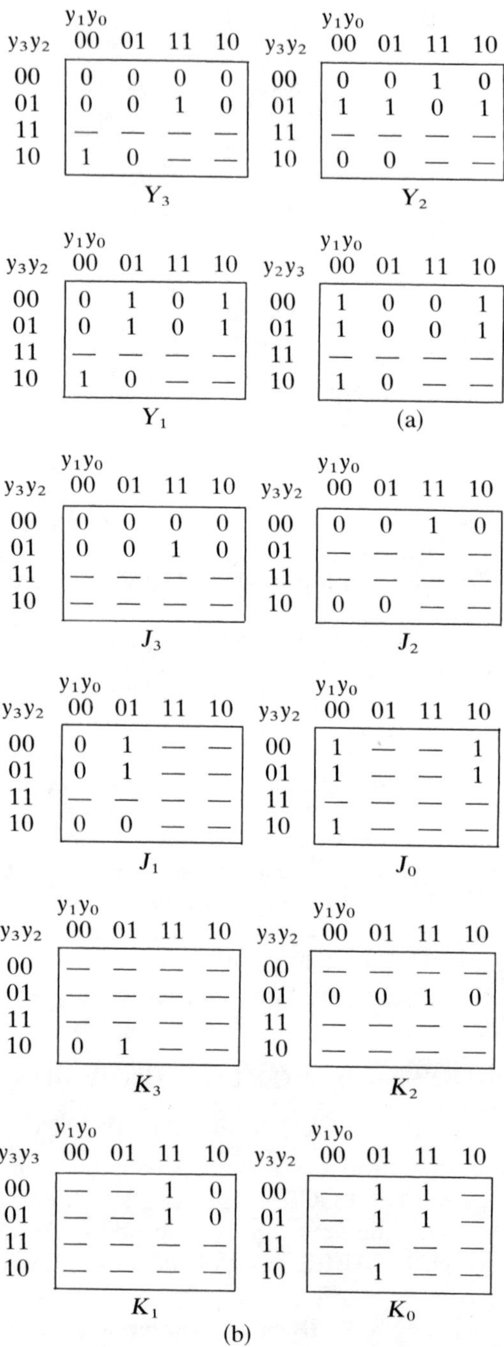

Figure 19. Excitation equations of a decimal counter

As an example, let us synthesize a circuit which counts in a binary coded decimal (BCD) code. The circuit specifications are gathered in the Karnaugh maps of Figure 19a. The input functions J_i and K_i (see Chapter VI) are given in Figure 19b. It thus remains to synthesize eight functions of four variables. Let us mention some solutions. Notice that they are equivalent to the number of gates:

$$J_3 = y_0 y_1 y_2,$$
$$K_3 = y_0,$$
$$J_2 = y_0 y_1,$$
$$K_2 = y_0 y_1 \quad \text{or} \quad y_0 y_1 y_2,$$
$$J_1 = y_0 \bar{y}_3,$$
$$K_1 = y_0 \quad \text{or} \quad y_0 y_1 \quad \text{or} \quad y_0 \bar{y}_3,$$
$$J_0 = 1 \quad \text{or} \quad \bar{y}_0,$$
$$K_0 = 1 \quad \text{or} \quad y_0.$$

The first solution leads to the logic diagram of Figure 20a.

The *a posteriori* analysis of the circuit allows its behaviour to be specified when lying in peripheral states. It leads to the connected transition graph of Figure 20b.

Connectivity is often considered as a desirable property when, for instance in industrial environment, external disturbances are able to put the counter in an undesirable peripheral state. The connectivity then assures

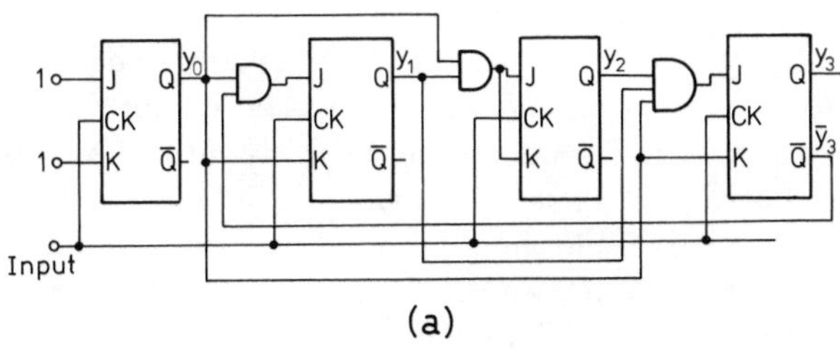

(a)

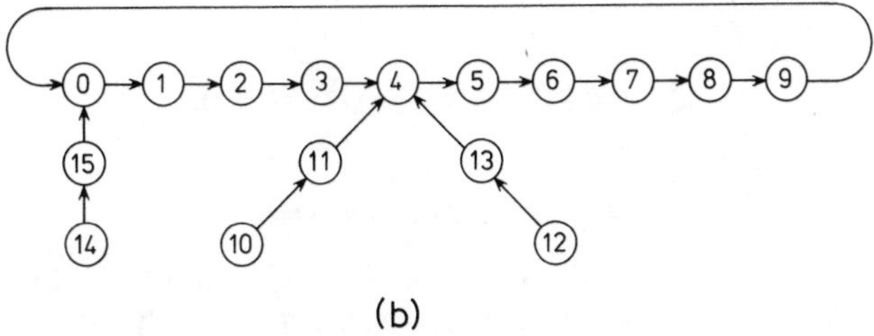

(b)

Figure 20. Decimal counter

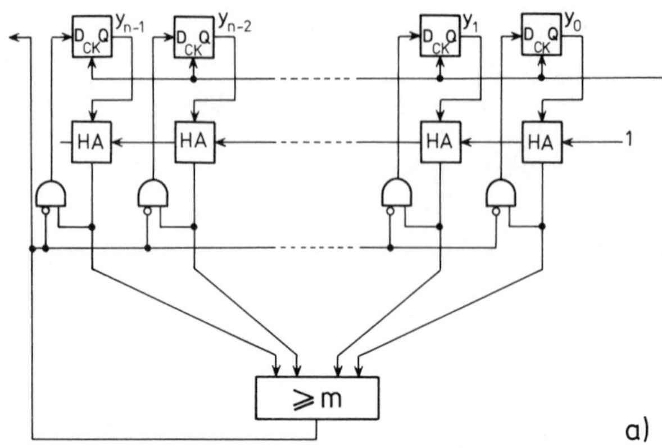

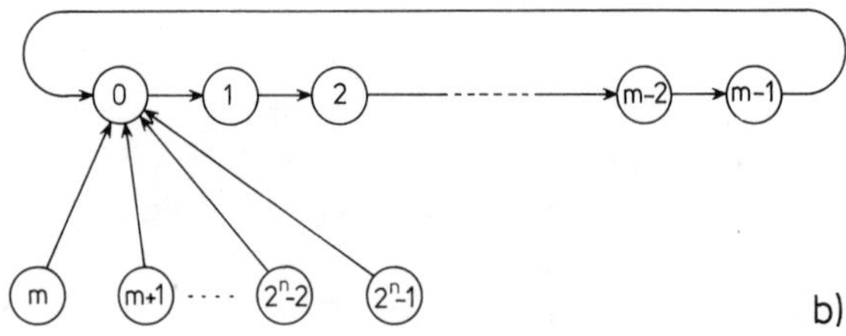

Figure 21. Modulo *m* counter

that the counter will return to the principal cycle, without any external action.

As before, we may develop *decimal down counters, decimal up–down counters,* and *decimal programmable counters.* The commercial versions of these circuits are always equipped with an enable control input and a ripple output allowing the counter to be extended.

More generally, we may build *modulo m counters,* with $m < 2^n$. The circuit of Figure 21a contains a comparator which generates an output value 1 if the next state of the counter is greater than or equal to *m*. The counting is performed by a series of half adders (HA). The corresponding transition graph is shown in Figure 21b.

4 FREQUENCY DIVIDERS

4.1 General discussion

We are now interested in the synthesis of irredundant dividers whose principal cycle has imposed length L ($2^{n-1} < L \leq 2^n$). The optimization of such systems is much more difficult than for counters since, as a rule, the designer has all the degrees of freedom associated with the encoding of the internal states. So it has been proved that all cycle lengths from two to thirty-one, except thirteen and sixteen, can be implemented by an irredundant circuit which includes only JK flip-flops, without additional gates (Manning, 1972). This type of argument indicates that it must always be possible to synthesize a divider including n flip-flops and a few logic gates. In fact, the cost optimization problem of dividers is far from being solved. Not much effort is devoted to this problem: this is due to the existence of simple and efficient design methods, yielding sufficiently cheap circuits.

4.2 Perturbation technique

The principal is illustrated by Figure 22: for transition

$$i \rightarrow i+1$$

we substitute transition

$$i \rightarrow j.$$

The length of the principal cycle then becomes

$$\text{for} \quad 0 \leqslant j \leqslant i: \qquad L = i - j + 1,$$
$$\text{for} \quad i < j \leqslant 2^n - 1: \qquad L = 2^n + i - j + 1.$$

Hence, in order to realize a cycle length $L \neq 0$, we choose

$$j = [\![i - L + 1]\!]_{2^n}. \tag{12}$$

If $i = i_{n-1} \ldots i_1 i_0$ in binary number system, let us denote by m_i the minterm

$$y_0^{(i_0)} y_1^{(i_1)} \ldots y_{n-1}^{(i_{n-1})}$$

which takes value 1 when the system lies in state i. On the other hand,

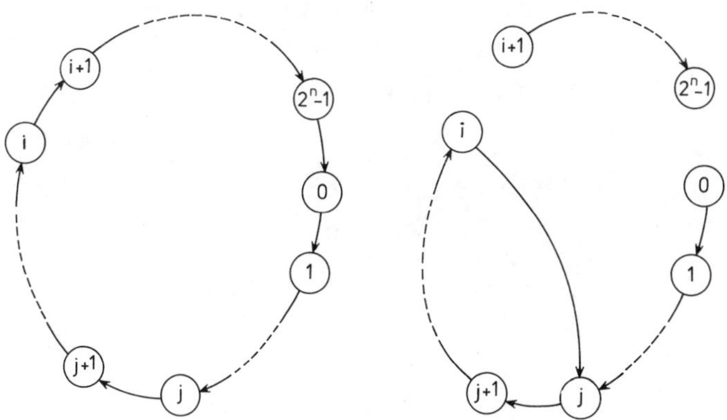

Figure 22. Principle of the perturbation technique

vector $\mathbf{p}_i=(p_{i0}, p_{i1}, \ldots, p_{in-1})$ is defined as follows: $p_{i\ell}=1$ if order ℓ bits of numbers $i+1$ and j, expressed in binary number system, are different; $p_{i\ell}=0$ in the contrary case. Let us briefly explain the working principle.

(1) So long as the system does not lie in state i, that is, so long as $m_i=0$, then it works as a circular counter.

(2) When the system gets to state i, that is, when $m_i=1$, the next state must be j instead of $i+1$; hence, it is necessary to modify the working of the order ℓT flip-flop if the corresponding bits of j and $i+1$ are different, that is, if $p_{i\ell}=1$.

The expected working is described by the next excitation equations:

$$\begin{aligned} y_0 &:= y_0 \oplus 1 \oplus m_i p_{i0}, \\ y_\ell &:= y_\ell \oplus y_0 y_1 \ldots y_{\ell-1} \oplus m_i p_{i\ell}, \qquad i=1,2,\ldots,n-1. \end{aligned} \tag{13}$$

The corresponding basic cell is shown in Figure 23.

The method can be applied in order to synthesize decimal counters, using either the natural binary code, or the excess-three code; both derive from the binary counter by an appropriate jump.

In the problem we are dealing with, namely the synthesis of frequency dividers, the only important parameter is the cycle length. Hence, we may choose the starting state i in order to simplify the input equations of the T flip-flops. Let us choose $i=2^n-1$, i.e. $11\ldots1$ in binary number system. Thus

$$m_i = y_0 y_1 \ldots y_{n-1},$$

$$y_0 y_1 \ldots y_{\ell-1} \oplus m_i = y_0 y_1 \ldots y_{\ell-1}(\bar{y}_0 \vee \bar{y}_1 \vee \ldots \vee \bar{y}_{n-1}),$$

and equations (13) become:

$$\begin{aligned} y_0 &:= y_0 \oplus (\bar{p}_{i0} \vee p_{i0}(\bar{y}_0 \vee \ldots \vee \bar{y}_{n-1})), \\ y_\ell &:= y_\ell \oplus y_0 y_1 \ldots y_{\ell-1}(\bar{p}_{i\ell} \vee p_{i\ell}(\bar{y}_0 \vee \ldots \vee \bar{y}_{n-1})), \quad i=1,2,\ldots,n-1. \end{aligned} \tag{14}$$

As an example, let us synthesize a divide by 60 circuit:

$$i=(63)_{\text{radix } 10}=(1\ 1\ 1\ 1\ 1\ 1)_{\text{radius } 2},$$

$$j=(4)_{10}=(0\ 0\ 0\ 1\ 0\ 0)_2,$$

$$i+1=(64)_{10}=(1\ 0\ 0\ 0\ 0\ 0\ 0)_2,$$

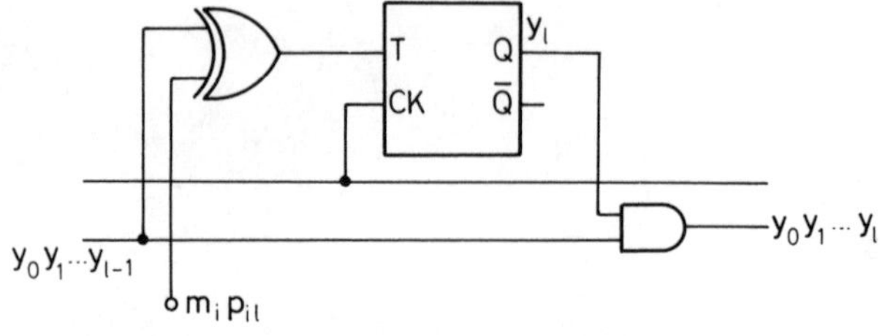

Figure 23. Basic cell of a divider using the perturbation technique

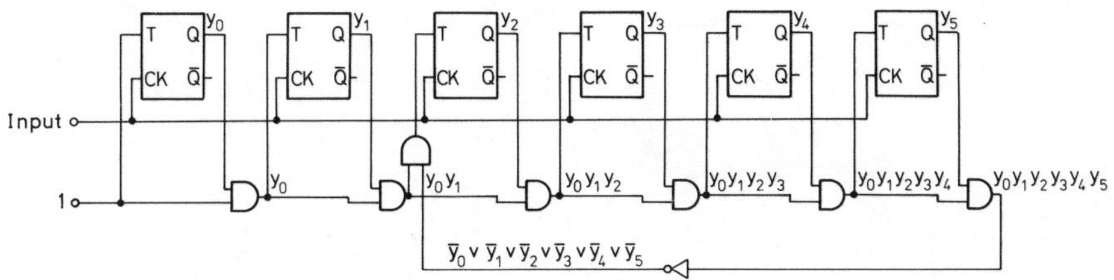

Figure 24. Divide by 60 circuit

and thus

$$\mathbf{p}_i = (0\ \ 0\ \ 1\ \ 0\ \ 0\ \ 0).$$

The corresponding logic diagram is shown in Figure 24.

4.3 A divider with programmable cycle length

An interesting variant of the preceding techniques leads to the synthesis of dividers with programmable cycle length. The circuit in question includes n flip-flops, and $(n-1)$ programming inputs

$$c_0, c_1, \ldots, c_{n-2},$$

allowing any cycle length between $2^{n-1}+1$ and 2^n to be realized.

Let us first consider a system whose excitation equations are

$$\begin{aligned} y_0 &:= y_0 \oplus 1 \oplus y_0 y_1 \ldots y_{j-1}\bar{y}_j, \qquad (j \geq 1), \\ y_i &:= y_i \oplus y_0 y_1 \ldots y_{i-1}, \qquad i = 1, 2, \ldots, n-1. \end{aligned} \tag{15}$$

They differ from equation (6) by the first equation. The normal transitions

$$\begin{aligned} (e_{n-1} \ldots e_{j+1}\ 0\ 1\ \ldots\ 1\ 1) &\to (e_{n-1} \ldots e_{j+1}\ 1\ 0 \ldots 0\ 0) \\ &\to (e_{n-1} \ldots e_{j+1}\ 1\ 0 \ldots 0\ 1) \end{aligned}$$

are replaced by a jump

$$(e_{n-1} \ldots e_{j+1}\ 0\ 1 \ldots 1\ 1) \to (e_{n-1} \ldots e_{j+1}\ 1\ 0 \ldots 0\ 1),$$

for every $(n-1-j)$-tuple $(e_{n-1} \ldots e_{j+1})$. Hence, there are 2^{n-1-j} such jumps and the length of the principal cycle is equal to

$$2^n - 2^{n-1-j},$$

instead of 2^n.

Let us now observe that, if $j \neq k$, the two products

$$y_0 y_1 \ldots y_{j-1}\bar{y}_j \quad \text{and} \quad y_0 y_1 \ldots y_{k-1}\bar{y}_k$$

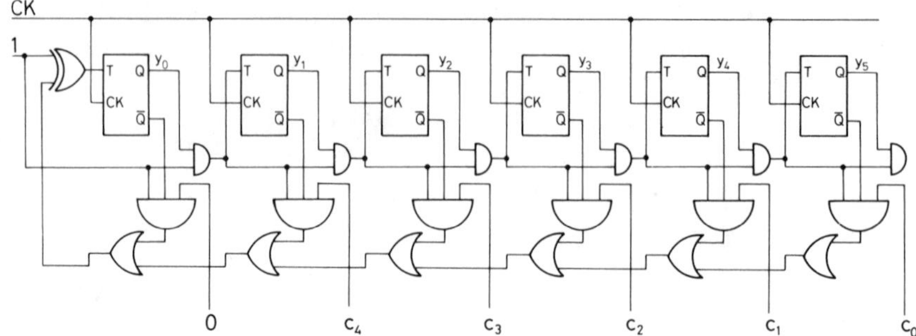

Figure 25. Divider with programmable cycle length

are disjoint. Therefore, the system whose excitation equations are

$$\begin{aligned} y_0 &:= y_0 \oplus 1 \bigvee_{\ell=1}^{n-1} c_{n-\ell-1} y_0 \dots y_{\ell-1} \bar{y}_\ell, \\ y_i &:= y_i \oplus y_0 y_1 \dots y_{i-1}, \qquad i = 1, 2, \dots, n-1, \end{aligned} \tag{16}$$

has a principal cycle whose length is equal to

$$2^n - \sum_{\ell=1}^{n-1} c_{n-\ell-1} 2^{n-\ell-1} = 2^n - \sum_{t=0}^{n-2} c_t 2^t.$$

The programming vector $(c_{n-2}, \dots, c_1, c_0)$ thus represents $2^n - L$, where L denotes the length of the principal cycle. The logic diagram corresponding to equations (16) is shown in Figure 25.

4.4 Asynchronous counters and dividers

Let us briefly mention the existence of asynchronous counters and dividers. As an example, the circuit of Figure 26, made up of a cascade interconnection of master–slave or edge-triggered flip-flops, is an asynchronous circular counter (*ripple-counter*): the clock input of flip-flop number i is driven by the output of flip-flop number $i-1$.

Usually, in asynchronous systems, the state changes of a given flip-flop are produced by state changes of other flip-flops; hence, these circuits often include closed loops. A complete timing analysis must be performed in order to check the design validity.

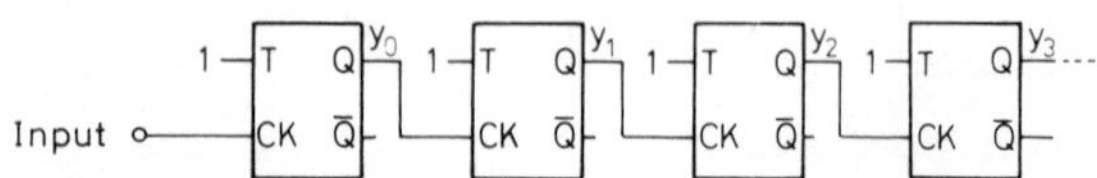

Figure 26. Asynchronous circular counter (ripple counter)

Notice also that, in some cases, an asynchronous counter will work correctly at higher frequencies than a synchronous counter, implemented in the same technology. Nevertheless, in an asynchronous counter, the state changes of the flip-flops do not simultaneously arise; at high working frequencies, the timing differences between flip-flops are no longer negligible with respect to the input signal period. The last feature prevents the use of asynchronous systems in some applications.

5 SEQUENCE GENERATORS

A *sequence generator* is an autonomous system which generates a prescribed periodic sequence of bits. If the period is equal to ℓ, it can be realized by means of a counter whose principal cycle length is ℓ, and a decoding combinational circuit. As an example, a three stage circular counter generates the periodic sequence

$$\ldots 1\ 1\ 1\quad 0\ 0\ 0\ 0\ 1\ 1\ 1\ 1\quad 0\ 0\ 0\ 0\ 1\ 1\ 1\ 1\quad 0\ 0\ 0 \ldots \tag{17}$$

whose length is equal to eight.

Very often, the term 'sequence generator' refers to a particular implementation method based on the use of a shift register and a combinational circuit. The logic diagram of the circuit is shown in Figure 27; it is called a *feedback shift register.* If f is a *linear function,* that is if

$$f = a_0 \oplus a_1 y_1 \oplus a_2 y_2 \oplus \ldots \oplus a_n y_n, \qquad a_i \in \{0, 1\},$$

it is a *linear feedback shift register.*

As an example, let us synthesize a circuit which generates the periodic sequence

$$\ldots 1\ 1\ 1\quad 0\ 0\ 0\ 1\ 0\ 1\ 1\ 1\quad 0\ 0\ 0\ 1\ 0\ 1\ 1\ 1\quad 0\ 0 \ldots.$$

Let us try to realize the circuit by means of a feedback shift register including three cells. In order to determine f, let us write the sequences

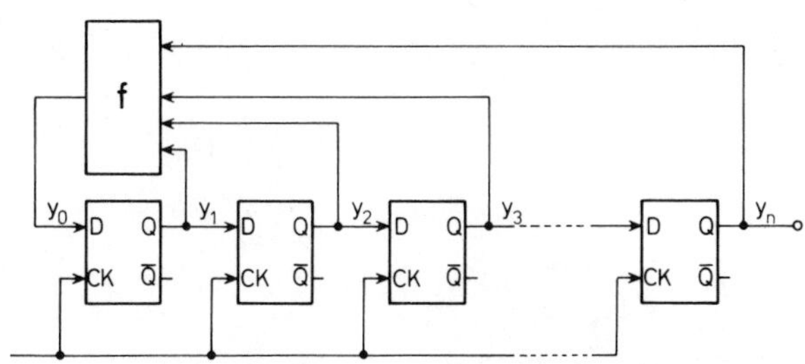

Figure 27. Feedback shift register

generated by y_0, y_1, y_2 and y_3, respectively:

$$y_3\colon \ldots 0\ 0\ 0\ 1\ 0\ 1\ 1\ 1 \ldots,$$
$$y_2\colon \ldots 0\ 0\ 1\ 0\ 1\ 1\ 1\ 0 \ldots,$$
$$y_1\colon \ldots 0\ 1\ 0\ 1\ 1\ 1\ 0\ 0 \ldots,$$
$$y_0\colon \ldots 1\ 0\ 1\ 1\ 1\ 0\ 0\ 0 \ldots.$$

It appears that y_0 is unambiguously determined by the knowledge of y_3, y_2 and y_1. The corresponding function f is described in Figure 28:

y_3	y_2	y_1	$y_0 = f(y_3, y_2, y_1)$
0	0	0	1
0	0	1	0
0	1	0	1
0	1	1	1
1	0	0	0
1	0	1	1
1	1	0	0
1	1	1	0

Figure 28. Feedback shift register

The *a posteriori* analysis of the circuit leads to the transition graph of Figure 29.

Notice, however, that it is not always possible to synthesize a sequence generator whose period is equal to ℓ with $\lceil \log_2 \ell \rceil$ cells of the shift register only. As an example, a circuit generating the sequence (17) cannot be realized by means of three cells:

$$y_3\colon \ldots 0\ 0\ 0\ 0\ 1\ 1\ 1\ 1 \ldots,$$
$$y_2\colon \ldots 0\ 0\ 0\ 1\ 1\ 1\ 1\ 0 \ldots,$$
$$y_1\colon \ldots 0\ 0\ 1\ 1\ 1\ 1\ 0\ 0 \ldots,$$
$$y_0\colon \ldots 0\ 1\ 1\ 1\ 1\ 0\ 0\ 0 \ldots.$$

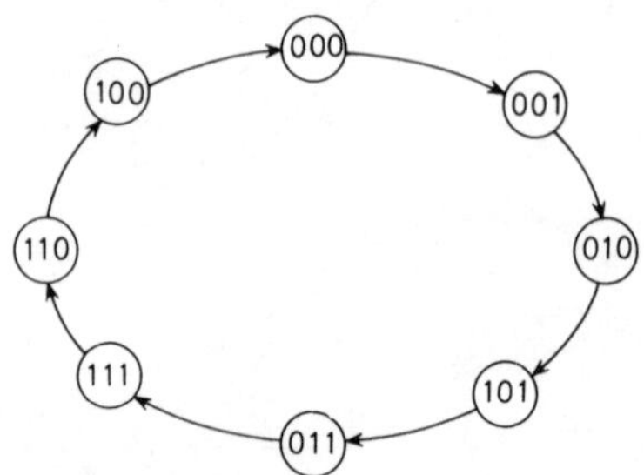

Figure 29. Transition graph of a feedback shift register

It is not possible to express y_0 as a function of y_3, y_2 and y_1. The sequence can be generated by a feedback shift register including four cells.

In any case, a periodic sequence, whose period is equal to ℓ, can always be generated by a ring counter, that is a linear feedback shift register including ℓ cells, with $y_0 = y_\ell$. The initial state of the ring counter represents one period of the sequence.

Let us recall that shift registers are basic synchronous sequential components which are easy to implement (dynamic shift-registers, charge coupled devices, etc.). This remark justifies the amount of work that has been devoted to feedback shift registers. A survey of the question, along with a long bibliography, may be found in Ronse (1980).

6 CONCLUSIONS

The practical applications presented in the preceding sections, along with the use of a transfer language as an external behaviour description tool, progressively led us to throw new light on synchronous systems. We first laid stress on the fact that data are serially introduced, and that results are progressively generated. Gradually, we emphasized the operations which are performed at every particular computation step.

Let us illustrate this comment by the *parallel accumulator A* of Figure 30. Its behaviour is described by the transfer

$$[A]:=[A]+B \qquad (0 \leqslant B \leqslant 2^k - 1). \tag{18}$$

Note, furthermore, that this evolution of our viewpoint was foreseeable from the moment we decided to delegate to a special control system tasks such as synchronization, feeding of the data, collecting of the results, and detection of end of computation (see Chapter V).

Another modification arose with respect to our starting viewpoint. Let us suppose that every register is able to perform several functional transfers, under the control of external variables **s**. The corresponding functional diagram is shown in Figure 31. There are both practical and conceptual consequences. The functional enrichment, which is due to the introduction

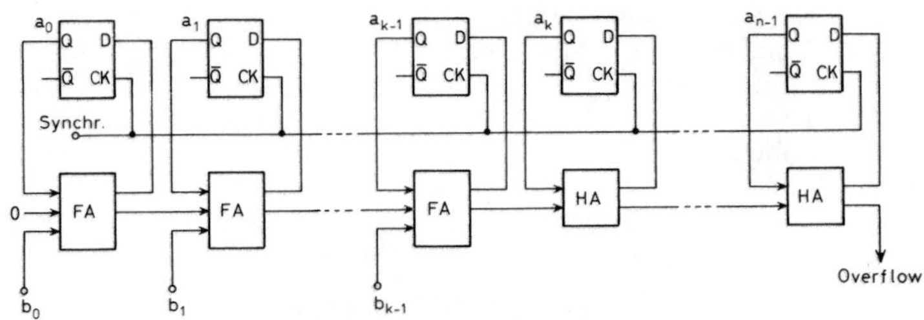

Figure 30. Parallel accumulator

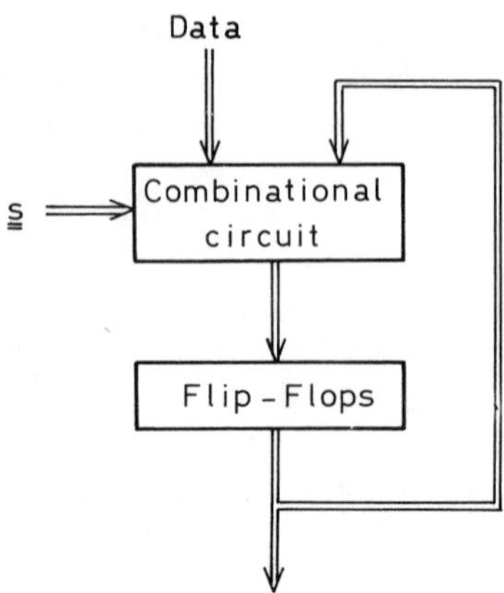

Figure 31. Functional diagram

of variables **s**, allows us to implement not only iterative computations, but also any computation scheme whose precedence graph is reduced to a single path. These notations will be generalized in Chapter VIII.

BIBLIOGRAPHICAL REMARKS

Further information concerning synchronous sequential components may be found in many textbooks dealing with switching circuits. Let us quote, for instance, Lee (1976), Mange (1978), Taub and Schilling (1977), and Oberman (1970). The theory of feedback shift registers is the subject matter of a number of textbooks: Golomb (1967), Gill (1966), Martin (1969), etc.

EXERCISES

E1. Design a counter for counting in the BCD excess-three code defined by:

$$0 \to 0011; \quad 1 \to 0100; \quad 2 \to 0101; \quad 3 \to 0110; \quad 4 \to 0111;$$
$$5 \to 1000; \quad 6 \to 1001; \quad 7 \to 1010; \quad 8 \to 1011; \quad 9 \to 1100.$$

Design variants of this counter (up counter, down counter programmable counter). Vary the type of combinatorial logic and the type of flip-flop.

E2. Assume a square wave signal is available at frequency 10 Hz. Discuss roughly the design of a 24-hour digital clock. Do not forget the possibility of setting time. Describe in detail the counters and/or frequency dividers you use in your design.

E3. Design the cell of a bidirectional shift-register with synchronous parallel load. At each clock-pulse, one of the four following behaviours is possible:

A: NO OPERATION,

B: LEFT SHIFT,

C: RIGHT SHIFT,

D: LOAD.

These four possibilities will be encoded by means of two binary variables (x_1, x_0) according to the scheme:

$$A \to (0,0); \quad B \to (1,0); \quad C \to (0,1); \quad D \to (1,1).$$

Describe a typical cell of that register if the flip-flop to be used is either a D flip-flop or a *J–K* flip-flop.

E4. Design a register R with three control variables having the following eight behaviours:

x_2	x_1	x_0		
0	0	0	NO OPERATION	$[R]:=[R]$
0	0	1	LOAD	$[R]:=A$
0	1	0	LEFT SHIFT	$[R]:=2[R]+x$
0	1	1	RIGHT SHIFT	$[R]:=\lfloor [R/2] \rfloor + 2^{n-1}y$
1	0	0	INCREM.	$[R]:=[R]+1$
1	0	1	DECREMENT	$[R]:=[R]-1$
1	1	0	ADD (mod 2^n)	$[R]:=[R]+A$
1	1	1	SUBSTRACT (mod 2^n)	$[R]:=[R]-A$

Think of an expandable design of that register and describe its typical cell. Discuss the choice of the encoding of the 8 operating modes by $x_2x_1x_0$.

E5. (Gray code.) To the integer $N(0 \leq N \leq 2^n - 1)$, we associate a binary n-tuple $\mathbf{z}^{(N)} = [z_{n-1}^{(N)}, \ldots, z_1^{(N)}, z_0^{(N)}]$ by the following rules:
(i) If $n=0$, $\mathbf{z}^{(0)}=[0]$ and $\mathbf{z}^{(1)}=[1]$.
(ii) Assume known the representation of any integer N' $(0 \leq N' \leq 2^{n-1}-1)$ by a binary $(n-1)$-tuple. The representation of N by a n-tuple is then defined by:

$$\mathbf{z}^{(N)} = [0, z_{n-2}^{(N)}, \ldots, z_0^{(N)}], \quad \text{if} \quad 0 \leq N \leq 2^{n-1}-1;$$
$$\mathbf{z}^{(N)} = [1, z_{n-2}^{(N')}, \ldots, z_0^{(N')}], \quad \text{with } N' = 2^n - 1 - N, \quad \text{if } 2^{n-1} \leq N \leq 2^n - 1.$$

Show:
(i) That when N runs cyclically along the sequence of integers $0, 1, \ldots, 2^n - 1, 0, 1, \ldots$ the Hamming distance of two consecutive n-tuples $d(\mathbf{z}^{(N)}, \mathbf{z}^{(N+1)})$ is always equal to 1.
(ii) That:

$$N < 2^{n-1} \Rightarrow \mathbf{z}^{(N+2^{n-1})} = (1, \overline{z_{n-2}^{(N)}}, z_{n-3}^{(N)}, \ldots, z_0^{(N)}).$$

Devise algorithms for passing from $\mathbf{z}^{(N)}$ to N and conversely show that, if $\mathbf{y}^{(N)} = [y_{n-1}^{(N)}, \ldots, y_1^{(N)}, y_0^{(N)}]$ is the binary representation of N, it is related to the corresponding Gray representation $\mathbf{z}^{(N)}$ by

$$y_i^{(N)} = \sum_{y=i}^{n-1} z_j^{(N)}; \qquad i = 0, 1, \ldots, n-1,$$
$$z_i^{(N)} = y_i^{(N)} + y_{i+1}^{(N)}; \qquad i = 0, 1, \ldots, n-2,$$
$$z_{n-1}^{(N)} = y_{n-1}^{(N)}.$$

Design circular counters in the Gray code.

E6. (Linear feedback shift registers.) A linear feedback shift register may be described by the following matrix equation over the field GF(2) of integer mod 2

$$\begin{bmatrix} y_0 \\ y_1 \\ y_2 \\ \cdots \\ y_{n-1} \end{bmatrix} = \left[\begin{array}{ccccc:c} c_0 & c_1 & c_2 & \cdots & c_{n-2} & 1 \\ \hdashline 1 & & & & & 0 \\ & 1 & & & & 0 \\ & & \cdots & & & \\ & & & & 1 & 0 \end{array}\right] \begin{bmatrix} y_0 \\ y_1 \\ y_2 \\ \cdots \\ y_{n-1} \end{bmatrix}$$

If the characteristic polynomial of the matrix

$$P = \lambda^n \oplus c_0\lambda^{n-1} \oplus c_1\lambda^{n-2} \oplus \ldots \oplus c_{n-2}\lambda \oplus 1$$

is primitive (see, for example, Peterson, 1961), the corresponding circuit runs cyclically along its $(2^n - 1)$ non-zero states. Describe and analyse the linear feedback shift-registers given by the polynomials

$$P_1 = \lambda^4 + \lambda^3 + 1$$
$$P_2 = \lambda^4 + \lambda^3 + \lambda^2 + \lambda + 1$$
$$P_3 = \lambda^5 + \lambda + 1$$

Show that, in a linear feedback shift register, the predecessor of the state

$(y_{n-1}, \ldots, y_1, y_0) = (0, \ldots, 0, 1)$ is necessarily $(1, 0, \ldots, 0, 1)$.

Show next that the replacement of

$$y_0^+ = y_{n-1} \oplus \sum_{j=0}^{n-2} c_j y_j$$

by

$$(y_0^+)^* = y_0^+ \oplus \bigwedge_{j=0}^{n-1} \bar{y}_j$$

allows one to insert the $(0, \ldots, 0, 0)$ state between the two aforementioned states.

E7. Use the ALU 74181 (Chapter IV), a register and, if necessary, an amplitude comparator to build up:
(i) an UP counter mod 2^4 $[C] := [C] + 1 \bmod 16$,
(ii) a DOWN counter mod 2^4 $[C] := [C] - 1 \bmod 16$,
(iii) an UP counter mod k ($k \leq 15$; k available as control input) $[C] := [C] + 1 \bmod k$,
(iv) a programmable UP–DOWN counter mod k.

Chapter VIII
Implementation of computation schemes

We pointed out at the beginning of this work that we consider digital systems as implementing algorithms. We also noticed that digital systems are hierarchically organized. At this stage, we have described the main building blocks: universal combinational components (gates, multiplexers, demultiplexers, ROMs, and PLAs), special purpose combinational components (ALUs, radix converters, and connecting networks), storage components (flip-flops, registers, and memories), and control components (counters and finite state machines). Chapters I to VII thus give an overall view of modern switching circuits.

We are now in a position to proceed one stage further of the hierarchy and look for algorithm implementation methods.

In this chapter we study the implementation of a particular class of algorithms: *computation schemes.* The main definitions are given in Section 3 of Chapter I.

The simplest way to implement a computation scheme is to design a combinational circuit (Chapter I, Section 3.3.4) in which the input wires correspond to constants (Chapter I, (24)) and variables (Chapter I, (25)), the combinational components correspond to computation primitives (Chapter I, (27)), and the interconnections are made according to (26) and (27) of Chapter I. An example was given in Section 4.4 of Chapter I.

The notions of cost and computation time of a computation scheme were introduced in Chapter I, Section 3.3.3; they are related to the precedence graph of the computations scheme. The *cost* gives the number of combinational components in the circuit. It corresponds to the actual cost if all combinational components have equal costs. In the contrary case, a weighted counting should be used.

The *computation time* gives the largest number of combinational components which can be encountered on a path linking an input wire to an output wire. It corresponds to the actual computation time if the components introduce equal delay values. This common value may be chosen as a time unit. Hereafter, we consider only such *unit execution time systems.* Indeed, digital systems are controlled by periodic clock signals. Hence, it is natural to decompose the whole system into components working in one clock period (see also comment (4) at the end of Chapter X, Section 1).

Nevertheless, direct implementation is not always the best solution. In the case where many combinational components are identical, the use of additional memory elements, along with a well-controlled resource and memory allocation policy, can greatly reduce the cost, while hardly increasing the computation time. On the other hand, pipelining can greatly increase the (apparent) computation time, without excessive cost increase.

1 LABELLING OF THE PRECEDENCE GRAPH

Let us consider the precedence graph of a unit execution time system $\{N_0, N_1, \ldots\}$. It is convenient to replace all source nodes by one BEGIN node, and to link all terminal nodes to an additional END node. This new graph is called a *closed precedence graph.*

Let us suppose that the input data are available at time 0. We may compute the *earliest* time $s_{0,j}$ at which step N_j can be performed. On the other hand, if we want the results to be available at time T, we may compute the *latest* time $s_{1,j}$ at which step N_j must be performed. The computation of $s_{0,j}$ and $s_{1,j}$, for every node N_j of the closed precedence graph, is performed by a classical critical path algorithm.

Algorithm 1.

A. $s_{0,j}$

Step 0. *Node BEGIN is labelled* 0.

Step k. All nodes which are not labelled, and all of whose predecessors are labelled, receive label k. Hence, $s_{0,j}$ *is the length of the longest path linking the BEGIN node to* N_j.

Let L be the label of the END node.

B. $s_{1,j}$

Step 0. *Node END is labelled T. It is necessary that* $T \geqslant L$.

Step k. All nodes which are not labelled, and all of whose successors are labelled, receive label $T-k$. *Hence,* $s_{1,j}$ *is the difference between T and the longest path linking* N_j *to the END node.* □

Figure 1 gives us two labelling examples of the same closed precedence graph: in the first case, $T=6$; in the second case, $T=L=4$.

We now consider the most natural case $T=L$. If N_j is labelled $(s_{0,j}, s_{1,j})$, the corresponding computation cannot be performed either before time $s_{0,j}$, or after time $s_{1,j}$. The time interval $[s_{0,j}, s_{1,j}]$ is the execution time interval of step N_j. If $s_{0j} = s_{1j} = t_j$, step N_j must be performed at time t_j: it is a *critical step.* A path which only includes critical steps is called a *critical path.* There always exists at least one critical path. In Figure 1b, the only critical path is (BEGIN, N_3, N_5, N_6, END).

Let us go back to the general case ($T \geqslant L$). For every computation step N_j,

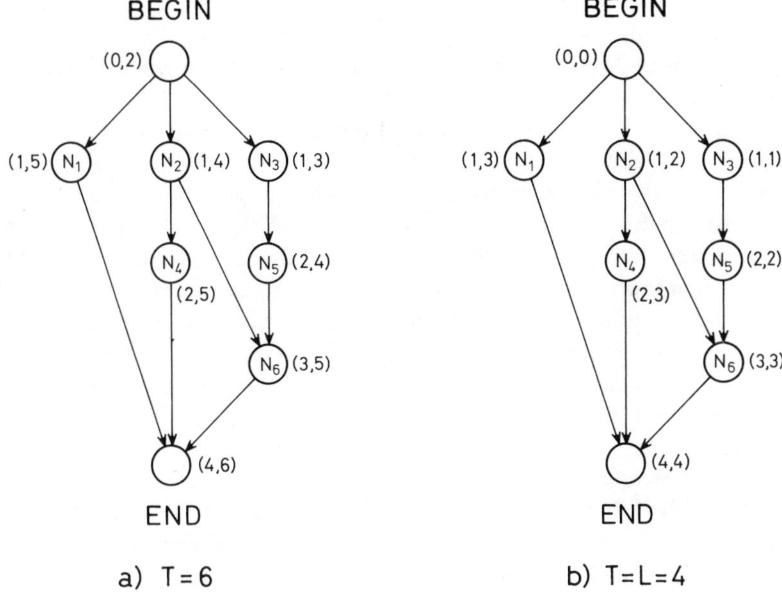

Figure 1. Labelling of a precedence graph

we may choose an execution time t_j. With this aim in view, we define an *acceptable labelling* ψ. It associates, with every node N_j of the closed precedence graph, the time $t_j = \psi(N_j)$. The two following conditions must be verified:

$$s_{0,j} \leq \psi(N_j) \leq s_{1,j}, \tag{1}$$

$$N_i \subset N_j \Rightarrow \psi(N_i) < \psi(N_j). \tag{2}$$

The second condition is necessary in order to respect the precedence relation. As an example, in the graph of Figure 1a, we may not choose $\psi(2) = 3$ and $\psi(4) = 2$.

2 COMPUTATION WIDTH AND INFORMATION FLOW WIDTH

Let us consider a circuit implementing some computation scheme, and suppose that its working timing corresponds to the acceptable labelling ψ of the precedence graph. At time k, the active resources correspond to the nodes N_j of the graph such that $\psi(N_j) = k$. The maximum number of simultaneously active resources is

$$W_c = \max_k |\{N_j \mid \psi(N_j) = k\}|. \tag{3}$$

It is the *computation width* of the computation scheme, with respect to the acceptable labelling ψ.

For the time being, we suppose that all resources are identical; this is a particular case of a unit execution time system.

The simultaneous working of several resources gives prominence to some kind of parallelism which is inherent in the algorithm and in the chosen labelling. Afterwards, we will design a circuit, implementing the computation scheme, which includes W_c resources, and thus respects the intrinsic algorithm parallelism. In order to achieve this aim, we must be able to supply the computing resources with the data they need every time: external data or intermediate results obtained during previous steps. If the same W_c resources are used at each step, intermediate results obtained at a particular step must be stored in order to be available during later steps. Hence, memory elements must be provided.

In order to estimate the minimum number of memory elements, let us define the *connection function* Λ of the computation scheme, with respect to the acceptable labelling ψ: $\Lambda(k)$ is the number of symbols which are generated during the steps N_i such that $\psi(N_i) \leq k$, and are next used as data during steps N_j such that $\psi(N_j) \geq k+1$. As an example, in the circuit of Figure 2, $\Lambda(k) = 12$.

The computation scheme can be implemented by means of W_c computation resources, and W_{fl} memory elements, where

$$W_{fl} = \max_k \Lambda(k); \tag{4}$$

each memory element is able to store one logic symbol (one bit in the binary

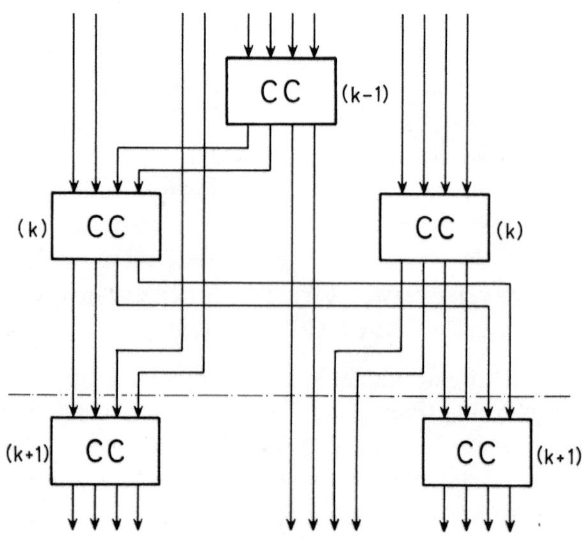

Figure 2. Information flow width

case). The parameter W_{fl} is the *information flow width* of the computation scheme, with respect to the acceptable labelling ψ.

3 IMPLEMENTATION OF COMPUTATION SCHEMES BY SYNCHRONOUS SEQUENTIAL NETWORKS

Let us consider a computation scheme which includes only one computation primitive and thus leads to a circuit, all of whose resources are identical. The choice of an acceptable labelling ψ of its precedence graph allows the parameters W_c and W_{fl} to be computed. We aim at implementing this computation scheme with a circuit including W_c identical resources CC_i, $i=1,2,\ldots,W_c$, and W_{fl} memory elements FF_j, $j=1,2,\ldots,W_{fl}$. FF stands for flip-flop: indeed, in most cases, the logic symbols belong to $\{0, 1\}$, and every memory element stores one bit. A first solution, illustrated by Figure 3, is based on a series of choices that we will call *resource allocation* and *memory allocation.*

(i) *Resource allocation at time k.* Let $\{N_i \mid \psi(N_j)=k\}$ be the set of nodes corresponding to computations performed at time k. The resource allocation at time k is a one-to-one mapping

$$\alpha_\rho(k):\{N_j\}\rightarrow\{CC_i\}, \tag{5}$$

by which we allocate one resource to every task performed at time k.

(ii) *Memory allocation at time k.* Let $\{s_{ij}\}$ be the set of symbols generated during the steps N_i such that $\psi(N_i)\leqslant k$, and next used as data during steps N_j such that $\psi(N_j)\geqslant k+1$. The memory allocation at time k is a one-to-one mapping

$$\alpha_\mu(k):\{s_{ij}\}\rightarrow\{FF_\ell\}$$

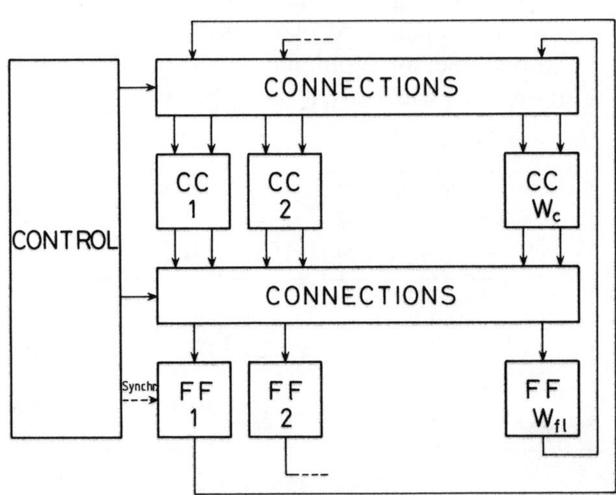

Figure 3. Implementation with two connection networks

by which we allocate one memory element to every logic symbol which must be stored between times k and $k+1$. In particular, $\alpha_\mu(0)$ gives the number of input symbols (external data).

The next observation will help to understand entirely the diagram of Figure 3. Once allocations $\alpha_\rho(k)$ and $\alpha_\mu(k)$ are chosen, the set of operations which must be performed at time k can be described by a series of transfers:

$$\mathrm{FF}_j(k) := f_{j,k}(\mathrm{FF}_{j1}, \mathrm{FF}_{j2}, \ldots), \qquad j = 1, 2, \ldots, W_{\mathrm{fl}}, \tag{6}$$

where functions $f_{j,k}$ are generated by the computation resources. The interposition of controlled interconnection networks will be sufficient to complete the algorithm implementation: at every time k, the control circuit positions the connection networks in accordance with the chosen resource and memory allocations $\alpha_\rho(k)$ and $\alpha_\mu(k)$.

A second solution is shown in Figure 4. It differs from the previous one by the disappearance of one interconnection network. In fact, we may decide always to store in the same memory elements the results generated by a given particular resource. The number of necessary memory elements is generally increased: in the circuit of Figure 4, a memory element FF_j which is connected to a computation resource CC_i which is not active at time k, cannot be used at time k. If every computation resource has n outputs, the minimum number of memory elements is

$$W' = nW_c + \max_k \,(\Lambda(k) - n\,|\{N_i \mid \psi(N_i) = k\}|). \tag{7}$$

Similarly, it is possible to delete the other connection network by requiring every resource always takes its data from the same memory elements.

Figure 5 illustrates the possibility of fixing rigidly the connections between resources and memory elements. Every step is broken down into two substeps: during the first one, the resources take data from fixed memory elements, compute new values, and place them in fixed memory elements; during the second one, memory contents are permuted in order to bring to appropriate positions data which are used at the next step. Loop A of Figure

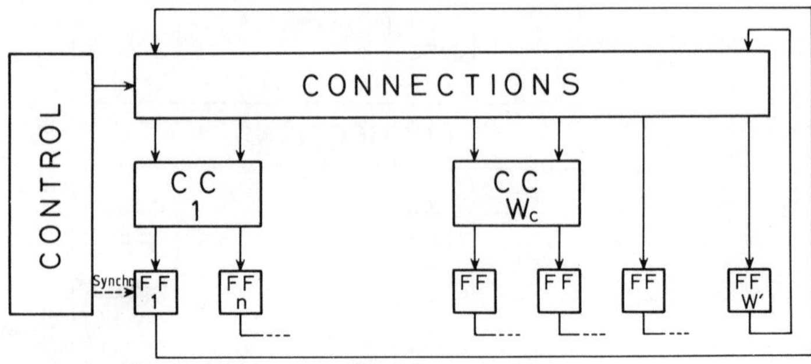

Figure 4. Implementation with one connection network

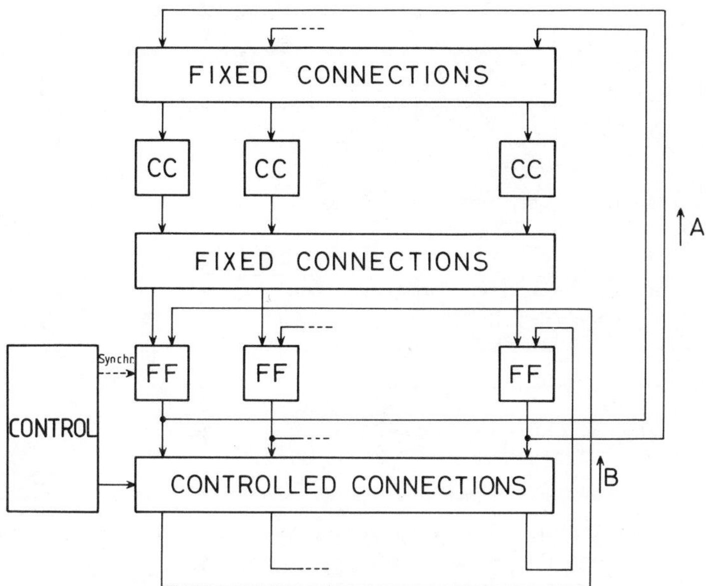

Figure 5. Implementation with fixed connections

5 is activated during the first substep, and loop B during the second one. The controlled connection network is used for permuting data during the second substep.

Let us generalize the method to the case of unit execution time systems. Let $\{\Pi_i \mid i=1, 2, \ldots, R\}$ be the set of different computation primitives appearing in the computation scheme. It is possible to design a programmable resource $\mathscr{P}$, which is controlled by r binary variables ($2^r \geq R$), and such that every primitive Π_i is implemented by $\mathscr{P}$, for (at least) one control r-tuple value. A slight modification of the circuit of Figure 4 leads to the cricuit of Figure 6, which is able to implement a unit execution time system. At every step, the control circuit positions the connection network and indicates which particular primitives are presently implemented by the programmable resources.

The implementation methods, which are developed above, are very important, and their study will be continued in Section 6 after examination of two examples. From now on we can derive some conclusions.

(1) The circuit organizations, depicted in Figures 3 to 6, give prominence to a natural decomposition of the system into two parts: the control unit and the processing unit. The control unit is an autonomous system which generates signal sequences intended for the connection networks and, if the case arises, for the programmable resources. It is an autonomous synchronous sequential machine (sequence generator and counter) which can be implemented by classical methods (Chapter VI and Chapter VII).

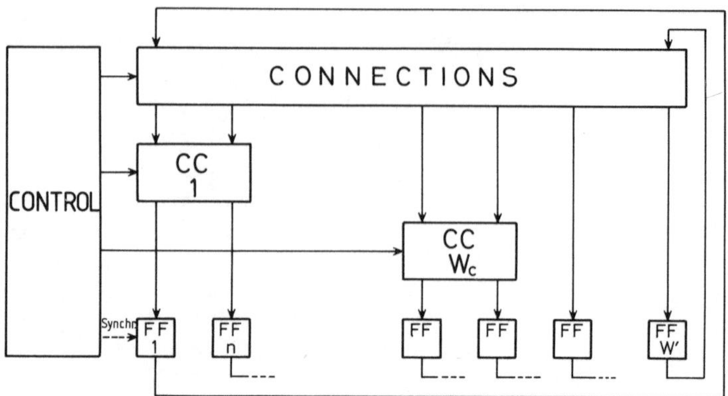

Figure 6. Implementation of a unit execution time system

The processing unit is described in a transfer language (see equation (6)). Let us notice that the memory elements can often be logically grouped together in order to form registers: the contents of a set of memory elements can be a vector which possibly represents significant alphanumeric quantities. Registers are denoted by capital letters such as $R, C, \ldots$, and their contents by $[R], [C], \ldots$. Transfers can be described, in condensed form, at the register level. For instance

$$[R]:=2[R]+1 \pmod{2^n},$$

is the characteristic equation of a shift register (see Chapter VII, equation (1)), and

$$[c]:=[c]+1 \pmod{2^n},$$

is the equation of a circular counter (Chapter VII, (equation (5)). The general form of a register transfer is

$$[R]:=f([R_1],[R_2],\ldots). \tag{8}$$

(2) We obviously want to compare the implementation methods described above, which we call *sequential implementations*, with the direct implementation by a combinational circuit (Chapter I, Section 3.3.4), which we call *combinational implementation*. Let us first compare the computation times. Both methods include the same number of steps since the sequential implementation respects the intrinsic algorithm parallelism: in both cases, the number of steps is equal to the length of the precedence graph. Nevertheless, the step durations are not equal: in the combinational implementation, the duration of a step is equal to the resource computation time while, in the sequential implementation, it is the sum of the same resource computation time and of the memory element storing time.

Let us now compare the computation costs. At this stage of our study, it is not possible to derive pertinent conclusions. The sequential implementation uses much less computation resources, but requires memory elements, connection networks, and a control unit. The cost of these circuits is greatly influenced by the choice of the acceptable labelling, and of the resource and memory allocations.

4 EXAMPLE: STUDY OF A SORTING ALGORITHM

Let us consider eight four-bit numbers which are stored in eight registers $R_0, \ldots, R_7$. We aim at designing a circuit which permutes these numbers, and eventually arranges them in increasing order. Hence, at the end of the

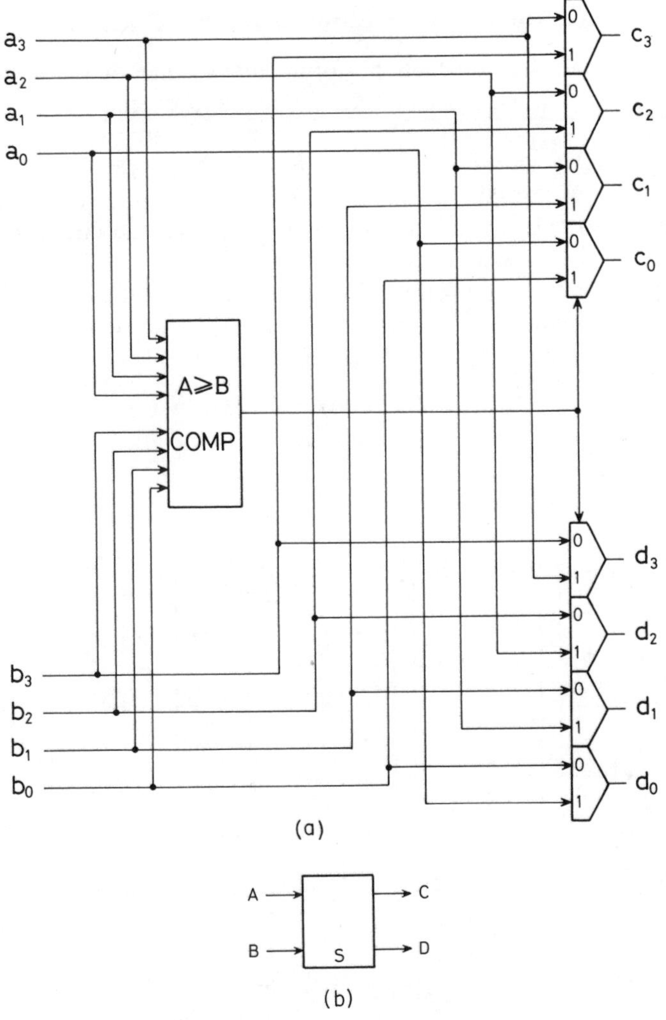

Figure 7. Two element sorting circuit

computation

$$[R_0] \leqslant [R_1] \leqslant \ldots \leqslant [R_7].$$

This is a classical sorting problem (Knuth, 1973). Sorting algorithms use, as their only computation primitive, a two element sorting circuit. This computation resource, and the corresponding logic symbol, are shown in Figure 7. In this basic cell, the operator COMP generates an output signal '0' if $A \geqslant B$, and '1' if $A < B$. The cell behaviour is described by the two following equations:

$$C = \max\{A, B\}; \qquad D = \min\{A, B\}. \tag{9}$$

4.1 The odd–even merge: algorithm description

A combinational circuit which performs the sorting of eight numbers is shown in Figure 8. As an example, let us suppose that a_3 is the largest number. Then, the hatched cells perform the indicated transpositions, and number a_3 is transmitted to output b_7. Furthermore, if we delete the hatched cells, we obtain a similar circuit acting on seven numbers. A recurrent use of this argument will convince us that the circuit of Figure 8 actually sorts numbers a_0 to a_7.

Let us denote by γ and δ the computation cost and the computation time of resource S (Figure 7). The computation cost Γ and the computation time Δ of a sorting circuit acting on N numbers are equal to

$$\Gamma(N) = (N(N-1)/2)\gamma, \tag{10}$$

$$\Delta(N) = N\delta. \tag{11}$$

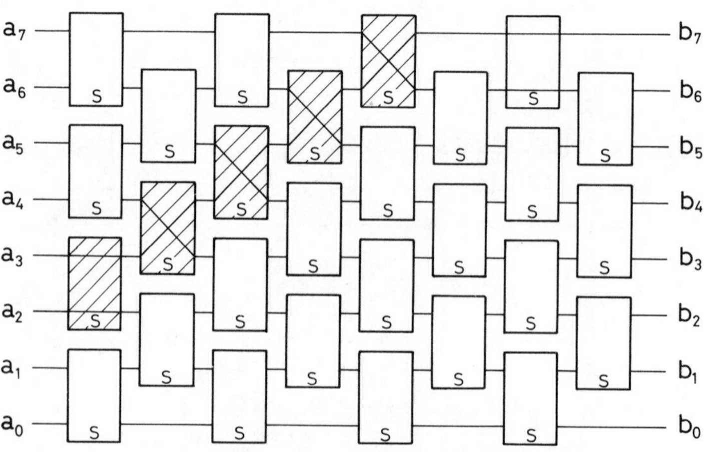

Figure 8. Eight element sorting circuit

4.2 Sequential implementation

The precedence graph corresponding to the circuit of Figure 8 is shown in Figure 9a. An acceptable labelling is indicated. The computation width is equal to 4, and the information flow width is equal to 32 since every edge of the graph corresponds to a group of 4 connections. The design can be started from this graph. Nevertheless, it is possible to increase the symmetry of the circuit by adding one register R_8: initially, it contains the maximum possible value (namely 15) and thus plays a fictitious role; the corresponding precedence graph is shown in Figure 9b. The computation width is still equal to 4, but now four computation resources are active during each step. The information flow width is equal to 36: this corresponds to the nine four-cell input registers.

The architecture of the whole circuit is depicted in Figure 10 and Figure 11 which only differ in the memory allocations. Let us observe that a trivial resource, namely a simple connection, has been introduced: it gives an account of jumps from an even step to an odd step, or from an odd step to an even step, which appear at the left and right sides of the precedence graph (Figure 9b). The four resources are denoted I, II, III, IV, and their outputs are m (minimum) and M (maximum). We only choose one resource allocation which is imposed in practice by the problem symmetry.

A first memory allocation is described by Figure 10a. It indicates, at each step, from which memory elements the resources draw their data, and to which memory elements they transfer the results. It is convenient to consider separately the odd and even steps; the corresponding states of the connection network are shown in Figures 10b and c.

We said above that it is possible to delete one of the two connection networks. For instance, we may force the resources always to take their data from the same memory elements. We so obtain the memory allocation of Figure 11a. The states of the connection networks, during odd and even steps, are given in Figures 11b and c. The deletion of the first connection network does not impose new memory elements. The second connection network can be implemented by the circuit of Figure 12. Control variable z is equal to 0 during the even steps, and to 1 during the odd steps. Cost and delay are equal to

$$\Gamma'(N) = \frac{N}{2}\gamma + (N+1)\gamma_{\mathrm{MUX}}, \tag{12}$$

$$\Delta'(N) = N(\delta + \delta_{\mathrm{MUX}} + \delta_{\mathrm{MEM}}), \tag{13}$$

(N even),

where γ_{MUX} and δ_{MUX} are the computation cost and time of a multiplexer, and δ_{MEM} is the storing time of a memory element. We may notice that Γ' is a linear function of N while Γ is a quadratic function of N (see (10) and (12)). On the other hand, both Δ and Δ' are linear functions of N((11) and (13)).

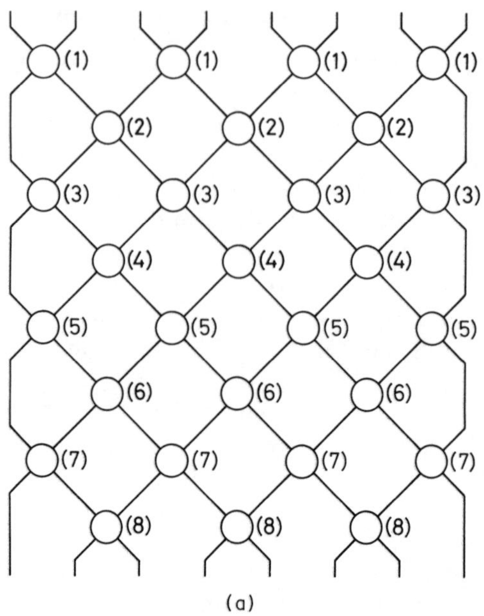

(a)

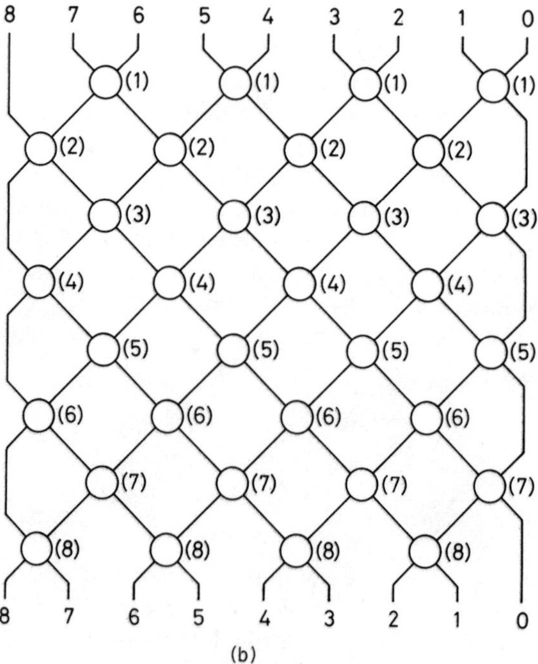

(b)

Figure 9. Precedence graphs

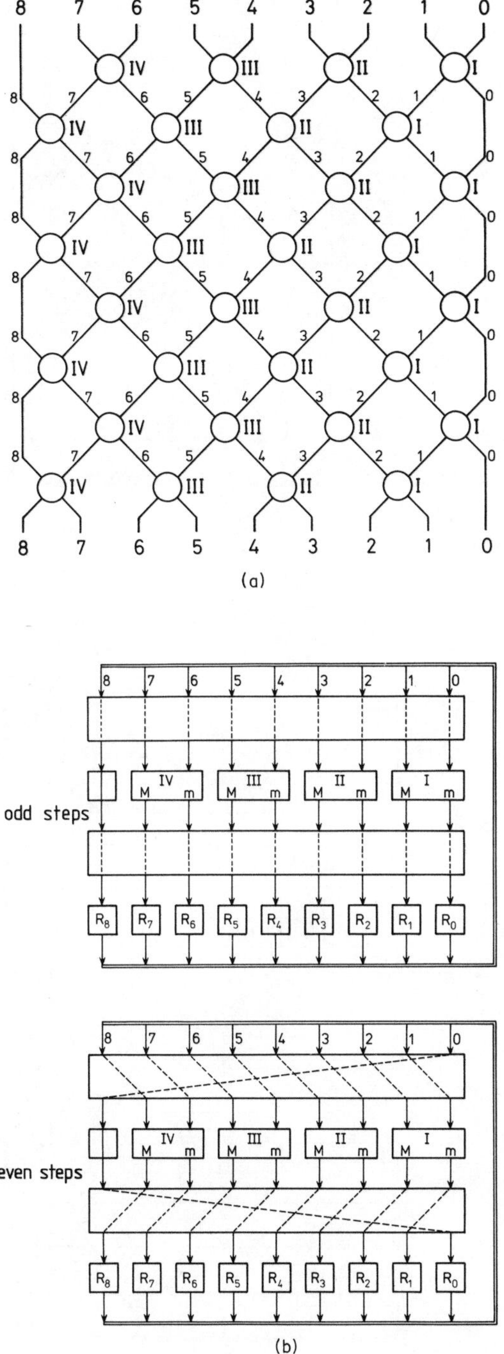

Figure 10. Sorting circuit 1

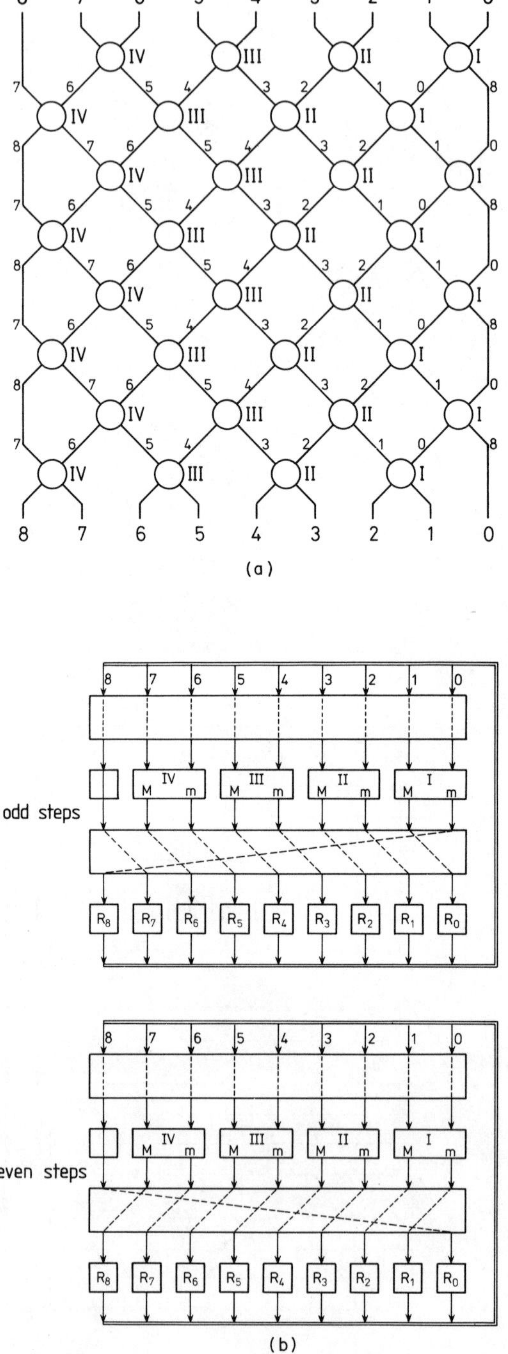

Figure 11. Sorting circuit 2

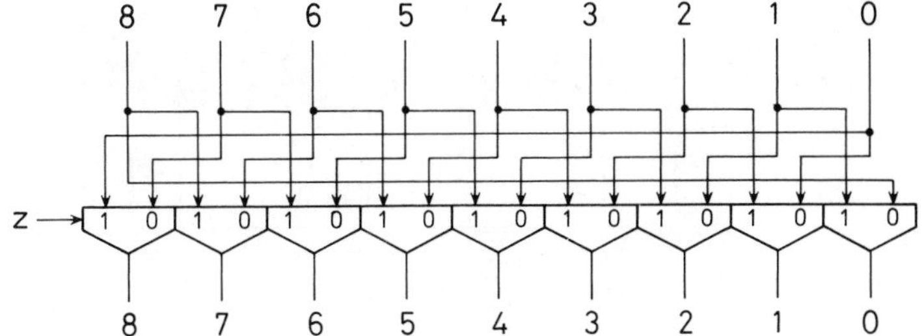

Figure 12. Connection network

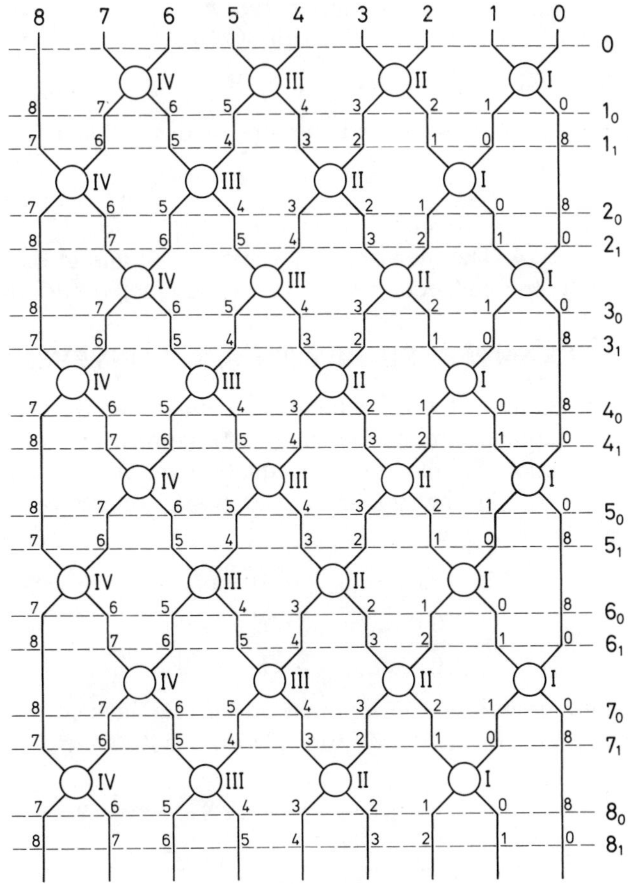

Figure 13. Precedence graph

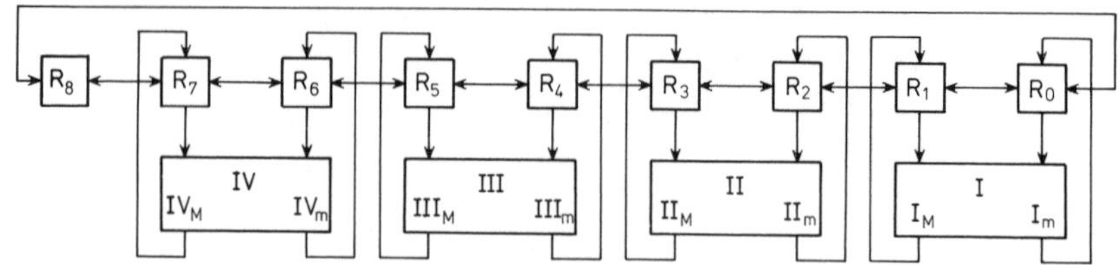

Figure 14. Sorting circuit 3

Let us observe that there is *memory reallocation* at every even step: the result, which is stored in R_8 during an odd step, is transferred to R_0 in the course of the next step, during which it is not used.

We may require that the resources take the data from fixed memory elements, and transfer the results to fixed memory elements. As it has been seen above, this result can be achieved by breaking down each step into two substeps. In this case, the contents of registers $R_8, R_7, \ldots, R_0$, are right shifted during the second substeps of the even steps:

$$R_0 := R_1, \qquad R_1 := R_2, \ldots, \qquad R_7 := R_8, \qquad R_8 := R_0, \tag{14}$$

and left shifted during the second substeps of the odd steps:

$$R_0 := R_8, \qquad R_1 := R_0, \ldots, \qquad R_7 := R_6, \qquad R_8 := R_7. \tag{15}$$

The corresponding precedence graph is shown in Figure 13; it leads to the circuit of Figure 14. The memory reallocations (14) and (15), performed at each step, allow the connection networks to be deleted entirely.

5 EXAMPLE: BINARY TO BCD CONVERSION

As a second example, let us implement a binary to decimal conversion circuit. Combinational circuits realizing radix conversions were studied in Chapter IV, Section 5. According to equation (111) of Chapter IV, with $b=2$ and $\beta=10$, the behaviour of the basic computation resource is described by the next equation:

$$\begin{aligned} \rho_j^2 + q_{j-1}^* &= q_j^* 10 + \rho_j^*, \\ 0 \leq \rho_j, &\quad \rho_j^* \leq \beta - 1 = 9, \\ 0 \leq q_{j-1}^*, &\quad q_j^* \leq b - 1 = 1. \end{aligned} \tag{16}$$

The ten-valued quantities will be encoded by four bits: $\rho_j \simeq (x_3, x_2, x_1, x_0)$, $\rho_j^* \simeq (y_3, y_2, y_1, y_0)$. If x_4 stands for q_{j-1}^* and y_4 for q_j^*, equation (16) becomes

$$(8x_3 + 4x_2 + 2x_1 + x_0)2 + x_4 = 10y_4 + 8y_3 + 4y_2 + 2y_1 + y_0. \tag{17}$$

Hence,

$$\begin{aligned} y_0 &= x_4, \\ 5y_4 + 4y_3 + 2y_2 + y_1 &= 8x_3 + 4x_2 + 2x_1 + x_0. \end{aligned} \tag{18}$$

x_3	x_2	x_1	x_0	y_4	y_3	y_2	y_1
0	0	0	0	0	0	0	0
0	0	0	1	0	0	0	1
0	0	1	0	0	0	1	0
0	0	1	1	0	0	1	1
0	1	0	0	0	1	0	0
0	1	0	1	1	0	0	0
0	1	1	0	1	0	0	1
0	1	1	1	1	0	1	0
1	0	0	0	1	0	1	1
1	0	0	1	1	1	0	0
1	0	1	0	—	—	—	—
1	0	1	1	—	—	—	—
1	1	0	0	—	—	—	—
1	1	0	1	—	—	—	—
1	1	1	0	—	—	—	—
1	1	1	1	—	—	—	—

Figure 15. Basic cell of a binary to decimal converter

The basic cell input–output behaviour is described by the table of Figure 15. The corresponding logic symbol is shown in Figure 16a. Figure 16b represents a conversion array whose input–output behaviour is described by

$$\sum_{i=0}^{n-1} a_i 2^i = \sum_{j=0}^{m-1} \alpha_j 10^j. \tag{19}$$

Let us denote by γ and δ the computation cost and time of the basic cell. As

$$m \cong \frac{n}{3},$$

the computation cost is roughly equal to

$$\Gamma(n) \cong \frac{n^2\gamma}{3}; \tag{20}$$

it is a *quadratic* function of n. Let us notice that some cells can be deleted, since their inputs and outputs are equal to zero, but this will not impair the asymptotic value of the computation cost.

The computation time is equal to

$$\Delta(n) = n\delta; \tag{21}$$

it is a *linear* function of n.

Let us now come to the sequential implementation of the circuit. The precedence graph is shown in Figure 17, in the case where $n = 10$, along with its natural acceptable labelling. The examination of the graph gives prominence to new properties of algorithms: the data access and result

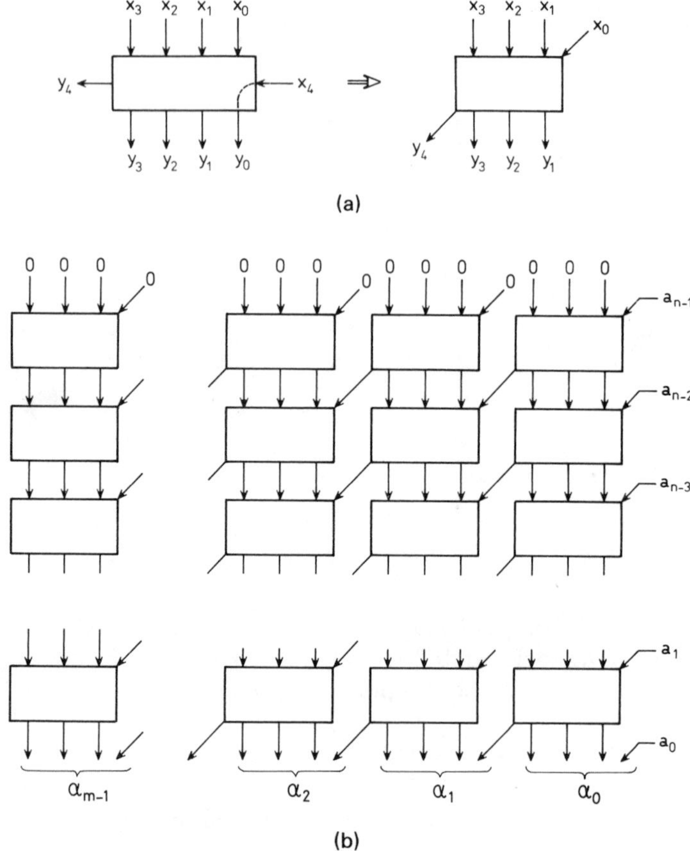

Figure 16. Conversion array

generation modes. In this case, every step requires only one data bit: the most significant bit, a_9, is used during the first step, the next bit, a_8, is used during the second step, and so on. It is a *serial-in* algorithm. On the other hand, the final result of the conversion process is generated during the last step, and no partial result is available before the last step. It is a *parallel-out* algorithm. According to this terminology, the sorting circuit of the previous section implements a *parallel-in, parallel-out algorithm.*

The computation width is equal to 4 and, if we do not take the input data into account, the information flow width is equal to 16. If we decide to fix rigidly the connections between the resources and the memory elements, we must make use of 26 flip-flops, among which 10 store the input data. The 16 flip-flops, which will eventually store the result, have fixed connections. On the other hand, the 10 flip-flops which store the input data can be interconnected so as to form a shift register. In this way, there is no connection

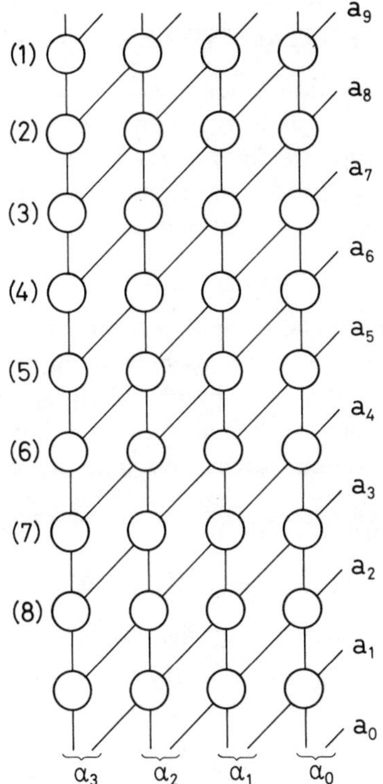

Figure 17. Precedence graph

network. Figure 18 shows the logic diagram of the sequential binary to decimal conversion circuit. Its computation cost is proportional to n, instead of n^2 in the combinational version. Let us again observe that the memory reallocations, which are implicitly performed by the shift register, allow the connection network to be deleted.

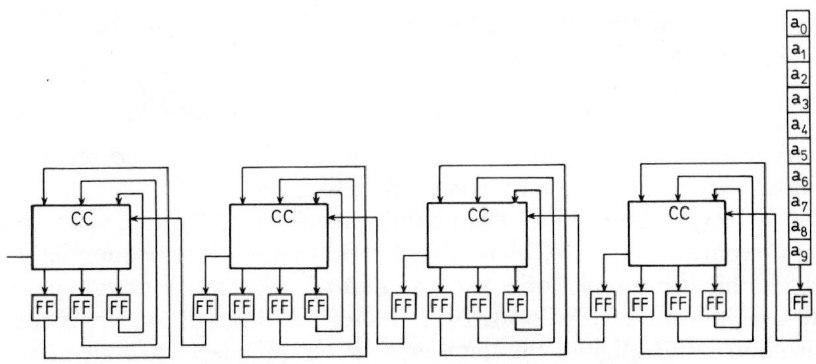

Figure 18. Binary to decimal conversion circuit

6 COMPARISON BETWEEN THE COMBINATIONAL AND THE SEQUENTIAL IMPLEMENTATIONS

Let us consider a computation scheme whose precedence graph includes N nodes, and whose longest path between a source and a terminal node contains L nodes. We suppose that all resources have computation cost γ_{cc} and computation time δ_{cc}. Let Γ_c and Δ_c stand for the computation cost and time of the corresponding combinational circuit:

$$\Gamma_c = N\gamma_{cc}, \qquad \Delta_c = L\delta_{cc}. \tag{22}$$

We may also implement the same algorithm by a synchronous sequential network including memory elements whose computation cost and time are γ_{FF} and δ_{FF}. The choice of an acceptable labelling allows W_c and W_{fl} to be determined. The computation scheme can be implemented by a circuit including W_c computation resources, W_{fl} memory elements, and two controlled connection networks: see Figure 3. The computation cost and time, Γ_s and Δ_s, of the circuit are equal to

$$\begin{aligned} \Gamma_s &= W_c\gamma_{cc} + W_{fl\gamma FF} + \Gamma_{cn}, \\ \Delta_s &= L(\delta_{cc} + \delta_{FF} + \Delta_{cn}), \end{aligned} \tag{23}$$

where Γ_{cn} is the whole connection cost (including the control cost), and Δ_{cn} is the whole connection delay time.

In order to compare both circuits, we make two assumptions:

$$W_{fi} = pW_c \quad \text{and} \quad W_c = \frac{N}{L}.$$

They are verified by a very regular computation scheme whose typical precedence graph is shown in Figure 19. Hence

$$\Gamma_c = LW_c\gamma_{cc} \quad \text{and} \quad \Gamma_s = W_c\gamma_{cc} + pW_c\gamma_{FF} + \Gamma_{cn}; \tag{24}$$

Γ_s is smaller than Γ_c if

$$p\gamma_{FF} + (\Gamma_{cn}/W_i) < (L-1)\gamma_{cc}. \tag{25}$$

If the connection cost is negligible, as it arose in the circuits of Figures 14 and 18, condition (25) becomes

$$p\gamma_{FF} < (L-1)\gamma_{cc}. \tag{26}$$

In many cases the first term of (26) is much smaller than the second one, because L or γ_{cc} are large by comparison with $p\gamma_{FF}$.

On the other hand, the computation times δ_{FF} and Δ_{cn} often are smaller than, or comparable to, δ_{cc}. The computation time of the sequential circuit thus has the same magnitude order as that of the combinational circuit.

In conclusion, if it is possible to define allocations which make the connection cost negligible, the sequential circuit often leads to a much smaller cost–delay time product than the combinational circuit.

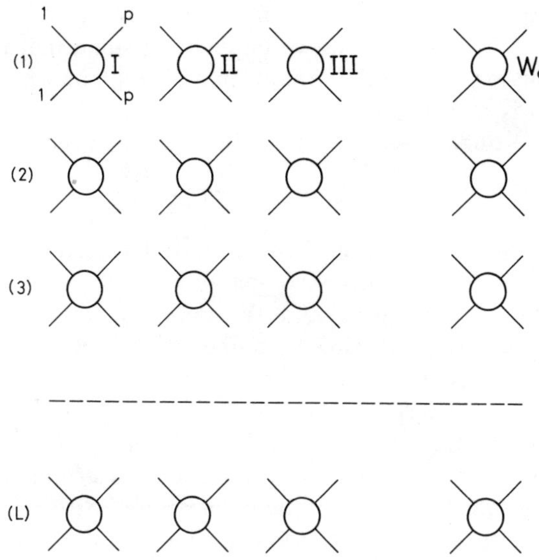

Figure 19. Precedence graph

The conclusion is obviously different if it is not possible to make the connection cost negligible. Let us again consider a computation scheme whose graph is given in Figure 19, and suppose that the allocation choice does not bring any connection network simplification. Figure 20 describes a connection network which includes N_{fl} multiplexers with pW_c data inputs and one perfect shuffle. As $W_{fl} = pW_c$ and since there are two connection networks,

$$\Gamma_{cn} \cong 2(pW_c)^2\gamma_{cn},$$

and

$$\Delta_{cn} = 2(\log_2 pW_c)\delta_{cn}, \qquad (27)$$

where γ_{cn} and δ_{cn} are the cost and delay time of a two data input

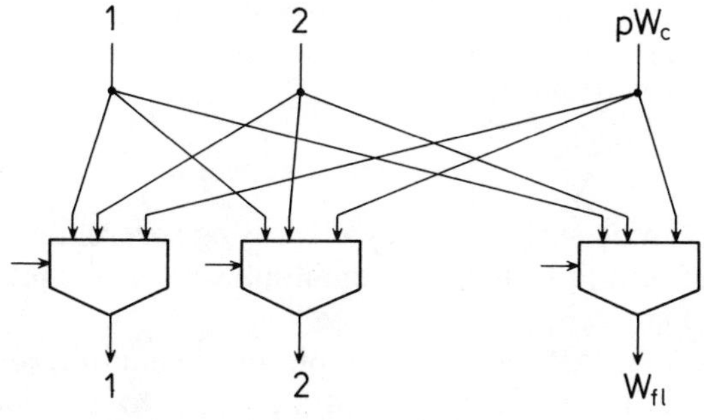

Figure 20. Connection network

multiplexer: see relations (3) and (4) in Chapter III with $2^n = pW_c$ and thus $n = \log_2 pW_c$. Let us observe that we implicitly assume that the control cost is distributed among all two data input multiplexers, and is included in γ_{cn}; this assumption is rather questionable.

Condition (25) under which Γ_s is smaller than Γ_c becomes

$$p\gamma_{FF} + (2p^2 W_c)\gamma_{cn} < (L-1)\gamma_{cc}. \tag{28}$$

On the other hand, computation times Δ_s and Δ_c are comparable if δ_{FF} and $(\log_2 pW_c)\delta_{cn}$ are smaller than, or comparable to, δ_{cc}.

Hence, for small values of p and W_c, that is in the case of algorithms having a low intrinsic parallelism, the sequential circuit could be better than the combinational circuit, even if the connection network cost cannot be made negligible.

The preceding discussion suggests two comments.

(1) The resource and memory allocation choice is a fundamental problem. According to the selected solution, the circuit complexity can vary widely. In particular, if an appropriate allocation choice makes the connection network cost negligible, the whole sequential circuit cost is often much smaller than the combinational circuit cost, without great delay increase. It would be unreasonable to implement the combinational version.

(2) The connection networks are best paid off if the implemented algorithm has a low intrinsic parallelism. Hence, it is sometimes rewarding to modify the initial algorithm, in order to lower its intrinsic parallelism (see Chapter XI). This operation decreases the cost, but increases the computation time.

7 THROUGHPUT INCREASE: CONCURRENT AND PIPELINE NETWORKS

Let us consider a combinational circuit whose computation cost and time are Γ_c and Δ_c. It can generate a new result every Δ_c time units, and its maximum *throughput* is equal to $1/\Delta_c$.

It may arise that input data arrive, and must be processed, at a higher rate than $1/\Delta_c$. Two ways to increase the throughput of a circuit are described in this section: concurrent and pipeline networks.

7.1 Concurrent networks

The throughput can be simply increased by using several copies of the same circuit, along with multiplexing and demultiplexing equipment. The method is illustrated by Figure 21.

If we neglect the cost of the multiplexer, demultiplexer, and control circuit, the cost of the concurrent network is equal to $k\Gamma_c$, and its apparent computation time, that is the inverse of the throughput, is equal to Δ_c/k.

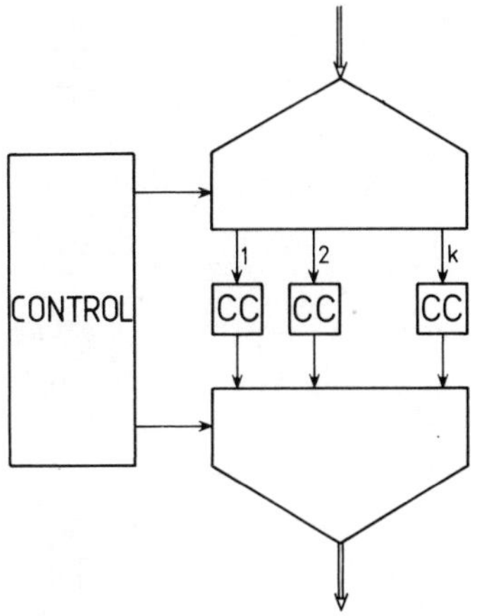

Figure 21. Concurrent networks

Hence, the cost–delay time product keeps a constant value

$$(k\Gamma_c)(\Delta_c/k) = \Gamma_c\Delta_c. \tag{29}$$

7.2 Pipeline networks

Let us consider a computation scheme, its precedence graph, and an acceptable labelling ψ of its nodes. We can define the connection function Λ with respect to ψ: $\psi(x)$ is the number of symbols which are generated during the steps N_i such that $\psi(N_i) \leqslant x$, and are next used during steps N_j such that $\psi(N_j) \geqslant x+1$ (see Section 2). As above (Section 6), L stands for the number of nodes encountered on the longest path linking a source node to a terminal node.

Let us suppose that $L = nk$, where n and k are positive integers. A *pipeline network*, with *latching grade k*, is deduced from the initial combinational network by adding a memory element on every connection wire linking a resource N_i such that $\psi(N_i) \leqslant qn$, to a resource N_j such that $\psi(N_j) \geqslant qn+1$, for every $q = 1, 2, \ldots, k$. Figure 22 shows an example, with $L=6$, $k=3$, $n=2$.

The pipeline circuit works as follows: when the results of computations performed at times $(q-1)n+1, (q-1)n+2, \ldots, qn$, are obtained, they are stored in the $\Lambda(qn)$ memory elements lying between levels qn and $qn+1$; these results are available for the next computations, while the resources which generated them are now inactive and available for new computations. If we neglect the computation time of the memory elements, the throughput is multiplied by k since new input data may be introduced every n time intervals instead of every L time intervals.

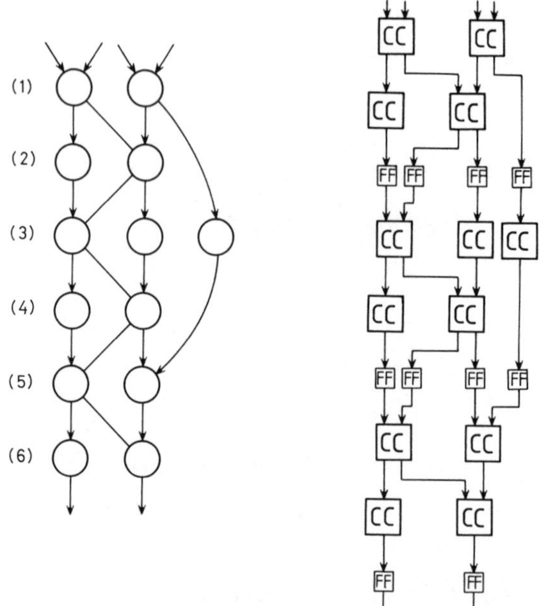

Figure 22. Pipeline network

In order to compare the pipeline network with both the concurrent network and the initial combinational network, we use the same notations N, γ_{cc}, δ_{cc}, γ_{FF}, δ_{FF}, as in Section 6. The computation cost and time of the pipeline network are equal to

$$\Gamma_p = N\gamma_{cc} + (\Lambda(n) + \Lambda(2n) + \ldots + \Lambda(kn))\gamma_{FF}, \tag{30}$$

and

$$\Delta_p = n\delta_{cc} + \delta_{FF} = \frac{L}{k}\delta_{cc} + \delta_{FF}. \tag{31}$$

If Λ stands for the maximum value of $\Lambda(x)$ over all x's, we may replace (30) by the upper bound

$$\Gamma_p = N\gamma_{cc} + k\Lambda\gamma_{FF}.$$

The cost–delay time product of the pipeline network is equal to

$$\Gamma_p\Delta_p = (N\gamma_{cc} + k\Lambda\gamma_{FF})\left(\frac{L}{k}\delta_{cc} + \delta_{FF}\right). \tag{32}$$

A simple derivative computation shows that (32) is minimum when

$$k \cong (LN\gamma_{cc}\delta_{cc}/\Lambda\gamma_{FF}\delta_{FF})^{1/2}; \tag{33}$$

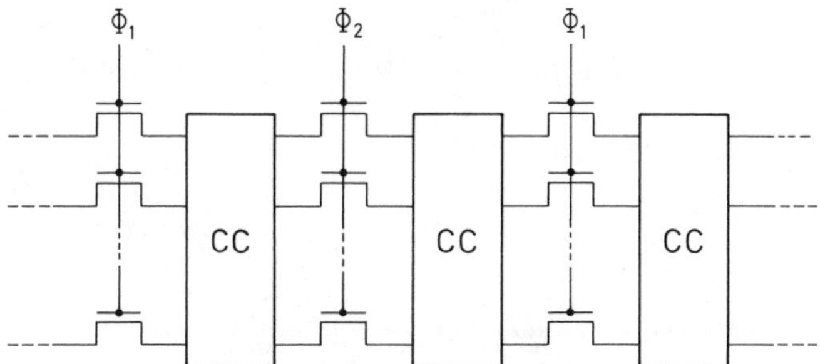

Figure 23. Pipeline circuit with dynamic memory elements

the minimum value is equal to

$$\Gamma_p\Delta_p \cong N\gamma_{cc}\delta_{FF} + L\Lambda\gamma_{FF}\delta_{cc} + 2(NL\Lambda\gamma_{cc}\delta_{cc}\gamma_{FF}\delta_{FF})^{1/2}$$
$$= \frac{\Gamma_c\Delta_c}{L}\left(\frac{\delta_{FF}}{\delta_{cc}} + \frac{L\Lambda}{N}\frac{\gamma_{FF}}{\gamma_{cc}} + 2\left(\frac{L\Lambda}{N}\frac{\gamma_{FF}}{\gamma_{cc}}\frac{\delta_{FF}}{\delta_{cc}}\right)^{1/2}\right), \tag{34}$$

where $\Gamma_c = N\gamma_{cc}$ and $\Delta_c = L\delta_{cc}$ are the computation cost and time of the combinational circuit.

In order to compare (29) and (34), we make the same assumptions as in Section 6: $\Lambda = W_{fl} = pW_c = pN/L$. Hence,

$$\Gamma_p\Delta_p = \frac{\Gamma_c\Delta_c}{L}\left(\frac{\delta_{FF}}{\delta_{cc}} + \frac{p\gamma_{FF}}{\gamma_{cc}} + 2\left(\frac{p\gamma_{FF}}{\gamma_{cc}}\frac{\delta_{FF}}{\delta_{cc}}\right)^{1/2}\right). \tag{35}$$

As long as δ_{FF} is smaller than or comparable with δ_{cc}, $p\gamma_{FF}$ smaller than or comparable with γ_{cc}, and L great enough, the pipeline network has a lower cost–delay time product than the concurrent network and the initial combinational network. Furthermore, if γ_{FF} and δ_{FF} are negligible, then, according to (32), $\Gamma_p\Delta_p \cong \Gamma_c\Delta_c/k$. More details may be found in Davio and Bioul (1978).

Let us notice that pipeline networks are examples of circuits in which a continuous flow of information circulates (hence the name). They are conveniently implemented by combinational resources and dynamic memory elements (Figure 23).

7.3 Example: algorithm to find the g.c.d. of two integers

Let (a, b) stand for the greatest common divisor of positive integers a and b. The algorithm is based on the three following rules:

(i) if a and b are even, then $(a, b) = 2(a/2, b/2)$;

(ii) If a is even and b odd, then $(a, b) = (a/2, b)$;

(iii) if a and b are odd, and if $a \geqslant b$, then $(a, b) = (a-b, b)$, and $(a-b)$ is even.

Algorithm 2. *Initial step* 0: $a_0 = a$, $b_0 = b$, $c_0 = 0$.
Step i: compute a_i, b_i, c_i, *in function of* a_{i-1}, b_{i-1}, c_{i-1}, *according to the five following rules:*
(1) *if* a_{i-1} *and* b_{i-1} *are even, then*

$$a_i = \frac{a_{i-1}}{2}, \qquad b_i = \frac{b_{i-1}}{2}, \qquad c_i = c_{i-1} + 1;$$

(2) *if* a_{i-1} *is even and* b_{i-1} *odd, then*

$$a_i = \frac{a_{i-1}}{2}, \qquad b_i = b_{i-1}, \qquad c_i = c_{i-1};$$

(3) *if* a_{i-1} *is odd and* b_{i-1} *even, then*

$$a_i = a_{i-1}, \qquad b_i = \frac{b_{i-1}}{2}, \qquad c_i = c_{i-1};$$

(4) *if* a_{i-1} *and* b_{i-1} *are odd, and if* $a_{i-1} \geqslant b_{i-1}$, *then*

$$a_i = a_{i-1} - b_{i-1}, \qquad b_i = b_{i-1}, \qquad c_i = c_{i-1};$$

(5) *if* a_{i-1} *and* b_{i-1} *are odd, and if* $a_{i-1} < b_{i-1}$, *then*

$$a_i = a_{i-1}, \qquad b_i = b_{i-1} - a_{i-1}, \qquad c_i = c_{i-1}.$$

According to rules (i), (ii), and (iii):

$$(a, b) = (a_0, b_0) = 2^{c_i}(a_i, b_i). \tag{36}$$

In particular, if $a_i = 0$, then $(a, b) = b_i 2^{c_i}$.

Let us prove that after a finite number of steps $a_i = 0$. Indeed, if a_{i-1} and b_{i-1} are both positive, then $a_i + b_i < a_{i-1} + b_{i-1}$, and as only rule (4) can generate a zero value, there exists an index ℓ such that $a_\ell = 0$ and b_ℓ is odd; furthermore, according to rule (2), $a_{\ell+j} = 0$, $b_{\ell+j} = b_\ell$, and $c_{\ell+j} = i$, for every $j = 1, 2, \ldots$.

It remains to compute the maximum value of index ℓ such that $a_\ell = 0$. Let us first prove that if $a_i \neq 0$, then $a_{i+2}b_{i+2} \leqslant a_i b_i/2$. Indeed, either $a_{i+1} = a_{i+2} = 0$, or $a_{i+1} \neq 0$ and at least one of the steps $i+1$ and $i+2$ contains a division by two. An iterative use of the property proves that, for every index i, either $a_{2i} = 0$ or

$$a_{2i}b_{2i} \leqslant a_0 b_0/2^i = ab/2^i.$$

If $a < 2^s$ and $b < 2^s$, then $a_{4s} = 0$. In the contrary case,

$$1 \leqslant a_{4s}b_{4s} \leqslant ab/2^{2s} < 1.$$

Figure 24 shows a computation resource which performs step i of the algorithm. It includes $2s + m$ binary inputs and outputs. Let us notice that $2^{c_i} \leqslant (a, b) < 2^s$, and thus $c_i < s$. Hence, we may choose $m = \lceil \log_2 s \rceil$.

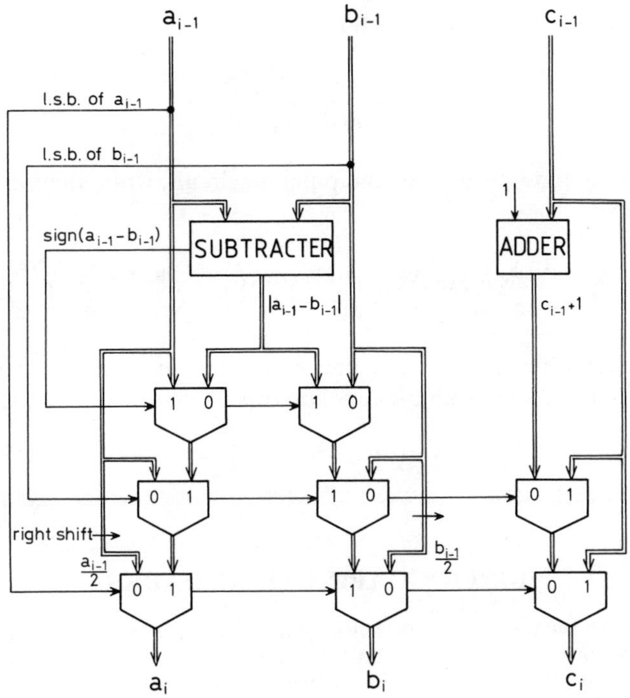

Figure 24. Computation resource

Figure 25 describes the whole circuit. It gives the g.c.d. of a and b under the form

$$(a, b) = B2^C.$$

The parameters of the algorithm are

$$N = L = 4s; \qquad W_c = 1; \qquad \Lambda = W_{fl} = p = 2s + m.$$

Figure 25. Computation of $(a, b) = B2^C$

The optimal value of k is given by (33):

$$k=\frac{4s}{(2s+m)^{1/2}}\left(\frac{\gamma_{cc}}{\gamma_{FF}}\frac{\delta_{cc}}{\delta_{FF}}\right)^{1/2}.$$

The cost–delay time product of the pipeline circuit, with latching grade k, is equal to (35):

$$\Gamma_p\Delta_p=\frac{\Gamma_c\Delta_c}{4s}\left(\frac{\delta_{FF}}{\delta_{cc}}+\frac{(2s+m)\gamma_{FF}}{\gamma_{cc}}+2\left(\frac{(2s+m)\gamma_{FF}}{\gamma_{cc}}\frac{\delta_{FF}}{\delta_{cc}}\right)^{1/2}\right).$$

The computation cost γ_{cc} is a linear function of s, and its computation time δ_{cc} is, at best, a logarithmic function of s. Hence, for large values of s:

$$\Gamma_p\Delta_p\cong\frac{\Gamma_c\Delta_c}{2}\frac{\gamma_{FF}}{\gamma_{cc}}\ll\Gamma_c\Delta_c.$$

BIBLIOGRAPHICAL REMARKS

Very few works seem to be devoted to computation schemes. Let us mention two papers by Howard (1975a, 1975b) and, from a more theoretical point of view, the paper by Savage (1972). The relation between precedence graph labelling and the critical path method does not seem to have been pointed out elsewhere. More details about critical path techniques may be found in Kelley and Walker (1959) and Kelley (1961). Pipeline circuits are discussed in Cotten (1965, 1969), Deverell (1975), Helln and Flynn (1972), Majithia (1976), and Davio and Bioul (1978).

EXERCISES

E1. In Section 5, we discussed the sequential implementation of a binary to BCD conversion. Discuss similarly the BCD to binary conversion. For the above two algorithms, discuss the efficiency of pipelining. The general study of radix conversion (Chapter IV) has pointed out the possibility of examining pairs of bits in the binary representation by using radix 4 to BCD conversion. Does this algorithm lend itself well to sequential implementation?

E2. Discuss the pipeline of the addition of two 64-bit integers, varying the cost criterion: choose as cost criterion
(i) the number of gates in the circuit, and
(ii) the number of commercial components (full-adders, hexadecimal adders).

E3. Consider the following program to find the g.c.d. of two integers a and b. The least significant bits of a and b are denoted by a_0 and b_0, respectively.

```
x:=a;     y:=b;     z:=0
while x≠0 do
   begin if (x0=0 and y0=0) then begin x:=x div 2;     y:=y div 2;
                                        z:=z+1
                                 end;
```

```
if (x0 = 0 and y0 = 1) then x := x div 2;
if (x0 = 1 and y0 = 0) then y := y div 2;
if (x0 = 1 and y0 = 1) then begin if x >= y
                                    then x := x - y
                                    else y := y - x
                               end;
                    g.c.d (a, b) := y * (2 ** z)
end
```

Prove that this algorithm actually computes the g.c.d. of a and b. Observe that the four *if* statements in the program body are independent and may be executed in parallel. Design a combinational circuit implementing the above algorithm. Discuss its sequential and its pipeline counterparts.

Chapter IX
Multiplication and division

1 INTRODUCTION

The two simplest arithmetic operations, namely addition and subtraction were studied in Chapter IV and we postponed until now a study of the two remaining basic arithmetic operations, namely multiplication and division. The reason for handling these two pairs of operations separately is to be found in their different complexity levels: the practical counterpart of this fact may be found in the fact that addition and subtraction are most often realized by combinational circuits, while multiplication and division are most often realized by sequential circuits; in this respect, the techniques developed in the last chapter will play a useful role. The above argument may be sharpened a little further. Let us, for example, discuss multiplication: we know that the usual pencil and paper multiplication algorithm reduces the problem of multiplying two n-digit numbers to the problem of adding n numbers. This means that a computation involving both multiplications and additions may be reformulated in terms of addition alone. On the one hand, the number of distinct computation primitives is reduced; on the other hand, all the primitives put at work in a given computation have the same duration. This important point will be discussed again in Chapter X.

2 MULTIPLICATION OF POSITIVE INTEGERS

2.1 The shift and add algorithms

Consider two integers X and Y given by their radix B representations:

$$X \simeq x_{n-1} \ldots x_1 x_0; \qquad Y \simeq y_{n-1} \ldots y_1 y_0; \qquad 0 \leq x_i, y_i < B. \tag{1}$$

In particular, one has:

$$X = \sum_{i=0}^{n-1} x_i B^i. \tag{2}$$

One wishes to compute the product $Z = XY$ of these two integers. It is easy to show that the radix B representation of Z consists of, at most, $2n$ digits, and we write accordingly:

$$Z \simeq z_{2n-1} \ldots z_1 z_0. \tag{3}$$

From (2), we immediately obtain

$$Z = \sum_{i=0}^{n-1} x_i Y B^i. \tag{4}$$

The product Z immediately appears as the *sum* of *shifted* copies of Y, viz. YB^i, weighted by the digits x_i of X representation. Hence the name 'shift and add' algorithms. There exist, in fact, two possible variants of the shift and add algorithms that we may relate to distinct interpretations of (4).

A. *The Hörner scheme algorithm.* Equation (4) may be written under the form:

$$Z = ((\ldots((0 \cdot B + x_{n-1}Y)B + x_{n-2}Y)B + \ldots)B + x_1 Y)B + x_0 Y. \tag{5}$$

Let

$$C_i = (\ldots((0 \cdot B + x_{n-1}Y)B + x_{n-2}Y)B + \ldots)B + x_i Y. \tag{6}$$

According to (5), we may thus compute $Z = C_0$ by the recurrence:

$$\begin{aligned} &C_n = 0, \\ &C_{i-1} = BC_i + x_{i-1}Y; \qquad i = n, n-1, \ldots, 2, 1. \end{aligned} \tag{7}$$

Thus, in each of the steps of the recurrence (7), we add to the partial result C_i, first shifted one position to the left, the weighted copy $x_{i-1}Y$ of Y. Figure 1 illustrates an implementation of the preceding algorithm for $B = 2$ and $n = 6$. We retained here an implementation by a synchronous sequential machine and used freely an appropriate memory reallocation (left shift register containing X) and an appropriate connection pattern (left shift of C)

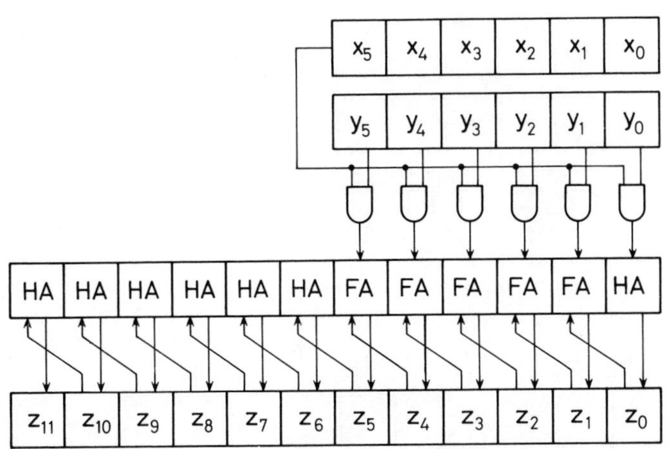

Figure 1. Shift and add multiplier (1st solution)

to simplify the involved hardware while performing the whole computation implied by (7) in a single clockpulse. It immediately appears from Figure 1 that this solution rests upon the use of a double-length adder. A second interpretation of (4), discussed below, will get us rid of that drawback.

B. Right to left factorization. From (4), we easily obtain:

$$\frac{Z}{B^{n-1}}=x_{n-1}Y+B^{-1}(x_{n-2}Y+B^{-1}(\ldots+B^{-1}(x_1Y+B^{-1}(x_0Y+B^{-1}0))\ldots)). \tag{8}$$

Let

$$D_i=x_iY+B^{-1}(x_{i-1}Y+B^{-1}(\ldots+B^{-1}(x_1Y+B^{-1}(x_0Y+B^{-1}0))\ldots)).$$

According to (8), we may now compute $Z/B^{n-1}=D_{n-1}$ by the recurrence

$$D_{-1}=0,$$
$$D_{i+1}=B^{-1}D_i+x_iY. \tag{9}$$

In this case at each step of the recurrence (9), we add to the partial result D_i, first shifted one position to the right, the weighted copy x_iY of Y. Figure 2 illustrates, again for $B=2$ and $n=6$, an implementation of the above algorithm as a synchronous sequential machine. Once again, we managed to realize within a single clock period the computation implied by (8).

The main advantage of the circuit in Figure 2 over that in Figure 1 is clearly that it only requires a simple length adder. This is due to the fact that, after the ith step of the recurrence (9), the i less significant bits of the product Z are known and are the i less significant bits of D_i. Observe that nothing similar may happen in Figure 1 as at step i, a carry from the ith position may propagate towards the more significant bit positions.

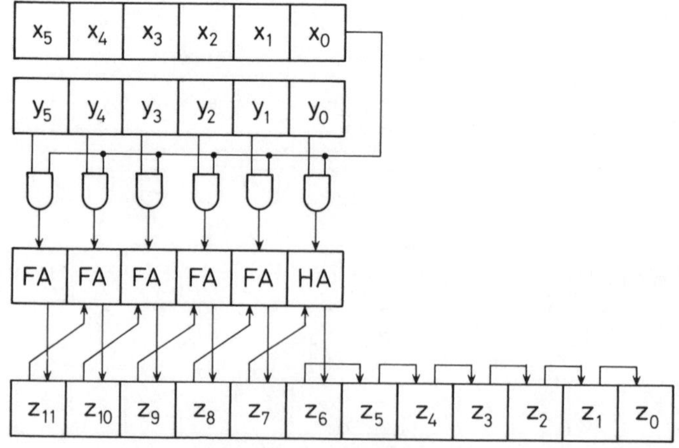

Figure 2. Shift and add multiplier (2nd solution)

Another interesting feature of the circuit in Figure 2 is that the n least significant flip-flops of the result register Z may be used to store initially the multiplier X: the X register becomes useless, once it is known that X may be lost after the computation.

It should be clear from the above discussion that the time required to carry out the multiplication of two n-bit integers is $0(n^2)$ if one uses a serial adder, and $0(n \log n)$ if one uses a fast adder, say of the carry-lookahead type. The cost of the sequentialized multiplier is $0(n)$.

2.2 Cellular multipliers

2.2.1 The ripple–carry multiplier

The use of combinational circuits to carry out multiplication may bring a significant displacement of the time–space trade-off. In this section, we present a combinational multiplier which is in fact a second implementation of the shift and add multiplication algorithm described in Section 2.1.B. Let x_i, x_j, w_k, z_1 represent four radix B digits. We thus have

$$0 \leqslant x_i, x_j, w_k, z_1 \leqslant B-1.$$

It is now possible to check that the expression $x_i y_j + w_k + z_1$ is bounded by

$$0 \leqslant x_i y_j + w_k + z_1 \leqslant B^2 - 1 \tag{10}$$

and may thus be represented by two radix B digits. As a matter of fact, the expression

$$x_i y_j + w_k + z_1 = Bc + s \tag{11}$$

defines the radix B digits c and s as functions of x_i, y_j, w_k, and z_1. The operator described by equation (1) will be our basic building block for the construction of cellular multipliers. It is sketched in Figure 3a and, for $B = 2$, it reduces to the combination of a AND-gate and a full-adder, as shown in Figure 3b. The combination (AND-gate, full-adder) was already present in Figure 2. In what follows, we shall use the symbol of Figure 3c to represent the cell of Figure 3a. Consider now the array of cells displayed in Figure 3d. It is easy to see that this array computes the quantity

$$R = XY + Z + W, \tag{12}$$

where X, Y, Z, and W are the 4-digit integers $x_3x_2x_1x_0$, $y_3y_2y_1y_0$, etc. More precisely, the set of vertical connections at the output of the row x_i carries the representation of D_i, as given by (8). The comparison of equations (11) and (12) leads us to the expandability of the function $xy + z + w$, and this also explains why the algorithm underlying Figure 3 is also a classical algorithm for multiprecision multiplication: in that case, x_i, y_j, w_k, and z_1 are r-bit integers.

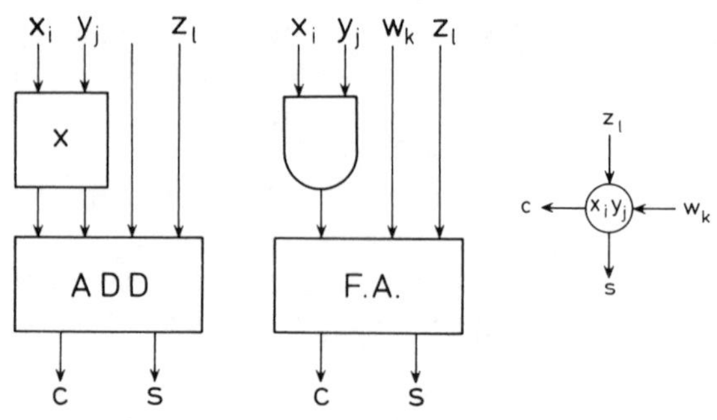

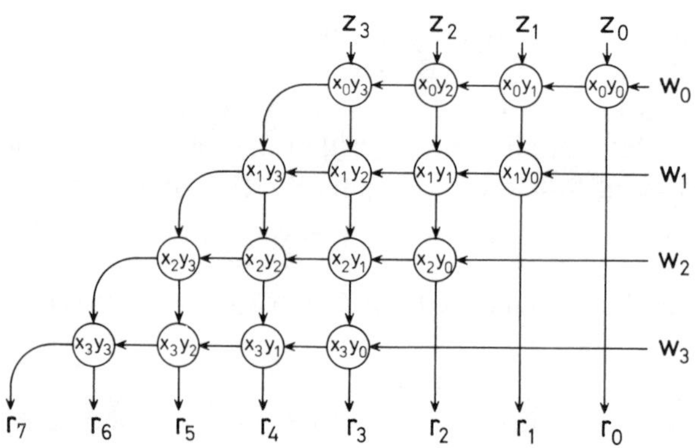

Figure 3. Ripple–carry multiplier

The multiplier illustrated by Figure 3 is known as the *ripple–carry multiplier.* The performances of the circuit are easily evaluated: its cost is $0(n^2)$ as there are n^2 cells of the type described in Figure 3a; its delay is $0(n)$. An easy count on Figure 3, viewed as a precedence graph, indeed shows that the computation time is $(3n-2)\tau$, where τ is the delay of the building block. The improvement with respect to the $0(n^2)$ delay obtained in Figure 2 is due to the parallelism involved in Figure 3: as a matter of fact all the cells located on the same descending diagonal in Figure 3 are simultaneously active.

2.2.2 The carry-save multiplier

To reach a new type of array multiplier, we consider again the circuit in Figure 3d. Consider the cell receiving the digits x_i and y_j. It is located in a column corresponding to the weight B^{i+j}. Its sum digit is summed within the column while the carry output is passed to the left neighbouring cell.

Obviously, the only important feature here is that the carry should be passed to the left neighbouring column and in particular, that it can be passed to the lower left neighbouring cell so as to avoid horizontal carry propagation. This process is illustrated by the four top rows of Figure 4. In the general situation where X, Y, Z and W have n digits, we are left, after n steps of the process, with two integers whose sum is equal to the numerical value $XY+Z+W$. These two integers are added up by a cascade of full-adders. It is not difficult, by labelling the circuit of Figure 4, viewed as a precedence graph, to conclude that the computation time of the circuit in Figure 4 is $2n\tau$, where τ is the (assumedly) common response time of the basic cell and of the full-adder: roughly speaking, the response time has been improved by a factor 2/3 with respect to Figure 3. The obtained multiplier is known as the *carry–save multiplier.*

It is particularly interesting to apply to the carry–save multiplier the sequential implementation techniques described in Chapter VIII. Before computing the various graph parameters associated with this multiplier, we first modify it slightly in order to eliminate the dissymmetry introduced in the algorithm by the cascade of full-adders in its bottom row. The principle of the modification is very simple: we just replace $X \simeq x_3x_2x_1x_0$ by the 8-digit representation $X \simeq 0\,0\,0\,0\;x_3x_2x_1x_0$ and we extend accordingly our multiplication array as illustrated by the precedence graph in Figure 5. In that figure, black and white dots stand for basic building cells and full-adders, respectively, As the four most significant digits of the result are now zero, the full-adders become useless and to reach the sequential counterpart of Figure 4 we may limit ourselves to the first eight rows of Figure 5. The computation width is 4 and the information flow width is 8 (neglecting the permanent storage of Y). The resource and memory allocations will take

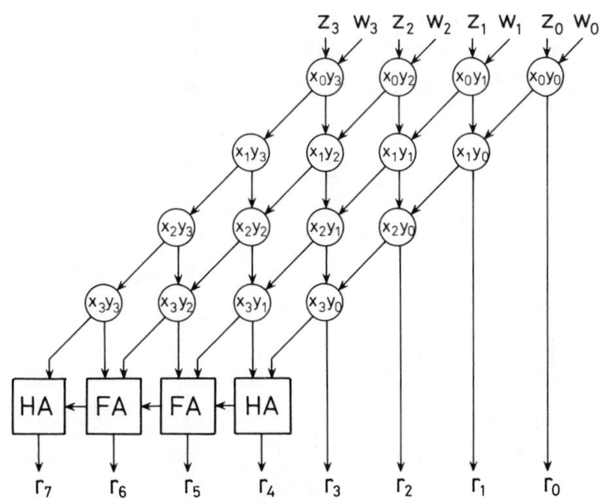

Figure 4. Carry–save multiplier

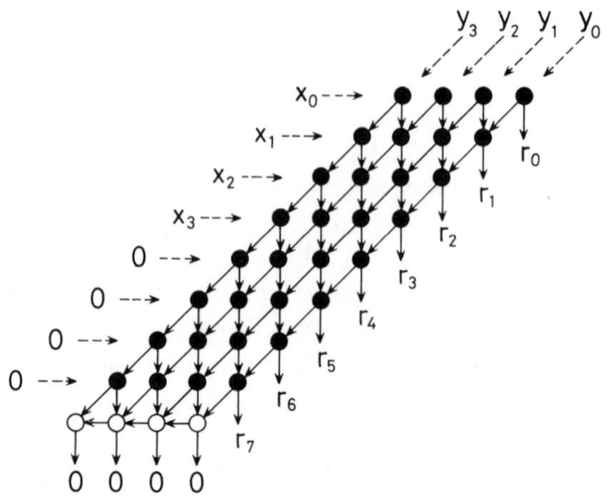

Figure 5. Sequentialization of the carry–save multiplier

into account the topological regularities of the array in Figure 8 and one naturally reaches the circuit shown in Figure 6. This circuit is known as the *parallel–serial multiplier* as the multiplicand Y is provided to the circuit in the parallel mode, while the multiplier X is provided to the circuit in the serial mode. The result R is also produced serially. Both X and R are handled by shift-registers which are not shown in Figure 6.

The performance improvement with respect to the multiplier shown in Figure 2 is noteworthy, as the computation time has dropped from $0(n^2)$ to $0(n)$: the multiplication in the circuit of Figure 6 indeed requires $2n$ clock periods. The origin of that improvement is obviously the algorithm choice: we indeed decided to postpone until the next clock period the handling of the carries produced during a particular clock period (hence the name carry–save) instead of letting these carries ripple through the combinatorial

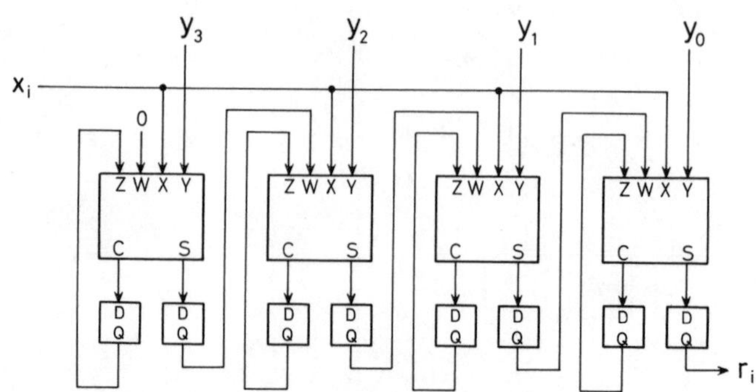

Figure 6. Serial–parallel multiplier

part of the circuit. Globally, the multiplier shown in Figure 6 has both linear time and linear cost (with respect to n).

Remark. Various other arrays for multiplying two integers have been proposed in the literature (Guild, 1969; De Mori 1969). Viewed as combinational circuits, these multipliers have performances very similar to the carry–save adder discussed above, viz. they have cost $0(n^2)$ and delay $0(n)$. We shall not discuss these circuits here.

2.2.3 Fast multipliers

In the foregoing section, we observed that, when one allows some parallelism in the computations, it is possible to carry out the multiplication of two n-bit numbers in linear time. In the present section, we show that, if one relaxes the requirement for a geometrical regularity and used more freely the possibilities of parallelism, it becomes possible to carry out the multiplication of two n-bit numbers in time $0(\log n)$.

The basic principle to be used for reaching this goal is again the carry–save principle. In its essence, this principle states that it is possible, within a single full-adder delay, to replace three integers A, B, and C by two integers F and G having the same sum, i.e. satisfying $A+B+C=F+G$ (observe that the computation time is one full-adder delay independently of the length of the involved integers). The principle is illustrated in Figure 7 for three 4-bit numbers A, B, and C.

Consider now again the multiplication of two n-bits numbers X and Y. During a first time unit, we form the n^2 bit by bit products $x_i y_j$. At this point, our problem reduces to the addition of n integers. Let $n=3q_1+r_1$. Then, applying the carry–save principle on q_1 set of three numbers, we shall, in a single time unit replace the n integers by $n_1=2q_1+r_1$ numbers. Then, if $n_1=3q_2+r_2$, we shall, in a second type, replace our n_1 summands by $n_2=2q_2+r_2$ summands, etc. Eventually, we shall reach a situation where two summands only are left over: at this state, we shall use a fast adder, such as a carry–lookahead adder to produce the final result.

It remains to give an estimate of the method performances. First, we evaluate the number of carry–save steps required to reduce the number of

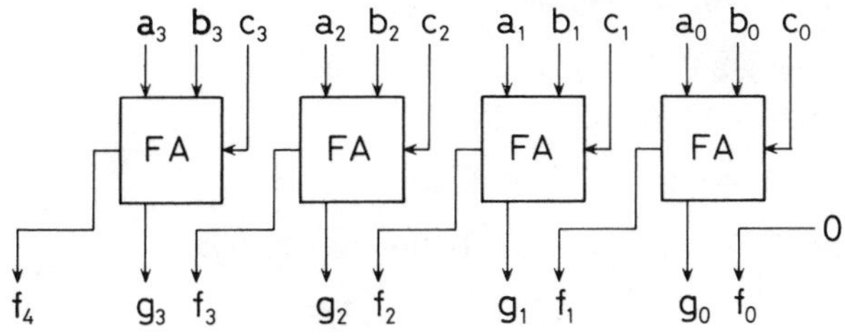

Figure 7. The carry–save principle

summands from n to 2. Let x_t be the maximum number of summands reducible to two in t steps. It is easy to characterize x_t by the recurrence:

$$\begin{aligned} &x_0 = 2, \\ &\text{if } x_{t-1} \text{ is even, then } x_t = 3x_{t-1}/2, \\ &\text{otherwise, } x_t = (3(x_{t-1}-1)/2)+1. \end{aligned} \tag{13}$$

We thus have

$$\tfrac{3}{2}x_{t-1} - \tfrac{1}{2} \leqslant x_t \leqslant \tfrac{3}{2}x_{t-1}. \tag{14}$$

This recurrence is easily solved and yields

$$\left(\frac{3}{2}\right)^t + 1 \leqslant x_t \leqslant 2 \cdot \left(\frac{3}{2}\right)^t. \tag{15}$$

In spite of its coarse character, the above equation will suffice for our purpose as it shows that a number n of summands such that $n \leqslant (3/2)^t + 1 \leqslant x_t$ is reducible to 2 summands in t steps, or, conversely, that $t = \lceil \log_{3/2}(n-1) \rceil$ steps suffice to carry out the reduction from n to two summands. Thus, the computation time for multiplying two n-bit numbers becomes $0(\log n)$ as the sum of two $2n$-bit numbers also requires a delay of the same order of magnitude. On the other hand, it is easy to prove that the cost required for this method of multiplication is $0(n^2)$.

3 MULTIPLICATION OF SIGNED INTEGERS

3.1 Sign-magnitude multiplication

The multiplication of two signed integers X and Y given by their sign-magnitude representation $(x_s, \mathbf{x})$ and $(y_s, \mathbf{y})$ is easily solved as the sign z_s of the result is given by $z_s = x_s \oplus y_s$ and as the magnitude $\mathbf{z}$ of the product may be computed by any of the methods discussed in the previous section.

3.2 2's complement multiplication

3.2.1 Shift and add algorithms

Consider two integers X and Y given by their 2's complement representations $x_{n-1} \ldots x_1 x_0$ and $y_{n-1} \ldots y_1 y_0$. Thus:

$$-2^{n-1} \leqslant X, Y \leqslant 2^{n-1} - 1. \tag{16}$$

The range of the product $Z = XY$ is thus given by:

$$-2^{n-1}(2^{n-1}-1) \leqslant Z \leqslant 2^{2n-2}. \tag{17}$$

The maximum value $Z = 2^{2n-2}$, obtained for $X = Y = 2^{n-1}$, is pathological in the sense that it is the only value requiring $2n$ bits for its 2's complement representation. The existence of that value leads us to express our results as $2n$-bit numbers; of course, except in the pathological case we have just mentioned, the two most significant digits of the result will be identical.

It is now easy to modify the shift and add algorithms of Section 2.1 so as to adapt them to the 2's complement case. The analog of equation (8) is, for example, written:

$$\frac{Z}{B^n}=B^{-1}(-x_{n-1}Y+B^{-1}(x_{n-2}Y+B^{-1}(\ldots+B^{-1}(x_1Y+B^{-1}(x_0Y+0))\ldots))). \tag{18}$$

The result is thus obtained at the end of the algorithm:

$$\begin{aligned}&D_{-1}=0\\ &\text{for}\quad i=0,1,\ldots,n-2:\qquad D_i=B^{-1}(x_iY+D_{i-1})\\ &D_{n-1}=B^{-1}(-x_{n-1}Y+D_{n-2}).\end{aligned} \tag{19}$$

Each step of (19) again consists of an addition followed by a shift, except for the last step where the addition is replaced by a subtraction to take into account the negative significance of the sign digit.

It should be clear that the additions and shifts involved in the above algorithm are 2's complement additions and shift, i.e. additions neglect the carries from the sign bit position and in the absence of overflow, shifts are sign bit preserving. If an overflow appears during an addition, care should be taken during the shift to recover the proper sign bit. These facts are illustrated by the numerical example

$$(1\ 0\ 1\ 0\ 1)\cdot(1\ 0\ 0\ 1\ 1):(-1\ 1)\cdot(-1\ 3)=143.$$

	x_i	z_s	overflow
D_{-1}	1	0 0 0 0 0	
		+1 0 1 0 1	
		1 0 1 0 1	0
D_0	Shift	1 1 0 1 0 1	
	1	+1 0 1 0 1	
		0 1 1 1 1 1	1
D_1	Shift	1 0 1 1 1 1 1	
	0	+0	
		1 0 1 1 1 1 1	0
D_2	Shift	1 1 0 1 1 1 1 1	
	0	+0	
		1 1 0 1 1 1 1 1	0
D_3	Shift	1 1 1 0 1 1 1 1 1	
	1		
	(subtr.)	+0 1 0 1 1	
		0 1 0 0 0 1 1 1 1	0
D_4	Shift	0 0 1 0 0 0 1 1 1 1	

Remark. 2's complement shift and add algorithms will be considered again in later chapters: in Chapter XI, the implementation of the above algorithm on the processing unit Am 2901 will allow us to describe more accurately the overflow preserving shift it involves. In Chapter XII, we shall describe a completely different 2's complement multiplication technique known as the modified Booth's algorithm.

3.2.2 2's complement multiplication arrays

As for shift and add multiplication, 2's complement multiplication arrays are obtained by suitable modifications of unsigned multiplication arrays. Consider the two operands X and Y written under the form:

$$X = -x_{n-1}2^{n-1} + X_0; \qquad X_0 = \sum_{i=0}^{n-2} x_i 2^i,$$

$$Y = -y_{n-1}2^{n-1} + Y_0; \qquad Y_0 = \sum_{i=0}^{n-2} y_i 2^i,$$

where X_0 and Y_0 represent positive integers. From the above expressions, one immediately deduces

$$Z = XY = X_0Y_0 + x_{n-1}y_{n-1}2^{2n-2} + (x_{n-1}(-Y_0) + y_{n-1}(-X_0))2^{n-1}.$$

The latter equation indicates that the result is the sum of two positive integers X_0Y_0 and $x_{n-1}y_{n-1}2^{n-2}$ possibly reduced by two correcting terms $x_{n-1}Y_02^{n-1}$ and $y_{n-1}X_02^{n-1}$. The process is illustrated below for our example multiplication $(-11)\cdot(-13)$. The position of the sign bit of the result is the position of weight 2^9, for the reason explained earlier: this fact should be kept in mind while preparing the copies of $-Y_02^{n-1}$ and of $-X_02^{n-1}$.

	9	8	7	6	5	4	3	2	1	0	weight position
							0	0	1	1	
						0	0	0	0		
					0	0	1	1			X_0Y_0
			0	0	0	0					
		1									x_4y_4
	1	1	1	1	0	1					$-Y_02^4$
+	1	1	1	0	1	1					$-X_02^4$
	0	0	1	0	0	0	1	1	1	1	

Practical implementations based on the above principles are numerous and differ in the geometrical positions of the summands and in variations of the summation mode. A complete survey of these multipliers is given by Hwang (1979). Basically, all these arrays have performances similar to the arrays of Section 2.2. The book by Hwang, together with the paper by Davio and Bioul (1977), give indications on the extension of the above principles to the general situation of radix complement multiplication.

4 DIVISION

4.1 Division of positive numbers: the basic algorithms

We complete in this section our study of the basic arithmetic operations with that of divison. We consider first positive rational numbers contained in the interval $[0, 1[$. Let us recall that a fraction $n/d \in [0, 1[$ has the representation:

$$n/d \simeq 0, q_1 q_2 \dots q_m; \qquad q_i \in \{0, 1\}$$

with the interpretation

$$n/d = \sum_{i=1}^{m} q_i 2^{-i}.$$

We first describe the two basic division techniques.

4.1.1 The restoring division algorithm

We first establish a convergence lemma.

Lemma 1. *If the integers a and b satisfy $a < b$, then there exist unique integers q, r satisfying the conditions*

$$2a = bq + r, \tag{20}$$

$$q \in \{0, 1\}, \tag{21}$$

$$0 \leqslant r \leqslant b - 1. \tag{22}$$

Proof. The integer division theorem immediately shows that there exist unique integers q and r satisfying (20) and (22). To show that (21) holds true, it suffices to observe that:

$$a < b \Rightarrow 2a < 2b \Rightarrow 2a \leqslant 2b - 1.$$

The right hand member of (20) indeed reaches this maximum value for $q = 1$ and $r = b - 1$. □

The interest of Lemma 1 lies in the fact that the inequality $a < b$ is forwarded to r and b under the form $r < b$. That property will justify the following division algorithm. We assume satisfied the initial condition $n < d$.

Algorithm A (Restoring division). *The representation* $0, q_1 q_2 \dots q_n$ *of the quotient* $q = n/d$ $(n < d)$ *is obtained by:*

$$p^{(0)} = n, \tag{23}$$

$$2p^{(i)} = q_{i+1} d + p^{(i+1)}; \qquad i = 0, 1, \dots, m-1. \tag{24}$$

Proof. Let us consider the set of equations

$$n = p^{(0)} \tag{25_0}$$

$$2p^{(0)} = q_1 d + p^{(1)} \tag{25_1}$$

$$2p^{(1)} = q_2 d + p^{(2)} \tag{25_2}$$

$$\cdots\cdots\cdots\cdots\cdots\cdots$$

$$2p^{(i)} = q_{i+1} d + p^{(i+1)} \tag{25_{i+1}}$$

$$\cdots\cdots\cdots\cdots\cdots\cdots$$

$$2p^{(m-1)} = q_m d + p^{(m)} \tag{25_m}$$

Taking Lemma 1 into acrount, and using the initial condition $p^{(0)} = n < d$, we obtain, for all i's, $p^{(i)} < d$. By elementary substitutions, we obtain, from the system (25)

$$n = dq + 2^{-m} p^{(m)}.$$

The remainder r of the division $n = dq + r$ appears as $2^{-m}p^{(m)}$ and it satisfies $r < 2^{-m}d$. The obtained bits $q_1 q_2 \ldots q_m$ are actually the m most significant quotient bits, as r/d has the form $0,0\ 0 \ldots 0\, q_{m+1}\, q_{m+2} \ldots$ □

The interpretation of equations (23) and (24) is straightforward. At step i of the algorithm, one receives as data $p^{(i)}$ and one has to compute q_{i+1} and

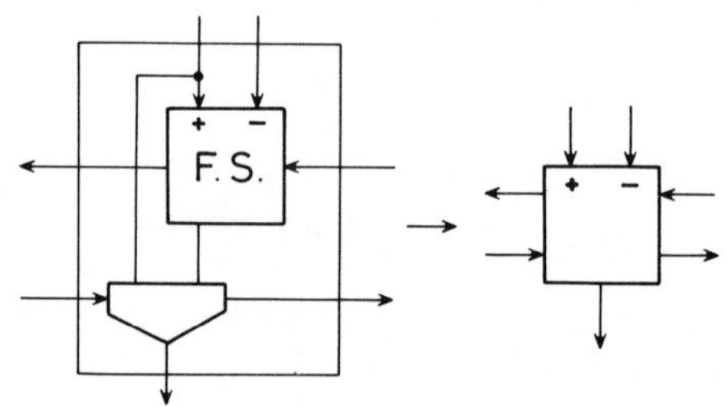

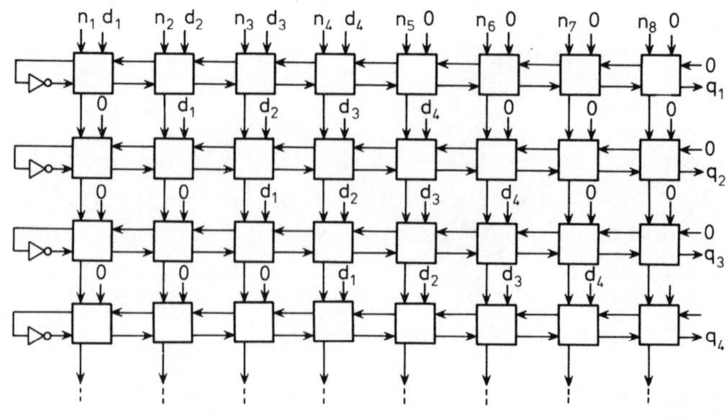

Figure 8. Restoring division array

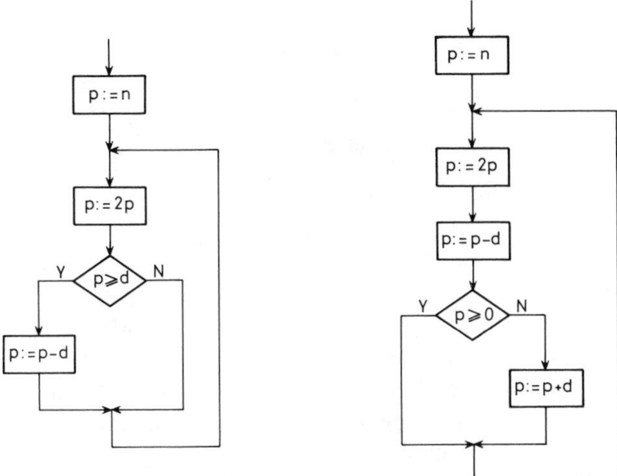

Figure 9. Flow charts for restoring division

$p^{(i+1)}$. This is done according to

$$2p^{(i)} - d \geq 0 \Rightarrow q_{i+1} = 1 \quad \text{and} \quad p^{(i+1)} = 2p^{(i)} - d,$$
$$2p^{(i)} - d < 0 \Rightarrow q_{i+1} = 0 \quad \text{and} \quad p^{(i+1)} = 2p^{(i)}.$$

The circuit shown in Figure 8b behaves according to the above principles. This array is built from cells displayed in Figure 8a. The circuit computes the quotient $0, q_1q_2q_3 \ldots$ of $n = 0, n_1n_2n_3 \ldots$ by $d = 0, d_1d_2d_3 \ldots$ under the assumption $n < d$. At each step, one uses the borrow of the difference $2p^{(i)} - d$ to select $2p^{(i)}$ or $2p^{(i)} - d$ as operand for the next step.

Figure 9a presents a flow chart equivalent to Algorithm A. This flow chart includes an amplitude comparison test '$p \geq d$?'. Such a test is not necessarily available as a facility in commercial processing units while tests of the form '$x \geq 0$?' are generally provided. When this is the case, the flowchart of Figure 9a may be replaced by that of Figure 9b. The latter is termed the restoring division algorithm: this connotes the addition $p := p + d$ required when $p < 0$ to restore the correct value of p.

If the restoring division algorithm is conceptually simple, one usually prefers its non-restoring counterpart. The latter, to be described in the following section, will then easily be modified to handle signed numbers.

4.1.2 The non-restoring division algorithm

Our discussion again starts with a convergence lemma.

Lemma 2. *Let a and b represent two integers such that $-b \leq a < b$. Then there* *exist unique integers q and r such that:*

$$2a = (2q-1)b + r, \tag{23}$$

$$q \in \{0, 1\}, \tag{24}$$

$$-b \leqslant r < b. \tag{25}$$

Proof. Equation (23) may be written:

$$2a + 2b = 2bq + b + r. \tag{26}$$

We know by the integer division theorem that, given the two integers $(2a+2b)$ and $2b$, we may find a unique quotient q and a unique remainder $(b+r)$ satisfying (26) and such that $0 \leqslant b+r < 2b$ or $-b \leqslant r < b$. Now, to show that (24) holds true, it suffices to observe that the maximum value attained by the left-hand member of (26) is $4b-1$ and that the latter value is represented in the right hand member by $q=1$ and $(b+r)=2b-1$, i.e. $r=b-1$. □

Once again, the interesting fact is that the property $a \in [-b, b[$ is transmitted to $r: r \in [-b, b[$. In the following Algorithm B we consider two positive numbers n and d satisfying: $0 \leqslant n < d$.

Algorithm B (Non-restoring division). *The representation* $0, q_1, q_2 \ldots q_m$ *of the quotient* $q = n/d$ $(0 \leqslant n < d)$ *is obtained by:*

$$2n - d = t^{(0)}, \tag{27}$$

$$2t^{(i)} = (2q_{i+1} - 1)d + t^{(i+1)}. \tag{28}$$

Proof. The initial condition $0 \leqslant n < d$ guarantees that $-d \leqslant t^{(0)} < d$. Hence, using recursively Lemma 2, we may assert, for all i, that: $-d \leqslant t^{(i)} < d$. Consider now the set of equations:

$$2n - d = t^{(0)} \tag{29}$$

$$2t^{(0)} = (2q_1 - 1)d + t^{(1)} \tag{30_1}$$

$$2t^{(1)} = (2q_2 - 1)d + t^{(2)} \tag{30_2}$$

$$\cdots\cdots\cdots\cdots\cdots\cdots\cdots\cdots\cdots \qquad \cdots\cdots$$

$$2t^{(i)} = (2q_{i+1} - 1)d + t^{(i+1)} \tag{30_{i+1}}$$

$$\cdots\cdots\cdots\cdots\cdots\cdots\cdots\cdots\cdots \qquad \cdots\cdots$$

$$2t^{(m-1)} = (2q_m - 1)d + t^{(m)} \tag{30_m}$$

Elementary substitutions yield:

$$n = dq + 2^{-m}\left(\frac{t^{(m)} + d}{2}\right),$$

and the remainder r appearing in $n = dq + r$ is now given by $r = 2^{-m}(t^{(m)} + d)/2$. As in Algorithm A, the remainder r satisfies $0 \leqslant r < 2^{-m}d$ so that the bits q_i obtained by the algorithm actually are the m most significant bits of q. □

The interpretation of the above algorithm is again straightforward; in step i (equation 30i), one receives as data the quantity $t^{(i)}$ and one has to determine $q_i \in \{0, 1\}$ and $t^{(i)}$ such that $-d \leq t^{(i)} < d$. Now,

$$q_i = 0 \Rightarrow t^{(i)} = 2t^{(i-1)} + d,$$
$$q_i = 1 \Rightarrow t^{(i)} = 2t^{(i-1)} - d.$$

Then, obviously, the choice has to be

$$t^{(i-1)} < 0 \Rightarrow q_i = 0 \quad \text{and} \quad t^{(i)} = 2t^{(i-1)} + d,$$
$$t^{(i-1)} \geq 0 \Rightarrow q_i = 1 \quad \text{and} \quad t^{(i)} = 2t^{(i-1)} - d.$$

The operation (add or subtract) to be carried out at step i is thus determined by the sign of $t^{(i-1)}$, *a result available at the outset of step* $(i-1)$: this is the main advantage of the non-restoring division algorithm over the restoring one. The core of a circuit based on the above principles is shown in Figure 10. In that circuit, the register containing initially n and subsequently the successive values $t^{(i)}$ has been provided with a sign bit and we assume that the adder–subtractor behaves in the 2's complement mode. The successive quotient bits appear on the output line q_i from the clockpulse on. The remainder r, if required, will have to be computed from the final value $t^{(m)}$ by $r = 2^{-m}(t^{(m)} + d)/2$.

Let us illustrate the above algorithm by a numerical example:

$$n = 9 \ (n \simeq 0\ 0\ 1\ 0\ 0\ 1); \qquad d = 21 \ (d \simeq 0\ 1\ 0\ 1\ 0\ 1);$$
$$-d \simeq (1\ 0\ 1\ 0\ 1\ 1).$$

$$\begin{array}{lr}
 & 0\ 0\ 1\ 0\ 0\ 1 \\
 & 0\ 1\ 0\ 0\ 1\ 0 \\
 & (-)\underline{1\ 0\ 1\ 0\ 1\ 1} \\
 & 1\ 1\ 1\ 1\ 0\ 1 \\
q_1 = 0 & 1\ 1\ 1\ 0\ 1\ 0 \\
 & +\underline{0\ 1\ 0\ 1\ 0\ 1} \\
 & 0\ 0\ 1\ 1\ 1\ 1 \\
q_2 = 1 & 0\ 1\ 1\ 1\ 1\ 0 \\
 & (-)\underline{1\ 0\ 1\ 0\ 1\ 1} \\
 & 0\ 0\ 1\ 0\ 0\ 1 \\
q_3 = 1 & 0\ 1\ 0\ 0\ 1\ 0 \\
 & (-)\underline{1\ 0\ 1\ 0\ 1\ 1} \\
 & 1\ 1\ 1\ 1\ 0\ 1
\end{array}$$

The result q is given by $0,0\ 1\ 1\ 0\ 1\ 1\ 0\ 1\ 1 \ldots = 3/7$.

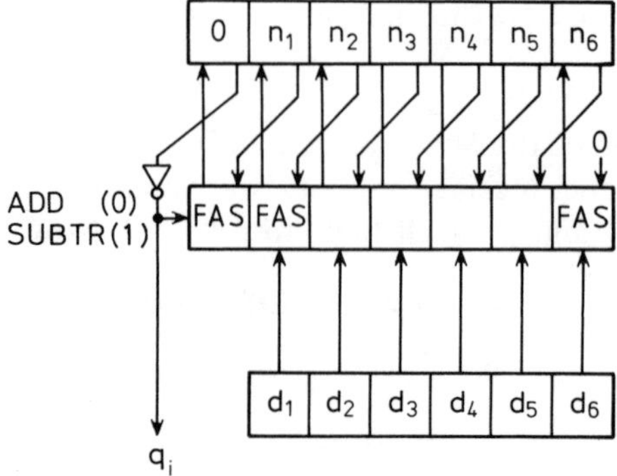

Figure 10. A sequential non-restoring divider

4.2 2's complement division

Even if the dividend and divisor are positive numbers, the non-restoring division algorithm discussed in the previous sections leads one to handle negative numbers. It is thus natural to consider the extension of the non-restoring division scheme to signed numbers. The problem is stated as follows: given two integers n and d such that:

$$0 \leqslant \|n\| < \|d\|,$$

obtain the quotient q and the remainder r such that:

$$n = dq + r, \tag{31}$$

$$2^m \|r\| < \|d\|, \tag{32}$$

$$\text{sign } r = \text{sign } n. \tag{33}$$

Observe in this definition the rather arbitrary nature of (33): we could have imposed $r \geqslant 0$. Let us denote by N and D the absolute values $\|n\|$ and $\|d\|$. By Algorithm B, we determine a quotient Q^* and a remainder R satisfying:

$$N = DQ^* + R, \tag{34}$$

$$0 \leqslant R < 2^{-m} D. \tag{35}$$

The successive quotient bits are obtained by equations (30). Let us now discuss the four possible cases:

(i) $n \geqslant 0$; $d > 0$: this is the case discussed in the previous section.

(ii) $n \geqslant 0$; $d < 0$: here $N = n$ and $D = -d$. Hence, by (34), $n = d(-Q^*) + R$, and we thus select $q = -Q^*$, $r = R$.

(iii) $n < 0$; $d > O$: $N = -n$, $D = d$. Using again (34), we obtain $-n = dQ^* + R$, i.e. $n = d(-Q^*) - R$. We choose $q = -Q^*$, $r = -R$.

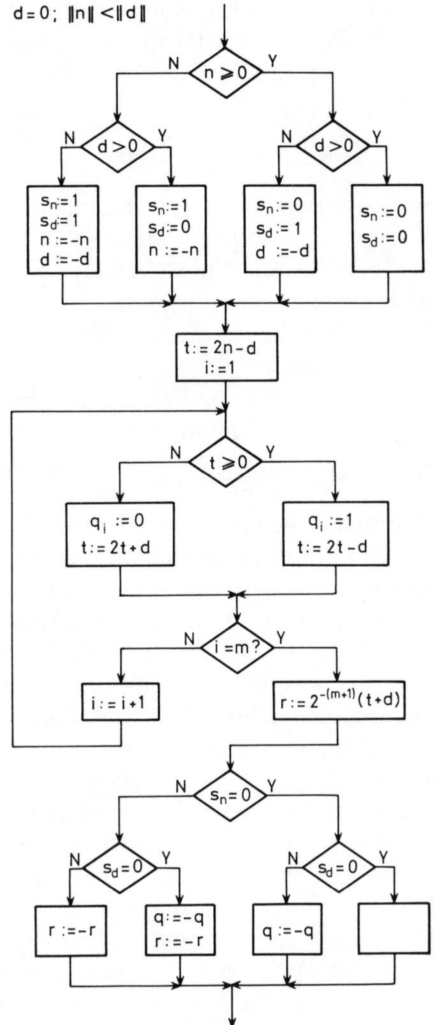

Figure 11. Flow chart for 2's complement division

(iv) $n<0$; $d<0$: $N=-n$, $D=-d$. By (34): $-n=-dQ^*+R$, $n=dQ^*-R$; $q=Q^*$, $r=-R$.

In all four cases, we obtain sign $r=$ sign n, as required by (33). Figure 11 summarizes the results of the previous discussion by presenting a complete flowchart of the non-restoring 2's complement division. The flowchart is divided into three parts: an initialization during which n and d are replaced by their absolute values N and D and their signs stored as the bits s_n and s_d;

the Algorithm B, as described by (30), completed by the initialization and updating of a counting index i and a termination phase during which the proper signs of q and r are restored according to the above discussion.

5 CONCLUSION

In the above sections, we presented some of the main algorithms in use for multiplication and division. The subject matter is far from being exhausted and entire books have been devoted to its study, since the early book by Flores (1963) to the previously mentioned book by Hwang (1979). The book by Knuth (1969), while being more software oriented, deserves a mention here as a source of algorithms.

Among the questions in arithmetic that we did not discuss in this chapter, we should mention arbitrary radix multiplication for signed numbers and arbitrary radix division both for unsigned and signed numbers. As a matter of fact, we tried to stick as far as possible to those algorithms which have had, and have, an influence on the hardware architecture of processing units.

So far, we are in position to draw two important conclusions:

(i) All the arithmetic operations we have studied may be decomposed in terms of the simple operations of addition, subtraction and shifts, so that these simple operations appear as natural primitives for arithmetic computations. Here we should stress once more that there is no point in introducing in an arithmetic unit additional facilities such as multipliers if their introduction leads one to choose a reduced clock rate: the overall performance could be jeopardized. The key rules are to select as primitive operations operations having similar execution times and, if parallelism is possible, to keep as many resources as possible simultaneously active at any computation step.

(ii) The second observation is that, even for simple operations such as sign-magnitude addition (Chapter IV) or 2's complement division, the algorithms to be implemented may become relatively complex and will in particular involve, apart from primitive operations, initialization, termination and conditional branchings. The need for a control mechanism clearly appears here and its discussion will be the body of the following chapter.

BIBLIOGRAPHICAL REMARKS

The subject of multiplication is extensively described in the literature. From our algorithmic point of view, the book by Knuth (1969) remains a fundamental reference. A number of cellular multipliers have been described by Guild (1968), De Mori (1969), Pezaris (1971), Baugh and Wooley (1973), and Davio and Bioul (1977). The sequential multiplier of Section 2.2.2 is known as a serial–parallel multiplier (Kostopoulos, 1975) but its relation to the carry–save multiplier does not seem to have been pointed out. Fast multiplication techniques are described in the early papers by Dadda (1965) and Wallace (1964). To this subject is also related the discussion of parallel counters: see Swartzlander (1973). At the other extreme, serial–serial multipliers have been considered by Atrubin (1965) and, more recently, by Chen and Willoner (1979). Division is far from being as well known as multiplication. A classical paper about serial division remains that of Robertson. Division arrays were studied by Dean (1968), Guild (1970), Cappa and Hamacher (1973), and Stefanelli (1972).

EXERCISES

E1. Design a pipelined version of the carry–save multiplier and optimize its merit factor.

E2. Design a synchronous multiplier based on the ripple–carry multiplier. Try to minimize the cost of the connection networks. *Hint:* Label carefully the precedence graph and observe the diagonal propagation of the information flow. Imbed your array in a larger rectangular array, the symmetry axis of which coincides with the direction of propagation. The latter array will appear as a succession of odd and even steps, much easier to sequentialize.

E3. (Parallel counters). A parallel counter of order n has 2^n data inputs $y_0, y_1, y_2, \ldots, y_{2^n-1}$ and n outputs $x_0 x_1 \ldots x_{n-1}$ such that

$$\sum_{j=0}^{n-1} x_j 2^j = \sum_{i=0}^{2^n-1} y_i.$$

Hence, the vector $\mathbf{x}$ is the binary representation of the number of ones in the vector $\mathbf{y}$. Use two parallel counters of order n and full-adders to build a parallel counter of order $(n+1)$. Discuss the cost and the delay of the obtained network. Discuss the possibility of sequentializing and of pipelining that network.

E4. (Squares). Let A be an integer having the binary representation $a_{n-1} \ldots a_1 a_0$. We wish to obtain the representation $s_{2x-1} \ldots s_1 s_0$ of the square S of A. A possible algorithm for computing S may be described as follows: let A_i represent the integer having the binary representation $a_{n-1} \ldots a_i$. The A_i's are clearly related by the recurrence relation

$$A_{i-1} = 2A_i + a_{i-1},$$

or

$$A_{i-1}^2 = 4A_i^2 + a_{i-1}(4A_i + 1).$$

Thus, if one chooses $A_n = 0$, one finally obtains $S = A_0^2$. Use that technique to devise various squares (combinational, sequential, and pipelined). Examine the interest of using a squarer for multiplication purposes (remember that $xy = ((x+y)^2 - (x-y)^2)/4$).

E5. (Square roots). Let X represent an integer having the binary representation $x_{2n-1} \ldots x_1 x_0$. Let $A \simeq a_{n-1} \ldots a_1 a_0$ be its square root. Assume A written under the form

$$A = A_1 2^p + a_{p-1} 2^{p-1} + A_0;$$

assume furthermore that

$$A_1 \quad \text{and} \quad X - (A_1 2^p) = R$$

are available at the outset of a typical algorithm step. We show that, from these data on, it is easy to compute

$$2A_1 + a_{p-1} \quad \text{and} \quad X - ((2A_1 + a_{p-1}) 2^{p-1})^2 = R',$$

which can be used again in a similar way at the next algorithmic step. Show that

$$R' = R - (4A_1 + 1)a_{p-1}2^{2p-2}.$$

Hence, the typical algorithm step (given R and A_1) is described by:
(i) compute $Q = R - (4A_1 + 1)2^{2p-2}$;
(ii) if $Q \geqslant 0$, then $a_{p-1} = 1$, i.e.

$$A_1 := 2A_1 + 1 \quad \text{and} \quad R' := Q$$

otherwise, $a_{p-1} = 0$, i.e.

$$A_1 := 2A_1 \quad \text{and} \quad R' = R.$$

Use the above description to design a combinational and a sequential square root extractor.

E6. Discuss the similarities between square-root extraction and division. Devise, in particular, restoring and non-restoring versions of the square-root extraction algorithms.

E7. Discuss the possibility of extending the above algorithms for multiplication, division, square, and square-root to radixes other than two. Discuss the interest of these techniques.

E8. (Remainder of an integer division). Consider the integers A and B given by their binary representations $a_{m-1} \ldots a_1 a_0$ and $b_{n-1} \ldots b_1 b_0$. One wishes to compute the binary representation $r_{n-1} \ldots r_1 r_0$ of the remainder R of the integer division
$A = BQ + R$; $\quad (0 \leqslant R \leqslant B - 1)$. Thus

$$R = \|A\|_B$$

Clearly

$$R = \left[\!\left[\sum_{i=0}^{m-1} a_i 2^i\right]\!\right]_B$$

or

$$R = \left[\!\left[\sum_{i=0}^{m-1} a_i [\![2^i]\!]_B\right]\!\right]_B.$$

Assume given at a typical step

$$P_i = [\![2^i]\!]_B \quad \text{and} \quad R_i = \left[\!\left[\sum_{j=0}^{i-1} a_j [\![2^j]\!]_B\right]\!\right]_B.$$

Clearly,

$$P_{i+1} = [\![2P_i]\!]_B \quad \text{and} \quad R_{i+1} = [\![R_i + a_i P_i]\!]_B.$$

Show that the computation resource required for the computation of P_{i+1} reduces to a conditional adder:

$$\text{if } 2P_i < B \text{ then } P_{i+1} = 2P_i \text{ otherwise } P_{i+1} = 2P_i - B.$$

Design a combinational network for computing R and the corresponding sequential network, and discuss their performances.

E9. (Integer division). The principles developed in exercise E8 may be extended to integer division, i.e. to the computation of Q and R such that $A = BQ + R$. Let indeed

$$2^i = S_i B + P_i.$$

Then

$$2^{i+1} = S_{i+1} B + P_{i+1} = 2S_i B + 2P_i,$$

and, if we put

$$2P^i = S_i^* B + P_{i+1},$$

we obtain

$$2^{i+1} = (2S_i + S_i^*)B + P_{i+1},$$

i.e.

$$S_{i+1} = 2S_i + S_i^*,$$
$$P_{i+1} = [\![2P_i]\!]_B,$$
$$S_i^* = 0 \quad \text{if} \quad 2P_i < B: \qquad S_i^* = 1 \quad \text{otherwise.}$$

Let now A_i be the integer represented by $a_i \ldots a_1 a_0$. Put

$$A_i = T_i B + U_i.$$

Then

$$\begin{aligned} A_{i+1} &= a_{i+1} 2^{i+1} + A_i \\ &= (T_i B + U_i) + a_{i+1}(S_{i+1}B + P_i) \\ &= (T_i + S_{i+1}a_{i+1})B + (U_i + a_{i+1}P_i). \end{aligned}$$

Finally, let

$$U_i + a_{i+1}P_i = \tilde{Q}_i B + I_{i+1}.$$

One obtains

$$A_{i+1} = (T_i + S_{i+1}a_{i+1} + \tilde{Q}_i)B + U_{i+1}.$$

Finally, either

(i) $U_i + a_{i+1}P_i < B$: in that case $\tilde{Q}_i = 0$ and $T_{i+1} = T_i + S_{i+1}a_{i+1}$; $U_{i+1} = U_i + a_{i+1}P_i$,
or
(ii) $U_i + a_{i+1}P_i \geqslant B$: in that case $\tilde{Q}_i = 1$ and $T_{i+1} = T_i + S_{i+1}a_{i+1} + 1$; $U_{i+1} = U_i + a_{i+1}P_i - B$.

From the above equations, it is thus possible to compute

$$S_{i+1}P_{i+1}T_{i+1} \quad \text{and} \quad U_{i+1} \quad \text{from} \quad S_i, P_i, T_i \quad \text{and} \quad U_i.$$

Use the equations to build a combinational circuit for integer division. Discuss its performance and compare your results with those obtained for classical division arrays.

E10. Consider two integers $X \simeq x_3x_2x_1x_0$ and $Y \simeq y_3y_2y_1y_0$. If these two numbers are shifted with respect to each other according to

$$\begin{array}{lllllll}
0) & x_3 & x_2 & x_1 & x_0 & & \\
 & & & & y_0 & y_1 & y_2 \quad y_3 \\
1) & x_3 & x_2 & x_1 & x_0 & & \\
 & & & y_0 & y_1 & y_2 & y_3 \\
2) & x_3 & x_2 & x_1 & x_0 & & \\
\ldots & & y_0 & y_1 & y_2 & y_3 &
\end{array}$$

they offer the following characteristic property: at step i, all the pairs of bits (x_j, y_k) such that $j + k = i$, i.e. all the pairs of bits of weight 2^i in the product XY, are facing each other. Use that property to design a multiplier.

Chapter X
Control unit and processing unit

1 INTRODUCTION EXAMPLE: EXPONENTIATION IN A FINITE FIELD

In Chapter VIII we considered the implementation of computation schemes by means of synchronous sequential sysems. A natural decomposition of the system into a *control unit* and a *processing unit* appeared. Let us recall that the control unit is an autonomous system.

We now consider a wider class of algorithms, namely those which can be implemented according to the block diagram of Figure 1. The main innovation, with respect to the preceding situation, consists in the introduction of a feedback connection from the processing unit to the control unit. We shall prove in this chapter that this elementary modification makes feasible the implementation of a much larger class of algorithms than computation schemes allow: as a matter of fact, the new structure yields systems whose computation power is comparable to that of high level languages. On the other hand, the model that we start studying will at the same time allow the operations of initiation and end of computation to be described.

Let us illustrate these points by an exponentiation algorithm. Let X be an element of a finite field F, and A a positive integer smaller than 2^n; in binary numeration system,

$$A = a_0 + a_1 2 + a_2 2^2 + \ldots + a_{n-1} 2^{n-1}, \qquad a_i \in \{0, 1\}. \tag{1}$$

We aim at computing X^A. According to (1)

$$\begin{aligned} X^A &= (X^{a_{n-1}})^{2^{n-1}} \cdot \ldots \cdot (X^{a_2})^{2^2} \cdot (X^{a_1})^2 \cdot X^{a_0} \\ &= (\ldots(((1)^2 X^{a_{n-1}})^2 X^{a_{n-2}})^2 \ldots X^{a_1})^2 X^{a_0}. \end{aligned} \tag{2}$$

Formula (2) leads to an exponentiation algorithm which we may formulate as follows (Pascal-like notation):

$$\begin{aligned} &P := 1; \\ &\textit{for} \quad i := n-1 \quad \textit{downto} \quad 0 \quad \textit{do} \quad P := P^2 X^{a_i}. \end{aligned} \tag{3}$$

It is an iteration. At each step, the partial result is squared, and then

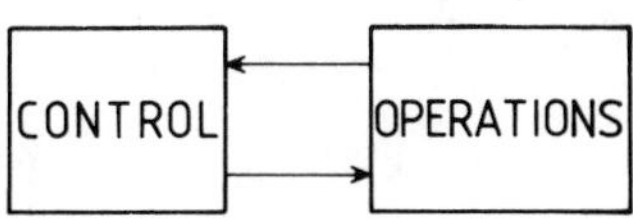

Figure 1. Basic model

multiplied by X if $a_i = 1$. Hence, we need computation resources able to perform products in the field F.

If $F = \mathrm{GF}(2^r)$, that is the *Galois field of order* 2^r, every element S of F is represented by a polynomial

$$s_0 + s_1 x + \ldots + s_{r-1} x^{r-1}, \qquad s_i \in \{0, 1\}, \tag{4}$$

whose degree is smaller than r (see, for instance, Davio, Deschamps, and Thayse, 1978, section 1.4.3).

The sum of two elements of F is represented by the sum of the corresponding polynomials. The product of two elements of F is represented by the product of the corresponding polynomials, modulo an irreducible polynomial

$$M = m_0 + m_1 x + \ldots + m_{r-1} x^{r-1} + m_r x^r, \qquad m_i \in \{0, 1\}, \qquad m_r = 1, \tag{5}$$

whose degree is equal to r.

Let us perform the product $P \cdot S$:

$$P \cdot (s_0 + s_1 x + \ldots + s_{r-1} x^{r-1}) = ((\ldots((0 \cdot x + s_{r-1} P)x + s_{r-2} P)x + \ldots + s_2 P)x + s_1 P)x + s_0 P. \tag{6}$$

Formula (6) leads to a multiplication algorithm:

$$\begin{aligned} &Q := 0 \\ &\textit{for} \quad j := r-1 \quad \textit{downto} \quad 0 \quad \textit{do} \quad Q := Qx + s_j P. \end{aligned} \tag{7}$$

It is again an iteration. At each step, the partial result is multipied by x, and then $s_j = 1$ is added to P. We thus need resources able to perform both the multiplication by x and the addition.

The addition of

$$P = p_0 + p_1 x + \ldots + p_{r-1} x^{r-1} \quad \text{and} \quad Q = q_0 + q_1 x + \ldots + q_{r-1} x^{r-1}$$

is performed in a componentwise way:

$$P + Q = (p_0 \oplus q_0) + (p_1 \oplus q_1)x + \ldots + (p_{r-1} \oplus q_{r-1})x^{r-1},$$

where $\oplus$ stands for the modulo 2 addition.

As regards the multiplication by x, we consider two cases:

(1) $q_{r-1} = 0$:

$$Qx = q_0 x + q_1 x^2 + \ldots + q_{r-2} x^{r-1}; \tag{8}$$

(2) $q_{r-1}=1$:

$$\begin{aligned} Qx &= q_0x+q_1x^2+\ldots+q_{r-2}x^{r-1}+q_{r-1}x^r, \bmod M \\ &= m_0+(q_0 \oplus m_1)x+(q_1 \oplus m_2)x^2+\ldots+(q_{r-2} \oplus m_{r-1})x^{r-1}, \end{aligned} \quad (9)$$

since $q_{r-1}=m_r=1$.

We are now in a position to conceive a sequential implementation of the exponentiation algorithm. Every element S of F is represented by the r-component vector $(s_{r-1}, \ldots, s_1, s_0)$ of the polynomial coefficients. In order to store data and intermediate results, we use six registers:

R_X is an r-bit register storing X;

R_A is an n-bit register storing A;

R_M is an r-bit register storing $(m_{r-1}, \ldots, m_1, m_0)$;

R_P, R_Q and R_S are r-bit registers storing intermediate results P, Q, and S.

In order to manage indices i and j, we use two down counters

$$C(i)\colon n \rightarrow 0,$$
$$C(j)\colon r \rightarrow 0.$$

Registers R_A, R_Q and R_S are left shift registers. This allows all bits of A and S to be successively accessed, and the multiplication of Q by x to be performed.

The next pieces of information are necessary to control the progress of the computation:

(a) the most significant bit (m.s.b.) a_{n-1} of R_A;
(b) the most significant bit q_{r-1} of R_Q;
(c) the most significant bit s_{r-1} of R_S;
(d) the terminal count down (t.c.d.) output of $C(i)$: it is equal to 1 if $[C(i)]=0$;
(e) the terminal count down output of $C(j)$.

Let us describe the list of transfers which are performed during the computation:

(1) Initiation:

$$\sigma_0\colon[R_p]:=1,$$
$$\sigma_1\colon[R_Q]:=0,$$
$$\sigma_2\colon[C(i)]:=n,$$
$$\sigma_3\colon[C(j)]:=r.$$

(2) Counting:

$$\sigma_4\colon[C(i)]:=[C(i)]-1,$$
$$\sigma_5\colon[C(j)]:=[C(j)]-1.$$

(3) Transfer of intermediate results:

$$\sigma_6 : [R_S] := [R_p],$$
$$\sigma_7 : [R_p] := [R_Q],$$
$$\sigma_8 : [R_s] := [R_X],$$
$$\sigma_9 : [R_Q] := 2[R_Q], \text{ mod } 2^r,$$
$$\sigma_{10} : [R_S] := 2[R_S], \text{ mod } 2^r,$$
$$\sigma_{11} : [R_A] := 2[R_A], \text{ mod } 2^n,$$
$$\sigma_{12} : [R_Q] := [R_Q] \oplus [R_M],$$
$$\sigma_{13} : [R_Q] := [R_Q] \oplus [R_p].$$

Notice that the presence of transfers σ_{12} and σ_{13} implies that R_Q is a master–slave register.

For every register we write the list of data it receives and of operations it performs, as well as an arbitrarily chosen encoding:

R_p

Data sources	z_0	Operations	z_1
constant 1	0	NO OPERATION	0
R_q	1	LOAD	1

R_q

Data sources	z_2	Operations	z_3	z_4	z_5
R_M	0	NO OPERATION	0	0	0
R_p	1	RESET	1	0	0
		SHIFT	0	1	0
		EXOR	0	0	1

R_S

Data sources	z_6	Operations	z_7	z_8
R_p	0	NO OPERATION	0	0
R_X	1	LOAD	1	0
		SHIFT	0	1

R_A

Operations	z_9
NO OPERATION	0
SHIFT	1

$C(i)$

Operations	z_{10}	z_{11}
NO OPERATION	0	0
RESET TO $n-1$	1	0
DECREMENT	0	1

$C(j)$

Operations	z_{12}	z_{13}
NO OPERATION	0	0
RESET TO $r-1$	1	0
DECREMENT	0	1

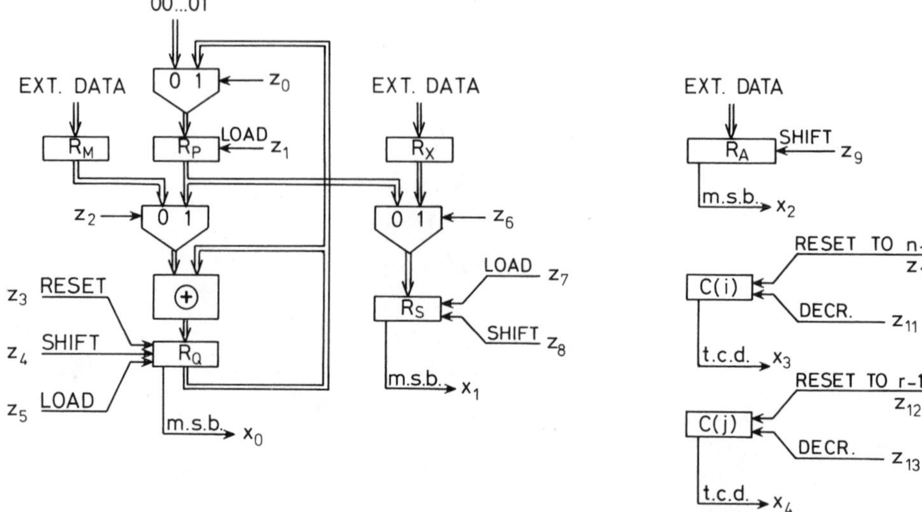

Figure 2. Processing unit

The whole processing unit is shown in Figure 2. It is controlled by fourteen *control signals* $z_0, \ldots, z_{13}$, plus the clock signal(s). It generates five *condition variables* $x_0, \ldots, x_4$.

Let us construct the complete algorithm. Instruction (3) leads to the flow chart of Figure 3a; the exponent of $[R_X]$ is always a_{n-1} thanks to the left shift of $[R_A]$ at the end of every iteration (σ_{11}). The transfer

$$[R_p]:=[R_p]^2[R_X]^{a_{n-1}}$$

can be decomposed according to the flow chart of Figure 3b. The multiplication instruction (7) leads to the flow chart of Figure 3c; the test always concerns $x_1 = s_{r-1}$ thanks to the left shift of $[R_S]$ at the end of every iteration (σ_{10}). The multiplication of $[R_Q]$ by x is performed according to the flow chart of Figure 3d: see formulae (8) and (9).

The assembly of the various algorithm subparts yields the algorithm shown in Figure 4.

It remains to translate the algorithm specification of Figure 4 into the input–output behaviour of a control circuit. Let us first observe that some transfers can be simultaneously performed: for instance, transfers $\sigma_0([R_p]:=1)$ and $\sigma_2([C(i)]:=n)$ do not interact. The whole algorithm can be formulated as follows:

Step 1. Execute σ_0 and σ_2.
Step 2. If $x_3=0$, then execute $\sigma_6, \sigma_1, \sigma_3$, and go to step 3. If $x_3=1$, then go to END.

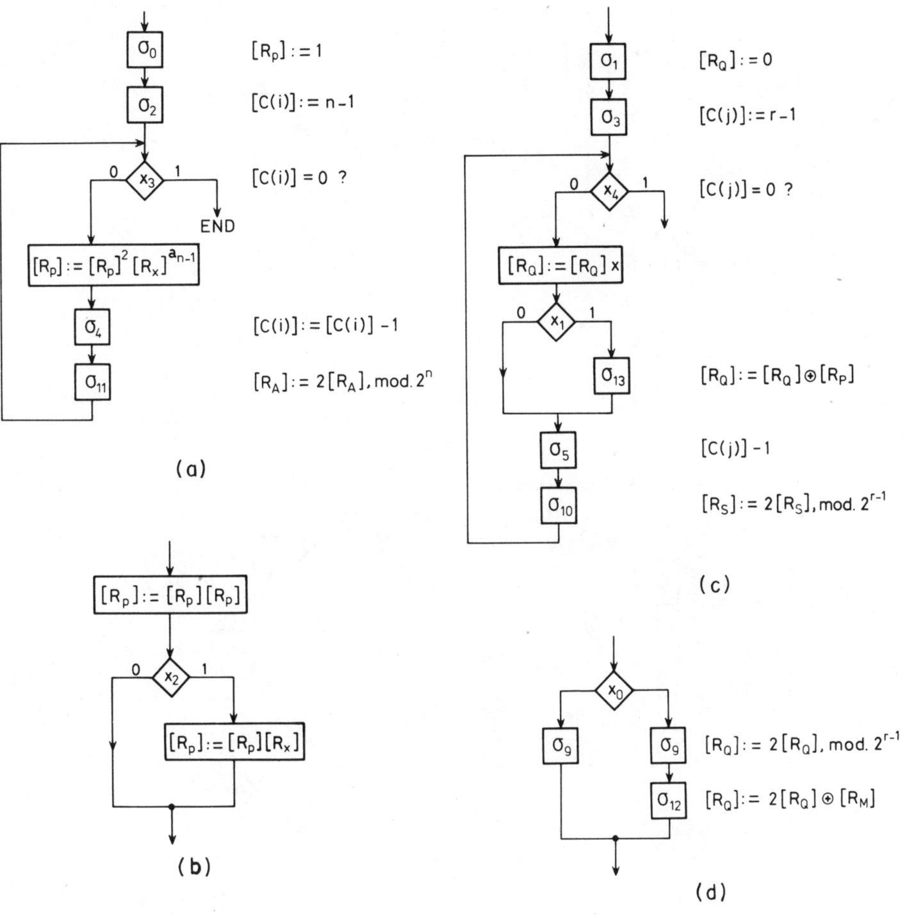

Figure 3. Algorithm subjects

Step 3. If $x_4 = 0$, then go to step 4. If $x_4 = 1$, then execute σ_7 and go to step 8.

Step 4. If $x_0 = 0$, then execute σ_9 and go to step 6. If $x_0 = 1$, then execute σ_9 and go to step 5.

Step 5. Execute σ_{12}.

Step 6. If $x_1 = 0$, then go to step 7. If $x_1 = 1$, then execute σ_{13} and go to step 7.

Step 7. Execute σ_5 and σ_{10} and go to step 3.

Step 8. If $x_2 = 0$, then execute σ_4, σ_{11} and go to step 2. If $x_2 = 1$, then execute $\sigma_8, \sigma_1, \sigma_3$, and go to step 9.

Step 9. If $x_4 = 0$, then go to step 10. If $x_4 = 1$, then execute $\sigma_7, \sigma_4, \sigma_{11}$, and go to step 2.

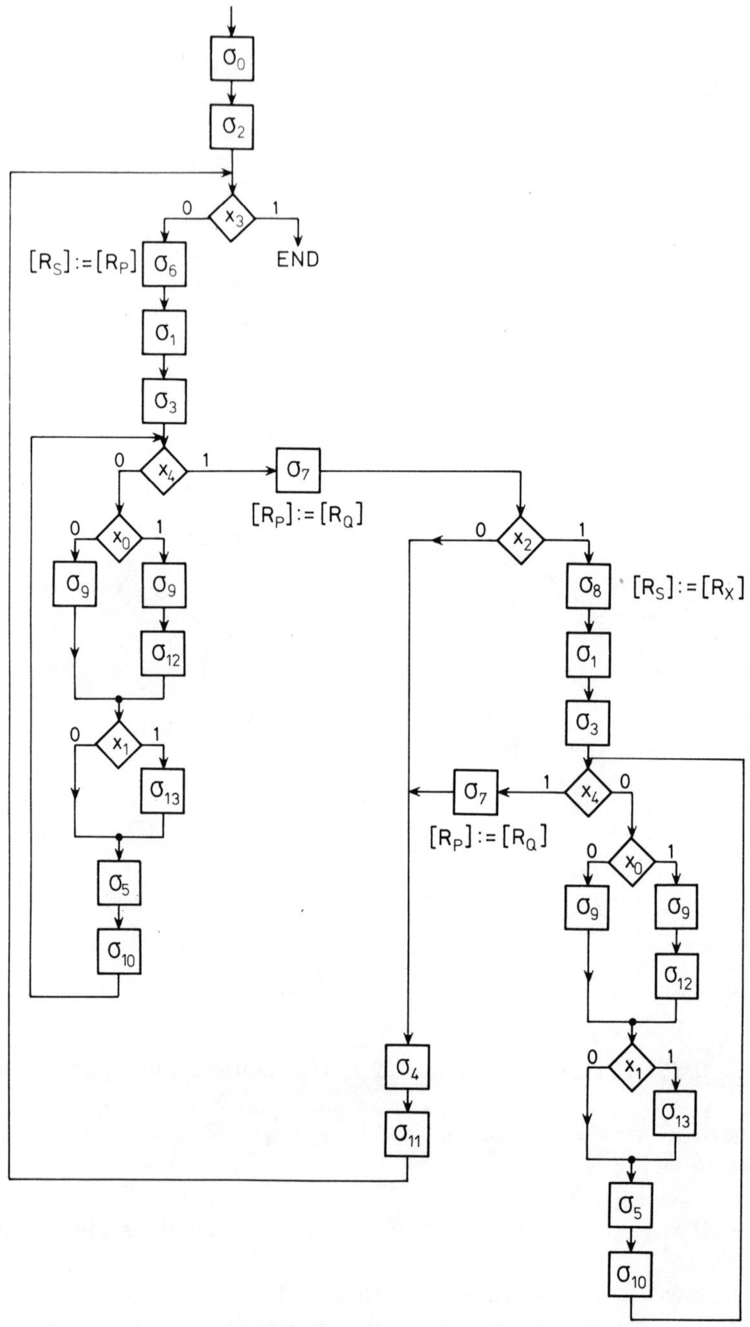

Figure 4. Exponentiation algorithm

Step 10. If $x_0=0$, then execute σ_9 and go to step 12. If $x_0=1$, then execute σ_9 and go to step 11.
Step 11. Execute σ_{12}.
Step 12. If $x_1=0$, then go to step 13. If $x_1=1$, then execute σ_{13} and go to step 13.
Step 13. Execute σ_5 and σ_{10} and go to step 9.

Every step includes:
(1) an identification number, and
(2) the possible test of a condition variable.
According to the result of the test (if any), it also includes
(3) the execution of some transfer(s), and
(4) the transition towards another step.
In condensed form, every step may be written as follows:

$$\begin{array}{llll} N_s & \bar{x}_k & \sigma_\ell \ldots \sigma_m & N_t, \\ & x_k & \sigma_p \ldots \sigma_q & N_u. \end{array} \tag{10}$$

This means:

Step s. If $x_k=0$, then execute $\sigma_\ell, \ldots, \sigma_m$, and go to step t. If $x_k=1$, then execute $\sigma_p, \ldots, \sigma_q$, and go to step u.
Let us introduce two notations:

t: true condition,

λ: no operation.

We also define composite transfers:

$$\sigma_{14}=\sigma_0\sigma_2, \qquad \sigma_{15}=\sigma_6\sigma_1\sigma_3, \qquad \sigma_{16}=\sigma_5\sigma_{10}, \qquad \sigma_{17}=\sigma_4\sigma_{11},$$
$$\sigma_{18}=\sigma_8\sigma_1\sigma_3, \qquad \sigma_{19}=\sigma_7\sigma_4\sigma_{11}.$$

The whole algorithm can be described by the following *program*:

$$\begin{array}{llll@{\qquad}llll} N_1 & t & \sigma_{14} & N_2 & N_8 & \bar{x}_2 & \sigma_{17} & N_2 \\ N_2 & \bar{x}_3 & \sigma_{15} & N_3 & & x_2 & \sigma_{18} & N_9 \\ & x_3 & \lambda & N_{14} & N_9 & \bar{x}_4 & \lambda & N_{10} \\ N_3 & \bar{x}_4 & \lambda & N_4 & & x_4 & \sigma_{19} & N_2 \\ & x_4 & \sigma_7 & N_8 & N_{10} & \bar{x}_0 & \sigma_9 & N_{12} \\ N_4 & \bar{x}_0 & \sigma_9 & N_6 & & x_0 & \sigma_9 & N_{11} \\ & x_0 & \sigma_9 & N_5 & N_{11} & t & \sigma_{12} & N_{12} \\ N_5 & t & \sigma_{12} & N_6 & N_{12} & \bar{x}_1 & \lambda & N_{13} \\ N_6 & \bar{x}_1 & \lambda & N_7 & & x_1 & \sigma_{13} & N_{13} \\ & x_1 & \sigma_{13} & N_7 & N_{13} & t & \sigma_{16} & N_9 \\ N_7 & t & \sigma_{16} & N_3 & N_{14} & & & \text{END} \end{array} \tag{11}$$

The correspondence between the transfers and the values of the control

	z_0	z_1	z_2	z_3	z_4	z_5	z_6	z_7	z_8	z_9	z_{10}	z_{11}	z_{12}	z_{13}
σ_7	1	1	—	0	0	0	—	0	0	0	0	0	0	0
σ_9	—	0	—	0	1	0	—	0	0	0	0	0	0	0
σ_{12}	—	0	0	0	0	1	—	0	0	0	0	0	0	0
σ_{13}	—	0	1	0	0	1	—	0	0	0	0	0	0	0
σ_{14}	0	1	—	0	0	0	—	0	0	0	1	0	0	0
σ_{15}	—	0	—	1	0	0	0	1	0	0	0	0	1	0
σ_{16}	—	0	—	0	0	0	—	0	1	0	0	0	0	1
σ_{17}	—	0	—	0	0	0	—	0	0	1	0	1	0	0
σ_{18}	—	0	—	1	0	0	1	1	0	0	0	0	1	0
σ_{19}	1	1	—	0	0	0	—	0	0	1	0	1	0	0
λ	—	0	—	0	0	0	—	0	0	0	0	0	0	0

Figure 5. Control variables

variables z is given in Figure 5. As an example, the composite transfer σ_{15} includes

$$[R_S]:=[R_p] \Rightarrow z_6=0,\ z_7=1,\ z_8=0;$$
$$[R_q]:=0 \Rightarrow z_3=1,\ z_4=0,\ z_5=0;$$
$$[C(j)]:=r \Rightarrow z_{12}=1,\ z_{13}=0.$$

There are no other transfers. Hence $z_1=0$, $z_9=0$, $z_{10}=0$, $z_{11}=0$.

Program (11) defines a synchronous sequential machine whose input variables are $x_0, \ldots, x_4$, output variables are $z_0, \ldots, z_{13}$, and internal states encode $N_1, \ldots, N_{14}$. All operations included in a given step (test of the condition variable value, execution of the transfer(s), and transition to the next state) are performed in one clock period. *Instructions* $N_1, \ldots, N_{14}$ are encoded by means of four internal state variables $y_0, \ldots, y_3$. The flow table of the sequential machine is shown in Figure 6.

The next step consists in implementing the sequential machine of Figure 6. This can be performed by methods described in Chapters V, VI, and VII. As an example, Figure 7 shows the logic diagram of an implementation based on the use of a read only memory: it includes $2^9=512$ rows and eighteen columns.

It remains to interconnect the control unit (*control automaton, finite state machine controller*, and *algorithmic state machine*) and the processing unit (*operational automaton* and *data path*). Appropriate care must be taken in order to avoid hazards. Among others, combinational closed loops must be avoided. In principle, it suffices that both the control and operational automata be realized by means of master–slave or edge triggered flip-flops, and that one of them, for instance the operational automaton, be a Moore automaton: the input state changes of a Moore automaton do not directly act upon its output state. A simple practical organization consists in using a two-phase ($\emptyset_1, \emptyset_2$) clocking scheme, controlling master-slave type registers. The corresponding logic diagram is shown in Figure 8. In the control unit,

$x_4x_3x_2x_1x_0$	000 00	000 01	000 10	000 11	001 00	001 01	001 10	001 11	010 00	010 01	010 10	010 11	011 00	011 01	011 10	011 11
N_1	N_2, σ_{14}	N_2, σ_{14}	N_2, σ_{14}	N_2, σ_{14}	N_2, σ_{14}	N_2, σ_{14}	N_2, σ_{14}	N_2, σ_{14}	N_2, σ_{14}	N_2, σ_{14}	N_2, σ_{14}	N_2, σ_{14}	N_2, σ_{14}	N_2, σ_{14}	N_2, σ_{14}	N_2, σ_{14}
N_2	N_3, σ_{15}	N_3, σ_{15}	N_3, σ_{15}	N_3, σ_{14}	N_3, σ_{15}	N_3, σ_{15}	N_3, σ_{15}	N_3, σ_{15}	N_{14}, λ	N_{14}, λ	N_{14}, λ	N_{14}, λ	N_{14}, λ	N_{14}, λ	N_{14}, λ	N_{14}, λ
N_3	N_4, λ	N_4, λ	N_4, λ	N_4, λ	N_4, λ	N_4, λ	N_4, λ	N_4, λ	N_4, λ	N_4, λ	N_4, λ	N_4, λ	N_4, λ	N_4, λ	N_4, λ	N_4, λ
N_4	N_6, σ_9	N_5, σ_9	N_5, σ_9	N_5, σ_9	N_5, σ_9	N_5, σ_9	N_5, σ_9	N_5, σ_9	N_5, σ_9	N_5, σ_9	N_5, σ_9	N_5, σ_9	N_5, σ_9	N_5, σ_9	N_5, σ_9	N_5, σ_9
N_5	N_6, σ_{12}	N_6, σ_{12}	N_6, σ_{12}	N_6, σ_{12}	N_6, σ_{12}	N_6, σ_{12}	N_6, σ_{12}	N_6, σ_{12}	N_6, σ_{12}	N_6, σ_{12}	N_6, σ_{12}	N_6, σ_{12}	N_6, σ_{12}	N_6, σ_{12}	N_6, σ_{12}	N_6, σ_{12}
N_6	N_7, λ	N_7, λ	N_7, σ_{13}	N_7, σ_{13}	N_7, λ	N_7, λ	N_7, σ_{13}	N_7, λ	N_7, λ	N_7, σ_{13}	N_7, σ_{13}	N_7, λ	N_7, λ	N_7, λ	N_7, σ_{13}	N_7, σ_{13}
N_7	N_3, σ_{16}	N_3, σ_{16}	N_3, σ_{16}	N_3, σ_{16}	N_3, σ_{16}	N_3, σ_{16}	N_3, σ_{16}	N_3, σ_{16}	N_3, σ_{16}	N_3, σ_{16}	N_3, σ_{16}	N_3, σ_{16}	N_3, σ_{16}	N_3, σ_{16}	N_3, σ_{16}	N_3, σ_{16}
N_8	N_2, σ_{17}	N_2, σ_{17}	N_2, σ_{17}	N_2, σ_{17}	N_9, σ_{18}	N_9, σ_{18}	N_9, σ_{18}	N_9, σ_{18}	N_2, σ_{17}	N_2, σ_{17}	N_2, σ_{17}	N_2, σ_{17}	N_9, σ_{18}	N_9, σ_{18}	N_9, σ_{18}	N_9, σ_{18}
N_9	N_{10}, λ	N_{10}, λ	N_{10}, λ	N_{10}, λ	N_{10}, λ	N_{10}, λ	N_{10}, λ	N_{10}, λ	N_{10}, λ	N_{10}, λ	N_{10}, λ	N_{10}, λ	N_{10}, λ	N_{10}, λ	N_{10}, λ	N_{10}, λ
N_{10}	N_{12}, σ_9	N_{11}, σ_9	N_{12}, σ_9	N_{11}, σ_9	N_{12}, σ_9	N_{11}, σ_9	N_{12}, σ_9	N_{11}, σ_9	N_{12}, σ_9	N_{11}, σ_9	N_{12}, σ_9	N_{11}, σ_9	N_{12}, σ_9	N_{11}, σ_9	N_{12}, σ_9	N_{11}, σ_9
N_{11}	N_{12}, σ_{12}	N_{12}, σ_{12}	N_{12}, σ_{12}	N_{12}, σ_{12}	N_{12}, σ_{12}	N_{12}, σ_{12}	N_{12}, σ_{12}	N_{12}, σ_{12}	N_{12}, σ_{12}	N_{12}, σ_{12}	N_{12}, σ_{12}	N_{12}, σ_{12}	N_{12}, σ_{12}	N_{12}, σ_{12}	N_{12}, σ_{12}	N_{12}, σ_{12}
N_{12}	N_{13}, λ	N_{13}, λ	N_{13}, λ	N_{13}, λ	N_{13}, λ	N_{13}, λ	N_{13}, λ	N_{13}, λ	N_{13}, λ	N_{13}, λ	N_{13}, λ	N_{13}, λ	N_{13}, λ	N_{13}, λ	N_{13}, λ	N_{13}, λ
N_{13}	N_9, σ_{16}	N_9, σ_{16}	N_9, σ_{16}	N_9, σ_{16}	N_9, σ_{16}	N_9, σ_{16}	N_9, σ_{16}	N_9, σ_{16}	N_9, σ_{16}	N_9, σ_{16}	N_9, σ_{16}	N_9, σ_{16}	N_9, σ_{16}	N_9, σ_{16}	N_9, σ_{16}	N_9, σ_{16}
N_{14}	N_{14}, λ	N_{14}, λ	N_{14}, λ	N_{14}, λ	N_{14}, λ	N_{14}, λ	N_{14}, λ	N_{14}, λ	N_{14}, λ	N_{14}, λ	N_{14}, λ	N_{14}, λ	N_{14}, λ	N_{14}, λ	N_{14}, λ	N_{14}, λ

$x_4x_3x_2x_1x_0$	100 00	100 01	100 10	100 11	101 00	101 01	101 10	101 11	110 00	110 01	110 10	110 11	111 00	111 01	111 10	111 11
N_1	N_2, σ_{14}	N_2, σ_{14}	N_2, σ_{14}	N_2, σ_{14}	N_2, σ_{14}	N_2, σ_{14}	N_2, σ_{14}	N_2, σ_{14}	N_2, σ_{14}	N_2, σ_{14}	N_2, σ_{14}	N_2, σ_{14}	N_2, σ_{14}	N_2, σ_{14}	N_2, σ_{14}	N_2, σ_{14}
N_2	N_3, σ_{15}	N_3, σ_{15}	N_3, σ_{15}	N_3, σ_{15}	N_3, σ_{15}	N_3, σ_{15}	N_3, σ_{15}	N_3, σ_{15}	N_{14}, λ	N_{14}, λ	N_{14}, λ	N_{14}, λ	N_{14}, λ	N_{14}, λ	N_{14}, λ	N_{14}, λ
N_3	N_8, σ_7	N_8, σ_7	N_8, σ_7	N_8, σ_7	N_8, σ_7	N_8, σ_7	N_8, σ_7	N_8, σ_7	N_8, σ_7	N_8, σ_7	N_8, σ_7	N_8, σ_7	N_8, σ_7	N_8, σ_7	N_8, σ_7	N_8, σ_7
N_4	N_6, σ_9	N_5, σ_9	N_6, σ_9	N_5, σ_9	N_6, σ_9	N_5, σ_9	N_6, σ_9	N_5, σ_9	N_6, σ_9	N_5, σ_9	N_6, σ_9	N_5, σ_9	N_6, σ_9	N_5, σ_9	N_6, σ_9	N_5, σ_9
N_5	N_6, σ_{12}	N_6, σ_{12}	N_6, σ_{12}	N_6, σ_{12}	N_6, σ_{12}	N_6, σ_{12}	N_6, σ_{12}	N_6, σ_{12}	N_6, σ_{12}	N_6, σ_{12}	N_6, σ_{12}	N_6, σ_{12}	N_6, σ_{12}	N_6, σ_{12}	N_6, σ_{12}	N_6, σ_{12}
N_6	N_7, λ	N_7, λ	N_7, σ_{13}	N_7, σ_{13}	N_7, λ	N_7, λ	N_7, σ_{13}	N_7, σ_{13}	N_7, λ	N_7, λ	N_7, σ_{13}	N_7, σ_{13}	N_7, λ	N_7, λ	N_7, σ_{13}	N_7, σ_{13}
N_7	N_3, σ_{16}	N_3, σ_{16}	N_3, σ_{16}	N_3, σ_{16}	N_3, σ_{16}	N_3, σ_{16}	N_3, σ_{16}	N_3, σ_{16}	N_3, σ_{16}	N_3, σ_{16}	N_3, σ_{16}	N_3, σ_{16}	N_3, σ_{16}	N_3, σ_{16}	N_3, σ_{16}	N_3, σ_{16}
N_8	N_2, σ_{17}	N_2, σ_{17}	N_2, σ_{17}	N_2, σ_{17}	N_9, σ_{17}	N_9, σ_{17}	N_9, σ_{17}	N_9, σ_{17}	N_2, σ_{17}	N_2, σ_{17}	N_2, σ_{17}	N_2, σ_{17}	N_9, σ_{18}	N_9, σ_{18}	N_9, σ_{18}	N_9, σ_{18}
N_9	N_2, σ_{19}	N_2, σ_{19}	N_2, σ_{19}	N_2, σ_{19}	N_2, σ_{19}	N_2, σ_{19}	N_2, σ_{19}	N_2, σ_{19}	N_2, σ_{19}	N_2, σ_{19}	N_2, σ_{19}	N_2, σ_{19}	N_2, σ_{19}	N_2, σ_{19}	N_2, σ_{19}	N_2, σ_{19}
N_{10}	N_{12}, σ_9	N_{12}, σ_9	N_{12}, σ_9	N_{12}, σ_9	N_{12}, σ_9	N_{12}, σ_9	N_{12}, σ_9	N_{12}, σ_9	N_{12}, σ_9	N_{12}, σ_9	N_{12}, σ_9	N_{12}, σ_9	N_{12}, σ_9	N_{12}, σ_9	N_{12}, σ_9	N_{12}, σ_9
N_{11}	N_{12}, σ_{12}	N_{12}, σ_{12}	N_{12}, σ_{12}	N_{12}, σ_{12}	N_{12}, σ_{12}	N_{12}, σ_{12}	N_{12}, σ_{12}	N_{12}, σ_{12}	N_{12}, σ_{12}	N_{12}, σ_{12}	N_{12}, σ_{12}	N_{12}, σ_{12}	N_{12}, σ_{12}	N_{12}, σ_{12}	N_{12}, σ_{12}	N_{12}, σ_{12}
N_{12}	N_{13}, λ	N_{13}, λ	N_{13}, σ_{13}	N_{13}, σ_{13}	N_{13}, λ	N_{13}, λ	N_{13}, σ_{13}	N_{13}, σ_{13}	N_{13}, λ	N_{13}, λ	N_{13}, σ_{13}	N_{13}, σ_{13}	N_{13}, λ	N_{13}, λ	N_{13}, σ_{13}	N_{13}, σ_{13}
N_{13}	N_9, σ_{16}	N_9, σ_{16}	N_9, σ_{16}	N_9, σ_{16}	N_9, σ_{16}	N_9, σ_{16}	N_9, σ_{16}	N_9, σ_{16}	N_9, σ_{16}	N_9, σ_{16}	N_9, σ_{16}	N_9, σ_{16}	N_9, σ_{16}	N_9, σ_{16}	N_9, σ_{16}	N_9, σ_{16}
N_{14}	N_{14}, λ	N_{14}, λ	N_{14}, λ	N_{14}, λ	N_{14}, λ	N_{14}, λ	N_{14}, λ	N_{14}, λ	N_{14}, λ	N_{14}, λ	N_{14}, λ	N_{14}, λ	N_{14}, λ	N_{14}, λ	N_{14}, λ	N_{14}, λ

Figure 6. Sequential machine

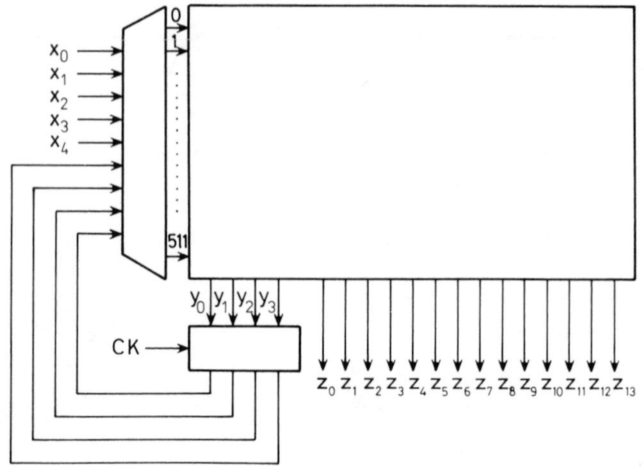

Figure 7. ROM implementation

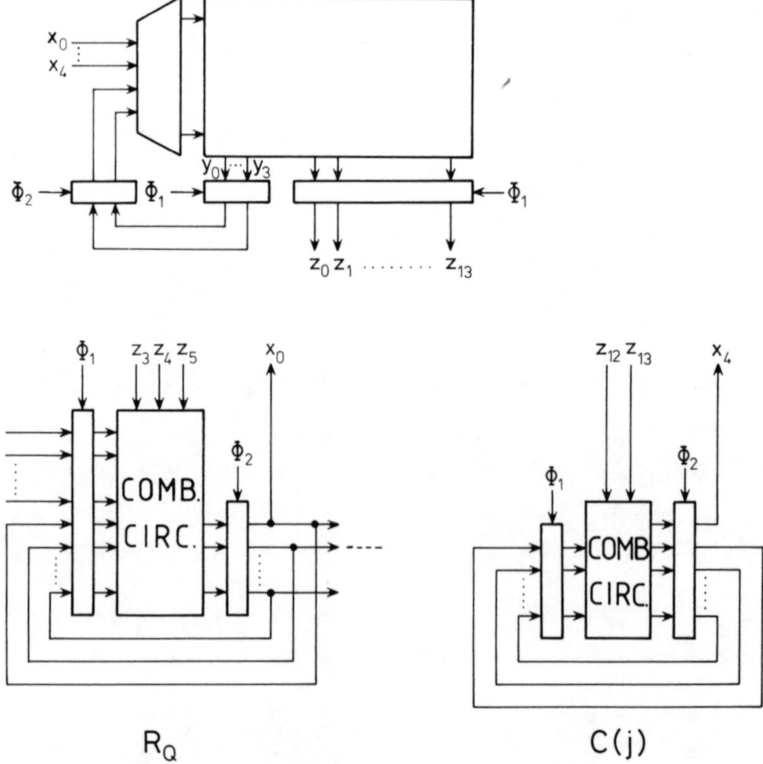

Figure 8. Two-phase clock implementation

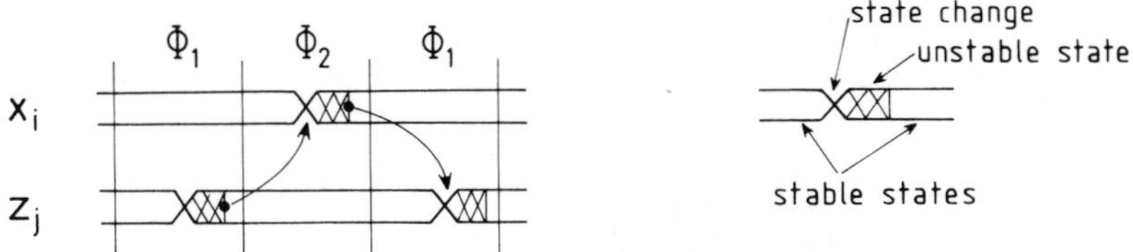

Figure 9. Timing diagram

the masters are activated during $\emptyset_1$ and the slaves during $\emptyset_2$; in the processing unit, the masters are activated during $\emptyset_2$ and the slaves during $\emptyset_1$. During $\emptyset_1$, condition variables $x_0, \ldots, x_4$ are stable, and control variables $z_0, \ldots, z_{13}$ can take new values. During $\emptyset_2$, control variables $z_0, \ldots, z_{13}$ are stable, and condition variables $x_0, \ldots, x_4$ can take new values. Hence, period $\emptyset_1$ (resp. $\emptyset_2$) must be long enough in order that variables $z_0, \ldots, z_{13}$ (resp. $x_0, \ldots, x_4$) reach their new values and be stable at the beginning of $\emptyset_2$ (resp. $\emptyset_1$). This fact is illustrated by Figure 9.

Comments.

(1) We arbitrarily decided to stop the system in state N_{14}(END). We can use this state N_{14} in order to link the exponentiation program to another program.

(2) In the flow chart of Figure 4, all tests concern one condition variable at a time. On the other hand, the implementation of Figure 7, based on the flow table of Figure 6, implicitly contains tests on all five condition variables. The question naturally arises whether a suitable circuit architecture can implement tests concerning one variable at a time. The affirmative answer is given in Chapters XII and XIII. A large cost saving can be obtained by taking into account the program structure, and by designing circuits which match this structure.

(3) The flow chart of Figure 4 contains twice the flow chart of Figure 10. We can again try to design a circuit matching this propety, and avoid storing the same subprogram several times. We will deal with that matter in Chapter XIII.

(4) It is worthwhile to recall the way we obtained the algorithm of Figure 4. The starting point was the algorithm of Figure 3a. We would have thought of a direct implementation of this algorithm by means of a circuit including a computation resource able to perform in one clock period the rather complicated transfer

$$[R_p]:=[R_p]^2[R_X]^{a_{n-1}}. \tag{12}$$

Nevertheless, this resource would have a very long computation time, compared with the time necessary for performing σ_0, σ_2, σ_4, and σ_{11}. As the minimum clock period depends on the resource having the longest computation time, the whole system would be very slow. A great improvement in

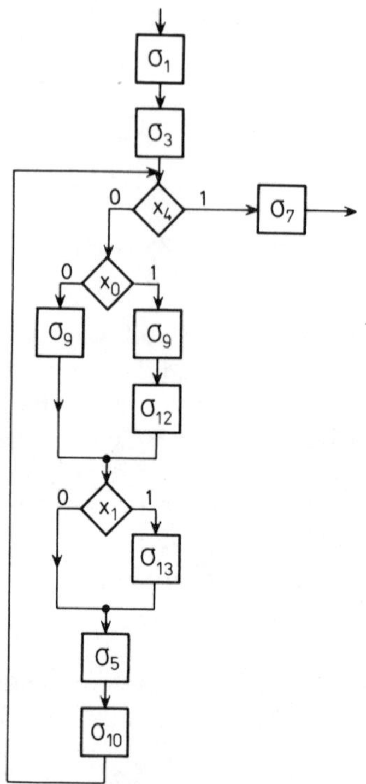

Figure 10. Subpart of the exponentiation algorithm

speed is obtained by sequentializing the transfer (12), in order to obtain an algorithm which only contains simple operations: load, reset, shift, count down, and modulo two sum. The general method thus consists in refining the algorithm in order to get primitives having comparable computation times. The clock signal generator then works at maximum rate, and every transfer appearing in the initial algorithm just receives the number of clock periods it needs. This method presents another advantage: every computation resource of the refined algorithm belongs to the list of standard circuits described in the preceding chapters. This point is obviously important if the system is made up of interconnected standard integrated circuits. It is also important within the frame of very large scale integrated circuit design: computer aided design facilities can include libraries of standard circuits; the whole circuit is then composed of well known, and *a priori* designed, structures (registers, counters, multiplexers, arithmetic and logic circuits, arrays, drivers, etc.) The design work mainly consists in devising the circuit architecture and interfacing the various parts, with the possibility of focusing attention on the geometrical aspects.

(5) The processing unit architecture (Figure 2) was chosen in a rather arbitrary way. Some decisions were quite natural. As an example, register R_p appears in the first number of transfers σ_0 ($[R_p]:=1$) and σ_7

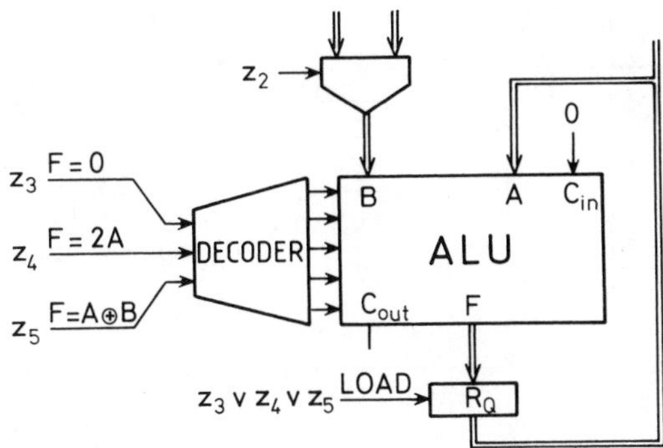

Figure 11. Alternative architecture

($[R_p]:=[R_Q]$). Hence, we choose to use a multiplexer, controlled by z_0, which allows both transfers to be performed. On the other hand, some decisions are more or less arbitrary. As an example, register R_Q appears in the first member of transfers

$$\sigma_1([R_Q]:=0), \qquad \sigma_9([R_Q]:=2[R_Q], \text{mod } 2^r),$$
$$\sigma_{12}([R_Q]:=[R_Q] \oplus [R_M])$$

and

$$\sigma_{13}([R_Q]:=[R_Q] \oplus [R_p]).$$

An alternative solution is shown in Figure 11. Our choice came from the fact that the shift and reset operations can easily be implemented inside the register. Nevertheless, the alternative solution is perhaps preferable if the exponentiation algorithm is part of a larger algorithm. Similarly, we could think of sequentializing the transfers and replace the multiplexers by buses (see Chapter XI).

2 BASIC MODEL

The aim of this section is to formalize the notions which are mentioned in the introductory example. Let us recall that we intend to describe a computing system, made up of a control unit and a processing unit (see Figure 1).

We get at once some idea of the processing unit architecture and of its working mode: it includes a set $\{R_1, R_2, \ldots, R_s\}$ of registers which store the processed data (initial data, intermediate, and final results), a set $\{cc_1, cc_2, \ldots, cc_t\}$ of computation resources which perform the actual processing, and an interconnection network capable of connecting registers and computation resources. Both the computation resources and the interconnection network are externally controlled. Examples of computation resources were given in Chapter IV.

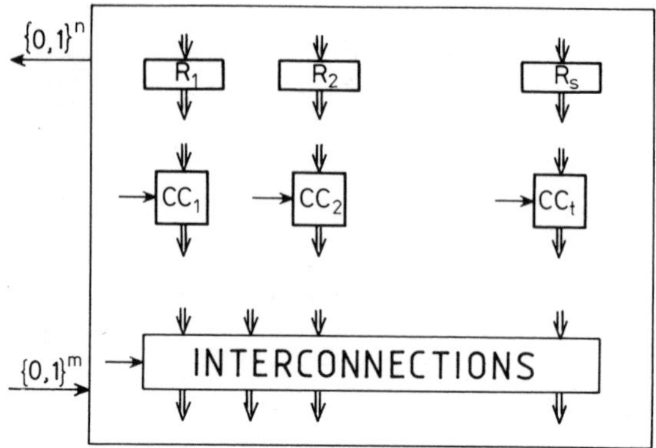

Figure 12. General structure of the processing unit

This rather general definition is illustrated by Figure 12.

The processing unit thus naturally appears as being a finite automaton. It is sometimes called an *operational automaton.* The elements of the input alphabet are *commands* which are usually denoted by small greek letters $\sigma, \tau, \ldots$. They are encoded by means of *m control variables* $z_0, z_1, \ldots, z_{m-1}$, each of which defines the state of a *control line*; examples of control lines are enable or inhibit lines of registers, programming lines of computation resources, and control lines of the interconnection network. One of the commands, namely the *empty command* λ, keeps the automaton in its present state (*no operation* command).

The elements of the output alphabet are encoded by means of *condition variables* (*qualifiers* and *status variables*) $x_0, \ldots, x_{n-1}$. They give information to the control unit about the present state of the operational automaton.

In view of the number of different input and internal states, the transition and output functions generally cannot be represented by flow tables. Besides, the general structure shown in Figure 12 suggests the use of a transfer language. To every command σ corresponds an internal state modification which can be described under the general form

$$\sigma:[R_1]:=g_1([R_{1,1}],\ldots,[R_{1,v_1}]); \qquad \ldots; \qquad R_s:=g_s([R_{s,1}],\ldots,[R_{s,v_s}]), \tag{13}$$

where $[R]$ stands for the contents of register R. Contents $[R_{i,j}]$, appearing in the right-hand members of (13), relate to the time which immediately precedes the clock pulse, while contents $[R_k]$, appearing in the left-hand members, relate to the time which follows the clock pulse.

A state modification, which only includes transfers such as

$$\ldots,[R_i]:=A_i,\ldots,[R_j]:=[R_j],\ldots,$$

by which some register contents $[R_i]$ are replaced by an explicitly mentioned value A_i, is called *value transfer.*

We suppose finally that the output function is a mapping from the set of internal states, that is the register states, to the output alphabet, encoded by condition variables. These condition variables are most often outputs of flip-flops (see x_0, x_1, and x_2 in Figure 2) or simple Boolean functions of flip-flops outputs (see x_3 and x_4 in Figure 2), which yield information concerning the register contents (overflow, sign, zero, contents, etc.).

According to the classical terminology, the operational automaton is a Moore automaton.

Let us now turn to the control unit. As in the introductory example (Section 1), we aim at presenting the computation algorithm in such a way that the control unit can also be implemented by a finite automaton. Hence, the whole system can be implemented by a synchronous sequential machine, according to classical methods presented in previous chapters.

According to Mischenko (1967), we describe computation algorithms by programs which we first give a formal definition. A *program* is made up of a series of *instructions*, labelled $N_0, N_1, \ldots, N_{p-1}$, including an *initial instruction* N_0 and a *final instruction* N_{p-1}. Every instruction has the general form

$$\begin{array}{llll} N_i & f_{i0}(\mathbf{x}) & \sigma_{i0} & N_{i0} \\ & f_{i1}(\mathbf{x}) & \sigma_{i1} & N_{i1} \\ & \cdots\cdots & \cdots & \cdots \\ & f_{ir-1}(\mathbf{x}) & \sigma_{ir-1} & N_{ir-1} \end{array} \tag{14}$$

in which $f_{i0}, \ldots, f_{ir-1}$ are Boolean functions depending on the condition variables $\mathbf{x} = x_0, x_1, \ldots, x_{n-1}$; as above, $\sigma_{i0}, \sigma_{i1}, \ldots, \sigma_{ir-1}$ denote commands and $N_{i0}, N_{i1}, \ldots, N_{ir-1}$ are instruction labels. A proper interpretation of instruction (14) will lead to the definition of a *control automaton.*

We first state that the set of tasks, which must be performed according to (14), will be completed in one clock period. This leads to the identification of the set of instructions with the set of internal states of the control automaton. The input letters of the control automaton are the minterms in $\mathbf{x}$. Functions f_{ij} define the transition and output functions. The values taken by functions f_{ij} must be computed according to the values taken by the condition variables $\mathbf{x}$ before execution of any command σ_{ij}. Indeed, the execution of a command σ_{ij} could modify the state of the operational automaton, and thus also the value of some condition variables. In fact, the instruction execution can be divided into two phases: an *evaluation phase* during which all functions f_{ij} are evaluated at point $\mathbf{x}$, and an *execution phase* during which some command(s) are executed, according to the result of the evaluation phase. Let us define

$$J = \{j \mid f_{ij}(\mathbf{x}) = 1\} \tag{15}$$

and

$$k = \max J. \tag{16}$$

The transition and output functions, **M** and **N**, of the control automaton are defined as follows:

$$N_iM_{\mathbf{x}} = N_{ik}, \qquad \text{if } J \neq \emptyset, \tag{17}$$

$$N_iM_{\mathbf{x}} = \emptyset, \qquad \text{if } J = 0, \tag{18}$$

$$N_iN_{\mathbf{x}} = \prod_{j \in J} \sigma_{ij}, \quad \text{if } J \neq \emptyset, \tag{19}$$

$$N_iN_{\mathbf{x}} = \emptyset, \qquad \text{if } J = \emptyset, \tag{20}$$

where Π stands for the command concatenation. We conventionally choose the increasing order of indices j of J for defining composite commands $\Pi_{j \in J} \sigma_{ij}$. Anyway, some order convention is necessary since commands generally do not commute. As an example, let us consider two commands acting on register R:

$$\sigma_1 : [R] := 2[R]; \qquad \sigma_2 : [R] := [R] + 1. \tag{21}$$

The corresponding commands $\sigma_1\sigma_2$ and $\sigma_2\sigma_1$ define two different transfers:

$$\sigma_1\sigma_2 : [R] := 2[R] + 1; \qquad \sigma_2\sigma_1 : [R] := 2[R] + 2. \tag{22}$$

We have reduced the control unit synthesis problem to a classical problem: the synthesis of a synchronous sequential machine implementing a given finite automaton. The control circuit will include p flip-flops, with

$$2^p \geqslant P, \tag{23}$$

and a combinational circuit. These p flip-flops form a register, whose contents may be considered as the address number of the current instruction; it is the *instruction address register.* The combinational circuit produces $(m+p)$ functions $Y_0, \ldots, Y_{p-1}$, $z_0, \ldots, z_{m-1}$, of $(n+p)$ variables $y_0, \ldots, y_{p-1}$, $x_0, \ldots, x_{n-1}$. It can be implemented by any of the methods described above. A read only memory implementation of the circuit leads to a structure introduced by Wilkes (1951); it was the basic idea for the development of microprogrammed systems. It is shown in Figure 7 ($p = 4$, $n = 5$, $m = 14$). The cost Γ of this circuit is assessed by the number of read only memory bits:

$$\Gamma = (m+p)2^{(n+p)}. \tag{24}$$

The relevance of this cost estimation method implicitly supposes that a two level addressing scheme of the read only memory is used: see Chapter III.

Below we will often adopt the graphical presentation of Figure 7 for representing a control unit, whatever the actual circuit might be. Classical alternatives are *random logic*, that is rather irregular interconnection of logic gates, and programmable logic arrays.

We conclude this section by stating some properties of the basic model. Let us first observe that the ability of the model to describe algorithms can be considered as satisfactory. As an example, the Pascal statements

while $f(\mathbf{x})$ *do* σ,

repeat σ *until* $f(\mathbf{x})$,

if $f(\mathbf{x})$ *then* σ_0 *else* σ_1,

correspond to

$$\begin{array}{llll} N_i & f(\mathbf{x}) & \sigma & N_i \\ & \overline{f(\mathbf{x})} & \cdots & \cdots, \\ N_i & t & \sigma & N_j \\ N_j & \overline{f(\mathbf{x})} & \lambda & N_i \\ & f(\mathbf{x}) & \cdots & \cdots, \\ N_i & f(\mathbf{x}) & \sigma_0 & \cdots \\ & \overline{f(\mathbf{x})} & \sigma_1 & \cdots. \end{array}$$

Pieces of program, equivalent to *for* statements, have also been written in the introductory example.

Let us return to general instruction (14). It will serve principally as a theoretical tool. In fact, a correct interpretation of (14) leads to an equivalent instruction

$$\begin{array}{llll} N_i & m_0(\mathbf{x}) & \sigma'_{i0} & N'_{i0} \\ & m_1(\mathbf{x}) & \sigma'_{i1} & N'_{i1} \\ & \cdots\cdots & \cdots\cdots & \cdots\cdots \\ & m_{2^n-1}(\mathbf{x}) & \sigma'_{i2^n-1} & N'_{i2^n-1} \end{array} \tag{25}$$

where functions m_j are minterms in $\mathbf{x}$, and where commands σ'_{ij} and instruction labels N'_{ij} are defined by (17) to (20). A program which is made up of such instructions is called a *canonical program*. The translation from (14) to (25) obviously corresponds to the definition of the control automaton flow table: as an example, let us compare program (11) and the flow table in Figure 6.

The change from the general form (14) to the canonical form (25) can raise some difficulty: according to (19), some σ'_{ij} can be composite commands. If the corresponding transfers, to perform inside the processing unit, do not interact, they may be simultaneously initiated, and the composite command is executed in one clock period. However, in some cases the corresponding transfers interact; as an example, commands σ_1 and σ_2 defined by (21) initiate transfers concerning the same register R. These two transfers cannot be executed in one clock period. We may think of modifying the processing unit in order that it can execute, in one clock period, a composite transfer such as $\sigma_1\sigma_2$ (or $\sigma_2\sigma_1$). In some cases, this is not possible: for instance, the processing unit can be made up of standard circuits, such as bit slice processors, whose internal working cannot be modified. Furthermore, it can be preferable to break down the execution phase into two clock periods rather than enlarge the computation or transfer capabilities of the processing unit (time–cost trade off): as an example, let us consider the canonical instruction ($n = 1$)

$$\begin{array}{cccc} N_i & \bar{x} & \sigma_1\sigma_2 & N_j \\ & x & \cdots & \cdots; \end{array}$$

it can be replaced by the next piece of canonical program

$$\begin{array}{cccc} N_i & \bar{x} & \sigma_1 & N_k \\ & x & \cdots & \cdots \\ N_k & \bar{x} & \sigma_2 & N_j \\ & x & \sigma_2 & N_j \end{array}$$

which execute $\sigma_1\sigma_2$ in two clock periods.

Let us call *data* the initial contents of registers $R_1, R_2, \ldots, R_s$, and *results* the final contents of these same registers. In other words, 'data' and 'results' are the initial and final state of the operational automaton. Two programs which associate identical results to identical data are called *equivalent program.*

Once the operational and the control automata have been defined, it remains to implement the corresponding synchronous sequential machines. It has been seen in the introductory example that precautions must be taken in order to avoid hazards. Figure 13 shows the logic diagram of the whole circuit. The operational automaton is a Moore automaton; hence, condition variables $\mathbf{x}$ depend on the operational automaton state, that is on the contents of registers $R_1, R_2, \ldots, R_s$, but not on control variables $\mathbf{z}$. If all

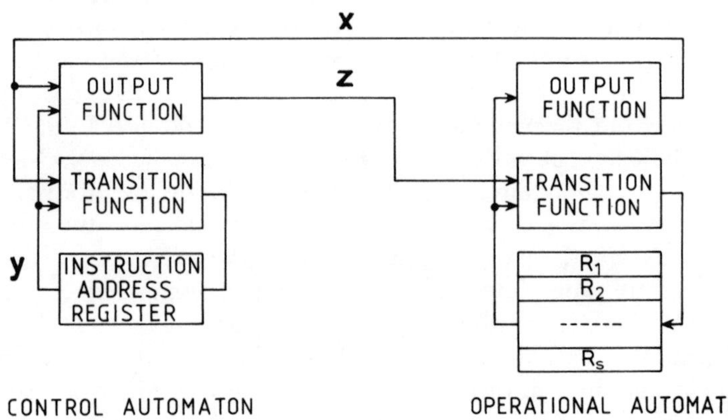

Figure 13. Logic diagram corresponding to the basic model

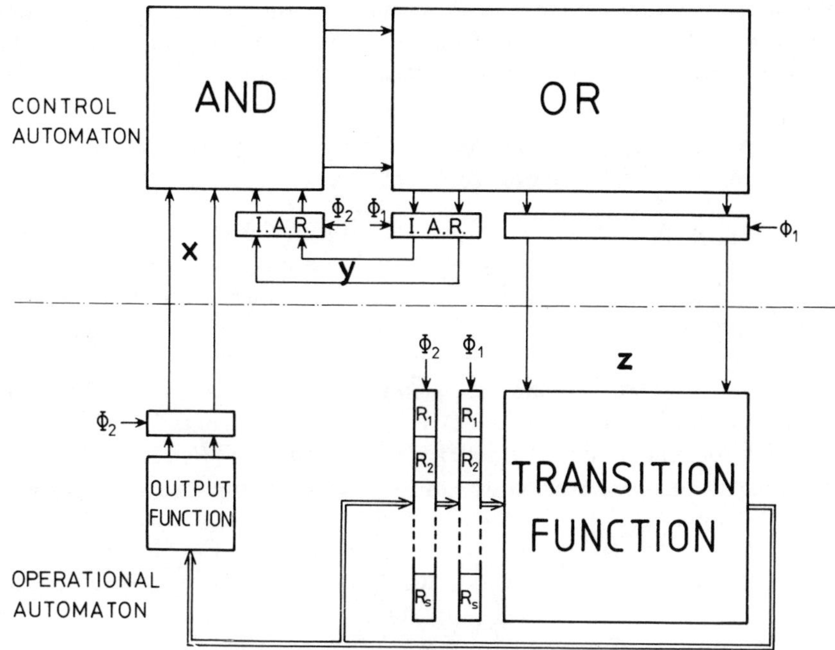

Figure 14. Two-phase clocking scheme organization

registers are made up of master–slave or edge triggered flip-flops, the system does not contain any closed loop and can work properly. A simple practical organization consists in using a two-phase clocking scheme: see Figure 14. The fact that the operational automaton is a Moore automaton is then irrelevant.

Remark. Figure 15 shows a combinational circuit which is equivalent to the basic model: A stands for the operational automaton, and B for the control automaton. In fact, this diagram does not give an account of an essential feature of the model, namely the possibility of guaranteeing computation completion in finite time, without knowing in advance the computation time. As an example, the number of program steps which are actually performed during execution of the exponentiation program (Figure 4) depends on the

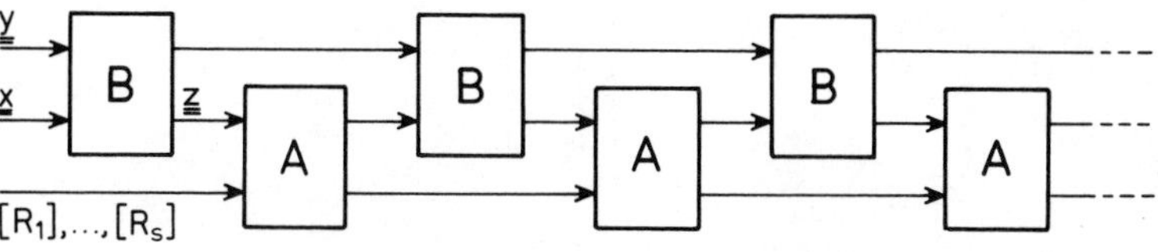

Figure 15. Combinational form of the basic model

value of the external data X and A, through condition variables x_0, x_1, and x_2. In the combinational circuit of Figure 15, this ignorance of the computation time leads to an uncertainty concerning the number of cells.

3 BOUNDARY BETWEEN THE CONTROL AND PROCESSING UNITS

Figure 15 gives prominence to the rather arbitrary character of the boundary between the control and operational automata. We first give two examples of boundary moving which directly come from the program structure observation.

Let us suppose that one of the control variables, for instance z_0, only depends on the condition variables $\mathbf{x}$ and not on the internal variables $\mathbf{y}$ of the control automaton. Variable z_0 is called a *local control variable.* The value of z_0 depends only on $\mathbf{x}$ which depends only on $[R_1], [R_2], \ldots, [R_s]$; hence, z_0 can be directly generated in the processing unit. Figure 16a shows the logic diagram of the modified processing unit, in the case where the two-phase clocking scheme is used.

It can also arise that the transition function of the control automaton does

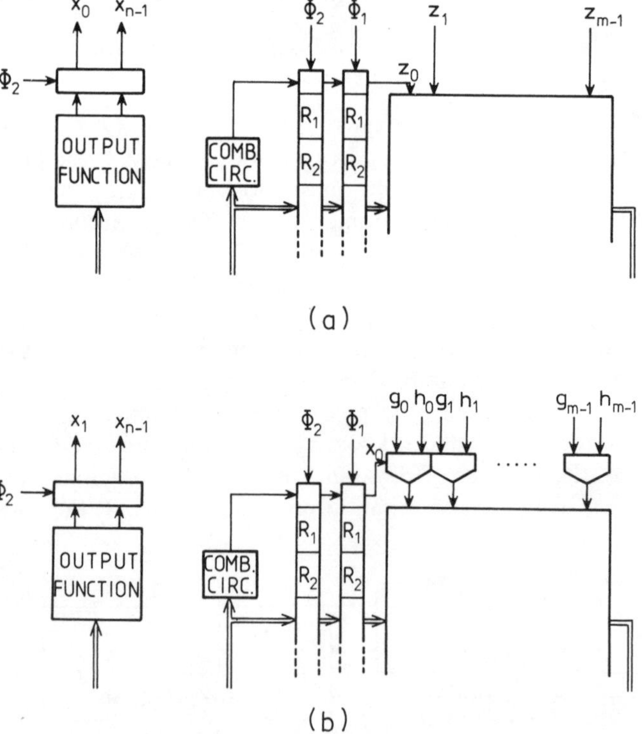

Figure 16. (a) Local control variable. (b) Local condition variable

not depend on one of the condition variables, for instance x_0. Variable x_0 is called a *local condition variable.* Every control variable z_j is a function of $\mathbf{x}$ and $\mathbf{y}$; it can be decomposed under the form

$$z_j = \bar{x}_0 g_j(x_1, \ldots, x_{n-1}, \mathbf{y}) \vee x_0 h_j(x_1, \ldots, x_{n-1}, \mathbf{y}).$$

Figure 16b shows the modified processing unit. As an example, variable x_1 of program (11) is a local condition variable: it appears in instructions N_6 and N_{12} in which the next address labels are N_7 and N_{13} respectively, whatever the value of x_1.

If all control variables are local control variables, then the control unit no longer issues commands to the processing unit. The control automaton receives information about the progress of the computation, but cannot act upon the working of the processing unit. This particular situation arises when the implemented algorithm is an iteration.

If all condition variables are local condition variables, the values of $\mathbf{y}$ and $\mathbf{z}$ do not depend on $\mathbf{x}$: the control automaton is an autonomous system. This occurs if the implemented algorithm is a computation scheme (Chapter VIII).

The existence of local variables allowed the system architecture to be slightly modified and to move the boundary between the control and processing units. We thus are faced with the next problem: does there exist a natural boundary between control and processing unit? A partial answer was given by Stabler (1970). He noticed that, very often, processing units have a great number of internal states and a regular structure, while the control units have less internal states and an irregular structure. According to this criterion, control units implement all functions which are not intrinsically regularly organized. The control units can be made up of random logic. As the number of internal states is not too large, they can also be made up of regular arrays such as read only memories or programmable logic arrays; the whole system then only includes regular circuits.

A partial answer can also be found in comments (4) and (5) at the end of Section 1: once an algorithm has been defined, a processing unit architecture is chosen. It must be able to implement the algorithm and includes standard circuits (registers, arithmetic and logic units, etc.); the algorithm is possibly refined in order to get basic operations which have equal computation times. Furthermore, in some cases, the processing unit is partially known in advance. This happens if it is decided to use standard circuits such as bit slice processing units, or to implement a previously designed processing unit.

Nevertheless, the choice of a processing unit architecture remains rather arbitrary, and mainly depends on the designer's ingenuity, on perhaps aesthetic criteria, but also on habits and tradition. The same remarks obviously apply to the boundary between control and processing units.

The general formalism we introduced in Section 2 allows, to some extent, boundary motions to be described. Let us consider instruction

$$N_i \quad t \quad \sigma \quad N_j, \tag{26}$$

and suppose that

$$\sigma : [R_1] := f([R_2]). \tag{27}$$

We denote by $A, B, C, \ldots, I, \ldots, K$, the various values that function f takes. Functions $f^{(I)}$ are defined by

$$\begin{aligned} f^{(I)}([R_2]) &= 1, \quad \text{if } f([R_2]) = I \\ &= 0, \quad \text{if } f([R_2]) \neq I. \end{aligned} \tag{28}$$

Commands σ_I are value transfers

$$\sigma_I : [R_1] := I. \tag{29}$$

Instruction (26) may obviously be replaced by

$$\begin{array}{llll} N_i & f^{(A)}([R_2]) & \sigma_A & N_j \\ & f^{(B)}([R_2]) & \sigma_B & N_j \\ & \cdots\cdots & & \\ & f^{(K)}([R_2]) & \sigma_K & N_j \end{array} \tag{30}$$

Instruction (30) only includes value transfers.

A processing unit able to implement a program including instruction (26) will likely include a resource which computes function f. On the other hand, if instruction (26) is replaced by (30), the computation of f is transferred from the processing unit to the control unit. In fact, this transformation can be performed for every instruction of a given program, and it is theoretically possible to transform any program into an equivalent program in which all commands are value transfers. The processing unit then amounts to a set of registers. At every step, they are loaded by values which are fed by the control unit. All computations are performed inside the control unit, while the processing unit is used for storing intermediate results. Such a system is called a *Boolean processor.*

It is also possible to move only part of the computation resources from the processing unit to the control unit. As an example, let us suppose that the command σ appearing in (26) controls the transfer

$$\sigma : [R] := f([R_1], [R_2]),$$

with

$$f([R_1], [R_2]) = \overline{h([R_1])}g_0([R_2]) \vee h([R_1])g_1([R_2]).$$

Instruction (26) may be replaced by

$$\begin{array}{lll} \overline{h([R_1])} & [R] := g_0([R_2]) & N_j \\ h([R_1]) & [R] := g_1([R_2]) & N_j. \end{array} \tag{31}$$

Another transformation consists in transferring all the storing capability of the circuit into the processing unit. The instruction address register is put in the processing unit, and the internal state variables $\mathbf{y}$ become condition variables. Let us consider a program whose general instruction is (14), and define

$$G_j^i(\mathbf{x},\mathbf{y}) = 1, \quad \text{if} \sum_{k=0}^{m-1} y_k 2^k = i \quad \text{and} \quad f_{ij}(\mathbf{x}) = 1,$$

$$= 0, \quad \text{otherwise.}$$

The whole program may be replaced by one instruction:

$$\begin{array}{llll} N_0 \cdots\cdots\cdots\cdots\cdots\cdots & & & \\ & G_0^i(\mathbf{x},\mathbf{y}) & \sigma_{i0}, [\text{I.A.R.}] := i0 & N_0 \\ & G_1^i(\mathbf{x},\mathbf{y}) & \sigma_{i1}, [\text{I.A.R.}] := i1 & N_0 \\ & \cdots\cdots\cdots & & \\ & G_{r-1}^i(\mathbf{x},\mathbf{y}) & \sigma_{ir-1}, [\text{I.A.R.}] := ir-1 & N_0 \\ & \cdots\cdots\cdots & & \end{array}$$

where I.A.R. stands for instruction address register.

The various transformations, which are described in this section, give prominence to the abundance of choices we are faced with during the implementation of an algorithm. In what concerns the boundary between control and operational automata, we feel the need to define objective criteria. The choice of the boundary is particularly important if the whole system is made up of two components, implementing the processing and control units, respectively. In this respect, a relevant objective is the minimization of the number of connecting wires (control and condition variables) which link the two components. Another decomposition method consists in performing inside the control unit all transfers whose execution depends on external data, in such a way that the processing unit remains more or less universal.

BIBLIOGRAPHICAL REMARKS

To the best of our knowledge, the decomposition of a computation system into a control unit and a processing unit may be traced back to Glushkov (1965, 1966, 1970). Observe that this model is slightly coarser than the familiar Von Neumann model. The model was later used by Mischenko (1967, 1968a, 1968b), Gerace (1968), Gerace *et al.* (1974). The model was initially seen as a microprogram model: hence, its relation to Wilkes (1951). The tutorial paper by Rauscher and Adams (1980) is quite representative of that trend. See also Davio and Thayse (1980). A more recent trend, probably first exemplified by Clare (1973) considers the discussed model as having its own value for program implementation. The paper by Swartzlander (1979) is a representative of that new trend.

EXERCISES

E1. The reader should now be able to design as synchronous sequential machines the control automata for all the algorithms studied in the previous chapters.

E2. A shift and add multiplication algorithm may be terminated in either one of the following ways.

(i) Examination of a condition variable z_c which reflects the state of a counter (assumed available in the processing unit); say $z_c = 1$ iff $[c] = n$.
(ii) Examination of a condition variable z_m which reflects the state of the multiplicator register; say $z_m = 1$ iff $[R_m] = 0$.
In the second case, the multiplication table becomes dependent on the value of the multiplicator. The variable z_m is called the *essential condition variable.* Discuss the significance of discovering essential condition variables. Does the presence of essential condition variables influence the possibilities of displacing the boundary between the control unit and the processing unit? Modify the odd–even merge algorithm (Chapter VIII) by introducing an essential control variable.

E3. The specification sheet of a machine tool contains the following requirements:
(1) The successive positions of the tool are described as 18-bit words in a ROM. (The code of a word is irrelevant here; we assume that the information in a word is decoded by three digital to analog converters and transmitted to the tool).
(2) There are four possible sequences of positions, each of which consists of 256 consecutive positions (words in the ROM). These sequences are denoted A_0, A_1, A_2, A_3. Thus, these sequences occupy the following ROM positions A_0: [0,255]; A_1: [256,511]; A_2: [512,767]; A_3: [768,1023].
(3) The system contains a position register PR (18 bit) loaded with the present tool position and feeding the D/A converters.
(4) The system contains a start–stop flip-flop SS.
(5) The system contains a two-bit sequence register SR whose contents, prior to the start of the execution, indicate the suited sequence.
(6) If the operator imposes SS:= 1, the system executes the sequence indicated by SR and completes its job by SS:= 0; PR:= 0.
(7) If, during the sequence, the operator imposes SS:= 0; the system interrupts its task and sets PR:= 0.

Describe in detail the organization of the processing unit, and the various possible modes of the registers involved. Produce a flow chart of the control algorithm and design the control unit using either a ROM or a PLA.

E4. The commercial component 7488 is a ROM containing 32 words of 8 bits. We denote by $x_4x_3x_2x_1x_0$ the address bits, and by $y_1y_2 \ldots y_8$ the output bits. This component is actually a sine table. If

$$x = \sum_{i=0}^{5} x_i 2^i$$

and if

$$\alpha = \frac{90}{32} \times \deg$$

the output

$$y = \sum_{j=1}^{8} y_j 2^{-j}$$

satisfies $y \cong \sin \alpha$. In particular, $x = 0 \Rightarrow y = 0$ and $x = 31 \Rightarrow y_j = 1$ ($j = 1 \ldots 8$). The component is provided with an additional ENABLE input. In normal use, ENABLE = 0. However, if ENABLE = 1, one obtains $y_j = 1$ ($j = 1 \ldots 8$), independently of x. Design a sine wave generator using the 7488 as the basic resource of the processing unit.

Chapter XI
Sequentialization: impact on the processing unit

1 BASIC PRINCIPLES

1.1 Partial or total sequentialization; acceptable chronologies

In Chapter VIII, we studied various implementations of computation schemes by synchronous sequential circuits. The type of implementation studied in that chapter respected the intrinsic parallelism of the given computation scheme; stated otherwise, two partial computations that were simultaneous in the starting algorithm remained simultaneous in the synchronous implementation. That method appears as a fruitful one in many practical situations such as arithmetic computations, where the iterative or recursive nature of the given algorithms often corresponds to highly regular combinational circuits and, accordingly, to particularly simple sequential implementations. Examples of that fact appeared in Chapter IX.

In the present chapter we carry on that discussion, starting from the following new idea: if two partial computations are independent, i.e. if they are not related by the precedence relation, they may be performed one after the other. For example, the precedence graph displayed in Figure 1a may be replaced by that in Figure 1b if we decide to postpone the execution of the partial computation B until completion of the partial computation A. As simple as it is, the technique will appear as a fundamental one: indeed, it allows us to reduce the computation width, and thus the number of computation resources, while increasing the graph length, i.e. the computation time. In fact, the type of algorithm transformation we have just introduced is one of the basic tools in the study of the time–cost tradeoff.

The above idea is easily formalized in the case of computation schemes. Let us indeed call *acceptable chronology* any irreflexive order stronger than the precedence relation ($\subset$). Let us denote such a chronology by the symbol [. We shall have:

$$N_i \subset N_j \Rightarrow N_i \, [\, N_j. \tag{1}$$

Clearly now, if we choose stronger and stronger acceptable chronologies, the

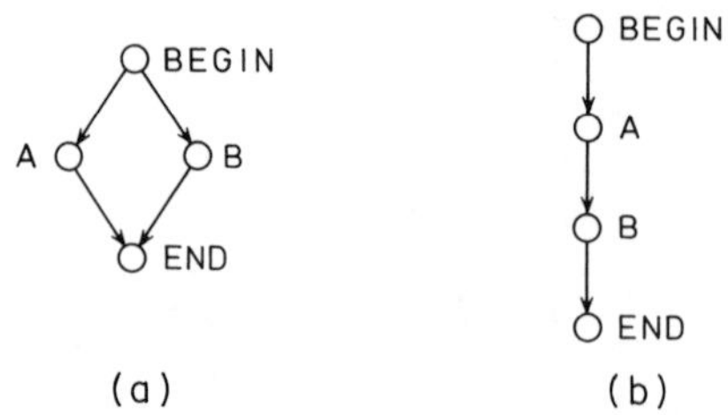

Figure 1. Sequentialization of a precedence graph

computation time increases and the computation width decreases. We call this process *sequentialization.* In the limit case, the acceptable chronology becomes a total order and we then speak of *total sequentialization.* The computation width is then equal to one and this shows that any algorithm may be implemented with a single computation resource.

In the case of unit execution time systems, the acceptable chronologies are easily deduced from the acceptable labellings. Let indeed ψ denote an acceptable labelling (Chapter VIII, (1)). The associated chronology will then be defined by

$$N_i \,[\, N_j \Leftrightarrow \psi(N_i) < \psi(N_j). \tag{2}$$

By equation (2) of Chapter VIII, we have

$$N_i \subset N_j \Rightarrow \psi(N_i) < \psi(N_j) \Rightarrow N_i \,[\, N_j,$$

and equation (1) is indeed satisfied. It should also be clear that the above ideas not only apply to computation schemes but are also valid for other algorithms.

1.2 Example: sequentialization of a decimal display

A classical system for displaying decimal digits has been studied in Section 4.2.5 of Chapter II. For a single decimal digit, given by the four bits $x_3x_2x_1x_0$ of its BCD representation, we use a seven-segment display and a BCD to seven-segment decoder. We now consider the following problem: given N 4-bit registers $R_0, R_1, \ldots, R_{N-1}$, realize the display of the corresponding N-digit number on N seven-segment displays. We shall choose $N = 2^p$ and discuss various solutions.

A first solution, illustrated by Figure 2a for $N = 8$, consists of using as many decoders as registers and displays. It corresponds to the maximum parallelism and its precedence graph is given in Figure 2b. If we denote by γ_S and δ_S the cost and delay of the decoder, we obtain, for the first solution

$$\Gamma_1 = N\gamma_S, \tag{3}$$

$$\Delta_1 = \delta_S. \tag{4}$$

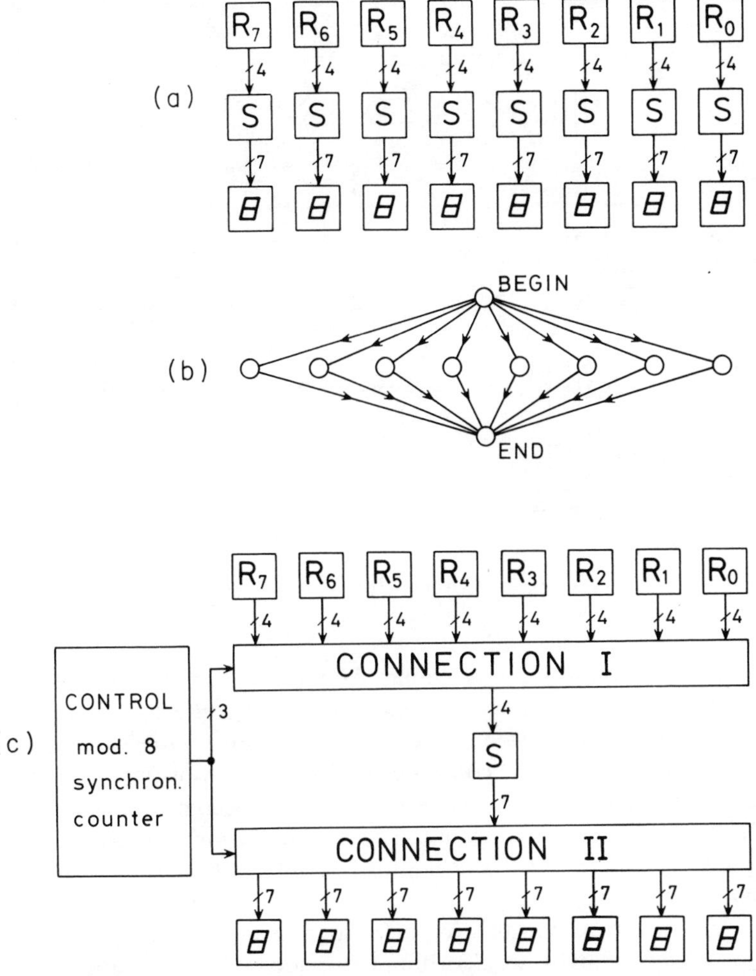

Figure 2. Sequentialization of a decimal display (A)

Let us now replace the underlying algorithm by a totally sequentialized algorithm in order to measure the cost decrease that may result from an asymptotically significant time increase. Observe that it is actually possible to replace the parallel algorithm in Figure 2a by a totally sequentialized algorithm: to convince oneself of this fact, one only has to compare the typical delay of a seven-segment decoder ($0, 1\ \mu s$) with the persistency of retinal images ($0, 1\ s$). In the sequentialized solution, the N registers $R_0, R_1, \ldots, R_{N-1}$ will in turn be connected to the corresponding displays through a unique copy of the decoder S and this operation will be repeated cyclically. A block-diagram corresponding to that solution is given in Figure 2c.

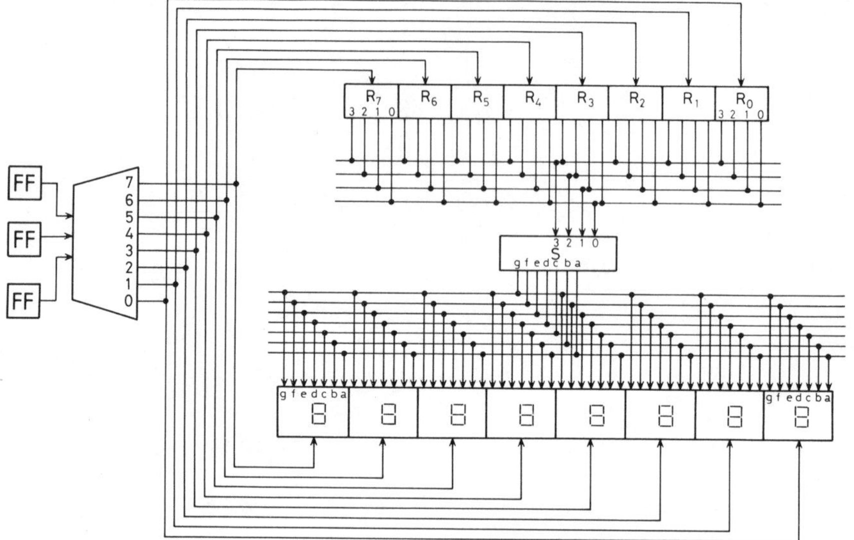

Figure 3. Sequentialization of a decimal display (B)

In this diagram, the connecting network I basically consists of four N-input multiplexers and connecting network II consists of seven N-output demultiplexers. Even if we use the refined technique described in Figure 10b of Chapter II, we have to conclude that the cost Γ_2 of that second solution remains proportional to N, while the delay Δ_2 becomes proportional to N. The return to be expected from the sequentialization process does not lie in the present example in the asymptotic trend of the cost but in the actual numerical value of the cost. The choice of an appropriate technology will play a prominent role in this respect. Figure 3 illustrates a practical realization of the sequentialized algorithm. Two technological features cooperate to achieve a cost-drop.

(i) The registers $R_0, R_1, \ldots, R_7$ are of the TRI-STATE type. Such registers are provided with an ENABLE control input, the 0 or 1 state of which determines a high or low output impedance. In the high impedance state, the registers may be considered as being disconnected from the remaining part of the circuit. In the low impedance state, the register outputs act on the circuit according to their 0 or 1 logical state: that action happens over a bundle of wires called *bus*; functionally, the bus plays the role of an OR-gate and, together with the TRI-STATE circuitry, it performs as the set of four multiplexers considered in the above discussion (see also Section 2).

(ii) The seven segments of the display are implemented as seven diodes having one common electrode. A particular display is activated or inhibited by setting up or by breaking the connection of that common electrode to the power supply. The behaviour of the circuit in Figure 3 is then easily understood: a register R_i and the corresponding display are simultaneously activated by a modulo 8 counter acting through an address decoder.

Let us conclude this example with two general remarks.

(i) The sequentialization process displaces the cost from the computation resources to the connection networks, allowing one to access the data and to dispatch the results. If we wish to preserve to a *sequentialized processor* a *general purpose* nature, one will have to be able, at each step of the computation process, to read the appropriate data and to write the obtained results from and to arbitrary memory locations. One should dispose, to achieve that result, of efficient and cheap random access memories (RAM's).

(ii) If, on the other hand, one adopts the philosophy of a *special purpose processor*, it will be rewarding to take advantage of the data access specificities to simplify as much as possible the connection networks. In the above example we could have connected the eight registers $R_0, R_1, \ldots, R_7$ as a ring to allow the serial presentation of their contents to the decoder. That observation already shows that it may become interesting, in order to decrease the connection cost, to modify the data location during the computation. A similar statement holds true for intermediate results, as it will appear from a second example.

1.3 Example: partial sequentialization of the carry–save multiplier

The carry–save multiplier has been studied in Section 2.2.2 of Chapter IX together with its synchronous realization, the serial–parallel multiplier. In the latter, one uses n cells $(xy+z+w=2c+s)$ and $2n$ flip-flops to perform the multiplication of two n-bit numbers in $2n$ clock periods. To be more specific, we assume in what follows that $n=8$. We intend to realize our multiplier with only four of the cells of the type displayed in Figure 3b of Chapter IX. We start by choosing an acceptable labelling compatible with the given flow width. Such a labelling is partially displayed in Figure 4. In this graph, the edges corresponding to the data and result are not represented. The remaining 15 interlevel edges correspond to intermediate results. Once our acceptable labelling has been chosen, we may proceed

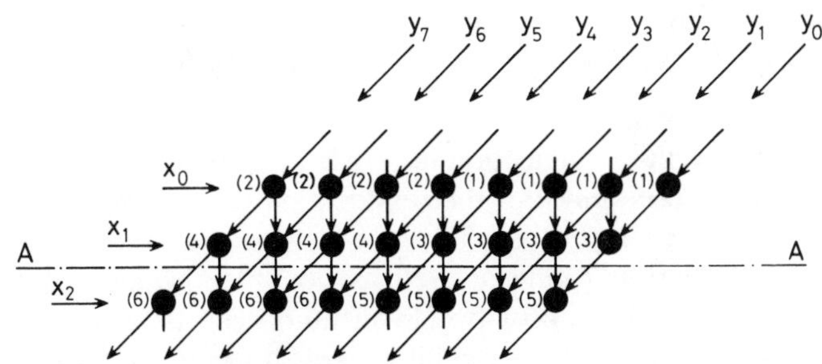

Figure 4. Sequentialization of the carry–save multiplier (A)

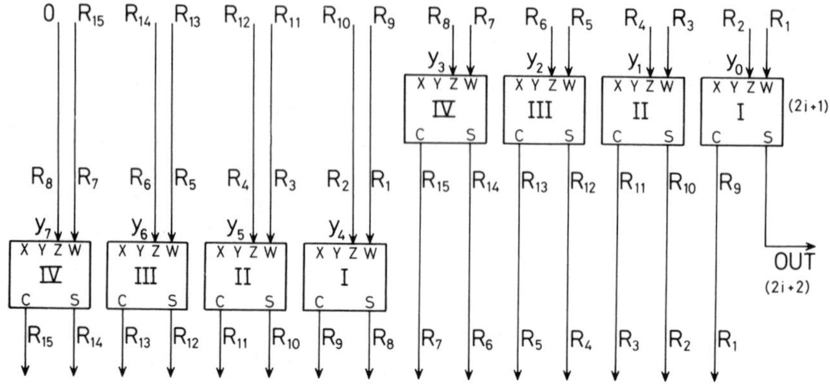

Figure 5. Sequentialization of the carry–save multiplier (B)

with the methods described in Chapter VIII and we have thus to decide on the resource and memory allocations. Obviously, we shall retain the same allocations for all the odd clock periods and for all the even clock periods. Figure 5 represents the succession of an odd clock period and of an even one. The resources are labelled I, II, III, and IV, and we retain the natural choice, as shown on the figure. The allocation of the 15 bistables deserves some more attention. The purpose is to modify at each step the smallest possible number of connections, by trying, for example, to feed a given computation resource from constant flip-flops. Figure 5 suggests a solution to the allocation problem: it shows that it is possible to feed permanently the four resources from the flip-flops $R_1, R_2, \ldots, R_8$ and to let them write their results into the flip-flops $(R_8), R_9, \ldots, R_{15}$. It suffices to realize, in parallel with the computation itself, the transfers

$$R_1 := R_9; \qquad R_2 := R_{10}; \qquad \ldots; \qquad R_7 := R_{15}.$$

The computation itself runs as follows:

(i) $t = 2i+1$

$$R_8 := 0; \qquad R_9 = c(x_i, y_0, R_1, R_2); \qquad R_{10} = s(x_i, y_1, R_3, R_4); \qquad \ldots;$$

(ii) $t = 2i+2$

$$R_8 := s(x_i, y_3, R_1, R_2); \qquad R_9 = c(x_i, y_3, R_1, R_2); \qquad \ldots.$$

The final scheme is given by Figure 6. The control unit reduces to a two-state counter C; the connection network consists of only 5 multiplexers: its simplicity is a consequence of the reallocation of the intermediate results by the transfers $R_1 := R_9$, $R_2 := R_{10}, \ldots$ which give to the processing unit memory an architecture fitting as well as possible the algorithm requirements. For the multiplication of two n-bit numbers, the circuit in Figure 6 will require $4n$ clock periods: the x_i bit is applied at the corresponding input at steps $2i+1$ and $2i+2$; the result bit z_i is obtained at step $2i+1$.

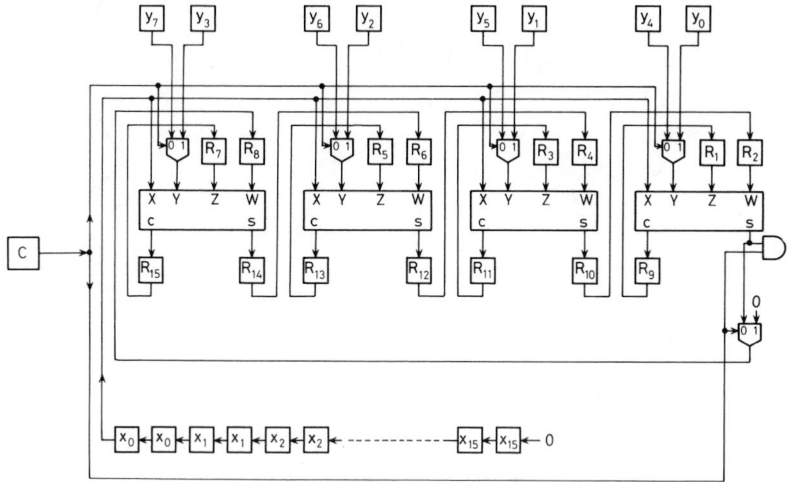

Figure 6. Sequentialization of the carry–save multiplier (C)

2 IMPACT OF THE SEQUENTIALIZATION PRINCIPLE ON THE CONNECTION NETWORK

In the preceding examples, we applied the sequentialization principle to the computation resources: more specifically, in the total sequentialization of an algorithm involving a single computation primitive π: $[R]:=f([R_1], [R_2], \ldots)$ we put to work at each step of the process the unique copy of the resource π. A typical step will thus be described by simultaneous transfers loading the input registers $R_1, R_2, \ldots$ of the computation resource with the appropriate values for the next step. Hence, if the resource π has n inputs, the connection network should be able to transmit simultaneously n data items.

Clearly, it is not necessary to realize simultaneously these n transfers: exactly as we did for independent partial computations, we may decide to load the input registers of the computation resource one after the other. The obvious purpose of this further sequentialization is to reduce the cost of the connection network. To illustrate the impact of this technique on the connection network, we shall consider the following problem: given N registers $R_1, R_2, \ldots, R_N$, describe a connection network allowing any transfer from register R_i to register R_j. Before discussing various solutions to this problem, we introduce two simplifying graphical conventions. First, the set of multiplexers in Figure 7a, allowing one to select an arbitrary word in a set, will be represented as shown in Figure 7b or even, if no confusion can occur, as shown in Figure 7c. Similarly, the sets of gates displayed in Figure 8a and Figure 8d will be replaced by the symbols in Figure 8 (b or c) and Figure 8 (e or f), respectively.

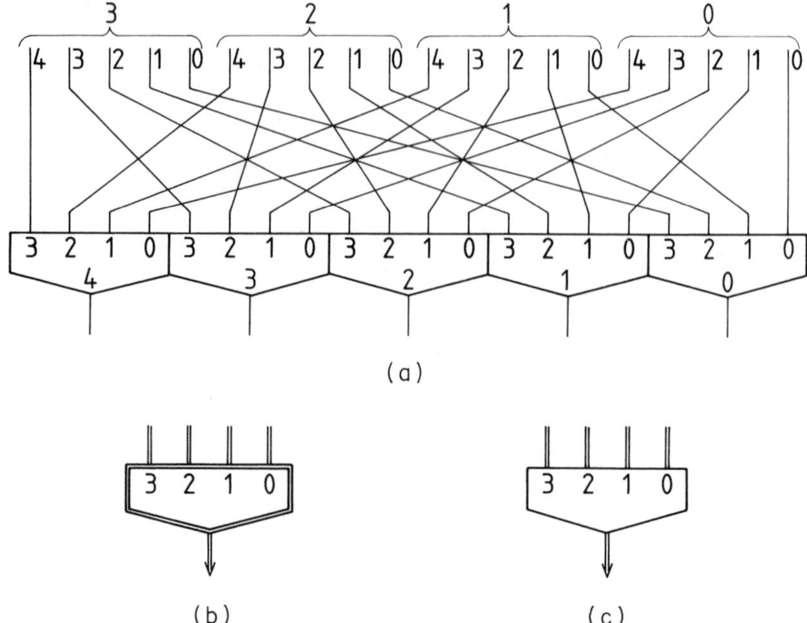

Figure 7. Graphical symbols for a set of multiplexers

We now return to our transfer problem. Two solutions are presented in Figure 9 and Figure 10. In the first solution, the connection network allows one to load simultaneously the N registers from arbitrary data. The cost of this network is proportional to N^2. Using more sophisticated techniques, we could reduce that cost to $N \log N$ (see Chapter IV), but the control of the

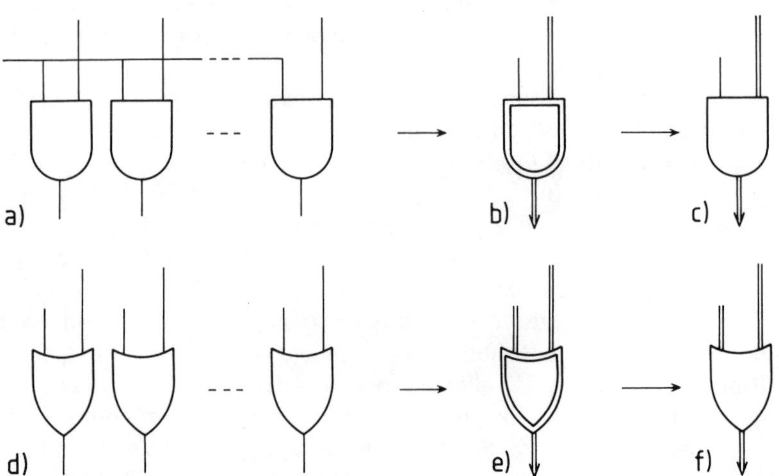

Figure 8. Graphical symbols for set of gates

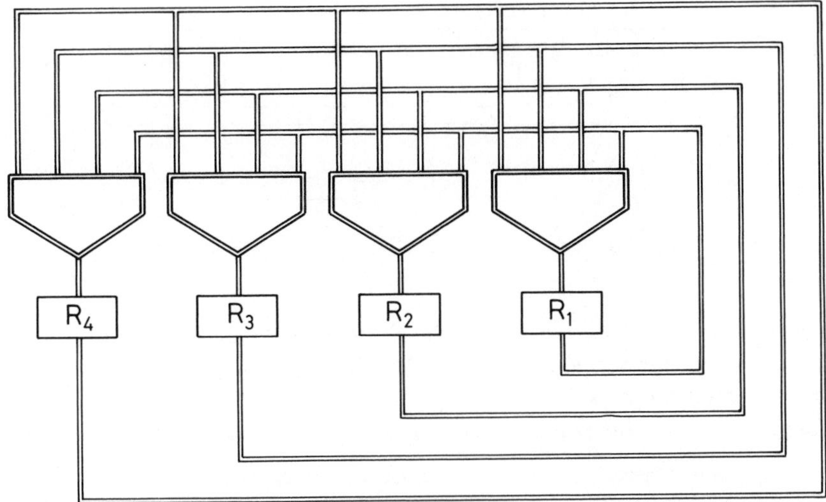

Figure 9. Parallel connection networks

connection network would become more complex. In the scheme of Figure 10, a single transfer per clock pulse is possible. Basically, the connection network consists of a multiplexer which we implemented by tristate registers connected to a bus: the output bundle of that multiplexer is the common path over which any transferred data is transmitted. The transferred data is then distributed to the appropriate destination by sets of AND gates. The cost of this second circuit is clearly proportional to *N*. The circuit of Figure 10a will be represented as shown in Figure 10b; if required, the output TRI-STATE control of a register will be represented as in Figure 10c.

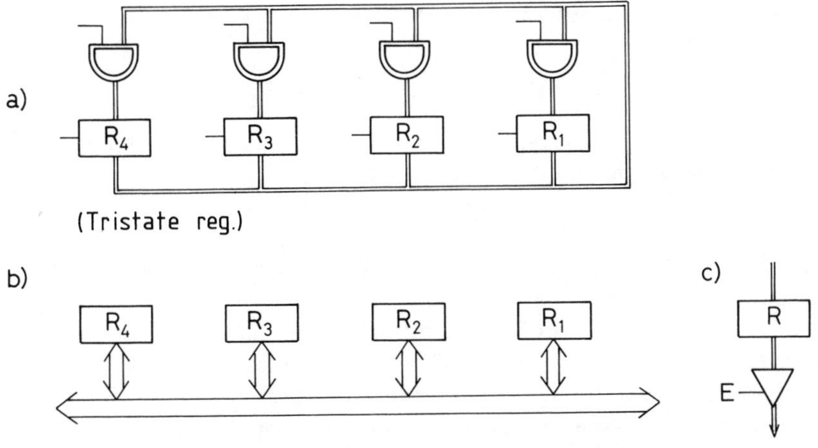

Figure 10. Serial connection networks

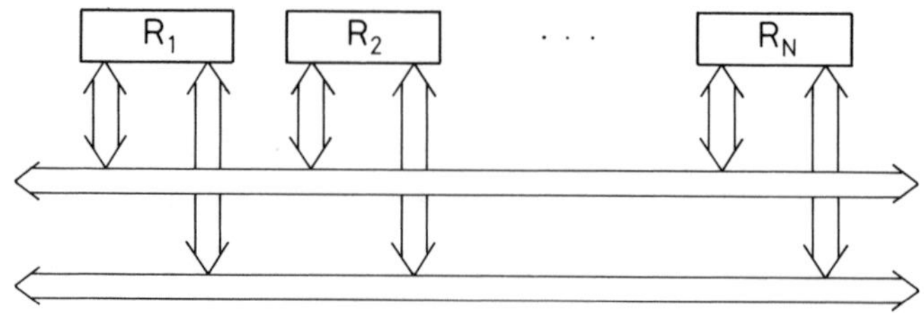

Figure 11. Registers connected on a two bus structure

The above discussion shows the cost decrease that may result from a systematic transfer sequentialization. The use of a single bus is in fact an extreme situation and intermediate situations will put to work a fixed number of buses. This is illustrated in Figure 11 by a two-bus organization.

3 THE PROCESSING UNIT OF A SEQUENTIALIZED SYSTEM

3.1 General discussion. The arithmetic and logic unit

The present section will discuss the application of the sequentialization principles to the design of the processing unit of a sequentialized general purpose processor. We first discuss the choice of the combinational resource.

From a theoretical point of view, any functionally complete logical primitive, such as a NAND gate, could be considered as the unique combinational resource to be put at work in a processing unit. Obviously, this choice is not a natural one for most practical situations and it would lead in most cases to unacceptable control complexity and computation time. In practice, one will select a number of functions such as arithmetic and logical operations which appear convenient for a target range of application: for example, addition and shift would be considered as candidates for arithmetic processors.

More precisely, let

$$\Pi = \{f_0(\mathbf{x}), f_1(\mathbf{x}), \ldots f_p(\mathbf{x})\}$$

be the set of suitable logical primitives. It is then always possible to define a programmable resource $\psi(\mathbf{x}, \mathbf{y})$ such that

$$\forall i, \quad \exists e_i : \psi(\mathbf{x}, \mathbf{e}_i) = f_i(\mathbf{x}).$$

The $\mathbf{y}$ inputs to that resource are its control inputs. We shall use, to represent such a resource, the symbol of Figure 12.

The choice of the set Π of logical primitives may clearly affect significantly the system performances. As a matter of fact, the choice has more quantitative than qualitative aspects: this results from the possibility of moving the boundary between control and processing units. (see Chapter X). The basic option is that of the complexity level of the logical primitives to be handled

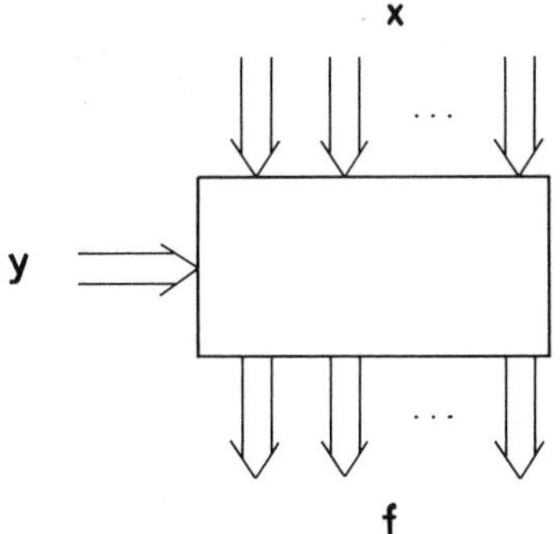

Figure 12. Graphic symbol for an ALU

by the processing unit and the choice may range from the absence of primitives in the processing unit to highly sophisticated primitives. One of the trends of today's technology seems to be the introduction of specialized hardware features oriented, for example, towards floating point arithmetic or to some form of interpretation of high level languages.

The choice of the word size is clearly, too, an important element of that discussion. In that respect, an important feature is that of *expandability*. Let us illustrate this by an example. In most general purpose processors, the combinational resource is an *arithmetic and logic unit* (ALU). Such an ALU has been described in Chapter IV and gives a typical example of the capabilities of commercial ALU's. Also, that ALU is expandable and is provided with the necessary 'hooks' for carry-lookahead generators.

3.2 Single bus and multibus organizations

Let us now discuss the organization of a processing unit built around an ALU of the above type. The ALU has two data inputs and one data output. The functional transfers will thus have the general form:

$$[R_i] := F([R_j], [R_k]) \tag{5}$$

with, possibly, $i = j$ or k. If we go back to our discussion of Section 2, we immediately realize that the functional transfer (5) may be completed in one or more clock periods depending on the number of buses or of multiplexers available as connection network. This problem is the subject matter of the present section.

We carry out that discussion under the assumption of a two phase clock. The two non-overlapping clock-signals are denoted ϕ_1 and ϕ_2 (see Chapter VI, Figure 8c). These two clock-signals control half-registers made of latches. The cascade connection of two half-registers, respectively controlled by ϕ_1 and ϕ_2, may be viewed as a register in the usual sense of the word.

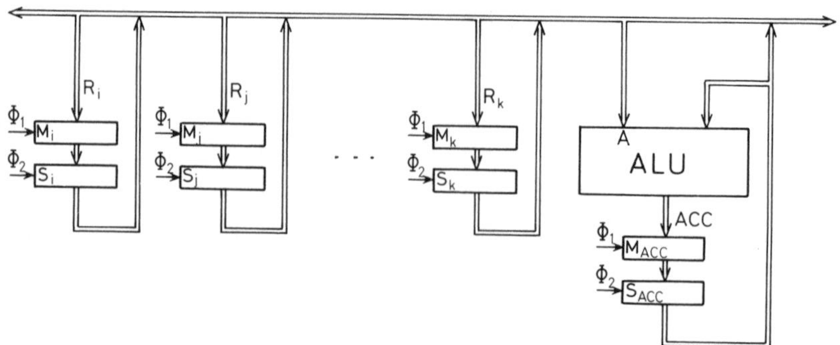

Figure 13. Single bus architecture (A)

Such a register could itself be built from master–slave or from edge triggered flip-flops. In the cascade connection of two half-registers, we call 'master' the half-register receiving its contents from the external world and 'slave' the half-register providing its contents to the external world. We make, however, no assumption on the actual technology supporting the architecture, so that when we display a set of register outputs connected to a bus, we may think either of a set of TRI-STATE registers or of a multiplexer.

We assume finally that the processing unit is provided with the control lines required to activate the load and read of registers.

A first single-bus operation is shown in Figure 13. The privileged register at the ALU output is called the *accumulator.* It appears as a standard sequential circuit: all the master registers are controlled by ϕ_1; all the slaves are controlled by ϕ_2. The execution of the transfer (5) spreads over three clock periods:

(i) $[\mathrm{ACC}]:=[R_j]$;
(ii) $[\mathrm{ACC}]:=F([\mathrm{ACC}],[R_k])$;
(iii) $[R_i]:=[\mathrm{ACC}]$.

Let us give a more detailed description of the circuit behaviour. We assume that, at the starting time, the data are present in S_j and S_k. During the half-periods $(\phi_1, \phi_2)=(1, 0)$, the contents of the slaves are stable and may be transferred over the bus to any selected master. During the half-periods $(\phi_1, \phi_2)=(0, 1)$, the masters transfer their information to the corresponding slaves and the bus is not used.

The inefficient use of the bus is the main drawback of the simple organization of Figure 13. In what follows we discuss various solutions allowing one to improve the situation. A first solution is given by the single bus organization of Figure 14. Here, the registers $R_i, R_j, \ldots, R_k$ are controlled in phase opposition with respect to the accumulator. To describe the behaviour, we assume the data initially present in M_j and M_k. The transfer

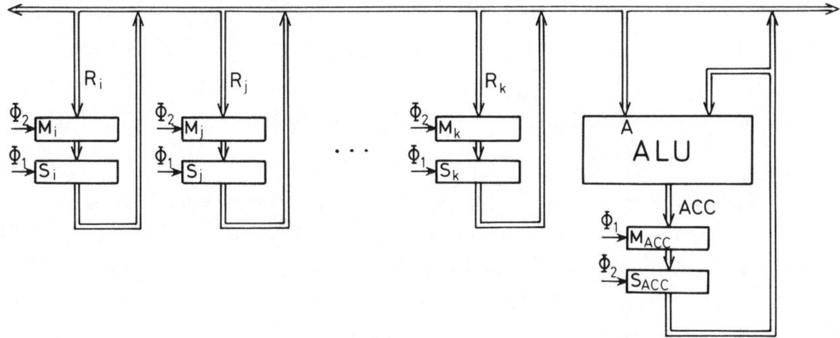

Figure 14. Single bus architecture (B)

(5) is realized in two clock periods:

(i) (a) $(\phi_1, \phi_2) = (1, 0) : [M_{ACC}] := [M_j]$ (via, S_j, bus),
(b) $(\phi_1, \phi_2) = (0, 1) : [S_{ACC}] := [M_{ACC}]$;
(ii) (a) $(\phi_1, \phi_2) = (1, 0) : [M_{ACC}] := F([S_{ACC}], [M_k])$ (via S_k, bus),
(b) $(\phi_1, \phi_2) = (0, 1) : [M_i] := [M_{ACC}]$ (via S_{ACC}, bus).

It should not, however, be concluded that the time required by the circuit of Figure 14 is given by the two-thirds of that shown in Figure 13. Indeed, in the circuit of Figure 14, a typical clock phase may contain up to three consecutive delays, for example S_j, ALU, M_{ACC}, while, in the circuit of Figure 13 one typical clock phase never contains more than two cumulated delays, for example ALU, M_{ACC}. The circuit in Figure 14 is thus particularly interesting in the case where the ALU delay is larger than the register delays.

Consider now the two-bus architecture displayed in Figure 15. The multiplexing function (B1) and demultiplexing functions (B2) are separated from each other. As in Figure 9, all the master flip-flops are controlled by ϕ_1 and all the slaves by ϕ_2. Assuming that the initial information is present in S_j

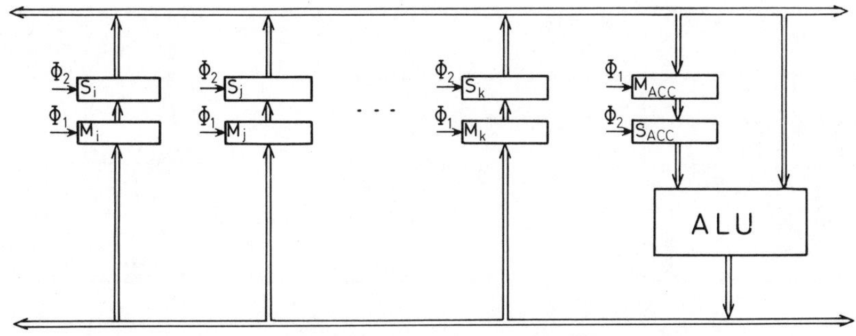

Figure 15. Two bus architecture

and S_k, we may detail the transfer (5) as follows:

$$
\begin{aligned}
&\text{(i) (a)}\ (\phi_1, \phi_2) = (1, 0) : [M_{\text{ACC}}] := [S_j] \quad (\text{via B1}),\\
&\qquad\quad (\phi_1, \phi_2) = (0, 1) : [S_{\text{ACC}}] := [M_{\text{ACC}}];\\
&\quad\ \text{(b)}\ (\phi_1, \phi_2) = (1, 0) : [M_i] := f([S_{\text{ACC}}], [S_k]) \quad (\text{via B1, ALU, B2}),\\
&\qquad\quad (\phi_1, \phi_2) = (0, 1) : [S_i] := [M_i].
\end{aligned}
$$

The completion of transfer (5) takes two clock phases which actually are of the same order of magnitude as in the case of Figure 13; indeed, in Figure 15, the critical phases involve only the ALU and one half-register. On the other hand, the efficiency of the bus is decreased with respect to Figure 14.

There are obviously many other possible architectures for processing units. Figure 16a illustrates a three-bus architecture: in that architecture, the transfer (5) may be considered as the basic type of circuit function, as it is realized in a single clock period. This point has already been discussed at the beginning of Section 2. A variant of the three-bus architecture is shown in Figure 16b. Here, one uses only two slaves S_A and S_B, connected to buses B1 and B2, which are shared by the masters M_i, M_j, . . . , M_k.

The above discussion exhibits the many degrees of freedom available in the design of a processing unit: number of buses, number of general purpose registers, and type of connections to the external world. So far, we have not

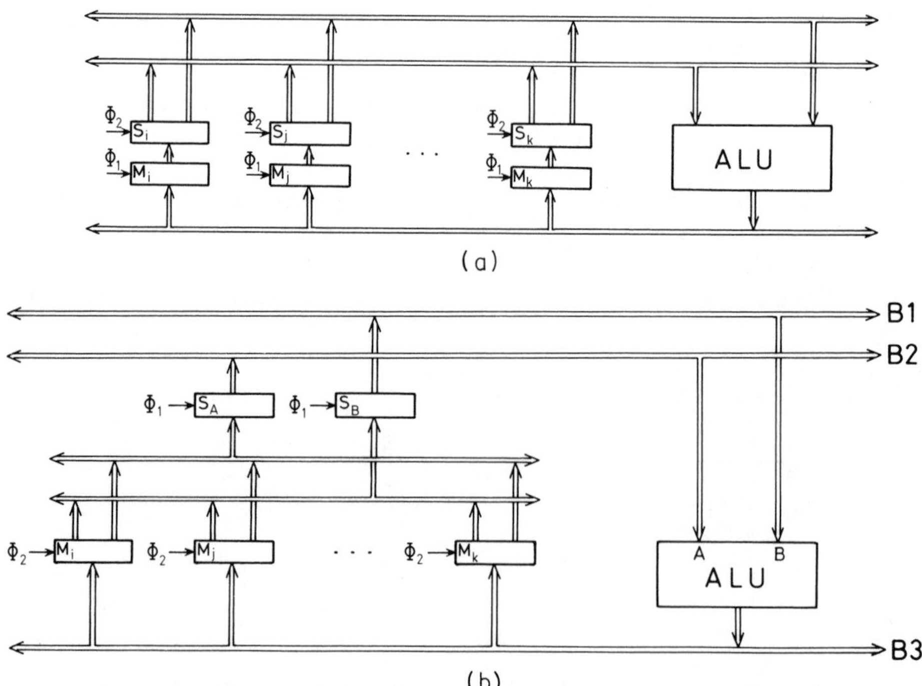

Figure 16. Three bus architecture

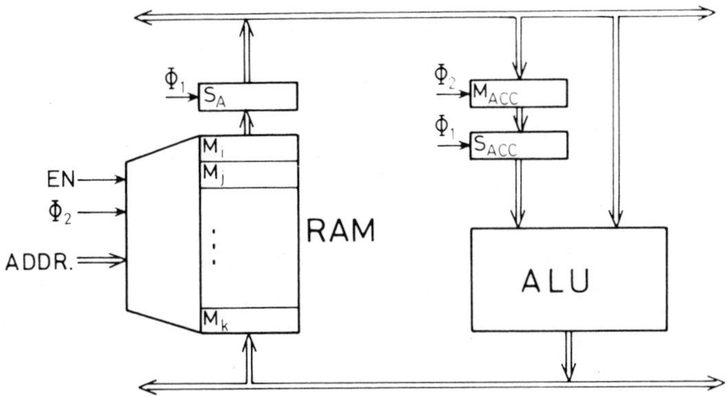

Figure 17. A simple processing unit architecture

taken into account the control requirements: the control should allow the various enable–disable functions, the bus accesses, etc. As far as the general purpose registers $R_i, R_j, \ldots, R_k$ are concerned, it will be interesting, if the number of connections to the control unit is an important parameter, to encode the corresponding control bits as an address. This amounts to restructuring the set of general purpose registers as a random access memory. This idea is illustrated by Figure 17 in the case of a two-bus architecture. The detailed analysis of that circuit is similar to the preceding ones and is left to the reader.

3.3 The bit-slice processing unit Am 2901 A

As an example of a processing unit, we describe in this section the Am 2901 A. The block diagram of this TTL circuit is shown in Figure 18. Basically, it consists of an ALU surrounded by a number of registers and multiplexers providing the interconnections.

The ALU comprises:

(i) Nine binary data inputs $\mathbf{R} = R_3R_2R_1R_0$, $\mathbf{S} = S_3S_2S_1S_0$, C_n.

(ii) Five binary data outputs C_{n+4}, $\mathbf{F} = F_3F_2F_1F_0$.

(iii) Three binary control inputs I_5, I_4, I_3 (or in short I_{543}). These control bits select the function carried out by the ALU according to the table in Figure 19a. The three first rows of the table correspond to arithmetic functions in the 2's complement representation. The five last rows correspond to logical bitwise operations.

(iv) Three status outputs or condition variables.

OVR is the overflow condition in 2's complement operations; F_3 is the sign bit in the same number system, and $F = 0$ indicates that the result at the ALU output is 0 0 0 0. The ALU also has outputs P and G (propagated and generated carries), not shown in Figure 18, for use with a companion carry-lookahead generator.

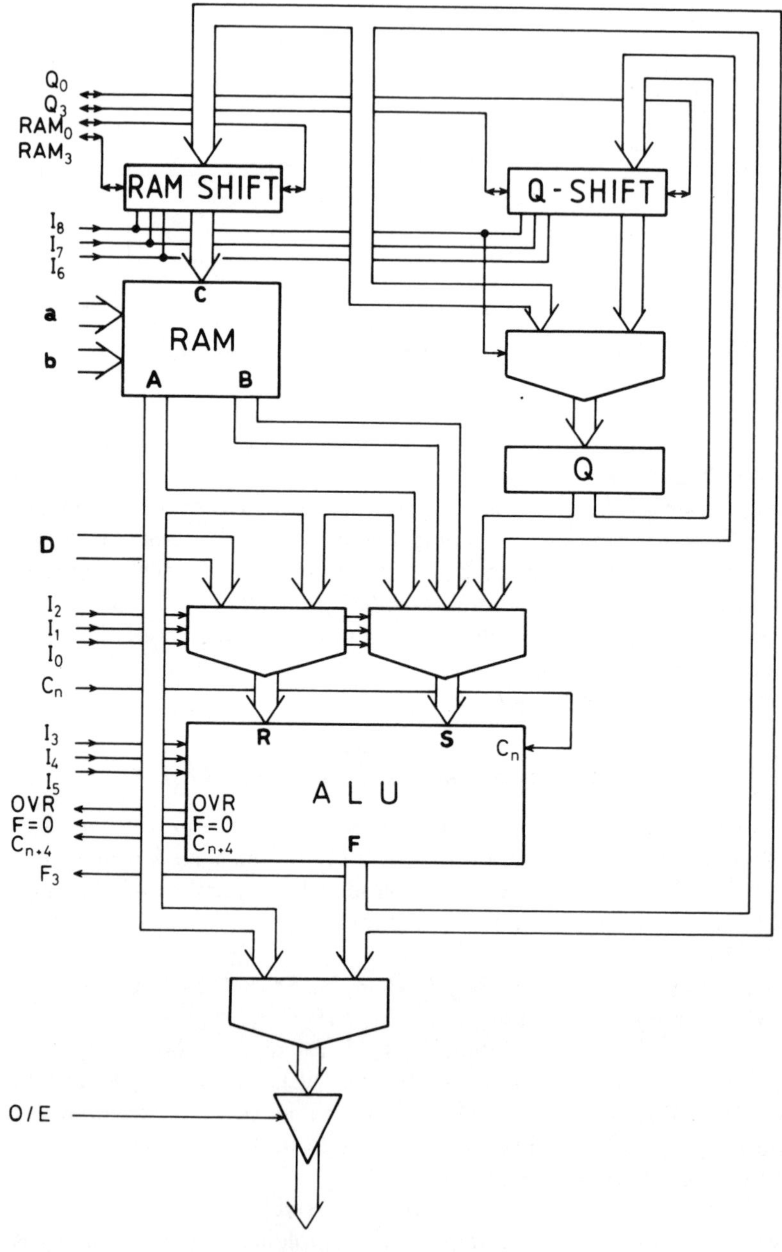

Figure 18. The bit-slice processing unit Am2901. (Copyright © 1978 Advanced Micro Devices, Inc. Reproduced with permission of copyright owner. All rights reserved)

a)

I_{543}	I_5	I_4	I_3	C_{n+4}, **F**
0	0	0	0	$R+S+C_n$
1	0	0	1	$S+\bar{R}+C_n$
2	0	1	0	$R+\bar{S}+C_n$
3	0	1	1	$R\vee S$
4	1	0	0	$R\wedge S$
5	1	0	1	$\bar{R}\wedge S$
6	1	1	0	$R\oplus S$
7	1	1	1	$\bar{R}\oplus S$

b)

I_{210}	I_2	I_1	I_0	R	S
0	0	0	0	A	Q
1	0	0	1	A	B
2	0	1	0	0	Q
3	0	1	1	0	B
4	1	0	0	0	A
5	1	0	1	D	A
6	1	1	0	D	Q
7	1	1	1	D	0

c)

I_{876}	I_8	I_7	I_6	B	Q	Y	RAM_0	RAM_3	Q_0	Q_3
0	0	0	0	—	F	F	—	—	—	—
1	0	0	1	—	—	F	—	—	—	—
2	0	1	0	F	—	A	—	—	—	—
3	0	1	1	F	—	F	—	—	—	—
4	1	0	0	$F/2$	$Q/2$	F	F_0	IN_3	q_0	IN_3
5	1	0	1	$F/2$	—	F	F_0	IN_3	q_0	—
6	1	1	0	$2F$	$2Q$	F	IN_0	$F3$	IN_0	q_3
7	1	1	1	$2F$	—	F	IN_0	F_3	—	q_3

Figure 19. The Am 2901: control variables

The processing unit contains 17 four-bit registers:

(i) The special purpose register Q ($q_3q_2q_1q_0$), and

(ii) 16 general purpose registers organized as a RAM. That memory has an input port **C** and two output ports **A** and **B**. The addresses of the two words simultaneously read in the memory are **a** and **b**, respectively. The write operation is always at address **b**. The organization of the RAM is very similar to that of Figure 16b and in particular involves two output latches A and B.

The ALU and registers are connected as indicated by Figure 18:

(i) three control variables I_2, I_1, I_0 (in short I_{210}) perform the selection of the ALU operand source. They act as indicated in Figure 19b.

(ii) three additional control variables I_8, I_7, I_6 (I_{876}) select the new contents of address **b** in the RAM, and of the Q register; they also select the source of the output data Y. The control variables I_{876} perform as indicated in Figure 19c. Four additional input–output pins allows one to handle the border bits in the case of up or down shift. In Figure 19c, the symbol IN means that the corresponding pin is used as an input.

Finally, the Am 2901 A exhibits four binary data inputs $\mathbf{D}=D_3D_2D_1D_0$ and one tristate control input OE.

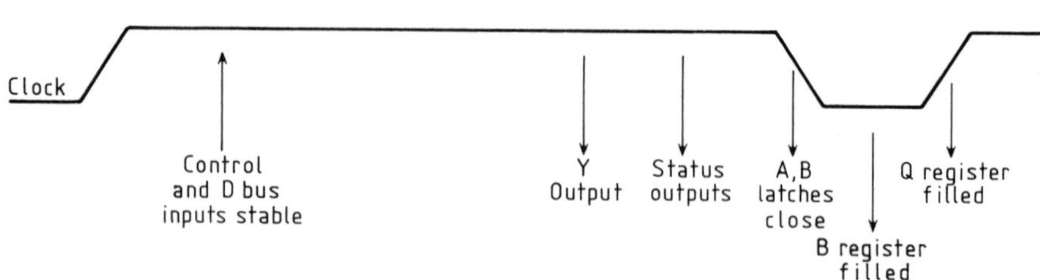

Figure 20. The Am2901: chronology

As such, the Am 2901 A appears as a rather simple processing unit or, more exactly, as a slice of processing unit. Indeed, by cascading m copies of the 2901 device, one obtains a $4m$-bit wide processing unit. That possibility originates in the linear expandability of the addition, subtraction, and shift, and has been implemented by the external connections C_n, C_{n+4}, Q_0, Q_3, RAM_0, RAM_3. This is the reason why the devices of the above type are called *bit-slice processing units.* A simplified view of the timing of the 2901 (after Myers, 1980) is shown in Figure 20.

To summarize, at each clock pulse, the control unit should provide the bit-slice with the following.

(i) Nine control bits $I_9 \ldots I_0$. In what follows, we shall use an octal notation for these control bits. Thus $I = 235$ represents $(I_8I_7I_6, I_5I_4I_3, I_2I_1I_0) = (0\,1\,0,\ 0\,1\,1,\ 1\,0\,1)$.

(ii) Eight address bits **a** and **b**.

(iii) If required, four data bits **D**.

(iv) One output control bit *OE*.

	I	
RESET	343 043	$B := 0$ $Q := 0$
LOAD	337 037	$B := D$ $Q := D$
TRANSF.	334 034 332 033	$B := A$ $Q := A$ $B := Q$ $Q := B$
INC DEC	$303^{(1)}$ $313^{(2)}$ $002^{(1)}$ $012^{(2)}$	$B := B+1$ $B := B-1$ $Q := Q+1$ $Q := Q-1$
$^{(1)}C_n = 1$	$^{(2)}C_n = 0$	

Figure 21. Elementary possibilities of the Am 2901

Thus we have a total of 22 bits. In counterpart, the processing unit will provide the control unit with the status bits C_{n+4}, OVR, F_3 and $F=0$.

Before examining more significant examples, we briefly review some of elementary possibilities of the processing unit: reset, load register by an external data, transfer from register to register, register increment and decrement. These possibilities are gathered in Figure 21. We now briefly discuss some other possibilities.

3.3.1 Simultaneous shift of B and Q; overflow preserving 2's complement addition

If we establish the connection $\text{RAM}_0 = Q_3$ and impose $I=433$, we realize the simultaneous right shift of B and Q, the least significant bit of B becoming the most significant bit of Q. That possibility allows one to store in B and in the most significant bit of Q a $(n+1)$-bit number, for example, the $(n+1)$-bit sum of two n bit 2's complement integers x and y. It suffices to introduce by RAM_3 the most significant bit of the sum $z = z_n z_{n-1} \ldots z_1 z_0$. The bit z_n is not known as such, but is easy to compute z_n from available quantities. Indeed

(a) $\text{OVR}=0 \Rightarrow z_n = z_{n-1}$ since, in this case:

$$-2^n z_n + 2^{n-1} z_{n-1} + \ldots = -2^n z_{n-1} + \ldots$$

(b) $\text{OVR}=1 \Rightarrow z_n = \bar{z}_{n-1}$ since we know that, in this case, we should have $z_n = x_{n-1} = y_{n-1}$, while the discussion in Chapter IV shows that, in an overflow situation, one has $z_{n-1} = \overline{x_{n-1}} = \overline{y_{n-1}}$.

Finally:

$$z_n = z_{n-1} \oplus \text{OVR},$$

and the instruction characterized by

$$I=401; \qquad C_n=0; \qquad \text{RAM}_0 = Q_3; \qquad \text{RAM}_3 = \text{OVR} \oplus F_3$$

places in B and in the most significant bit of Q the sum of the contents of registers A and B.

3.3.2 Multiplication

The discussion in Chapter IX has shown the usefulness of a double length register for multiplication. In the 2901, this role is again played by the concatenation of B and Q. The main instructions of a multiplication program are given in Figure 22. We assume that the operands are initially stored at addresses 0 and 1 of the RAM and that the most and least significant halves of the result will appear at addresses 2 and 3 of the RAM, respectively. The instructions are easily understood in the light of the discussion of shift and add multiplication. The multiplier is initially stored in Q and, during steps 3 and 4, it is shifted right, allowing one to observe its

	a	b	I	C_a	Q_3	RAM_3	
1	0	—	034	—	—	—	$Q := R_0$
2	—	3	243	—	—	—	$R_3 := 0$
3	1	3	401	0	RAM_0	$F_3 \oplus OVR$	$R_3Q := (R_1 + R_2)/2$
			403	0	RAM_0	$F_3 \oplus OVR$	$R_3Q := R3/2$
4	1	3	411	1	RAM_0	$F_3 \oplus OVR$	$R_3Q := (R_3 - R_1)/2$
			413	1	RAM_0	$F_3 \oplus OVR$	$R_3Q := R_3/2$
5	—	2	332	—	—	—	$R_2 := Q$

Figure 22. Multiplication on the Am 2901

successive bits at Q_0. Instruction 3 is repeated $(n-1)$ times (for n bits two's complement operands); at each execution, the observation of Q_0 determines the choice $I = 401$ or $I = 403$; i.e. the addition of the multiplicand or of zero to the partial result. This is easily accomplished by setting $I_1 = \bar{Q}_0$. Instruction 4 is similar to instruction 3 but it handles the negatively weighted most significant bit of the multiplier.

4 TRADE OFFS AND OPTIMIZATION PROBLEMS

4.1 General discussion

In the preceding sections, we reached a rather precise idea of the hardware architecture of a sequentialized processing unit. Most often, such a processing unit will contain:

(a) a programmable computation resource (ALU);

(b) a privileged register (ACC);

(c) a number of general purpose registers, possibly organized as a RAM; and

(d) a number of multiplexers or buses acting as a connection network. Many variants of this architecture are possible, in particular with respect to the general purpose registers: these could, for instance, be organized as a stack, as is the case in some pocket calculators.

Nothing, however, requires the designer to stick in all circumstances to a totally sequentialized architecture. For example, in many real time signal processing applications, the computation speed achievable by a custom microprocessor may become unacceptably low, while, for those applications, efficient algorithms involving an intrinsic parallelism are known. The designer's concern will then probably be to deduce, from the given algorithm (or algorithms), some partially sequentialized architecture allowing him to reach his speed specification at acceptable cost. The design problem thus becomes a typical trade-off. The technological evolution makes probable the development of new domains of application and it is likely that these new domains will suggest the design of mass-produced special purpose processors. This motivates the present discussion.

The trade-off and optimization problems raised here are by no means simple. This is easily perceived by considering the number of degrees of freedom left to the designer. For instance, in the case of unit execution time systems (see Chapter VIII), we have seen that the freedom degrees are:

(a) the choice of an acceptable labelling;
(b) the choice of the resource allocation, and
(c) the choice of the memory allocation.

These choices influence the system performances in the following respects:

(a) number of combinational resources;
(b) number of memories;
(c) complexity of the connection network, and
(d) computation time or throughput.

In general, the trade-off problem appears as a complex one. To our knowledge, no attempt to cover it as a whole has been performed to date. In the following subsections, we describe two partial optimization problems known as *scheduling* and *pebbling.*

4.2 Scheduling

4.2.1 Problem statement

Consider a computation scheme $\langle A, C\rangle$ with a set of tasks $A = \{N_1, N_2, \ldots, N_p\}$ and precedence relation $\subset$. *Scheduling theory* studies the execution of such a computation scheme on a finite number m of computation resources (processors). Accordingly, that theory discusses methods for obtaining acceptable labellings ψ such that $W_c(\psi) \leq m$. The retained acceptable labelling is called the *schedule.*

To keep to the highlights of scheduling theory, we restrict our presentation by the two following assumptions (for unit execution time systems):

Assumption 1. No upper bound is imposed on the number of memory places required to store intermediate computation results.

Assumption 2. The scheduling is *non-pre-emptive,* i.e. any partial task, once initiated, has to be completed.

Under these assumptions it becomes possible to represent a schedule by a two-dimensional table whose (i, j)-entry specifies the task that has to be performed on processor i during the jth time unit.

Note furthermore that scheduling theory focuses on the partial trade-off:

(a) number of computation resources, and
(b) computation time,

and pays no attention to the memory requirements and to the cost of the connection networks. Let us mention at once that this trade-off is 'well-behaved' in the sense that, if one divides by k the number of computation resources, one simultaneously multiplies by k the computation time. Similarly, scheduling theory raises various optimization problems, namely:

(i) Obtain a minimum length acceptable labelling such that $W_c(\psi) \leq m$;

(ii) Given a target length L, obtain a minimum width acceptable labelling having length T at most.

In the remaining part of this discussion, we discuss only the first of these two problems.

4.2.2 List scheduling

The principle of a *list scheduling procedure* may be formulated as follows: given a computation scheme $\langle A, C\rangle$ impose a total order on the set of partial computations in such a way that the schedule obtained by applying the following rule has minimum length:

Rule. Whenever a resource becomes available, assign it the unexecuted ready task with the highest rank for the imposed total order.

The actual problem of list scheduling is in fact the building up of the list itself. We shall not enter here a detailed discussion of these techniques. They are extensively described and validated in the book written by Coffman (1976) on the subject. We shall only mention, for illustration purposes, two methods in use for building up an adequate list.

Algorithm 1 (Level scheduling).
Setp 0. Apply part A of Algorithm 1 *(Chapter VIII) from the END node on. After this step, all the nodes have received a preliminary label.*
Step $i+1$. *Assume that all the nodes with preliminary labels* $1, 2, \ldots, i$ *have received the final labels* $1, 2, \ldots, p$. *Then, the nodes with preliminary label* $(i-1)$ *receive the final labels* $(p+1)$, $(p+2), \ldots$ *in an arbitrary order. The process is carried out until all the nodes in A have been reached.*

A refinement of the above procedure is based on the concept of lexicographical order of sequences. This concept has been introduced in Chapter I. It is extended to sequences of unequal length by completing the most significant digits of the shorter sequence with zeroes.

Algorithm 2 (Lexicographical scheduling).
Step 0. *Same as step* 0 *in Algorithm* 1.
Step i. Assume that all the nodes with preliminary labels $1, 2, \ldots, i-1$ *have received the final labels* $1, 2, \ldots, p$. *Consider now the nodes with preliminary label* i; *to each of these nodes, associate an integer sequence formed by the*

final labels of the immediate successors written by decreasing order. Now, the final labels $p+1$, $p'+2, \ldots$ *are associated to the nodes with preliminary label i ordered by increasing lexicographical order of the associated sequences.*

Algorithms 1 and 2 do not, in general, provide optimal schedules. It may be shown that Algorithm 1 provides an optimal schedule if the precedence graph is a forest (union of trees) and that Algorithm 2 provides an optimal solution if the number of resources is two.

4.2.3 Restricted enumeration scheduling

Consider the precedence graph $\langle A, C\rangle$ of a computation scheme. The basic enumeration procedure consists in forming step by step the list of acceptable labellings whose width remains smaller than the number m of available processors. That procedure is described by the following Algorithm 3 whose body (steps 2 to 5) accepts as data a family of task sets:

$$\mathcal{A}(t)=\{A_i\}; \qquad A_i \subseteq A,$$

each of which may be realized on the m processors within t units of time and produces as result a new family $\mathcal{A}(t+1)$ of task sets, each of which may be realized on the m processors within $(t+1)$ units of time.

Algorithm 3 (Enumerative scheduling).
Step 1. *Let* $t:=0$ *and* $\mathcal{A}(0)=\{\{\text{BEGIN}\}\}$
Step 2. *Let* $\mathcal{A}(t)=\{A_i\}$

(FORK) *For each* $A_i \in \mathcal{A}(t)$:

Step 3 *Compute the set* $H(A_i)$ *of all the nodes all of whose predecessors belong to* A_i.
Step 4. *If* $\mu=|H(A_i)|\leq m$, *then* $A_i:=A_i\cup H(A_i)$, *else form the* $\binom{\mu}{m}$ *subsets* $H_j(A_i)$ *of* $H(A_i)$ *with cardinality m and replace* A_i *by the family of sets*:

$$A_i:=\left\{A_i\cup H_j(A_i)\,\middle|\, j=1,2,\ldots,\binom{\mu}{m}\right\}.$$

Step 5. $\tilde{\mathcal{A}}(t+1):=\bigcup A_i$, *where the* A_i *are the families of task sets obtained during Step* 4. *Form* $\mathcal{A}(t+1)$ *by deleting in* $\tilde{\mathcal{A}}(t+1)$ *task sets that are included in other task sets.*
Step 6. *If* $A_0\in\mathcal{A}(t+1)$, *end, else go to step* 2 *with* $t:=t+1$.

The above algorithm will clearly yield all the optimal solutions. The price paid for reaching this result is a high computational complexity: as many optimization problems encountered in this book, the research of optimal schedules is a NP-complete problem. Various techniques may be used to reduce the enumeration involved in Algorithm 3. We shall not enter here into the details of that discussion and instead refer the reader to the already mentioned book by Coffman (1976) and to Sethi (1976), Coffman and Graham (1972), and Davio and Thayse (1979). The main feature to be remembered about the scheduling optimization problem is its well-behavedness which immediately provides the designer with a family of solutions of constant cost–delay product.

Once the schedule and the corresponding acceptable chronology, has been decided upon, one remains free of choosing the resource allocation to minimize, for instance, the cost of the connection networks. This problem has been taken tackled by Torng and Wilhelm (1977) who attempt to minimize the cost of input–output interface between computation resources and a system of buses.

4.3 Pebbling

Another partial optimization problem, known as *pebbling*, studies the trade-off between memory requirement and computation time. One assumes that there is a single computation resource, i.e. that the computation is totally sequentialized. To understand the problem, consider the precedence graph in Figure 23. We interpret such a graph as follows: when two arrows originate from the same node, say node A, this means that the partial result obtained at that node has to be used twice (in our example, for nodes B and E). Now, once the computation A is completed, we have two choices: either we store the produced result in some memory location until computation E is completed, which freezes a memory location until that time, or we use the result A for computation B and free the corresponding memory location which may, for instance, be used for result B; in that case, the result A will have to be computed again before starting E and the computation time increases.

The situation is modelled as a game on a graph. The game, a one-player game, consists in placing pebbles on the nodes of the graph according to certain rules. Placing a pebble on a node is interpreted in our context as

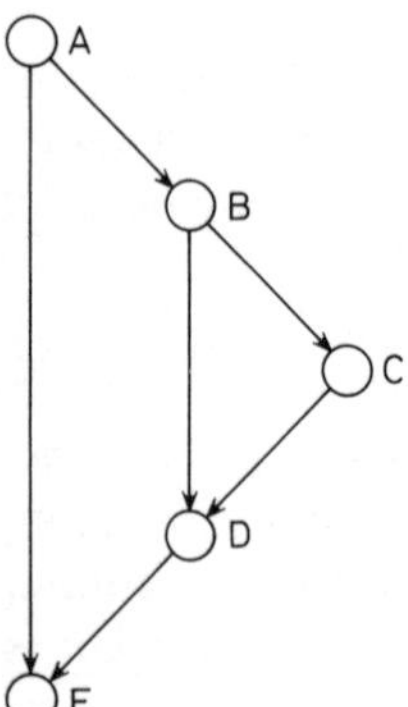

Figure 23. A precedence graph

performing the computation associated with that node and storing the result in some memory location. The rules are the following ones:

Rule 1. If all the immediate predecessors of a node have pebbles on them, a pebble may be placed on that node. In our terminology, a partial computation may be started once all the required data are available.

Rule 2. A pebble may be removed from any vertex, i.e. intermediate partial results may be destroyed.

Clearly, if the precedence graph of a computation scheme has N nodes, the computation may be completed in N units of time using N memory locations. The results of the study of pebbling are extremely striking. We shall only mention here these results and refer the reader to the recent and complete survey by Pippenger (1980) for additional references. We keep assuming that the number of nodes in the graph is N.

(a) The required number of memory locations has an upper bound of the form $0(N/\log N)$ and there exist graphs requiring $\Omega(N/\log N)$ pebbles.

(b) In the range $S=\Omega(N/\log N)$ to $S=0(N)$, time

$$T=S \exp\{\exp[0(N/S)]\}$$

is sufficient for any graph, while for any N and S in this range, there exist graphs for which time

$$T=S \exp\{\exp[\Omega(N/S)]\}$$

is required.

Thus, if we attempt to approach the lower bound $\Omega(N/\log N)$ for the number of memory locations, we may have to pay a more than polynomial increase in computation time. The trade-off memory-time thus appears as much more critical than the one studied by scheduling theory.

4.4 Conclusion

To conclude our discussion of the trade-off and optimization problems, we should stress the incomplete nature of the set of tools presently available. From the point of view of the VLSI circuit designer, the cost factor is the silicon area and, from this point of view, one should obviously take into account the area occupied by the connections. This explains the new trends of that branch of complexity theory, illustrated by the recent paper of Vuillemin (1980).

BIBLIOGRAPHICAL REMARKS

The principle of sequentialization introduced in this chapter is obviously of primordial importance to the understanding of modern computing devices, such as the microprocessor. Rather curiously, its presentation generally remains more or less implicit in the discussion of computer organizations: see, for example, Burks, Goldstine, and Von Neumann (1946), Stone *et al.* (1975), and Hamacher, Vranesic, and Zaky (1978). It is indeed generally assumed that the processing unit has one ALU, two buses, etc. The discussion of Section 2 is due to Booth (1971). The bibliographical remarks about optimization appear in Section 4 above.

EXERCISES

E1. Consider the block diagram of the bit-slice processor Am 2901. Draw a functional equivalent of that device using buses instead of multiplexers. What do the control variables become?

E2. Assume that you have to perform the addition of three 16 bit numbers. Devise various schemes for reaching that objective, using 1, 2, or 4 Am 2901. Draw the flow charts of the corresponding control algorithms.

E3. The multiplication algorithm outlined in Section 3.3 takes advantage of the possibility of simultaneous shift and add on the Am 2901. Observe that this processor in fact contains three computation resources, namely the ALU and the two shifters. Would it be possible to obtain the same performance with the ALU as a single computation resource? Discuss in particular the computation time and the expandability.

E4. Imagine the main steps of the two's complement non-restoring division algorithm implemented on the Am 2901. (*Hint.* To use efficiently the possibilities of the processor, a slightly different version of the algorithm in Chapter IX will be required. One indeed uses in Figure 11 of Chapter IX the variable t, involved in the transfer

$$t := 2t + d$$

Use instead the variable $t' = 2t$. The above transfer is then replaced by

$$t' := 2(t' + d)$$

which is more easily implemented.)

E5. The following piece of Pascal program describes a sorting algorithm known as the 'bubble sort'

```
for i:=1 to (n-1) do
   begin for j:=1 to (n-i) do
            begin if a[j+1]<a[j]
                  then
                    begin b:=a[j];
                          a[j]:=a[j+1];
                          a[j+1]:=b
                    end
            end
   end.
```

Implement that algorithm on the Am 2901. Now design a dedicated processor for the same purpose, allowing in particular the simultaneous transfers

$$(a[j], a[j+1]) := (a[j+1], a[j]).$$

Chapter XII
Sequentialization. Impact on the control unit

1 INTRODUCTION

In Chapters X and XI we studied implementations of algorithms based on automata. The basic model includes a control automaton, which actually implements the algorithm, and an operational automaton which performs the data processing.

This chapter describes canonical and optimal implementations deduced from the basic model, and gives cost upper bounds for every realization.

Sections 2 to 4 describe different program transformations and the corresponding circuit realization. Cost upper bounds are derived for every realization. The starting point is a program with disjoint predicates (Section 2). It can be transformed into an equivalent conditional program (Section 3) or into an incremental program (Chapter XIV). The basic architectures are described in Section 4 where emphasis is put on the control part of the circuit: this is due to our choice of considering universal realizations and to the fact that the architecture of the operational automaton must be adapted to the particular algorithm it implements.

2 PROGRAM TRANSFORMATION

The initial program is a list of instructions of the form (14) of Chapter X i.e.

$$\begin{array}{llll} N & f_0(\mathbf{x}) & \sigma_0 & N_0 \\ & f_1(\mathbf{x}) & \sigma_1 & N_1 \\ & \cdots & \cdots & \cdots \\ & f_{r-1}(\mathbf{x}) & \sigma_{r-1} & N_{r-1} \end{array} \tag{1}$$

where every predicate f_i is a Boolean function of the n variables $\mathbf{x}=(x_{n-1}, \ldots, x_1, x_0)$ and where every command σ_i belongs to a set of 2^m different commands.

The definition (1) of an instruction tries to achieve two simultaneous goals:

(a) to cover as broad a class of instructions as possible so as to provide the model with a maximum number of capabilities, and

(b) to keep consistent with the automaton model and, more precisely, to preserve the identity of the instructions and of the internal states (see Chapter X, Section 2). This will allow us to design the control automaton by the standard procedures used for sequential synchronous circuits (see Chapter VI).

Remember (see Chapter X, Section 2) that the interpretation of instruction (1) is the following one:

(a) While starting the execution of N, we first compute the values of all the functions $f_i(\mathbf{x})$. This is the *evaluation phase.*

(b) We then perform, by increasing order of indices, all commands σ_i corresponding to functions $f_i(\mathbf{x})$ equal to 1. This is the *execution phase.* The need to keep consistent with the order of indices is imposed by the non-commutativity of the composition of commands.

(c) Finally we select a next instruction label N_i with the highest index i such that $f_i(\mathbf{x})=1$.

These rules allow us to define without ambiguity the transition and output functions of the control automaton (see Chapter X, Section 2).

We can distinguish two parts in instruction (1):

(a) search of the indices i for which $f_i(\mathbf{x})=1$ (*evaluation phase*), and

(b) execution of the corresponding commands (*execution phase*).

The main purpose of this chapter lies in the development of synthesis methods for programs made up of instructions (1) by means of instruction primitives to be defined further on.

First of all we introduce an elementary construction which allows us to transform an instruction of the type (1) into an evaluation instruction and a set of execution instructions. This transformation reflects the separation of instruction (1) into an evaluation phase and an execution phase. This transformation is, for example, for $r=3$ (λ: no operation, t: always true condition):

$$\left.\begin{array}{llll} N & g_0(\mathbf{x})=\bar{f}_2\bar{f}_1 f_0 & \lambda & N_1' \\ & g_1(\mathbf{x})=\bar{f}_2 f_1\bar{f}_0 & \lambda & N_2' \\ & g_2(\mathbf{x})=\bar{f}_2 f_1 f_0 & \lambda & N_3' \\ & g_3(\mathbf{x})=f_2\bar{f}_1\bar{f}_0 & \lambda & N_4' \\ & g_4(\mathbf{x})=f_2\bar{f}_1 f_0 & \lambda & N_5' \\ & g_5(\mathbf{x})=f_2 f_1\bar{f}_0 & \lambda & N_6' \\ & g_6(\mathbf{x})=f_2 f_1 f_0 & \lambda & N_7' \end{array}\right\}\ \text{evaluation phase} \qquad (2a)$$

$$\left.\begin{array}{llll} N_1' & \mathrm{t} & \sigma_0 & N_0 \\ N_2' & \mathrm{t} & \sigma_1 & N_1 \\ N_3' & \mathrm{t} & \sigma_0 & N_2' \\ N_4' & \mathrm{t} & \sigma_2 & N_2 \\ N_5' & \mathrm{t} & \sigma_0 & N_4' \\ N_6' & \mathrm{t} & \sigma_1 & N_4' \\ N_7' & \mathrm{t} & \sigma_0 & N_6' \end{array}\right\}\ \text{execution phase} \qquad (2b)$$

The type of transformation: instruction (1) → program (2a, 2b) is clearly of general nature. Let us call $P(r)$ the execution phase corresponding to instruction (1). This program should have 2^r inputs corresponding to the 2^r values (0) (1) . . . (2^r-1) of the vector $(f_{r-1}, \ldots, f_1, f_0)$. It should have $(r+1)$ outputs corresponding to the r instruction labels $N_0, N_1, \ldots, N_{r-1}$ and to the don't-care output required whenever all the functions have the value 0. We use, for representing $P(r)$, the symbol of Figure 1a. A recurrent formation method for $P(r)$ is shown in Figure 1. The starting point $P(1)$ is displayed in Figure 1b. The formation of $P(r+1)$ from $P(r)$ is illustrated in Figure 1c. The result of the construction for $r=4$ is shown in Figure 4b.

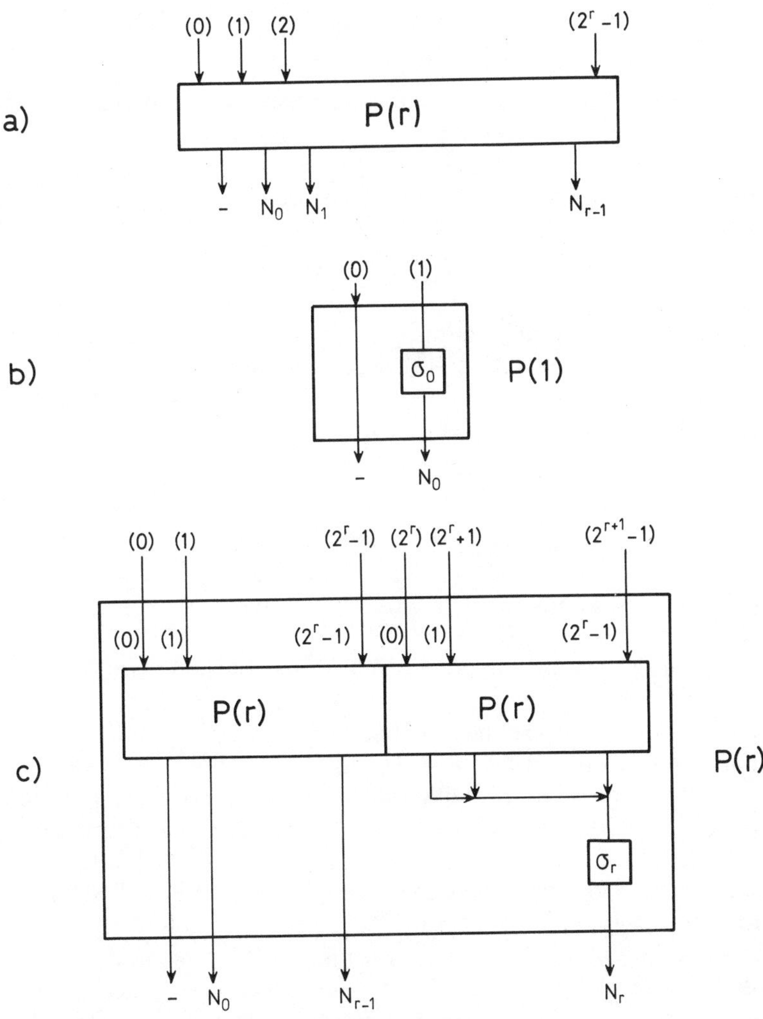

Figure 1. Organization of the execution program

Further on we shall always assume that $\forall f_i(\mathbf{x}) \equiv 1$ and thus $\forall g_j(\mathbf{x}) \equiv 1$, which eliminates the don't-care output in the scheme of Figure 1.

We can partially merge the evaluation phase (2a) and the execution program (2b) by introducing an execution instruction in every row of the evaluation instruction (2a), i.e.

$$\begin{array}{llll} N & g_0(\mathbf{x}) & \sigma_0 & N_0 \\ & g_1(\mathbf{x}) & \sigma_1 & N_1 \\ & g_2(\mathbf{x}) & \sigma_0 & N_1'' \\ & g_3(\mathbf{x}) & \sigma_2 & N_2 \\ & g_4(\mathbf{x}) & \sigma_0 & N_2'' \\ & g_5(\mathbf{x}) & \sigma_1 & N_2'' \\ & g_6(\mathbf{x}) & \sigma_0 & N_3'' \end{array} \tag{3a}$$

$$\begin{array}{llll} N_1'' & \mathrm{t} & \sigma_1 & N_1 \\ N_2'' & \mathrm{t} & \sigma_2 & N_2 \\ N_3'' & \mathrm{t} & \sigma_1 & N_2'' \end{array} \tag{3b}$$

Instructions of the type (2a) and (3a) are both instructions with disjoint predicates, i.e.

$$g_i(\mathbf{x})g_j(\mathbf{x}) = 0, \qquad \forall i \neq j. \tag{4}$$

As a conclusion, any instruction (1) may be replaced by an equivalent program made up of instructions (5) with disjoint predicates:

$$\begin{array}{llll} N & g_0(\mathbf{x}) & \sigma_0 & N_0 \\ & g_1(\mathbf{x}) & \sigma_1 & N_1 \\ & \cdots & \cdots & \cdots \\ & g_{q-1}(\mathbf{x}) & \sigma_{q-1} & N_{q-1} \end{array} \tag{5}$$

A program whose instructions are of the type (5) defines a *control automaton* (see Chapter X):

(1) the input letters are the 2^n possible values of the condition variables $\mathbf{x}$;

(2) the output letters are the commands σ_j;

(3) the states are the instructions N_j, and

(4) with the n-tuple $\mathbf{e}$ and with the state N we associate the state N_j corresponding to the index j for which $g_j(\mathbf{e}) = 1$. The automaton generates the output letter σ_j.

The next state and output functions are unambiguously defined since the predicates are disjoint.

After the encoding of the 2^p instructions and 2^m commands, one gets a sequential machine with n input variables $\mathbf{x} = (x_{n-1}, \ldots, x_1, x_0)$, p state variables $\mathbf{y} = (y_{p-1}, \ldots, y_1, y_0)$, and m output variables $\mathbf{z} = (z_{m-1}, \ldots, z_1, z_0)$.

Note that the automaton is complete since we assume that:

$$\bigvee_i g_i(\mathbf{x}) \equiv 1.$$

The starting point of the further sections of this chapter is a program made up of instructions with disjoint predicates and one command per row. Making certain assumptions concerning the commands and the condition variables, we have associated with this program a circuit including a control part and an operational part.

The control part can be implemented under the form of a finite state machine, i.e. by a programmable logic array or a read only memory cooperating with the register. On the other hand, the implementation of the operational part depends on the actual operations it must perform.

In Chapter X we gave prominence to circuit transformations affecting both the control part and the operational part. In other words, these transformations modify the program, and also the operational part.

The next sections are devoted to program transformations that do not affect the operational part of the corresponding circuit. As was the case for the finite automata, distinct programs may execute the same computation. As a matter of fact, various concepts of equivalence may be introduced. In what follows, we use the following terminology (see also Chapter X):

(1) Two programs P_1 and P_2 are *functionally equivalent* iff, for every input, they perform the same computation.

(2) Two programs P_1 and P_2 are *semantically equivalent* iff they are functionally equivalent and use the same sets of (elementary) instructions.

3 SEQUENTIAL EXAMINATION OF THE CONDITION VARIABLES. CONDITIONAL PROGRAMS

3.1 Concept of conditional program

In this section we examine some consequences of the idea of sequentialization: this concept has been mentioned earlier, but it has mainly been considered from the point of view of the operational automaton (see Chapter XI, Section 1). Intuitively, to sequentialize an algorithm essentially consists in performing, one after the other, two otherwise unrelated partial computations. Similar considerations apply to the control automaton: observe indeed that, in the control automaton considered so far, the condition variables were examined simultaneously (see Chapter X, equation (25)). We shall now investigate the possibility of examining these variables one at a time.

This is obviously quite an extreme situation. A more general problem would be to question the possibility of examining k condition variables at a time, the parameter k being independent of the number n of variables in a particular problem. As a matter of fact, the principles that will be developed below would be easily extended to that general situation. There are other reasons for considering only the simple case. Remember first that the basic idea lying behind the concept of sequentialization is to realize an economy of hardware at the price of an increased computation time. If we now take into account the speed reached by digital integrated electronics, we may expect that a totally sequentialized system will be a relatively cheap system with a time performance acceptable in a number of applications. This is indeed the case with microprocessors.

On the other hand, it is often observed in practical applications that the most interesting designs are obtained when both the operational and the control automata are sequentialized simultaneously at well-matched sequentialization levels. These are the reasons why we consider only the extreme situation of conditional programs.

A *conditional program* is a program all of whose instructions (except the last one) belong to either one of the following two types:

$$\begin{matrix} N & \bar{x}_j & \sigma_0 & N_k \\ & x_j & \sigma_1 & N_\ell \end{matrix} \quad (\textit{branching instruction}), \tag{6a}$$

$$N' \quad t \quad \sigma \quad N_j \quad (\textit{execution instruction}). \tag{6b}$$

Observe that an execution instruction is a degenerate case of a branching instruction. The affinity of the concept of conditional program with that of program schemata (Manna, 1974) and with that of flow chart currently used in programming, is clear and is illustrated by Figure 2. The flow chart representation contains, however, too little information about the time evaluation of the computation. We shall most often use instead the concept

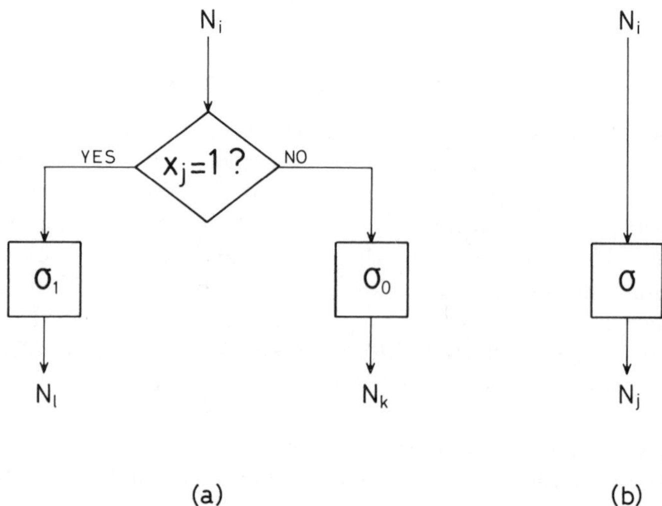

Figure 2. Elementary instruction representation

of *conditional automaton:* we call a conditional automaton a control automaton having as input alphabet:

$$\tilde{\Sigma}=\{x_0, \bar{x}_0, x_1, \bar{x}_1, \ldots, x_{n-1}, \bar{x}_{n-1}, \mathrm{t}\}$$

consisting of $(2n+1)$ literals (instead of 2^n minterms for the canonical automaton accepting instructions (25) of Chapter X).

Theorem 1. *Any program P formed by a set of instructions* (1) *may be replaced by a* (*sementically*) *equivalent conditional program.*

Proof. Because of the result of Section 2 of Chapter X, we only have to give the proof for a cononical program. The proof consists in replacing the typical instruction, i.e. equation (25) of Chapter X, of a canonical program by an equivalent piece of conditional program. This type of replacement is, for example, for $n=2$:

$$\begin{array}{llll} N & \bar{x}_1\bar{x}_0 & \sigma_0 & N_0 \\ & \bar{x}_1 x_0 & \sigma_1 & N_1 \\ & x_1\bar{x}_0 & \sigma_2 & N_2 \\ & x_1 x_0 & \sigma_3 & N_3 \end{array} \rightarrow \begin{array}{llll} N & \bar{x}_1 & \lambda & M_0, \\ & x_1 & \lambda & M_1, \\ M_0 & \bar{x}_0 & \sigma_0 & N_0, \\ & x_0 & \sigma_1 & N_1, \\ M_1 & \bar{x}_0 & \sigma_2 & N_2, \\ & x_0 & \sigma_3 & N_3, \end{array} \tag{7}$$

and it is clearly of a general nature. □

It is interesting to give an estimate of the transformation impact on the program characteristics. Let us call P_{EV} the number of instructions in the obtained evaluation phase of the conditional program. The transformation used in the proof of Theorem 1 immediately shows (see also Figure 4a for $n=3$) that:

$$P_{\mathrm{EV}} \leqslant 2^n - 1. \tag{8}$$

As a matter of fact, the bound given by (8) is not asymptotically optimal. It has been shown by Lee (1959) that there exists a constant K such that

$$P_{\mathrm{EV}} \leqslant K\frac{2^n}{n}. \tag{9}$$

For large values of n, the bound (9) obviously becomes stronger than (8).

It is possible to reduce the upper bound (8) by using a construction very similar to Muller's (see Shannon, 1949; Muller, 1956, and Lee, 1959). This construction is classical in complexity theory: it proves that every r-tuple of n-variable Boolean functions can be realized with, at most:

$$K\frac{2^s}{s} \tag{10}$$

multiplexers with one control variable, where K does not depend on

$$s = n + \log r.$$

(The output functions of a multiplexer and of a branching instruction are the same.) We consider a decision tree similar to that of Figure 4a but where only $(n-k)$ condition variables are considered. Such a decision tree contains $(2^{(n-k)}-1)$ instructions and $2^{(n-k)}$ outputs, each of which evaluates a r-tuple of k-variable functions. Note that there are:

$$T = (2)^{r2^k}$$

r-tuples of k-variable Boolean functions. Assume that we have a series of T subprograms each of which evaluates an r-tuple of functions. It is then possible to connect the outputs of the program to the adequate imputs of the series of subprograms. We then prove that the number of instructions needed to generate the series of subprograms is at most T. Indeed, for $k = 1$, 4^r instructions are sufficient (since there are 4 different 1-variable functions). Moreover, if we have a series of subprograms which generates any r-tuples of $(k-1)$ variable functions, only one additional instruction of the form (6a) is sufficient to generate each r-tuple of k-variable functions. We thus have:

$$P_{\mathrm{EV}} \leqslant (2)^{r2^k} + 2^{n-k} - 1. \tag{11}$$

It remains to choose the value of k in order to minimize the right part of inequality (11). Let us define $\tilde{k}$ as follows:

$$\tilde{k} = k + \log r. \tag{12}$$

Hence:

$$P_{\mathrm{EV}} \leqslant (2)^{2^{\tilde{k}}} + r2^{n-\tilde{k}} - 1. \tag{13}$$

We verify that the right member of (13) is minimum for:

$$\tilde{k} = \lfloor \log (s - \log s) \rfloor, \tag{14}$$

with

$$s = n + \log r. \tag{15}$$

We then obtain:

$$2^{2^{\tilde{k}}} = \frac{2^s}{s},$$

$$r2^{n-\tilde{k}} = 2^{s-\tilde{k}} = \frac{2^s}{2^{\lfloor \log(s - \log s) \rfloor}} \leqslant 2 \frac{2^s}{2^{\log(s-\log s)}}$$

$$= 2 \frac{2^s}{s - \log s}.$$

Hence,

$$P_{\mathrm{EV}} \leqslant \frac{2^s}{s}\left(1 + \frac{2^s}{s - \log_2 s}\right) - 1. \tag{16}$$

The study of the function $2s/(s-\log s)$ shows that in (10) K may be chosen equal to 5. For large values of s, the number of instructions is approximately equal to

$$3\frac{2^s}{s}. \tag{17}$$

As a conclusion of the complexity computations, we see that any instruction of the type (1) may be realized by at most $K2^s/s$ branching instructions and 2^r-1 execution instructions. Hence the number of instructions P of a conditional program equivalent to instruction (1) satisfies the inequality:

$$P \leqslant K\frac{2^s}{s}+(2^r-1). \tag{18}$$

Example 1. We present an example of a direct transformation of an arbitrary instruction (1) into a conditional program, thus avoiding the intermediate of the canonical program of Chapter X, Section 2. The transformation to be described has itself been termed canonical in view of its universal and systematic nature. Related optimization problems will be discussed in Section 5.

We shall describe a transformation yielding from an arbitrary instruction a (semantically) equivalent piece of conditional program. We shall illustrate the transformation by the example

$$\begin{array}{llll} N & S_0 = A \oplus B \oplus C & \sigma_0 & N_0, \\ & C_0 = AB \vee AC \vee BC & \sigma_1 & N_1, \\ & S_1 = A \oplus \bar{B} \oplus C & \sigma_2 & N_2, \\ & C_1 = A\bar{B} \vee AC \vee \bar{B}C & \sigma_3 & N_3. \end{array} \tag{19}$$

The three variable Boolean functions $f_i(A, B, C)$ describe the input–output relationship of a controlled Add/Subtract cell (CAS) with the following meanings: S_0 = *sum output*, C_0 = carry output for the sum, S_1 = subtraction output, C_1 = carry output for the subtraction. A one-bit CAS adds two binary digits A, B, and one carry-input C to produce either simultaneously the outputs S_0, C_0, S_1, C_1 or, according to the value (0 or 1) of a parameter $\mathbf{y}$, simultaneously the outputs (S_0, C_0) or (S_1, C_1). Such CAS cells are used in construction iterative arithmetic logic arrays (see Hwang, 1979). These functions are represented by the Karnaugh map in Figure 3. We also assume

								A								
	0	0	1	0	1	0	0	1	0	1	1	0	1	0	0	0
C	1	0	0	1	0	1	1	1	1	1	0	1	0	1	1	0
												B				

Figure 3. Karnaugh map for the functions (19)

as earlier that, in general, the instruction to be transformed contains r functions of n variables.

During the evaluation phase the n variables are examined in turn by branching instructions of the type (6a). The evaluation part of the program is a tree of depth n containing $(2^n - 1)$ instructions and ending in 2^n outputs numbered from 0 to $2^n - 1$ and corresponding to the 2^n points of the domain. To each of these outputs (domain points) is associated one and only one of the 2^r values of the vector $(f_{r-1}, \ldots, f_1, f_0)$ (see Figure 4 for our running example). The output values are shown by their decimal equivalent (between brackets). Identically valued outputs are tied together.

The execution phase $P_{EX}(r)$ corresponding to an instruction of type (5) is illustrated in Figure 1. Such a part of a program should have 2^r inputs corresponding to the 2^r values $(0)(1), \ldots, (2^r - 1)$ of the vector $(f_{r-1}, \ldots, f_1, f_0)$.

It remains to tie the two parts of the program. This is done on a value equality basis. The useless parts of the program may finally be pruned away as shown by the dotted lines at the bottom of Figure 4. Obviously the evaluation phase should still be completed before starting the execution phase. However, some instruction mergings are possible at the border between the two phases: more precisely, one execution instruction, all of

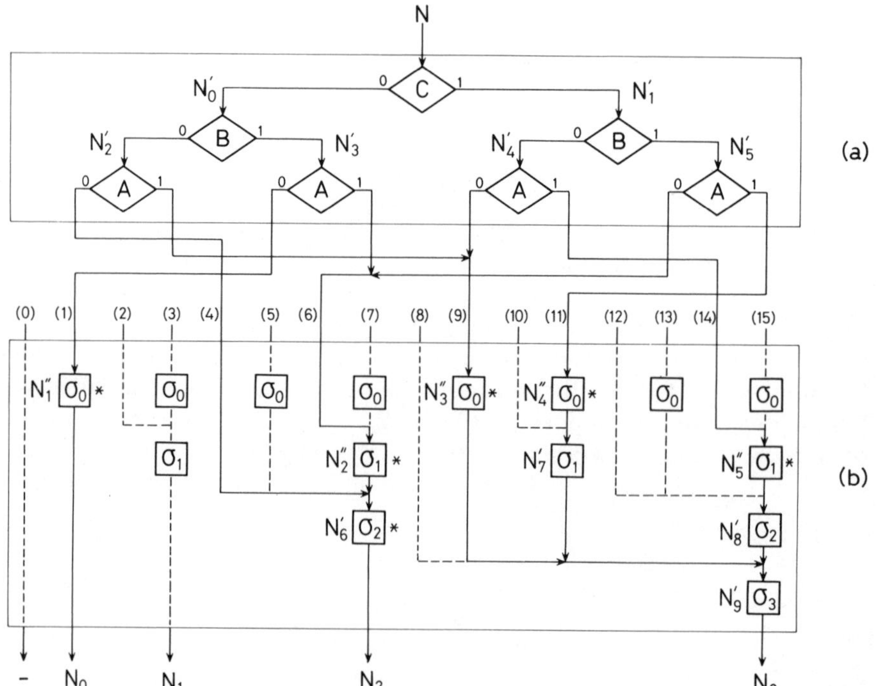

Figure 4. Implementation of a CAS cell by a conditional program

whose predecessors are of the branching type, may simultaneously be imbedded in all the predecessors. This is the case for the starred boxes in the lower part of Figure 4.

From the program flow chart of Figure 4 we deduce the following conditional program equivalent to instruction (19):

$$
\begin{array}{llllll}
(0\;0\;0\;0) & N & \bar{C} & \lambda & N_0' & \\
 & & C & \lambda & N_1' & \\
(0\;0\;0\;1) & N_0' & \bar{B} & \lambda & N_2' & \\
 & & B & \lambda & N_3' & \\
(0\;0\;1\;0) & N_1' & \bar{B} & \lambda & N_4' & \\
 & & B & \lambda & N_5' & \\
(0\;0\;1\;1) & N_2' & \bar{A} & \sigma_2 & N_2 & \\
 & & A & \sigma_0 & N_9' & \\
(0\;1\;0\;0) & N_3' & \bar{A} & \sigma_0 & N_0 & (1\;0\;1\;1) \\
 & & A & \sigma_1 & N_6' & \\
(0\;1\;0\;1) & N_4' & \bar{A} & \sigma_0 & N_9' & \\
 & & A & \sigma_1 & N_8' & \\
(0\;1\;1\;0) & N_5' & \bar{A} & \sigma_1 & N_6' & \\
 & & A & \sigma_0 & N_7' & \\
(0\;1\;1\;1) & N_6' & \mathrm{t} & \sigma_2 & N_2 & (1\;1\;0\;0) \\
(1\;0\;0\;0) & N_7' & \mathrm{t} & \sigma_1 & N_9' & \\
(1\;0\;0\;1) & N_8' & \mathrm{t} & \sigma_2 & N_9' & \\
(1\;0\;1\;0) & N_9' & \mathrm{t} & \sigma_3 & N_3 & (1\;1\;0\;1)
\end{array}
\qquad (20)
$$

3.2 Implementation of a conditional program

It is not difficult to reflect the characteristic features of a conditional program in the architecture of the control automaton. To do this, we first classify the accepted types of instructions. Referring to equations (6a) and (6b) we accept two types of instructions, i.e. the branching instruction and the execution instruction.

The second observation is that an instruction of a given type may contain up to four distinct items, namely two commands σ_0 and σ_1 and two next step identifiers or addresses N_k and N_ℓ. These observations suggest that, for the control automaton, the ROM structure displayed in Figure 5 should be used. Each word in the control memory is divided into five subwords or fields: the first field contains, under coded form, the name of a condition variable. It allows one to identify the instruction type. During the execution of a typical instruction, say (6a), that information is used by an input identification circuit to select the variable x_j among the set of the condition variables. The value of the latter is in turn to select the appropriate next step address N_k or

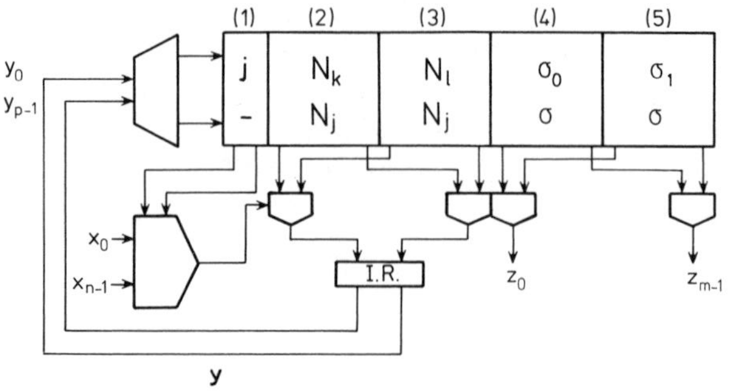

Figure 5. Implementation of a conditional program

N_ℓ and the appropriate output command σ_0 or σ_1. For instructions of the type (6b) both address fields are filled in with N_j while both execution fields are filled in with σ.

The implementation of the example of Section 3.1, by means of the architecture of Figure 5, is shown in Figure 6.

The coding adopted in the read only memory is:

(a) for the control variables:

$$A \simeq (0\ \ 0),$$
$$B \simeq (0\ \ 1),$$
$$C \simeq (1\ \ 0);$$

(b) for the commands:

$$\sigma_0 \simeq (0\ \ 0),$$
$$\sigma_1 \simeq (0\ \ 1),$$
$$\sigma_2 \simeq (1\ \ 0),$$
$$\sigma_3 \simeq (1\ \ 1);$$

(c) for the next instructions:

see the coding indicated in (20).

We compare from a cost point of view the realizations of programs by means of canonical instructions (see Chapter X, Section 2) and by means of conditional instructions.

In Chapter X we associated with any instruction (1) an equivalent canonical program. The transformation method used in the proof in this property (Section X.2.) while constructive, is not very attractive from a practical point of view. This is due to the prohibitive increase of the number of instructions. If the initial program contains 2^p instructions and if r is an upper bound of the number of predicates per instruction, we show that the number P_0 of

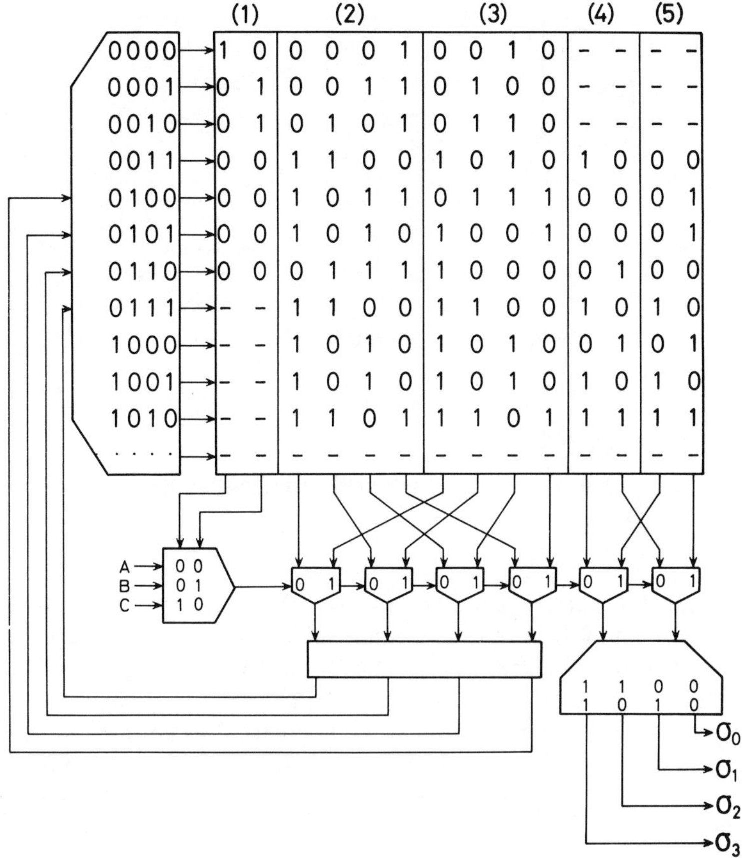

Figure 6. Implementation of the functions (19) by means of the architecture of Figure 5

instructions in the equivalent canonical program has the upper bound:

$$P_0 \leqslant [2^n(r-1)+1]2^p.$$

Indeed, a compund command σ_i' (see equation (25) of Chapter X) is the product of, at most, r elementary commands σ_j. The elimination of a single compound command σ_i' replaces an instruction of the general type (1) by r instructions. This elimination operation thus introduces $(n-1)$ canonical instructions without compound command.

A possible implementation of a canonical program is shown in Figure 7. In Figure 7, the memory part of the system is realized by means of clocked D flip-flops (synchronous unit delays); the combinational part of the system is realized by a read only memory provided with a two-level addressing scheme; the internal state variables (of the synchronous sequential circuit

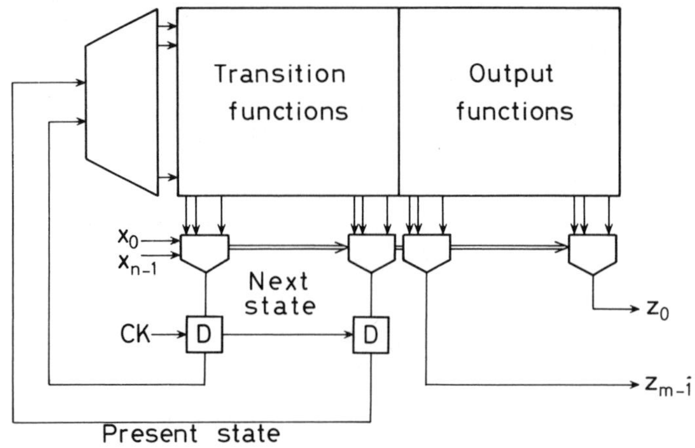

Figure 7. Implementation of a canonical program

materializing the control automation) feed the input addressing level (address decoder), and the external inputs (condition variables) feed the output selection level (multiplexers). Presented in this form, the control automaton has all the main features of the model proposed by Wilkes (1951) for microprogrammed systems. It is easy from Figure 7 to compute an estimate of the system cost if the cost factor is the number of ROM bits. Let, as before, P_0 represent the number of instructions in the given canonical program and put

$$p_0 = \lceil \log P_0 \rceil.$$

Assume furthermore that there are m binary output lines (commands to the operational automaton). The system cost is then given by

$$\Gamma_0 = 2^{(n+p_0)}(m+p_0). \tag{21}$$

Assume, moreover, that the number of instructions of the equivalent conditional program implemented by the structure of Figure 5 is P_1 and put

$$p_1 = \lceil \log P_1 \rceil.$$

According to Figure 5 the cost of this new (conditional) realization is:

$$\Gamma_1 = 2^{p_1}(\log n + 2p_1 + 2m). \tag{22}$$

In order to allow a simple comparison between (21) and (22) we shall assume that $r \ll n$ and so, in view of (15), $s = n$. This allows us to write (18) in the following approximate form:

$$P_1 \leqslant K \frac{2^n}{n} P_0 \leqslant K \frac{2^{n+p_0}}{n}. \tag{23}$$

From this last inequality we derive by retaining only the most significant terms:

$$p_1 \leqslant n + p_0 - \log n,$$

so that relation (22) involves the inequality:

$$\Gamma_1 \leqslant 2K \frac{2^{n+p_0}}{n}(n+m+p_0). \tag{24}$$

The comparison between (21) and (24) allows us to state the following important conclusion: if we may assume, as usually occurs in practical situations, that $m \gg n$, then the asymptotic evolutions of Γ_1 and Γ_0 with respect to n are as follows:

$$\Gamma_0 = 0(2^n) \quad \text{and} \quad \Gamma_1 = 0\left(\frac{s^n}{n}\right). \tag{25}$$

Under the above hypothesis the synthesis by means of a conditional implementation presents a cost advantage with respect to the canonical implementation. Note, however, that under the same hypothesis the time computation is multiplied by n for the conditional version since the n variables are tested one at a time. Roughly speaking, the transformation from a canonical program into a conditional program holds the merit factor constant. In practical cases, formulae (21) and (23) allow a precise cost comparison; it remains that the value of the ratio m/n is of first importance.

3.3 Conditional programs in the restricted sense

It has been observed in Section 3.2 that the maximum number of distinct information items (output commands and next labels) in an instruction is an important design parameter since it determines the number of fields in a ROM word. In this section let us restrict the accepted instruction types to:

$$\begin{array}{cccc} N & \bar{x}_i & \lambda & N_k, \\ & x_i & \lambda & N_\ell; \end{array} \tag{26a}$$

$$\begin{array}{cccc} N' & t & \sigma & N_j. \end{array} \tag{26b}$$

Such a restriction is always possible since instruction (6a) may be replaced by the (semantically) equivalent piece of program

$$\begin{array}{cccc} N & \bar{x}_i & \lambda & N'_k, \\ & x_i & \lambda & N'_\ell; \\ N'_k & t & \sigma_0 & N_k, \\ N'_\ell & t & \sigma_1 & N_\ell. \end{array} \tag{27}$$

Instructions of the types (26a) and (26b) are decision instructions (or branching instructions) and execution instructions respectively. Programs containing only instructions of the types (26a,b) will be called *conditional programs in the restricted sense.*

The striking feature of these programs is that they contain no instruction with more than two information items: N_k and N_ℓ in the branching instruction (26a), N_j and σ in the decision instruction (26b). In such a situation, the control ROM word would be divided in three fields: the type identification field and two information fields. According to the instruction type (26a) or (26b), the first of these two fields would contain a next instruction label, say N_k or N_j, while the last field would contain either a next instruction label N_ℓ or a command σ. A possible implementation structure is shown in Figure 8. Essentially this is the structure proposed by Mange (1978) for binary decision programs. The behaviour of this structure is quite similar to that of Figure 5. Observe, however, that the identification field (the left part of the ROM) has been itself decomposed to exhibit a special bit t whose value allows one to distinguish between branching instructions ($t=0$) and execution instructions ($t=1$). If $t=0$, the output lines are invalidated and the D flip-flops are fed by N_k or N_ℓ according to the value of the appropriate condition variable selected by the identification subfield v. If $t=1$, the output lines are validated and the D flip-flops are fed by N_j.

To sum up, the ROM structure associated with a conditional program in the restricted sense (Figure 8) differs from the ROM structure associated with a conditional program (Figure 5) in two essential points.

(a) The introduction of a one-bit field which characterizes the type of instruction (branching or execution instruction).

(b) The replacement of the four fields $N_0, N_1, \sigma_0, \sigma_1$ by the pairs of fields N_0, N' or N_1, σ according to the type of instruction; this replacement will be called *field merging.*

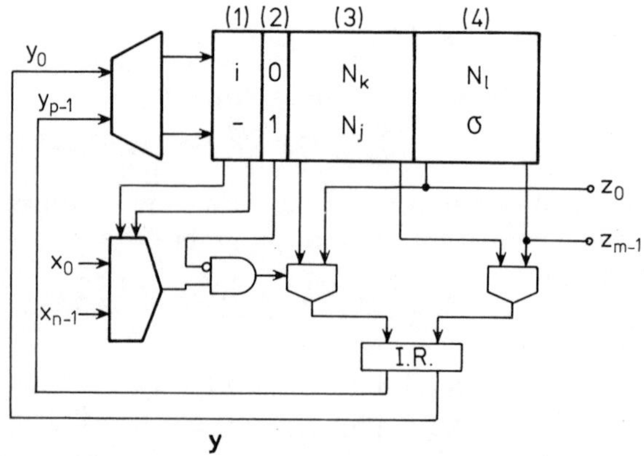

Figure 8. Implementation of a conditional program in the restricted sense

From Figure 8 we deduce that the bit cost of the circuit is:

$$\Gamma_1' = 2^{p_1}(\log n + 1 + p_1 + \max(p_1, m)), \tag{28}$$

which is clarly lower than Γ_1 (see (22); for identical p_1's). The cost reduction of the conditional (in the restricted sense) implementation with respect to the conditional implementation derives essentially from the field merging.

Example 1 (continued). The conditional program in the restricted sense equivalent to instruction (19) may be deduced either from the conditional program (20) or directly from the flow chart of Figure 4. We obtain accordingly:

$$\begin{array}{cccc@{\qquad\qquad}cccc}
N & \bar{C} & \lambda & N_0' & N_1'' & \mathrm{t} & \sigma_0 & N_0 \\
 & C & \lambda & N_1' & N_2'' & \mathrm{t} & \sigma_1 & N_6' \\
N_0' & \bar{B} & \lambda & N_2' & N_3'' & \mathrm{t} & \sigma_0 & N_9' \\
 & B & \lambda & N_3' & N_4'' & \mathrm{t} & \sigma_0 & N_7' \\
N_1' & \bar{B} & \lambda & N_4' & N_5'' & \mathrm{t} & \sigma_1 & N_8' \\
 & B & \lambda & N_5' & N_6' & \mathrm{t} & \sigma_2 & N_2 \\
N_2' & \bar{A} & \lambda & N_6' & N_7' & \mathrm{t} & \sigma_1 & N_9' \\
 & A & \lambda & N_3'' & N_8' & \mathrm{t} & \sigma_2 & N_9' \\
N_3' & \bar{A} & \lambda & N_1'' & N_9' & \mathrm{t} & \sigma_3 & N_3 \\
 & A & \lambda & N_2'' & & & & \\
N_4' & \bar{A} & \lambda & N_3'' & & & & \\
 & A & \lambda & N_5'' & & & & \\
N_5' & \bar{A} & \lambda & N_2'' & & & & \\
 & A & \lambda & N_4'' & & & & \\
\multicolumn{4}{c}{\text{(evaluation phase)}} & \multicolumn{4}{c}{\text{(execution phase)}}
\end{array} \tag{29}$$

As a conclusion for this section we consider the synthesis of the control unit for the non-restoring division algorithm, studied in Chapter XI, by means of the processing unit Am 2901. We assume that the dividend N is initially loaded in R_0; this direct access memory word also contains the $\mathrm{t}'^{(i)}$ and $\mathrm{t}^{(m)}$.

The divisor D is initially loaded in R_1. The register R_3, initially loaded with the value m, will be used as a counter. The organization of the processing unit is represented in Figure 9. Besides the Am 2901, the processing unit contains:

(1) two multiplexers allowing the command of the inputs I_r and C_n to be either in local mode or in external mode;

(2) four bistable flip-flops allowing the recording of the condition variables before their local use or their test; these flip-flops are selectively activated by the signals E_0, E_1, E_2, E_3, and

(3) a general activation input which acts on the synchronization of the system (input CK).

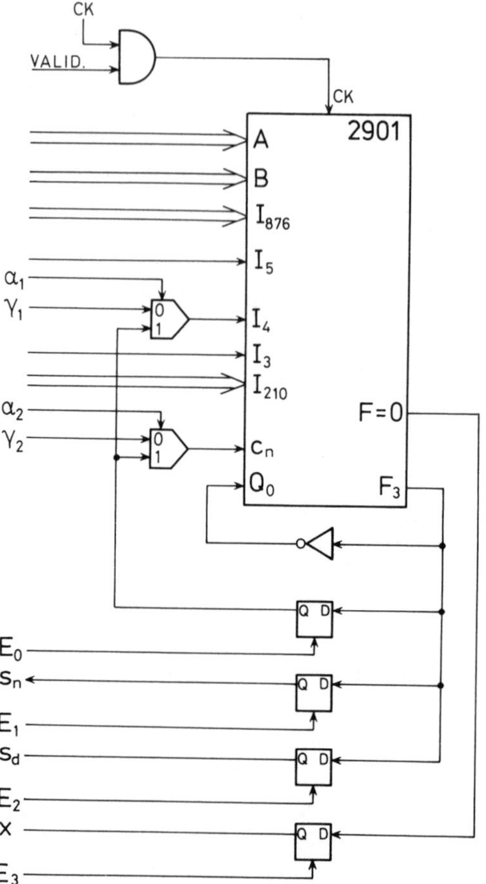

Figure 9. Organization of a processing unit

The program corresponding to the non-restoring division algorithm for two signed numbers is represented in Figure 10. It is made up with 19 instructions; there are 25 command lines and four condition variables. The read only memory of the control part will thus have an area of

$$(2+1+1+25) \cdot 32 = 1056 \text{ bits.}$$

3.4 Ultimate sequentialization. Boolean processors

We introduce in this section a quite different type of program transformation, namely the *ultimate sequentialization*. We say that a program is an ultimate sequantialized program if the only commands it contains are

n^i	test	Val	A	B	I_{876}	I_{543}	I_{210}	α_1	γ_2	α_2	E_0	E_1	E_2	E_3	N_0	N_1	Comments
1	—	1	—	0	3	3	3	0	—	—	0	1	0	0	2	—	registration of s_n
2	s_n	0	—	—	—	—	—	—	—	—	—	—	—	—	4	3	—
3	—	1	—	0	3	1	3	0	1	0	0	0	0	0	4	—	sign change of n
4	—	1	—	1	3	3	3	0	—	—	0	0	1	0	5	—	registration of s_d
5	s_d	0	—	—	—	—	—	—	—	—	—	—	—	—	7	6	—
6	—	1	—	1	3	2	3	0	1	0	0	0	0	0	7	—	sign change of d
7	—	1	—	0	7	3	3	0	—	—	0	0	0	0	8	—	computation of 2^n
8	—	1	1	0	7	1	1	0	1	0	1	0	0	0	9	—	computation of $t'^{(0)}=2(2n-d)$; memorization sign $t' \rightarrow y$
9	—	1	1	0	6	0–0	1	1	—	1	1	0	0	0	10	—	typical step
10	—	1	—	3	3	1	3	0	0	0	0	0	0	1	11	—	decrementation of R_3; memorization: $F:0 \rightarrow x$
11	x	0	—	—	—	—	—	—	—	—	—	—	—	—	9	12	test of x: $x=0$ means $[R_3] \neq 0$
12	—	1	—	0	5	3	3	0	—	—	0	0	0	0	13	—	computation of $t=t'/2$
13	—	1	1	0	5	0	1	0	0	0	0	0	0	0	14	—	computation of $r=(t+d)/2$
14	s_n	0	—	—	—	—	—	—	—	—	—	—	—	—	15	16	—
15	s_d	0	—	—	—	—	—	—	—	—	—	—	—	—	19	18	—
16	—	1	—	0	3	2	3	0	1	0	0	0	0	0	17	—	sign change of r
17	s_d	0	—	—	—	—	—	—	—	—	—	—	—	—	18	19	—
18	—	1	—	—	0	2	2	0	1	0	0	0	0	0	18	19	sign change of q
19	—																

Figure 10. Program corresponding to the non-restoring division algorithm

transfers of values. We shall state the following proposition:

Proposition. *To each program P it is possible to associate a (semantically) equivalent ultimately sequentialized program.*

In order to establish this proposition we first assume that the program P is written in conditional form. A particular instruction of this program is thus written

$$\begin{array}{llll} N & \bar{x} & \sigma_0 & N_0, \\ & x & \sigma_1 & N_1, \end{array} \tag{30}$$

the commands σ_0 and σ_1 representing functional transfers that we write in the following condensed form:

$$\begin{aligned} \sigma_0 &: [\mathbf{R}] := g_0([\mathbf{R}]), \\ \sigma_1 &: [\mathbf{R}] := g_1([\mathbf{R}]). \end{aligned} \tag{31}$$

Let $\{A, B, \ldots, M\}$ be the set of values which may be taken by the functions g_0 and g_1. For $i = 0, 1$ we define the Boolean functions $G_i^{(A)}, G_i^{(B)}, \ldots, G_i^{(M)}$ by the rules:

$$\begin{aligned} G_i^{(A)} &= 1, \quad \text{iff} \quad g_i = A, \\ G_i^{(A)} &= 0, \quad \text{iff} \quad g_i \neq A. \end{aligned} \tag{32}$$

The instruction (30) may be replaced by the following general instruction:

$$\begin{array}{llll} N & \bar{x}G_0^{(A)} & [\mathbf{R}] := A & N_0 \\ & \bar{x}G_0^{(B)} & [\mathbf{R}] := B & N_0 \\ & \cdots & \cdots & \cdots \\ & \bar{x}G_0^{(M)} & [\mathbf{R}] := M & N_0 \\ & xG_1^{(A)} & [\mathbf{R}] := A & N_1 \\ & xG_1^{(B)} & [\mathbf{R}] := B & N_1 \\ & \cdots & \cdots & \cdots \\ & xG_1^{(M)} & [\mathbf{R}] := M & N_1 \end{array} \tag{33}$$

which only contains transfers of values. This proves the proposition.

This proposition allows us to remove from the operational automaton all the combinatorial circuits, including the interconnection networks; the operational automaton then contains only registers.

Besides this, if we substitute by a conditional program an ultimate sequentialized program, the resulting conditional program is itself ultimately sequentialized. This shows that, if we add registers to the scheme of Figure 5, we obtain an universal computation model: the transformation of an ultimately sequentialized program into a strictly conditional program defines a *Boolean processor.* Its general organization is shown in Figure 11.

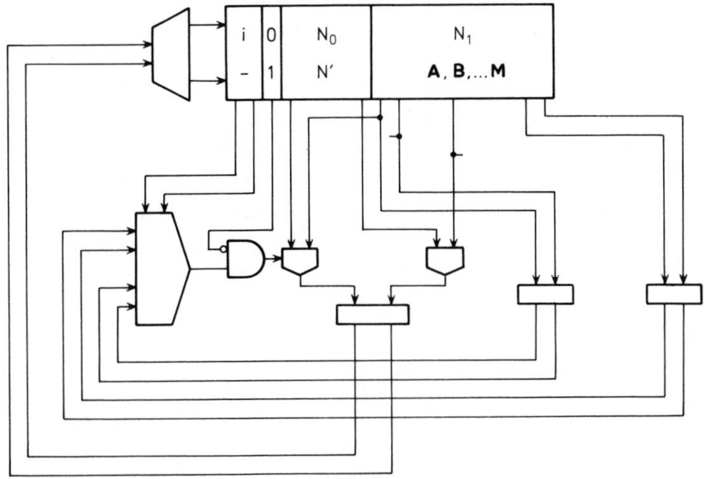

Figure 11. General organization of a Boolean processor

Boolean processors correspond to a personalized computation structure which has found several applications in the domain of automatic control (Boute, 1976). The main reason for using Boolean processors in this field of applications is of a technological nature: the transformations described above allow us to replace a conventional logic element by a read only memory of reduced cost and area (see also Sanchez and Stauffer, 1979, who use Boolean processors in the architecture of clock-maker microprocessors).

Example. We apply the above technique to the design of a decoder allowing the synthesis of a hexadecimal character (x_3, x_2, x_1, x_0) by means of a seven segment display using the convention of Figure 12.

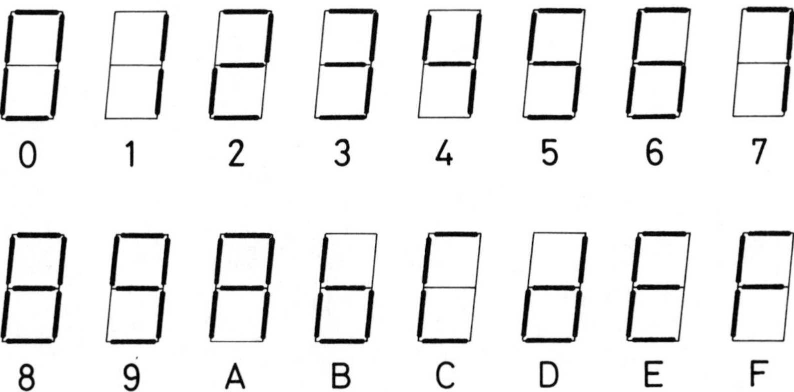

Figure 12. Seven-segment display for the synthesis of hexadecimal characters

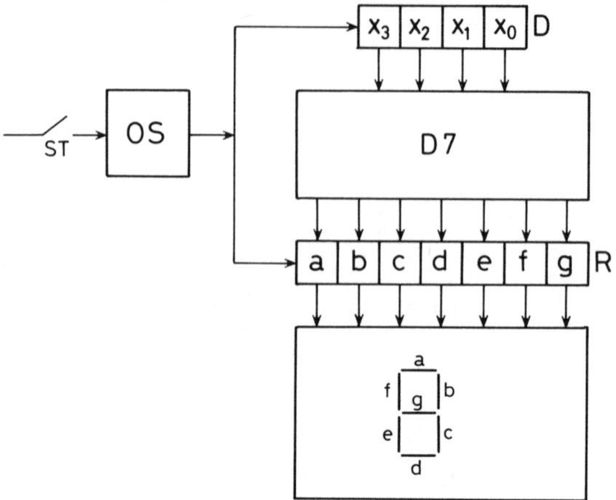

Figure 13. Initial enacting part of the seven-segment display

The initial enacting part of the seven segment display is shown in Figure 13; it is constituted by:

(a) a four bits register D: (x_3, x_2, x_1, x_0);

(b) a seven segment decoder $D7$; it is a 4-input, 7-output combinatorial network;

(c) a seven bits register R: (a, b, c, d, e, f, g) which is used as an interface between the decoder and the seven segment display; and

(d) a monostable multivibrator OS (i.e. one-shot) generating a single

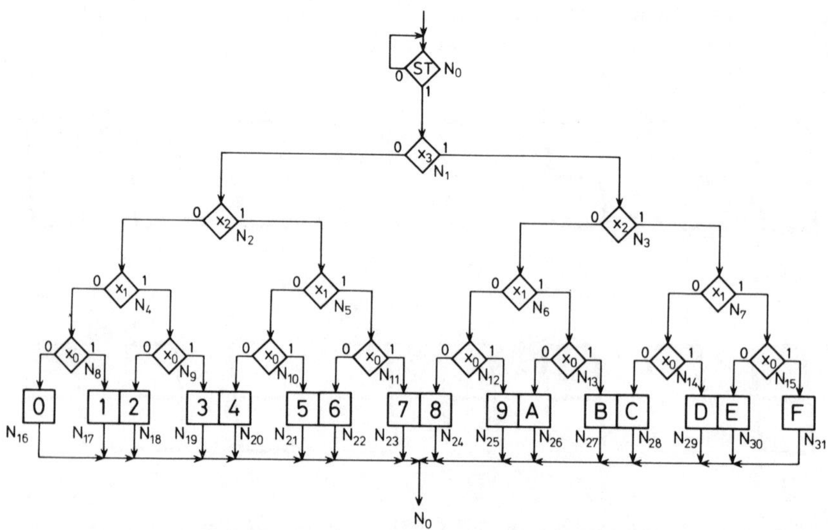

Figure 14. Ultimately sequentialized and conditional program

clock-pulse which commands the registers D and R: the clock-pulse is generated by the intermediate of a manual switch ST.

The scheme of Figure 14 realizes the program

$$
\begin{array}{llll}
N_0 & \overline{ST} & \lambda & N_0 \\
 & ST & \lambda & N_1 \\
N_1 & \mathrm{t} & \sigma & N_0
\end{array}
\qquad (34)
$$

$$\sigma : [\mathbf{R}] = f[\mathbf{D}].$$

We shall not write explicitly the program (34) under the form (33) which may be considered as a computation means; the instruction N_1 of (34) should be replaced by an instruction of the form (33) with 16 rows corresponding to the 16 characters $0, 1, \ldots, 9, A, B, \ldots, F$. In the present

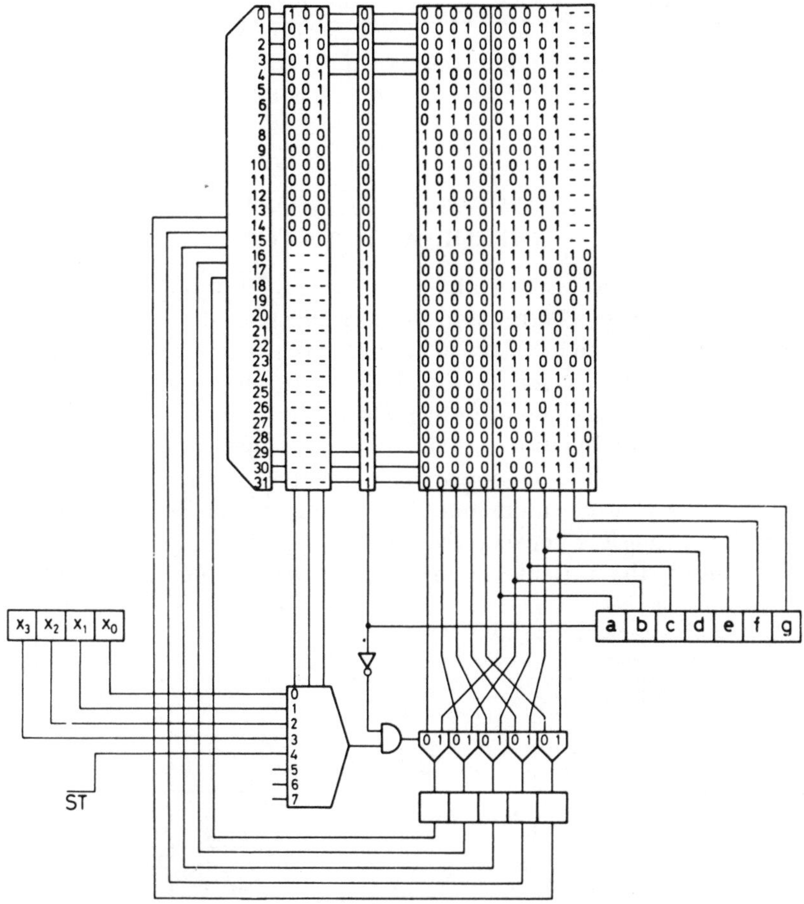

Figure 15. Microprogrammed structure corresponding to the general scheme of Figure 14

example a program which is at the same time ultimately sequentialized and conditional in the restricted sense may easily be written. Such a program based on the examination of the variables $x_3x_2x_1x_0$ by means of a canonical tree is described by the flow chart of Figure 14. It contains 32 instructions. The control automaton thus has 5 internal state variables which are denoted $y_4y_3y_2y_1y_0$.

The structure corresponding to the general scheme of Figure 14 is represented in Figure 15. The z variable which indicates that the output is available also activates the register R. Observe that the circuit of Figure 15 is far from being optimal; it will be discussed again in the following paragraph.

Remark. Note that the transformation discussed in the present section (namely the ultimate sequentialization) may be considered as a step by step process which has not necessarily to be conducted until its ultimate phase. This was already pointed out in Chapter X, Section 3 and may be considered with a new insight now that the conditional programs have been introduced. For example, instruction (31) of Chapter X is clearly a partially sequentialized instruction.

4 VLSI. MATERIALIZATION OF THE CONTROL AUTOMATON

Section 3 of Chapter X already gives some indications concerning the implementation of the control automaton. Remember that the whole system (control automaton and operational automaton) is realized by a circuit which has the configuration given in Figure 16. The instruction register (I.R.) includes the memory elements which store the present value of the state variables y. The combinational circuit A realizes the output function of the control automaton. The combinational circuit B realizes the next state function of the control automaton. The combinational circuit D generates the functional transfers between the registers $R_1, R_2, \ldots$, under the control of the value of the variables z immediately before the clock pulse. The

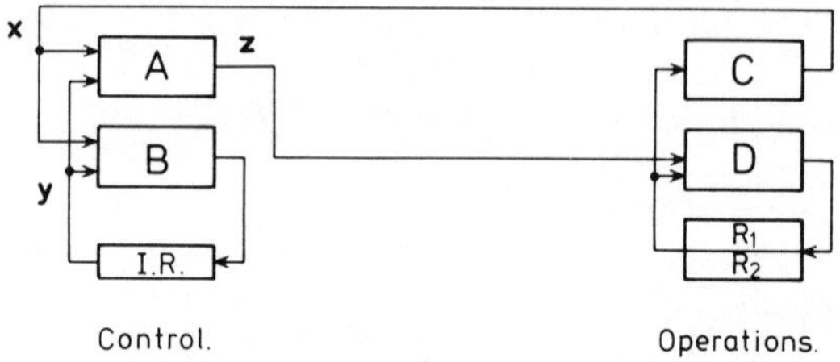

Figure 16. Configuration of the control automaton and operational automaton system

combinational circuit C computes the values of the condition variables $\mathbf{x}$ according to the contents of registers $R_1, R_2, \ldots$.

The right part of the circuit (operations) implements a Moore automaton since the output state $\mathbf{x}$ depends on the internal state (contents of $R_1, R_2, \ldots$) but not on the input state $\mathbf{z}$. Hence, the whole circuit does not contain any combinational loop: the changes of value of $\mathbf{x}$ occur as a result of a change of value of the contents of registers $R_1, R_2, \ldots$, i.e. immediately after the clock pulse.

The left part of the circuit (control) can be implemented under the form of a finite state machine, i.e. by a programmable logic array or a read only memory.

4.1 Program with disjoint predicates and one command per row

In Figure 17 the control part is implemented by a programmable logic array. The control PLA realizes $T=p+m$ Boolean functions of $N=n+p$ variables.

According to the asymptotic value of equation (31) of Chapter III, its cost is equal to

$$C_4 2^S, \tag{35}$$

with $S=N+\log T$; hence the cost of the control part is:

$$C_4(p+m)2^{n+p}. \tag{36}$$

We have also to evaluate the interconnection cost, i.e. the cost of interconnection between the two registers I.R and i.r (see Figure 17) and the cost of the interconnection between the control part and the operational part.

In the case of discrete component realizations one often considers the wiring cost as being negligible. This is however no longer true in the case of

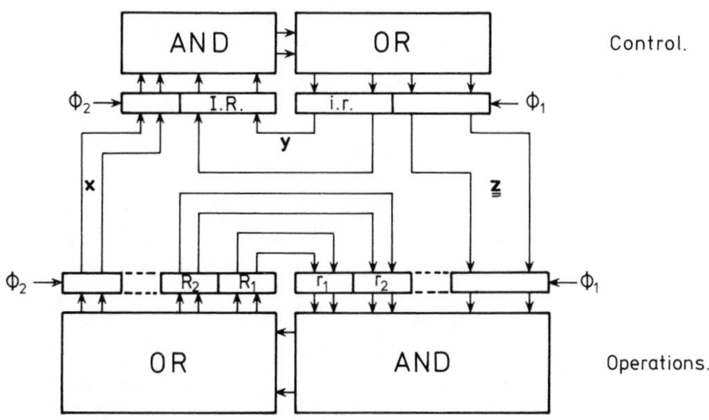

Figure 17. PLA implementation of the control and operational automata

circuits implemented on one chip. In some cases the interconnection cost of two subcircuits is actually negligible, in other cases the surface area occupied by the interconnection of two subcircuits may be important. We may consider as a worst case when the connections between n output wires of a subcircuit and m input wires of another subcircuit can be implemented by a $n \times m$ matrix whose cost is:

$$C_5 nm. \tag{37}$$

If we choose (37) as the interconnection cost, the cost of the interconnections linking the two registers I.R and i.r. of Figure 17 is $C_5 p^2$ while the cost of the interconnections linking the control part to the operational part is $C_5(m^2+n^2)$.

We have thus the following upper bound of the control automaton: this cost includes the internal interconnection cost and the cost of the interconnections with the operational automaton:

$$C_4(p+m)2^{n+p}+C_5(p^2+m^2+n^2). \tag{38}$$

At this point we introduce a synthesis method with two programmable logic arrays; this synthesis method presents important advantages, from a cost point of view, with respect to the synthesis method with one programmable logic array.

Let us designate by 2^R an upper bound of the number of actually different T-tuples of Boolean functions which must appear at the circuit output. If R is much smaller than T and if 2^R is much smaller than 2^n, it may be advantageous to break down the programmable logic array of Figure 17 into two parts as shown by Figure 18.

According to (35), the cost of part **a** is approximately equal to $C_4 R 2^N$; the cost of part **b** is approximately equal to $C_4 T 2^R$; the cost of the connections linking the R outputs of part **a** to the R inputs of part **b** is supposed to be proportional to R^2. The total cost is equal to:

$$C_4 R 2^R + C_4 T 2^R + C_5 R^2. \tag{39}$$

In the case where

$$R \ll T \quad \text{and} \quad 2^R \ll 2^N,$$

and thus

$$R^2 \ll R2^N \quad \text{and} \quad R^2 \ll T2^R,$$

the cost is approximately equal to

$$C_4 2^S\left(\frac{R}{T}+\frac{2^R}{2^N}\right), \tag{40}$$

which is much smaller than $C_4 2^S$.

Let us now come back to the synthesis of the control automaton for a program with disjoint predicates and one command per row. The number of

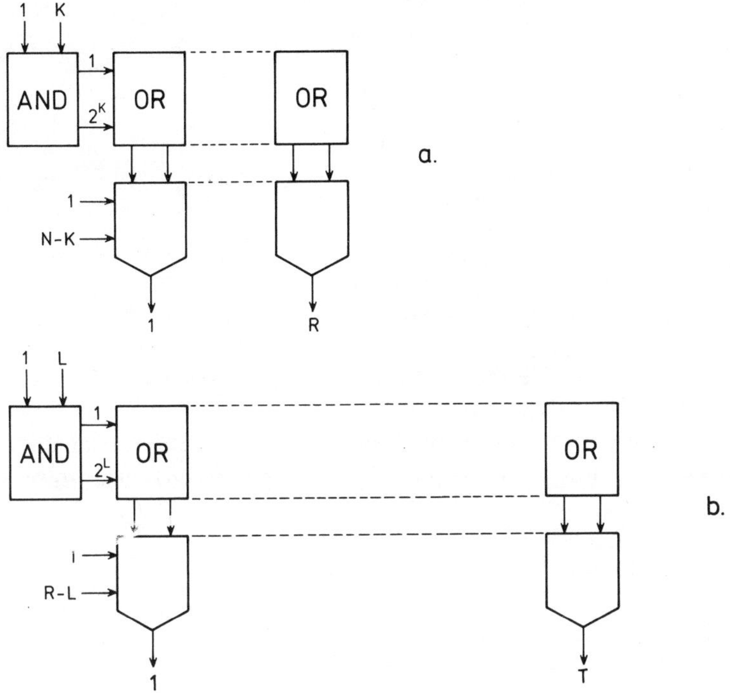

Figure 18. Decomposition of the PLA into two parts

different T-tuples which can actually appear at the PLA output is the number of the different pairs (N_i, σ_i) in the program. This number is upwardly bounded by

$$2^{p+m} \quad \text{and} \quad q2^p,$$

where q stands for the maximum number of predicates per instruction. Let us choose:

$$R = p + \log q \text{ and thus } 2^R = q2^p. \tag{41}$$

From the cost formula (40) and from (41) we deduce that the cost of the control part is equal to:

$$C_4(p+m)2^{n+p}\left(\frac{p+\log q}{p+m}+\frac{q}{2^n}\right). \tag{42}$$

If we add the interconnection costs we have the following upper bound cost of the control automaton cost:

$$C_4(p+m)2^{n+p}\left(\frac{p+\log q}{p+m}+\frac{q}{2^n}\right)+C_5(p^2+n^2+m^2). \tag{43}$$

4.2 Conditional program

Consider a conditional program containing 2^{p_1} instructions of the following types:

$$\begin{matrix} N & \bar{x}_i & \sigma_0 & N_0 \\ & x_i & \sigma_1 & N_1 \end{matrix} \tag{44a}$$

$$N \quad \mathrm{t} \quad \sigma \quad N' \tag{44b}$$

Figure 5 shows a realization of the corresponding control automaton using a read only memory divided into five fields.

(1) A first field containing $\lceil \log n \rceil$ columns indicates the index of the condition variable under test.

(2), (3) Two fields give the next instruction numbers N_0 and N_1 or N'.

(4), (5) Two fields define the two commands σ_0 and σ_1 or σ.

Figure 19 shows an equivalent realization using a programmable logic array. Note that the columns of the OR-array can be arranged in order to simplify the interconnection with the multiplexers. Let us estimate the cost of this last circuit.

(1) The programmable logic array realizes:

$$T = 2p_1 + 2m + \lceil \log n \rceil .$$

According to (35), its cost is equal to:

$$C_4(2p_1 + 2n + \lceil \log n \rceil)2^{p_1}. \tag{45}$$

(2) There are $p_1 + m$ multiplexers with $\lceil \log n \rceil$ control variable. According to Section 2 of Chapter III, the corresponding cost is equal to:

$$C_1^*(p_1 + m) + C_1^* \lceil \log n \rceil 1^{(\lceil \log n \rceil - 1)};$$

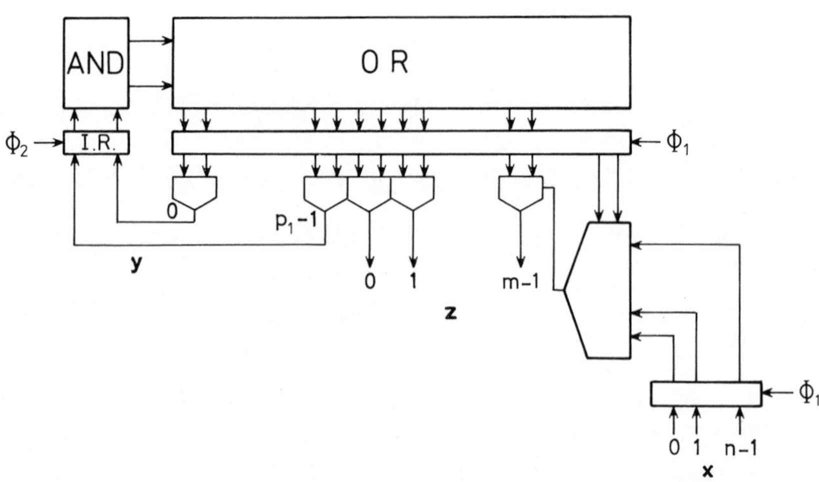

Figure 19. PLA implementation of a conditional program

it is bounded by

$$C_1^*(p_1+m+n\lceil\log n\rceil). \tag{46}$$

If we estimate the interconnection cost as above, the cost of the whole circuit is equal to:

$$C_4(2p_1+2m+\lceil\log n\rceil)2^{p_1}+C_1^*(p_1+m+n\lceil\log n\rceil)+C_5(p_1^2+n^2+m^2). \tag{47}$$

4.3 Conditional program in the restricted sense

Let us now consider a strictly conditional program containing 2^{p_2} instructions. It includes instructions of the following types:

$$\begin{matrix} N & \bar{x}_i & \lambda & N_0 \\ & x_i & \lambda & N_1 \end{matrix} \tag{48a}$$

$$\begin{matrix} N & \text{t} & \sigma & N_1' \end{matrix} \tag{48b}$$

Its implementation is based on the use of a type variable v which takes the value 0 for the branching instructions and the value 1 for the execution instructions; it constitutes also the enable signal for the operational part.

Figure 8 shows a realization using a read only memory divided into four fields.

(1) A field containing $\lceil\log n\rceil$ columns indicates the index of the condition variable under test.

(2) A one column field defines the type of instruction (branching or execution).

(3) The third field gives one of the next instruction numbers N_0 or N'.

(4) The last field gives the other next instruction number N_1 or the command σ.

A series of multiplexers realizes the information transfers under the control of the variable x_i selected by field (1).

Figure 20 shows an equivalent realization based on the use of a programmable logic array. Let us estimate the cost of this circuit.

(1) The programmable logic array realizes

$$T=p_2+\max(p_2, m)+\lceil\log n\rceil+1$$

functions of $N=p_1$ variables. According to equation (15) of Chapter III, its cost is equal to

$$C_4(p_2+\max(p_2, m)+\lceil\log n\rceil+1)2^{p_2}. \tag{49}$$

(2) There are p_2 multiplexers with one control variable, and one multiplexer with $\lceil\log n\rceil$ control variables. According to Section 2 of Chapter III, the corresponding cost is equal to

$$C_1^*p_2+C_1^*\lceil\log n\rceil 2^{(\lceil\log n\rceil-1)}$$

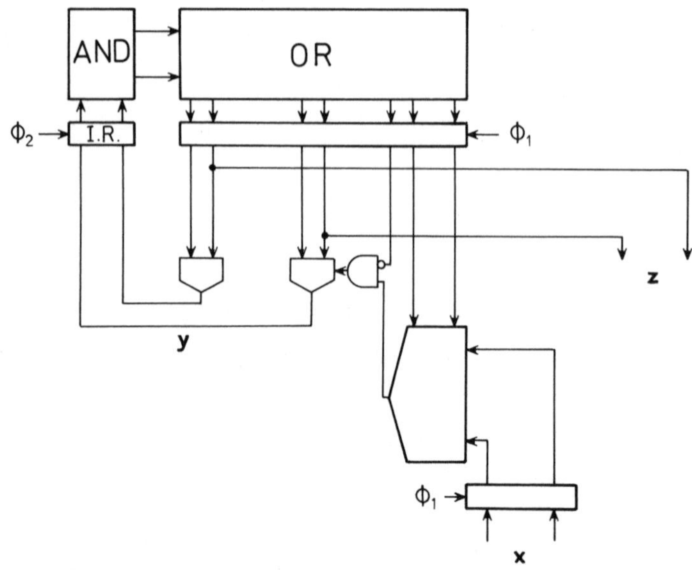

Figure 20. Implementation of a strictly conditional program

which is bounded by:

$$C_1^*(p_2 + n\lceil \log n \rceil). \tag{50}$$

If we estimate the interconnection cost as above, the cost of the whole circuit is equal to

$$C_4(p_2 + \max(p_2, m) + \lceil \log n \rceil + 1)2^{p_2} + C_1(p_2 + n\lceil \log n \rceil) + C_5(p^2 + n^2 + m^2). \tag{51}$$

4.4 Cost comparisons

Suppose that we must implement an initial program, with disjoint predicates and one command per row, which contains 2^p instructions. It may be realized by the circuit of Figure 17 with a cost of the control automaton given by (38) or (43). It can also be transformed into a strictly conditional program containing at most 2^{p_2} instructions. This program may be implemented by the circuits of Figure 20 with a cost given by (51).

In order to compare the two techniques we first express p_2 in terms of n (condition variable number) and q (upper bound of the number of predicates per instruction).

We have proved (see (18)) that instruction (5) can be replaced by a strictly conditional program including at most

$$K\frac{2^s}{s} - 1, \quad \text{branching instructions}$$

and

$$2^r - 1 = q, \quad \text{execution instructions,}$$

with $s = n + \log r = n + \log\log(q+1) \cong n + \log\log q$. If the initial program contains 2^p instructions, and if q represents further on an upper bound of the number of predicates per instruction then the number 2^{p_2} of instructions of the strictly conditional program is bounded by

$$\left(K\frac{2^s}{s} + q - 1\right)2^p = \left(\frac{K2^n \log q}{n + \log\log q} + q - 1\right)2^p. \tag{52}$$

In order to compare the two techniques we have later to make some approximations based on the following hypothesis

$$n \gg \log q.$$

Hence

$$n \gg \log\log q; \qquad \frac{s^n}{n} \gg \frac{q}{\log q}$$

and

$$2^{p_2} \cong \log qK\frac{2^{n+p}}{n}. \tag{53}$$

By taking into account the following implications

$$\log q \ll n \Rightarrow \frac{K \log q}{n} < 1 \Rightarrow \log\left(\frac{K \log q}{n}\right) < 0,$$

we deduce the following approximate value for p_2:

$$p_2 \cong n + p. \tag{54}$$

Making the following assumptions:

$$\lceil \log n \rceil + 1 \ll n + p + \max(n+p, m),$$

$$C_1^*(n + p + n\lceil \log n \rceil) \ll C_4(n + p + \max(n+p, m))\frac{K \log q}{n}2^{n+p}$$
$$+ C_5((n+p)^2 + n^2 + m^2),$$

we obtain in view of (52), (53), and (54) the following cost estimate:

$$C_4\frac{K \log q}{n}2^{n+p}(n + p + \max(n+p, m)) + C_5((n+p)^2 + n^2 + m^2). \tag{55}$$

Summarizing the above results, we obtained three cost expressions for the following types of architecture.

Direct implementation with one PLA:
  Cost c_1 given by expression (38).

Direct implementation with two PLA's:
Cost c_2 given by expression (43).
Implementation by the intermediate of a conditional program in the restricted sense:
Cost c_3 given by expression (55).

Let us study the dependence of the various costs upon p and m, while assuming n and q constant:

(1) $c_1 = (C_4 2^n)p2^p + (C_4 2^n)m2^p + C_5(p^2 + m^2 + n^2)$.

Neglecting p^2 and 1 with respect to $p2^p + m2^p$ we obtain:

$$C_1 \simeq \alpha p 2^p + \alpha m 2^p + C_5 m^2, \tag{56}$$

with

$$\alpha = C_4 2^n.$$

(2) $c_2 = (C_4 2^n)p2^p + (C_4 2^n \log q)2^p + (C_4 q)p2^p + (C_4 q)m2^p + C_5(p^2 + m^2 + n^2)$.

Neglecting 2^p, p^2, and 1 with respect to $p2^p$ we obtain:

$$c_2 \simeq \alpha p 2^p + \beta p 2^p + \beta m 2^p + C_5 m^2 \tag{57}$$

with

$$\beta = C_4 q.$$

(3) (We assume $n + p < m$):

$$c_3 = C_4 K \frac{\log q}{n} 2^{n+p}(n + p + m) + C_5((n+p)^2 + n^2 + m^2)$$
$$= \gamma n 2^p + \gamma p 2^p + \gamma m 2^p + C_5 p^2 + C_5 m^2 + (2C_5 n)p + (2C_5 n^2)$$

with

$$\gamma = C_4 K \log q \frac{2^n}{n}.$$

Hence:

$$C_3 \cong \gamma p 2^p + \gamma m 2^p + C_5 m^2. \tag{58}$$

If we take into account the fact that $n \gg \log q$, we deduce:

$$\alpha \gg \beta; \qquad \frac{\alpha}{\gamma} = \frac{n}{K \log q} \gg 1 \text{ and thus } \alpha \gg \gamma.$$

Therefore $c_1 > c_3$ which shows that for large values of m and p the solution of cost c_1 (direct implementation with one PLA) must be rejected.

As regards the comparison between c_2 and c_3, the conclusions are not so obvious; the choice between the solution of cost c_2 and the solution of cost c_3 depends on the relative values of p and m. If $m \gg p$, $c_2 < c_3$ and it is preferable to use the direct implementation with two programmable logic arrays. The implementation of the equivalent conditional program in the restricted sense could be the best solution when p and m have the same order of magnitude. If $m \ll p$ we show in next chapter that the best solution consists in the implementation by means of an incremental equivalent program.

BIBLIOGRAPHICAL REMARKS

For more references on the subject of this chapter, the reader is referred to the Bibliographical remarks in Chapters X and XIII.

EXERCISES

E1. Most of the programs described hitherto are already conditional by their own nature. In that case, the complexity estimates of Section 4 are unfair to the conditional implementation. The reader should select some multiplication or division algorithm and, for a given processing unit, compare the control ROM required for
(i) An automaton realization of the control unit according to Chapter X.
(ii) A conditional realization of the control unit according to Chapter XII.

E2. Consider two 4-bit integers $A \simeq a_3a_2a_1a_0$, $B \simeq b_3b_2b_1b_0$. Design a conditional program for an amplitude comparison with results $A > B$, $A = B$, and $A < B$ (see Chapter IV). Use this program in the control unit of a sorting network (odd–even merge (Chapter VIII) or bubble sort (Chapter XI, E5) to avoid the use of comparators in the computation resource of Figure 7a, Chapter VIII.

E3. Consider the two following instruction types

$$\begin{array}{lllll} N_i & \bar{x}_1\bar{x}_0 & \lambda & N_0 & \\ & \bar{x}_1 x_0 & \lambda & N_1 & \text{(four way branch)} \\ & x_1\bar{x}_0 & \lambda & N_2 & \\ & x_1 x_0 & \lambda & N_3 & \\ N_i & t & \sigma & N_j & \text{(execution)} \end{array}$$

Is it always possible to replace a general program by an equivalent program using the above two instruction types only? Is the equivalence functional or semantic? What would be an organization of the control unit appropriate to such a kind of program? Discuss its interest. (See also Chapter XIV, Section 3.5.)

E4. Consider the following three instruction types:

$$\begin{array}{llllll} \text{(i)} & N_i & f(x) & \Sigma_0 & N_{i+1} & (\textit{if} \ldots \textit{then else}), \\ & & f(x) & \Sigma_1 & N_{i+1}; & \\ \text{(ii)} & N_i & f(x) & \Sigma & N_i & (\textit{while do}), \\ & & f(x) & \lambda & N_{i+1}; & \\ \text{(iii)} & N_i & t & \Sigma & N_{i+1}; & \end{array}$$

where Σ, Σ_0, and Σ_1 represent programs and where $f(x)$ is a Boolean function of the condition variables. A program written with only the three above instruction types is called a *regular program*: it is a particular case of a *structured program* (goto less program: Dijkstra, 1968; Böhm and Jacopini, 1966; Aschroft and

Manna, 1972; Ito, 1973; Knuth and Floyd, 1961, and Ledgard and Marcotti, 1975). It will appear in the course of this exercise that any program may be replaced by a functionally equivalent regular program. The interest of the regular programs is that they do not require the explicit mention of the instruction label; as a matter of fact, these programs have a textual representation if one adopts the following notations for the three instruction types:

(i) $\underset{f(x)}{(}\Sigma_1 \vee \Sigma_0);$

(ii) $\underset{f(x)}{[}\ \Sigma\ ];$

(iii) Σ.

For example, the writing

$$\rho_i\rho_j\underset{f}{[}\,\underset{g}{[}\,\underset{x}{(}\sigma\vee\lambda)\tau_j]\tau_i]$$

represents the sorting program of Exercise E5, Chapter XI. If

(i) $\rho_i : i := 0;\qquad \rho_j : j := 0;$
(ii) $f = 1$ iff $i \geqslant n;\qquad g = 1$ iff $j > n - i;$
(iii) $x = 1$ iff $a[j+1] < a[j];$
(iv) $\sigma : (a[j], a[j+1]) := (a[j+1], a[j]);$
(v) $\tau_i : i := i+1;\qquad \tau_j : j := j+1.$

We furthermore use the notation $f \circ \Sigma$ to represent a function which takes the value 1 iff f were to take that value after execution of Σ. Prove the following properties:

(i) $\underset{f}{(}\Sigma\vee T) = \underset{\bar f}{(}T\vee\Sigma);$

(ii) $\underset{f}{(}\Sigma\vee T)Y = \underset{f}{(}\Sigma Y\vee TY);$

(iii) $\Sigma\underset{f}{(}T\vee Y) = \underset{f\circ\Sigma}{(}\Sigma T\vee\Sigma Y);$

(iv) $\underset{f}{(}\underset{g}{(}\Sigma_1\vee\Sigma_2)\vee\underset{h}{(}T_1\vee T_2)) = \underset{a}{(}\underset{f}{(}\Sigma_1\vee T_1)\vee\underset{f}{(}\Sigma_2\vee T_2)),$

where $a = (f\wedge g)\vee(\bar f\wedge h);$

(v) $\underset{f}{(}\underset{g}{(}\Sigma\vee T)\vee Y) = \underset{a}{(}\Sigma\vee\underset{f}{(}T\vee Y)),$

where $a = f\wedge g;$

(vi) $\underset{f}{(}\Sigma\vee\underset{g}{(}T\vee Y)) = \underset{a}{(}\underset{f}{(}\Sigma\vee T)\vee Y);$

(vii) $\Sigma = \underset{f}{(}T\vee Y\Sigma) \Rightarrow \Sigma = \underset{f}{[}Y]T$

(intuitive proof only);

(viii) $\Sigma = \Xi\underset{f}{(}T\vee Y\Sigma) \Rightarrow \Sigma = \Xi\underset{f}{[}Y\Xi]T;$

(ix) $\Sigma = \underset{f1}{(}T_1\vee\underset{f2}{(}T_2\vee\underset{f3}{(}T_3\vee Y\Sigma))) \Rightarrow$

$\Sigma = \underset{a}{[}Y]\underset{f1}{(}T_1\vee\underset{f_2}{(}T_2\vee T_3)),$

where $a = f_1\vee f_2\vee f_3$

(Generalize this formula to an arbitrary number of T_i's and f_i's);

(x) $\Sigma = \underset{f1}{\Xi_1}(T_1 \vee \underset{f2}{\Xi_2}(T_2 \vee \underset{f3}{\Xi_3}(T_3 \vee Y\Sigma))) \Rightarrow$

$\Sigma = \underset{a}{[}\Xi_1\Xi_2\Xi_3 Y]\underset{f1}{\Xi_1}(T_1 \vee \underset{f2}{\Xi_2}(T_2 \vee \Xi_3 T_3))),$

where $a = (\Xi_1 \circ f_1) \vee (\Xi_1\Xi_2 \circ f_2) \vee (\Xi_1\Xi_2\Xi_3 \circ f_3)$

(Generalize this formula to an arbitrary number of T_i's, of Ξ_i's, and of f_i's). Consider now the typical instruction of a conditional program

N_i	$\bar{x}$	σ_0	N_j
	x	σ_1	N_k

To that instruction N_i, we associate the program Σ_i (program executed if the computation is started in N_i) and the equation:

$$\Sigma_i = \underset{x}{(}\sigma_1\Sigma_k \vee \sigma_0\Sigma_j).$$

Hence, to a conditional program, we may associate a system of equations of the above type: this system is called the characteristic system of the program. Using properties (i) to (x), show that this system may always be solved for the program Σ_{init} corresponding to the initial instruction. Use the above regular program algebra to relate the *while do* and *repeat until* control structures. Replace the following (non-regular) program by a regular program. (*Hint*: Use right factorization (ii) whenever possible.)

N_1	$\bar{x}$	λ	N_2
	x	λ	N_3
N_2	$\bar{y}$	σ_1	N_4
	y	σ_2	N_1
N_3	$\bar{z}$	σ_3	N_4
	z	σ_4	N_1
N_4	END		

Chapter XIII
Optimization problems in the control unit

1 INTRODUCTION

One of the results of Section 2 of Chapter XII is to show that an instruction of the type (1) of Chapter XII may always be partitioned into two phases: an evaluation phase an execution phase (see instruction (2) of Chapter XII). The evaluation phase is described by an instruction with disjoint predictes of the form:

$$\begin{array}{llll} N & g_0(\mathbf{x}) & \lambda & N_0 \\ & g_1(\mathbf{x}) & \lambda & N_1 \\ & \cdots & \cdots & \cdots \\ & g_{q-1}(\mathbf{x}) & \lambda & N_{q-1} \end{array} \tag{1}$$

In Section 1 of Chapter XII we showed that an instruction (1) could always be transformed into a conditional program in the restricted sense. Such a conditional program may be described by means of a flow chart schematized in Figure 1. The conditional program of Figure 1 has one input (or initial instruction) labelled N and q outputs (or final instructions) labelled N_i, $0 \leq i \leq q-1$. It is made up of a loop free interconnection of instructions of the type (26a) of Chapter XII, each of which is represented by its usual flow chart symbol. Besides to being loop free, the only restriction to which the interconnection of instructions is constrained is that any instruction output is connected to one and only one instruction input: this constraint prevents any possibility of parallelism and of non-determinism in the program.

There are clearly a high number of conditional programs which may represent the instruction (1). Different programs are, e.g. obtained by taking different choices for the successive condition variables to be tested and by considering the possibility of having *reconvergent instructions*. In the scheme of Figure 1 the instruction M_c is called *reconvergent* since it may be reached through the intermediate of at least two (in this case M_a and M_b) instructions. There exists thus an optimization problem so far as the synthesis of conditional programs is concerned. The main optimization criteria are the following ones.

(a) The maximal duration of a computation (*time criterion*). This duration

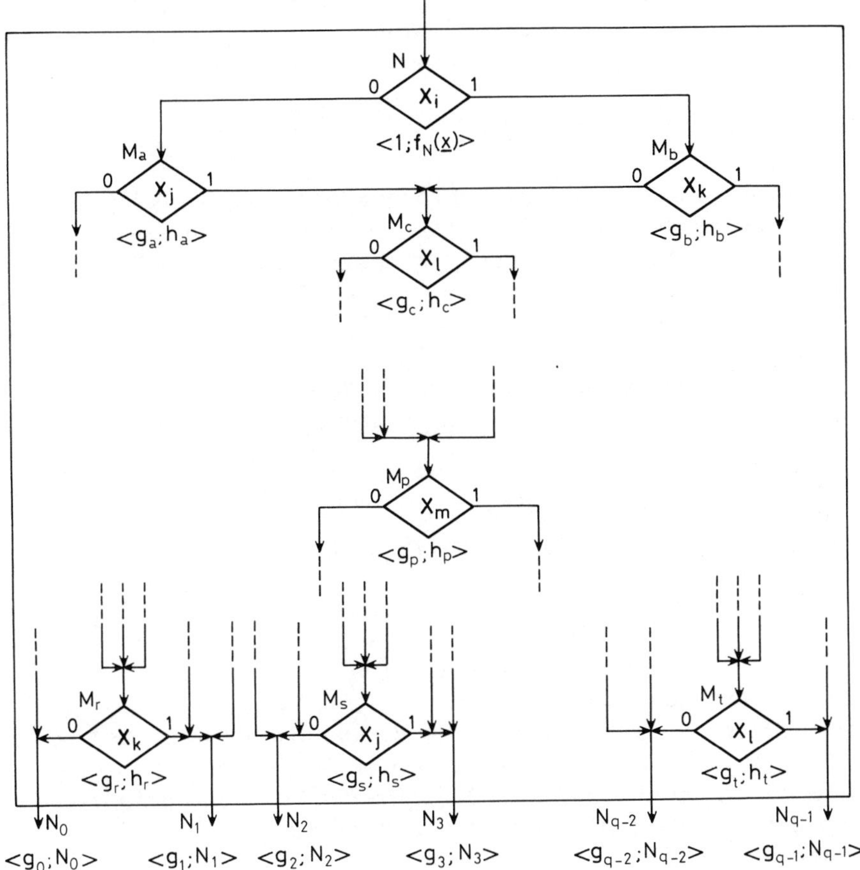

Figure 1. Flow chart representing a conditional program in the restricted sense

is reflected in the scheme of Figure 1 by the length of the longest path between N and any N_i, $0 \leq i \leq q-1$ (the length of a path is the number of its edges).

(b) The number of instructions (*cost criterion*).

The following criterion will also be considered.

(c) Between all the conditional programs having the shortest maximal computation time find those having a minimum number of instructions (*relative cost criterion*).

Consider the flow chart of Figure 1 representing a conditional program realizing the instruction (1). This instruction is also represented by means of the formal Boolean expression $f(\mathbf{x})$:

$$N \simeq f_N(\mathbf{x}) = \bigvee_{i=0,\, q-1} g_i(\mathbf{x}) N_i \qquad (f_N : \{0, 1\}^n \to \{N_0, N_1, \ldots, N_{q-1}\}). \qquad (2)$$

Now any of the internal instructions in the program (e.g. M_c in Figure 1), may be considered as an initial instruction to a subprogram; it may be represented by a formal Boolean expression of the same type as (2), i.e.:

$$M_c \simeq h_c(\mathbf{x}) = \bigvee_{i=0,\, q-1} g_i^c(\mathbf{x}) N_i. \tag{3}$$

The synthesis and optimization of conditional programs will be stated by introducing the concept of a *P-function* associated with an instruction.

Consider first the conditional program of Figure 1 as a black box with one input terminal N and q output terminals $N_0, N_1, \ldots, N_{q-1}$. To the input terminal N we associate the pair of functions $\langle 1; f_N(\mathbf{x})\rangle$; to each output terminal N_i we associate the pair of functions $\langle g_i(\mathbf{x}); N_i\rangle$. These functions will be called the *P-functions associated with the corresponding instruction label.* We verify that any one of the P-functions $\langle g; h\rangle$ associated with a terminal satisfies the relation

$$f_N(\mathbf{x})g(\mathbf{x}) = h(\mathbf{x})g(\mathbf{x}). \tag{4}$$

This relation is interpreted as follows: in the domain of the n-cube characterized by the relation

$$g(\mathbf{x}) = 1,$$

the function $f_N(\mathbf{x})$ reduces to the function $h(\mathbf{x})$ (f_N and h are both formal Boolean expressions of the form: $\{0, 1\}^n \to \{N_0, N_1, \ldots, N_{q-1}\}$). Hence the Boolean function $g(\mathbf{x})$ will be called the *domain function* while the function $h(\mathbf{x})$ will be called the *codomain function* of the P-function $\langle g; h\rangle$ respectively. To any internal instruction M_c represented by the formal expression (3) we associate the P-function $\langle g_c; h_c\rangle$ with g_c a Boolean function satisfying the relation (4), i.e.

$$f_N(\mathbf{x})g_c(\mathbf{x}) = h_c(\mathbf{x})g_c(\mathbf{x}).$$

The interest of the concept of P-function lies in the following observations (see also Figure 2):

(a) The instruction labelled M with x as condition variable and to which is associated the P-function $\langle g; h\rangle$ has the two following instructions labelled M_0 and M_1, and has $\langle \bar{x}g; h(x=0)\rangle$ and $\langle xg; h(x=1)\rangle$ as P-functions respectively (see Figure 2a).

(b) The instructions M_0 and M_1 having $\langle g_0; h_0\rangle$ and $\langle g_1; h_1\rangle$ as P-functions respectively are obtained from an instruction M having $\langle \bar{x}g_0 \vee xg_1;\ \bar{x}h_0 \vee xh_1\rangle$ as P-function (see Figure 2b).

We see that there exist transformation laws allowing us either, knowing the P-function of an instruction, to obtain the P-functions of its two following instructions, or knowing the P-functions of these two last instructions to find the P-functions of the instruction from which they were derived.

The synthesis of a conditional program may then be stated in the following way: starting from the instructions $\{N_i\}$ represented by their

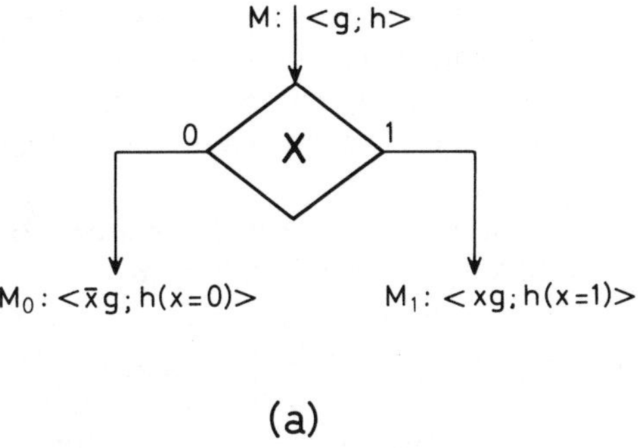

(a)

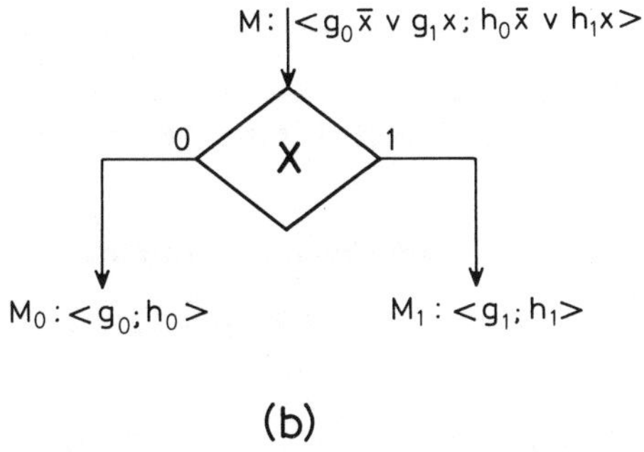

(b)

Figure 2. P-functions associated with a decision instruction

P-functions $\{\langle g_i\,; N_i\rangle\}$ find a composition law T which, when acting on a pair of P-functions, generates a new P-function, i.e.

$$\langle g_0;\, h_0\rangle T\langle g_1;\, h_1\rangle = \langle g;\, h\rangle.$$

It is required that an iterative use of this law T produces finally the P-function $\langle 1; f_N\rangle$ and that the intermediate P-functions are in one-to-one correspondence with the intermediate instructions of the conditional program. Thus the minimization of the number of instructions in a conditional program will be obtained by minimizing the number of intermediate P-functions in the transformation process which produces the P-function $\langle 1; f\rangle$ from the set of P-functions $\{\langle g_i\,; N_i\rangle\}$. Further on this transformation process will be represented as follows:

$$\{\langle g_i\,; N_i\rangle\} \xrightarrow{T} \langle 1\,; f_N\rangle. \tag{5}$$

In Section 2 we shall define two types of composition laws which will be able to generate two types of conditional programs, namely the simple conditional program and the non-simple conditional program.

A conditional program is said to be *simple* if any variable x_i may be tested once at most during a computation.

The most general type of conditional program is the *non-simple program* where the same variable may be tested several times during a given computation.

When it is not necessary to distinguish between these two types of programs the term *conditional program* will be used as before.

The following Section 2.2 introduces the algebra of P-functions and the composition laws in a more rigorous and complete way. In Section 2.3 we introduce a general type of algorithm for generating optimal conditional programs. Section 2.4 is devoted to a detailed study of a tabular method which transforms any instruction into an optimal conditional program. Section 2.5 considers the extension of this tabular method to the optimization of programs.

2 THE ALGEBRA OF *P*-FUNCTIONS

2.1 Introductory concepts

Let L be a finite non-empty set; we consider functions:

$$f\colon \{0, 1\}^n \to L. \tag{6}$$

It will generally be assumed that the set L is formed by a list of instruction labels, i.e. $L = \{N_0, N_1, \ldots, N_{q-1}\}$; the function f may then be represented by means of a formal Boolean expression of the form (2). We will also consider for L the set of integers: $\{0, 1, \ldots, q-1\}$ and its particular important case $L = \{0, 1\}$. The function f reduces then to the well-known *pseudo-logic* and *Boolean functons* respectively. One will moreover assume that the function f may be incompletely defined.

Let $f(\mathbf{x})$ and $h(\mathbf{x})$ be two functions of the form (6) and let $g(\mathbf{x})$ be a Boolean function with $\mathbf{x} = (x_{n-1}, \ldots, x_1)$; the pair of functions $\langle g; h\rangle$ will be called a P-function of f (and one will write $\langle g; h\rangle \nabla f$) if and only if:

$$fg = hg. \tag{7}$$

The condition $g = 1$ defines a domain where $f = h$ (see Section 1); the P-function $\langle g; h\rangle$ constitutes a *partial description* of f in the domain characterized by $g = 1$. The set of P-functions

$$\{\langle g_i; h_i\rangle \mid \langle g_i; h_i\rangle \nabla f, i \in I\} \tag{8}$$

constitutes a *total description* of f if and only if the solutions of

$$\bigvee_{i \in I} g_i(\mathbf{x}) = 1 \tag{9}$$

characterize the domain where f is defined; in particular if f is a completely defined function the condition (9) becomes:

$$\bigvee_{i \in I} g_i(\mathbf{x}) \equiv 1. \tag{10}$$

Clearly $\langle 1; f\rangle$ and $\{\langle g_i(\mathbf{x}); N_i\rangle\}$ constitute two total descriptions of:

$$f(\mathbf{x}) = \bigvee_{i=0,\, q-1} g_i(\mathbf{x}) N_i. \tag{11}$$

For $f(\mathbf{x})$ a Boolean function $\{\langle f; 1\rangle, \langle \bar{f}; 0\rangle\}$ constitutes another total description.

Further on, no distinction will be made between completely and incompletely defined functions; this unified treatment may be reached by modifying slightly the definition of the Boolean function $g_i(\mathbf{x})$ as follows: the solutions of the equation $g_i(\mathbf{x}) = 1$ will characterize the *maximal domain* in which f may take the label N_i. Since a non-specification may be viewed as the possibility for f to take several labels N_i on some subdomains of the n-cube, the g_i do not satisfy the orthogonality conditions (4) of Chapter XII when f is incompletely defined.

Let us prove some elementary properties of the algebra of P-functions.

Lemma 1.
(a) $\langle g; h\rangle \nabla f$ *and* $g' \leq g \Rightarrow \langle g'; h\rangle \nabla f$.
(b) $\langle g_0; h\rangle \nabla f$ *and* $\langle g_1; h\rangle \nabla f$
$\Rightarrow \langle g_0 g_1; h\rangle \nabla f$ *and* $\langle g_0 \vee g_1; h\rangle \nabla f$.

Proof.
(a) $(g; h) \nabla f$ and $g' \leq g \Rightarrow fg = hg$ and $gg' = g'$
$\Rightarrow fgg' = hgg'$ and $fg' = hg' \Rightarrow \langle g'; h\rangle \nabla f$.
(b) $\langle g_0; h\rangle \nabla f$ and $\langle g_1; h\rangle \nabla f \Rightarrow fg_0 = hg_0$ and $fg_1 = hg_1$
$\Rightarrow fg_0 g_1 = hg_0 g_1$ and $f(g_0 \vee g_1) = h(g_0 \vee g_1)$
$\Rightarrow \langle g_0 g_1; h\rangle \nabla f$ and $\langle g_0 \vee g_1; h\rangle \nabla f$. □

From the preceding lemma one deduces that the set:

$$\{g_i \mid \langle g_i; h\rangle \nabla f\}$$

is closed for the operations of conjunction and of disjunction. It constitutes thus a sublattice of the lattice of Boolean functions and has consequently a maximum element denoted $[f, h]$. The relation

$$[f, h] = 1$$

is thus satisfied on each vertex where $f = h$ and on these vertices only. Let us denote by $f \oplus h$ a Boolean function equal to 0 on the vertices where $f = h$ and equal to 1 otherwise. One has thus:

$$[f, h] = \overline{f \oplus h}. \tag{12}$$

The theorems that will be proven below are valid for formal Boolean expressions of the form (6). The proof will, however, be given for Boolean functions, since the form of $[f, h]$ in this last case allows us to use the usual Boolean formalism. For example, one proves that

$$\langle [f, h]; h\rangle \nabla f$$

by observing that:

$$f(\overline{f \oplus h}) = fh = h(\overline{f \oplus h}).$$

2.2 Theorems on composition laws

The formalism of proof for the theorems below and for f, a formal Boolean expression, is left to the reader.

Theorem 1 (Theorem on the composition law T^0).

$\langle g_0; h_0\rangle \nabla f$ *and* $\langle g_1; h_1\rangle \nabla f \Rightarrow \langle g_0; h_0\rangle T_x^0 \langle g_1; h_1\rangle = \langle g_0 \bar{x} \vee g_1 x;\ h_0 \bar{x} \vee h_1 x\rangle \nabla f.$

Moreover:

$$g_0 = [f, h_0] \quad \textit{and} \quad g_1 = [f, h_1] \Rightarrow \bar{x} g_0 \vee x g_1 = [f, \bar{x} h_0 \vee x h_1].$$

Proof. One has immediately:

$$f(\bar{x} g_0 \vee x g_1) = \bar{x} f g_0 \vee x f g_1,$$
$$(\bar{x} h_0 \vee x h_1)(\bar{x} g_0 \vee x g_1) = \bar{x} h_0 g_0 \vee x h_1 g_1.$$

The first part of the proposition follows then from the relations

$$f g_0 = h_0 g_0 \quad \text{and} \quad f g_1 = h_1 g_1.$$

Let us now prove the preservation of the maximality character of the domain function $\bar{x} g_0 \vee x g_1$. One has successively:

$$\begin{aligned} \overline{f \oplus (\bar{x} h_0 \vee x h_1)} &= \bar{f} \oplus (\bar{x} h_0 \oplus x h_1) \\ &= (\bar{x} \bar{f} \oplus x \bar{f}) \oplus (\bar{x} h_0 \oplus x h_1) \\ &= \bar{x}(\bar{f} \oplus h_0) \oplus x(\bar{f} \oplus h_1) = \bar{x} g_0 \vee x g_1. \end{aligned}$$

□

Theorem 2 (Theorem on the composition law T^1).

$$\langle g_0; h_0\rangle \nabla f \quad \textit{and} \quad \langle g_1; h_1\rangle \nabla f,$$
$$\langle g_0; h_0\rangle T_x^1 \langle g_1; h_1\rangle = \langle g_0(x=0) g_1(x=1);\ h_0 \bar{x} \vee h_1 x\rangle \nabla f.$$

Moreover, if $g_0 = [f, h_0]$ *and* $g_1 = [f, h_1]$, $g_0(x=0) g_1(x=1)$ *is the greatest function independent of x and contained in* $[f, (\bar{x} h_0 \vee x h_1)]$.

Proof. Taking into account the fact that

$$\bar{x} g_0(x) \vee x g_1(x) = \bar{x} g_0(0) \vee x g_1(1)$$

and applying Theorem 1, one obtains:

$$\langle \bar{x}g_0(0) \vee xg_1(1); \qquad \bar{x}h_0 \vee xh_1 \rangle \nabla f.$$

The property then results from the fact that

$$g_0(0)g_1(1) \leqslant \bar{x}g_0(0) \vee xg_1(1) \tag{13}$$

and from the application of Lemma 1(a). Moreover, one knows that if g_0 and g_1 are maximal domain functions $\bar{x}g_0 \vee xg_1$ is also a maximal domain function. Since $g_0(x=0)g_1(x=1)$ is the meet difference (see Davio, Deschamps, and Thayse, 1978) of this last function, it is the greatest function degenerated in x and satisfying the inequality (13). □

Theorems 1 and 2 presented properties of two laws that were called *composition laws*. Remember (see Section 2.1) that composition laws were introduced in order to obtain the total description $\langle 1; f\rangle$ of a program given by its total description $\{\langle g_j; N_j\rangle\}$. Let us define the composition laws in a more complete way.

A composition law T_x^ℓ is a law acting on pairs of P-functions and satisfying the following properties.

(a) To the instruction labels M_0, M_1 and M of

$$\begin{array}{cccc} M & \bar{x} & \lambda & M_0 \\ & x & \lambda & M_1 \end{array}$$

one associates the P-functions $\langle g_0; h_0\rangle$, $\langle g_1; h_1\rangle$ and $\langle g_\ell; h\rangle$ respectively; then:

$$\begin{aligned} \langle g_0; h_0\rangle T_x^\ell \langle g_1; h_1\rangle &= \langle g_\ell(g_0, g_1, x); \qquad h_0\bar{x} \vee h_1 x\rangle \\ &= \langle g_\ell; h\rangle. \end{aligned} \tag{14}$$

(b) An iterated use of the law $T_{x_i}^\ell$, $x_i \in \mathbf{x}$, acting on $\{\langle g_{jk}; N_j\rangle\}$ with g_{jk}, Boolean functions satisfying

$$\bigvee_k g_{jk} = g_j, \qquad j = 0, 1, \ldots, q-1 \tag{15}$$

produces in at least one way a P-function have 1 as domain function.

Comments.

(a) A composition law produces a unique codomain function, i.e. $h_0\bar{x} \vee h_1 x$, since this function is uniquely determined by the interconnection of instructions as shown by the Figure 2b; a composition law may produce several domain functions since the function f may reduce to $h_0\bar{x} \vee h_1 x$ in several subdomains of the n-cube. One requires the law T_x^ℓ to be able to produce a domain function equal to 1 since a total description of f implies that the complete n-cube must be covered by a unique domain function.

(b) If f is a completely defined function, to the domain function 1 is necessarily associated the codomain function f, i.e. the final P-function obtained is $\langle 1; f\rangle$. If f is incompletely defined, the final P-functions are of the form $\langle 1; f_j\rangle$ where the f_j are completely defined functions compatible with f.

The above definition of composition law T_x^ℓ allows us to state the following general theorem on composition laws.

Theorem 3 (Theorem on composition law T_x^ℓ). *A composition law T_x^ℓ associates to a pair of P-functions $\langle g_0; h_0\rangle$ and $\langle g_1; h_1\rangle$ a P-function $\langle g_\ell, h\rangle$ with:*

$$\langle g_\ell, h\rangle = \langle g_0(x=0)g_1(x=1) \leq g_\ell \leq \bar{x}g_0 \vee xg_1; \qquad \bar{x}h_0 \vee xh_1\rangle. \tag{16}$$

Proof. The domain function g_ℓ must be smaller than its maximal element, i.e.: $\bar{x}g_0 \vee xg_1$ (see Theorem 1); moreover, any composition law should be able to produce a domain function equal to 1. The law T_x^ℓ must be such that:

$$\langle \bar{x}\alpha_0 \vee x\beta_0 \vee \gamma_0; \quad h_0\rangle T_x^\ell \langle \bar{x}\alpha_1 \vee x\beta_1 \vee \gamma_1; \quad h_1\rangle = \langle 1; \quad \bar{x}h_0 \vee xh_1\rangle$$

if and only if $\alpha_0\beta_1 = 1$. One verifies that the smallest composition law acting on x and which effectively produces the term $\alpha_0\beta_1$ is T_x^1. □

In summary, we defined two particular composition laws T_x^0 and T_x^1 (see Theorems 1 and 2):

$$T_x^0\colon \langle g_0; h_0\rangle T_x^0 \langle g_1; h_1\rangle = \langle g_0\bar{x} \vee g_1 x; \qquad h_0\bar{x} \vee h_1 x\rangle$$

$$T_x^1\colon \langle g_0; h_0\rangle T_x^1 \langle g_1; h_1\rangle = \langle g_0(x=0)g_1(x=1); \qquad h_0\bar{x} \vee h_1 x\rangle$$

and we proved (see Theorem 3) that any acceptable composition law T_x^ℓ should satisfy the relation:

$$T_x^\ell\colon \langle g_0; h_0\rangle T_x^\ell \langle g_1; h_1\rangle = \langle g_\ell; \qquad h_0\bar{x} \vee h_1 x\rangle,$$

with

$$g_0(x=0)g_1(x=1) \leq g_l \leq g_0\bar{x} \vee g_1 x.$$

From a practical point of view only the laws T_x^0 and T_x^1 present an interest. Moreover, when computing P-functions it appears that the use of the law T_x^0 leads to much more intricate computations than the use of the law T_x^1. In a computation process for generating optimal conditional programs we show (see Thayse, 1980) that the use of the law T_x^1 instead of the law T_x^0 results in possibly eliminating some (optimal) non-simple conditional programs. This fact may, for example, easily be deduced from Theorem 2 which shows that the law T_x^1 produces a domain function which no longer depends on x; hence this variable may not be tested again in a computation process.

2.3 Algorithms for generating optimal conditional programs

A general scheme for algorithms which have to generate optimal (the optimality criterion will be stated precisely further on) conditional programs may be stated in terms of P-functions as follows.

Step 1. *Give the list of all the P-functions:*

$$\langle g_i(\mathbf{x}); N_i \rangle$$

Step 2. *Compute by using the composition laws* T_{x_j} $\forall x_j$ *all the P-functions which may be obtained from the list of step* 1.

Step k. Compute by using the composition laws T_{x_j} $\forall x_j$ *all the P-functions which may be obtained from those obtained at any preceding step.*

If the relative cost criterion is considered the computation ends as soon as a step k' is reached where the P-function $\langle 1; f_N \rangle$ appears. If the P-function $\langle 1; f_N \rangle$ is obtained in several ways at the step k', to each of these ways is associated a conditional program which is optimal in time (since the step k' is the earliest step where $\langle 1; f_N \rangle$ appears and that the computation time of the program is $k'-1$). An elementary computation allows us then to detect those of these programs which are also optimal in cost. If the cost criterion is considered, the computation ends as soon as a step k'' is reached which contains only the P-function $\langle 1; f_N \rangle$. Again an elementary cost comparison between all the programs leading to the ultimate P-function $\langle 1; f_N \rangle$ allows us to detect those having a minimal cost. If an optimal program is, e.g., obtained at a step $k'' > k'$, this program is optimal in cost but has a longer computation time than the programs derived at the step k'.

Let us give some justification for these algorithms. First the generation of algorithms which satisfy either the cost criterion or the relative cost criterion requires that the reconvergent nodes should be detected at each step of the computation. This goal is reached by starting the computation with the P-functions $\langle g_i(\mathbf{x}); N_i' \rangle$. Indeed, the $g_i(\mathbf{x})$ are the greatest domain functions where f_N takes the value P_i and we know (see Theorems 1 and 2) that the maximal character of the domain function is preserved during the computations. Since in a program instructions may be merged only if they have the same domain functions, programs having the maximum number of reconvergent instructions are obtained from algorithms generating P-functions have the greatest domain functions.

The algorithm(s) ends at the obtention of the P-function $\langle 1; f_N \rangle$. Such a P-function may always be obtained after a finite number of steps since we easily check that the laws T_{x_i} produce a domain function 1 providing the initial domain functions g_i satisfy:

$$\bigvee_i g_i(\mathbf{x}) \equiv 1.$$

The use of the laws $T^0_{x_i}$ produces optimal conditional programs while the use of the laws $T^1_{x_i}$ produces optimal simple conditional programs. In order to illustrate the above algorithm we first give an elementary example.

Example 1. Consider the function:

$$f:\{0, 1\}^4 \to \{a, b, c, d, e, g\}$$

given by means of its truth table of Figure 3. The following set of P-functions is a total description of f:

$$A_0 = \langle \bar{x}_1\bar{x}_2;\, a\rangle, \qquad A_1 = \langle x_0x_2;\, b\rangle, \qquad A_2 = \langle \bar{x}_0x_3(x_1 \vee x_2);\, e\rangle,$$

$$A_3 = \langle x_1\bar{x}_3(\bar{x}_0 \vee \bar{x}_2);\, d\rangle, \qquad A_4 = \langle \bar{x}_0\bar{x}_1x_2\bar{x}_3;\, c\rangle, \qquad A_5 = \langle x_0x_1\bar{x}_2x_3;\, g\rangle.$$

By using the composition laws $T^1_{x_i}$ one obtains successively (T^1_i stands for $T^1_{x_i}$):

$$\begin{aligned}
B_0 &= A_4T^1_1A_3 = \langle \bar{x}_0x_2\bar{x}_3;\; c\bar{x}_1 \vee dx_1\rangle,\\
B_1 &= A_2T^1_0A_5 = \langle x_1\bar{x}_2x_3;\; e\bar{x}_0 \vee gx_0\rangle;\\
C_0 &= B_0T^1_3A_2 = \langle \bar{x}_0x_2;\; ex_3 \vee \bar{x}_3(c\bar{x}_1 \vee dx_1)\rangle,\\
C_1 &= A_3T^1_3B_1 = \langle x_1\bar{x}_2;\; x_3(e\bar{x}_0 \vee gx_0) \vee \bar{x}_3d\rangle;\\
D_0 &= C_0T^1_0A_1 = \langle x_2;\; \bar{x}_0[ex_3 \vee \bar{x}_3(c\bar{x}_1 \vee dx_1)] \vee x_0b\rangle,\\
D_1 &= A_0T^1_1C_1 = \langle \bar{x}_2;\; \bar{x}_1a \vee x_1[x_3(e\bar{x}_0 \vee gx_0) \vee \bar{x}_3d]\rangle;\\
E &= D_1T^1_2D_0 = \langle 1; f\rangle.
\end{aligned}$$

By starting from the same set of functions A_i and by using the composition laws T^0_i one obtains successively:

$$\begin{aligned}
B'_0 &= A_4T^1_1A_3 = \langle \bar{x}_3(x_1\bar{x}_0 \vee x_1\bar{x}_2 \vee \bar{x}_0x_2);\; c\bar{x}_1 \vee dx_1\rangle,\\
B'_1 &= A_2T^1_0A_5 = \langle x_3(x_1\bar{x}_0 \vee x_1\bar{x}_2 \vee \bar{x}_0x_2);\; e\bar{x}_0 \vee gx_0\rangle,\\
C' &= B'_0T^1_3B'_1 = \langle x_1\bar{x}_0 \vee x_1\bar{x}_2 \vee \bar{x}_0x_2;\; \bar{x}_3(c\bar{x}_1 \vee dx_1) \vee x_3(e\bar{x}_0 \vee gx_0)\rangle,\\
D'_0 &= C'T^1_0A_1 = \langle x_2 \vee \bar{x}_0x_1;\; \bar{x}_0[\bar{x}_3(c\bar{x}_1 \vee dx_1) \vee x_3e] \vee bx_0\rangle,\\
D'_1 &= A_0T^1_1C' = \langle \bar{x}_2 \vee \bar{x}_0x_1;\; x_1[\bar{x}_3d \vee x_3(e\bar{x}_0 \vee gx_0)] \vee a\bar{x}_1\rangle,\\
E &= D'_1T^1_2D'_0 = \langle 1; f\rangle.
\end{aligned}$$

The programs resulting from the use of the laws $T^1_{x_i}$ and $T^0_{x_i}$ are depicted by their corresponding flow charts given in Figures 4a and 4b respectively. We verify that the program of Figure 4a is an optimal simple conditional

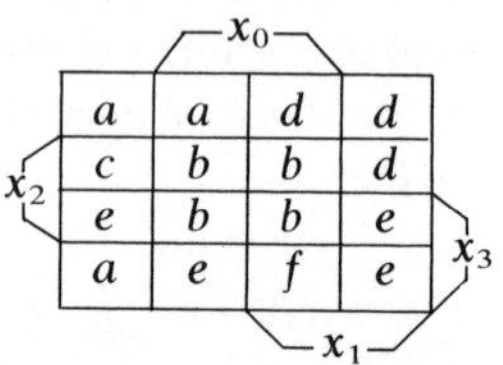

Figure 3. Truth table

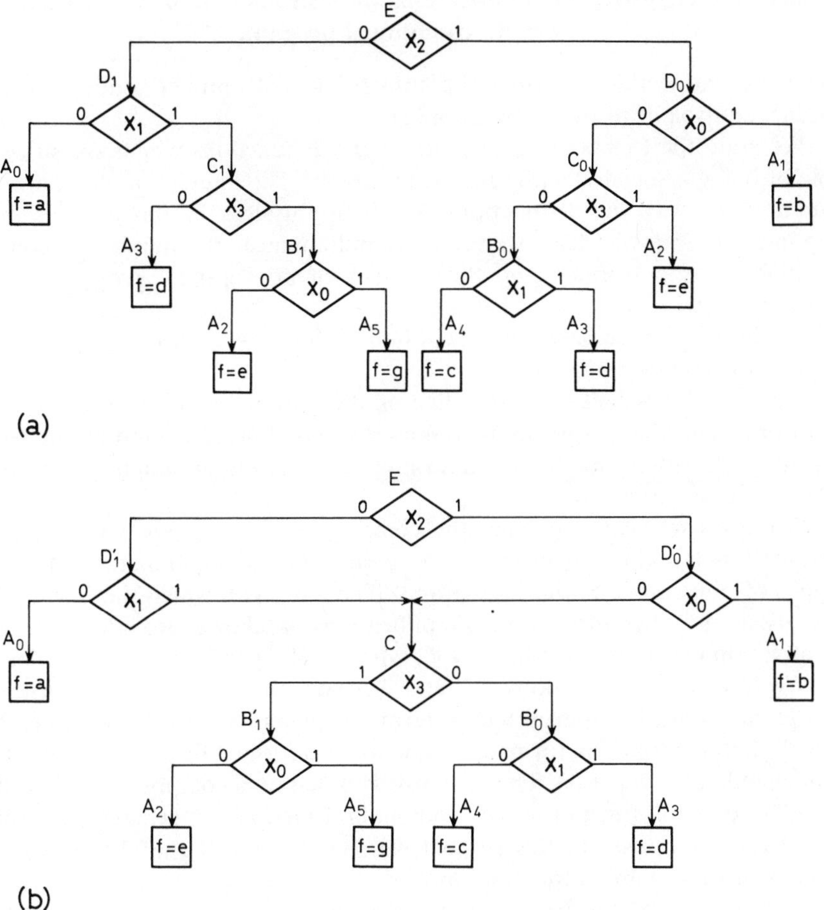

Figure 4. Optimal programs

program while the program of Figure 4b is an optimal non-simple conditional program.

We computed for the above example the only *P*-functions which led to the generation of the two optimal programs of Figure 4. Computing all the *P*-functions, as suggested by the algorithm, and retaining only those leading to optimal programs leads generally to an overwhelming number of computations.

In Section 2.4 we present a tabular type of algorithm for detecting optimal simple programs; this algorithm allows us to eliminate as soon as possible the *P*-functions which do not contribute to the generation of optimal programs. Hence the number of computations is generally drastically reduced with respect to the general type of algorithm presented in this section.

2.4 An algorithm for transforming an instruction into an equivalent simple conditional program

The present section describes the construction of optimal conditional programs equivalent to the instruction (1).

We state the functional properties of the P-functions which are susceptible of being associated with optimal programs. The interest of eliminating in an *a priori* way the P-functions which are useless in the generation of optimal programs is clear: we reduce simultaneously the number of computations to be performed and the number of non-optimal programs to be handled.

We impose on ourselves several kinds of limitations in the generation of optimal evaluation programs.

First of all we shall generally take as the optimality criterion the *relative cost criterion:* this allows us to reduce the number of computations to be performed, without modifying in a significant way the optimality property of the programs obtained.

We also exclusively consider the composition law T_x^1. We know that the use of this law (instead of the more general law T_x^0) eliminates only some types of non-simple optimal programs. The ease with which the law T_x^1 can be used and the number of simplifications it allows provide a definite motivation for eliminating the law T_x^0 when dealing with practical problems.

We have seen in Sections 2.2 and 2.3 how the synthesis of conditional programs could be interpreted in terms of generation of P-functions. We recall that a P-function (or pair of functions) is formed by a domain function and a codomain function. The synthesis is based on a composition law acting on the domain functions, the codomain function being used for some verification purpose. In this respect we shall represent the P-functions by means of their domain function only.

The purpose of the present section is to derive a computation method allowing us to obtain the domain functions associated with an optimal evaluation program in the simplest possible way. This method will consider in particular the following items.

(a) Elimination in an *a priori* way of the domain functions that cannot generate an optimal program.

(b) Elimination in the remaining domain functions of some terms that cannot generate any optimal program.

For the sake of clarity an example will be developed extensively in the course of this section.

Consider the two domain functions g_0 and g_1; let us write each of these functions as the disjunction of all its prime implicants and let us explicitly consider one of the variables x_i of $\mathbf{x}$:

$$g_j = \bar{x}_i g_j^0 \vee x_i g_j^1 \vee g_j^2, \qquad j = 0 \cdot 1, \tag{17}$$

where g_j^2 = disjunction of the prime implicants of g_j independent of x_i;

$x_i^{(k)}g_j^k$ = disjunction of the prime implicants of g_i dependent on $x_i^{(k)}$, $k=0, 1$ (remember that $x_i^{(0)}=\bar{x}_i$ and $x_i^{(1)}=x_i$).

We know (see Section 2.2) that the most general domain function obtained by composing g_0 and g_1 according to a law T_i^0 and with respect to the variable x_i is (for sake of simplicity $T_{x_i}^0$ and $T_{x_i}^1$ will be written T_i^0 and T_i^1, respectively):

$$\begin{aligned} g &= g_0 T_i^0 g_1 \\ &= \bar{x}_i g_0 \vee x_i g_i \vee g_0(x_i=0) g_1(x_i=1) \\ &= \bar{x}_i(g_0^0 \vee g_0^2) \vee x_i(g_1^1 \vee g_1^2) \vee, & \text{(a)} \\ &\quad g_0^0 g_1^2 \vee g_0^2 g_1^1 \vee g_0^2 g_1^2 \vee, & \text{(b)} \\ &\quad g_0^0 g_1^1. & \text{(c)} \end{aligned} \tag{18}$$

Part (a) of (18) is the contribution of the composition law T_i^0; if we restrict ourselves to parts (b) and (c) of (18) the composition law considered is T_i^1 which is able to generate any optimal simple program. We show that the restriction of the law T_i^1 to part (c) of (18) also generates all the simple optimal programs.

The P-function technique for generating (optimal) simple programs is based on the principle of obtaining new cube-functions at each step of the algorithm described below (see also Sections 4.2 and 4.3). These new cube-functions are contained in part (c) of (18): the cubes of g_0^2 and g_1^2 are already contained in g_0 and g_1.

The new cube-functions that may be generated by composition of g_0 with g_1 according to the law T_i^1 are the prime implicants of the functions:

$$g_0^0 g_1^1\text{: deriving from } g_0 T_i^1 g_1,$$
$$g_1^0 g_0^1\text{: deriving from } g_1 T_i^1 g_0.$$

The above rule is systematically used in Figure 7 where each of the variables x_i of $\mathbf{x}$ is succesively considered.

Example 2. Multiplication of two functional numbers written in the 2's complement notation by the modified Booth algorithm (Booth (1976)).

Let us first describe the proposed algorithm. Any fractional number $A=A'\varepsilon$ in which A' is an integer satisfying:

$$(-2^{2n-1}) \leqslant A' \leqslant 2^{2n-1}-1, \tag{19}$$

and where ε is given by

$$\varepsilon = 2^{-(2n-1)} \tag{20}$$

is represented in the 2's complement notation as

$$A \simeq a_0, a_1 a_2 \ldots a_{2n-1}, \tag{21}$$

with the interpretation

$$A = -a_0 + \sum_{i=1}^{2n-1} a_i 2^i. \tag{22}$$

An elementary arithmetic manipulation shows that it is always possible to represent A by

$$\begin{gathered} A \simeq \alpha_0 \alpha_1 \ldots \alpha_{n-1} \\ (\alpha_i \in \{-1, -\tfrac{1}{2}, 0, \tfrac{1}{2}, 1\}) \end{gathered} \tag{23}$$

with the interpretation:

$$A = \sum_{j=0}^{n-1} \alpha_j 2^{-2j}. \tag{24}$$

We have to choose:

$$\alpha_j = -a_{2j} + a_{2j+1} 2^{-1} + a_{2j+2} 2^{-1} \qquad (a_{2n} = 0). \tag{25}$$

The number A may also be written:

$$A = ((\ldots(\alpha_{n-1} 2^{-2} + \alpha_{n-2}) 2^{-2} + \ldots) 2^{-2} + \alpha_1) 2^{-2} + \alpha_0 \tag{26}$$

and the product AB is put in the form:

$$\begin{aligned} AB = ((\ldots(((0.2^{-2} + \alpha_{n-1} B) 2^{-2} + \alpha_{n-2} B) 2^{-2} + \alpha_{n-3} B) 2^{-2} \\ + \ldots) 2^{-2} + \alpha_1 B) 2^{-2} + \alpha_0 B. \end{aligned} \tag{27}$$

The multiplication algorithm grounded on the formula (27) is called a modified Booth algorithm. If R is a partial result, we may represent this algorithm by the diagram of Figure 5.

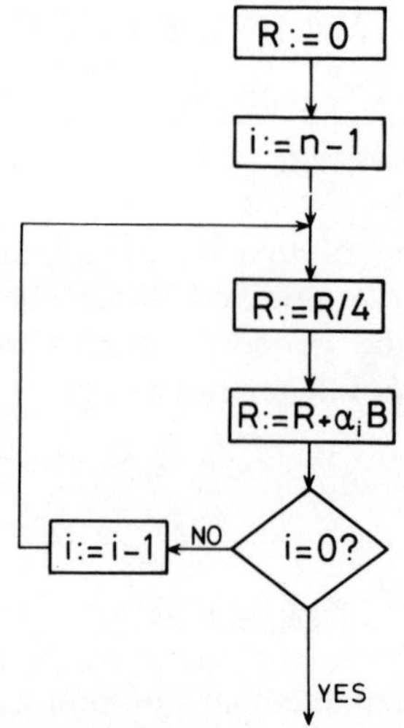

Figure 5. Modified Booth algorithm

We must first consider the realizability of the operation $R := R + \alpha_i B$, taking into account the fact that α_i may take five distinct values, namely $0, \pm\frac{1}{2}, \pm 1$. This shows that we have either $\pm B$ or $\pm B/2$. Assume that we have at our disposal ordinary shift registers realizing the usual shift operations, i.e. $R := 2R$ or $R := \lfloor R/2 \rfloor$. We must thus be able to realize the following sequences of commands:

$$R := R/2; \qquad R := R/2; \qquad R := R \pm B,$$
$$R := R/2; \qquad R := R/2; \qquad R := R \pm B/2,$$

which are respectively equivalent to:

$$R := R/4 \pm B,$$
$$R := R/4 \pm B/2.$$

These sequences can be realized as follows respectively:

$$R := R/2; \qquad R := R/2; \qquad R := R \pm B,$$
$$R := R/2; \qquad R := R \pm B; \qquad R := R/2.$$

Let us call x_2, x_1, and x_0 the three bits corresponding to a_{2j}, a_{2j+1}, and a_{2j+2}, respectively. The values of the α_i's and the corresponding operations are given in Figure 6. We define the commands as follows:

$$\tau_0\colon R := 0; \qquad \tau_1\colon i := n-1; \qquad \tau_2\colon i := i-1,$$
$$\sigma_0\colon R := R/2; \qquad \sigma_+\colon R := R + B; \qquad \sigma_-\colon R := R - B.$$

The modified Booth algorithm may then be described by the program:

$$
\begin{array}{llll}
N_0 & t & \tau_0\tau_1 & N_1 \\
N_1 & t & \sigma_3 & N_2 \\
N_2 & \bar{x}_2\bar{x}_1\bar{x}_0 \vee x_2x_1x_0 & \sigma_0 & N_3 \\
 & \bar{x}_2(\bar{x}_1x_0 \vee x_1\bar{x}_0) & \sigma_+\sigma_0 & N_3 \\
 & \bar{x}_2x_1x_0 & \sigma_0\sigma_+ & N_3 \\
 & x_2\bar{x}_1x_0 & \sigma_0\sigma_- & N_3 \\
 & x_2(\bar{x}_1x_0 \vee x_1\bar{x}_0) & \sigma_-\sigma_0 & N_3 \\
N_3 & \bar{z} & \lambda & N_4 \\
 & z & \tau_2 & N_1 \\
N_4 & \text{END} & &
\end{array}
\tag{28}
$$

The next step is the replacement of the program (28) by an optimal conditional program; the only instruction which has to be transformed is instruction N_2.

x_2	x_1	x_0	α	Operations		
0	0	0	0	$R := R/2;$	$R := R/2;$	λ
0	0	1	1/2	$R := R/2;$	$R := R+B;$	$R := R/2$
0	1	0	1/2	$R := R/2;$	$R := R+B;$	$R := R/2$
0	1	1	1	$R := R/2;$	$R := R/2;$	$R := R+B$
1	0	0	−1	$R := R/2;$	$R := R/2;$	$R := R-B$
1	0	1	−1/2	$R := R/2;$	$R := R-B;$	$R := R/2$
1	1	0	−1/2	$R := R/2;$	$R := R-B;$	$R := R/2$
1	1	1	0	$R := R/2;$	$R := R/2;$	λ

Figure 6. Value of the α_i's in the modified Booth algorithm

Algorithm.

Step 1. The five domain functions given by the disjunction of their prime implicants are:

$$A_0 = \bar{x}_2\bar{x}_1\bar{x}_0 \vee x_2x_1x_0,$$

$$A_1 = \bar{x}_2\bar{x}_1x_0 \vee \bar{x}_2x_1\bar{x}_0,$$

$$A_2 = \bar{x}_2x_1x_0,$$

$$A_3 = x_2\bar{x}_1\bar{x}_0,$$

$$A_4 = x_2\bar{x}_1x_0 \vee x_2x_1\bar{x}_0.$$

In Figure 7, $A_0, \ldots, A_4$ are successively written down with respect to the pairs of literals $\{\bar{x}_i, x_i\}\ \forall i = 0, 1, 2$. In the entry $\{A_j, \bar{x}_i\}$ of Figure 7 we place the coefficients of $\bar{x}_i$ of the prime implicants of A_j ($j = 0, 1$). These coefficients correspond to the functions g_0^0 and g_1^1 of (18c).

The successive domain functions are generated in the following way: the label B_0 in the entries $\{A_0, \bar{x}_0\}$ of Figure 7 means that the domain function B_0 is the product of the functions of these two entries. This function B_0 derives from the use of the control variable x_0 applied to A_0 and A_1.

The evaluation of each of the possible products between the entries corresponding to the rows $A_0 \ldots A_4$ produces the domain function B_i, $0 \leq i \leq 11$ of Figure 7.

Step 2. Obtain (as in step 1) all the possible products corresponding to the rows $A_0 \ldots A_4$, $B_0 \ldots, B_{11}$ of Figure 7. The products produce the domain functions labelled C_i, $0 \leq i \leq 5$.

Step 3. Obtain all the possible products corresponding to the rows A_i, B_j, and C_k; these products lead to the domain function $D = 1$. The tabular part of the algorithms ends when a domain function equal to 1 is obtained:

(a) in at least one pair $\{x_i, \bar{x}_i\}$ of columns if the relative cost criterion is adopted, and

(b) in all the pairs of columns $\{x_i, \bar{x}_i\}$ if the cost criterion is adopted.

Final step: Cost comparison. It remains to compare the costs (number of

	$\bar{x}_0$	x_0	$\bar{x}_1$	x_1	$\bar{x}_2$	x_2
$A_0=\bar{x}_2\bar{x}_1\bar{x}_0 \vee x_2x_1x_0$	$\bar{x}_2\bar{x}_1$; B_0	x_2x_1; B_3	$\bar{x}_2\bar{x}_0$; B_5	x_2x_0; B_7	$\bar{x}_1\bar{x}_0$; B_9	x_1x_0; B_{11}
$A_1=\bar{x}_2(\bar{x}_1x_0 \vee x_1\bar{x}_0)$	$\bar{x}_2x_1$; B_1	$\bar{x}_2\bar{x}_1$; B_0	$\bar{x}_2x_0$; B_6	$\bar{x}_2\bar{x}_0$; B_5	$\bar{x}_1x_0 \vee x_1\bar{x}_0$; B_{10}	
$A_2=\bar{x}_2x_1x_0$		$\bar{x}_2x_1$; B_1		$\bar{x}_2x_0$; B_6	x_1x_0; B_{11}	
$A_3=x_2\bar{x}_1\bar{x}_0$	$x_2\bar{x}_1$; B_4		$x_2\bar{x}_0$; B_8			$\bar{x}_1\bar{x}_0$; B_9
$A_4=x_2(\bar{x}_1x_0 \vee x_1\bar{x}_0)$	x_2x_1; B_3	$x_2\bar{x}_1$; B_4	x_2x_0; B_x	$x_2\bar{x}_0$; B_8		$\bar{x}_1x_0 \vee x_1\bar{x}_0$; B_{10}
$B_0=\bar{x}_2\bar{x}_1$			$\bar{x}_2$; C_2		$\bar{x}_1$; C_0	
$B_1=\bar{x}_2x_1$				$\bar{x}_2$; C_2	x_1; C_1	
$B_3=x_2x_1$				x_2; C_3		x_1, C_1
$B_4=x_2\bar{x}_1$			x_2; C_3			$\bar{x}_1$; C_0
$B_5=\bar{x}_2\bar{x}_0$	$\bar{x}_2$; C_2				$\bar{x}_0$; C_4	
$B_6=\bar{x}_2x_0$		$\bar{x}_2$; C_2			x_0; C_5	
$B_7=x_2x_0$		x_2; C_3				x_0; C_5
$B_8=x_2\bar{x}_0$	x_2; C_3					$\bar{x}_0$; C_4
$B_9=\bar{x}_1\bar{x}_0$	$\bar{x}_1$; C_0		$\bar{x}_0$; C_4			
$B_{10}=\bar{x}_1x_0 \vee \bar{x}_1x_0$	x_1; C_1	$\bar{x}_1$; C_0	x_0; C_5	$\bar{x}_0$; C_4		
$B_{11}=x_1x_0$		x_1; C_1		x_0; C_3		
$C_0=\bar{x}_1$			1; D			
$C_1=x_1$				1; D		
$C_2=\bar{x}_2$					1; D	
$C_3=x_2$						1; D
$C_4=\bar{x}_0$	1; D					
$C_5=x_0$		1; D				
$D=1$						

Figure 7. Optimization algorithm for the modified Booth algorithm

instructions or equivalently of domain functions) of the different programs leading to the domain function 1.

We verify that for the example there exist two programs with a minimal cost of six branching instructions. One of these programs is depicted in Figure 8.

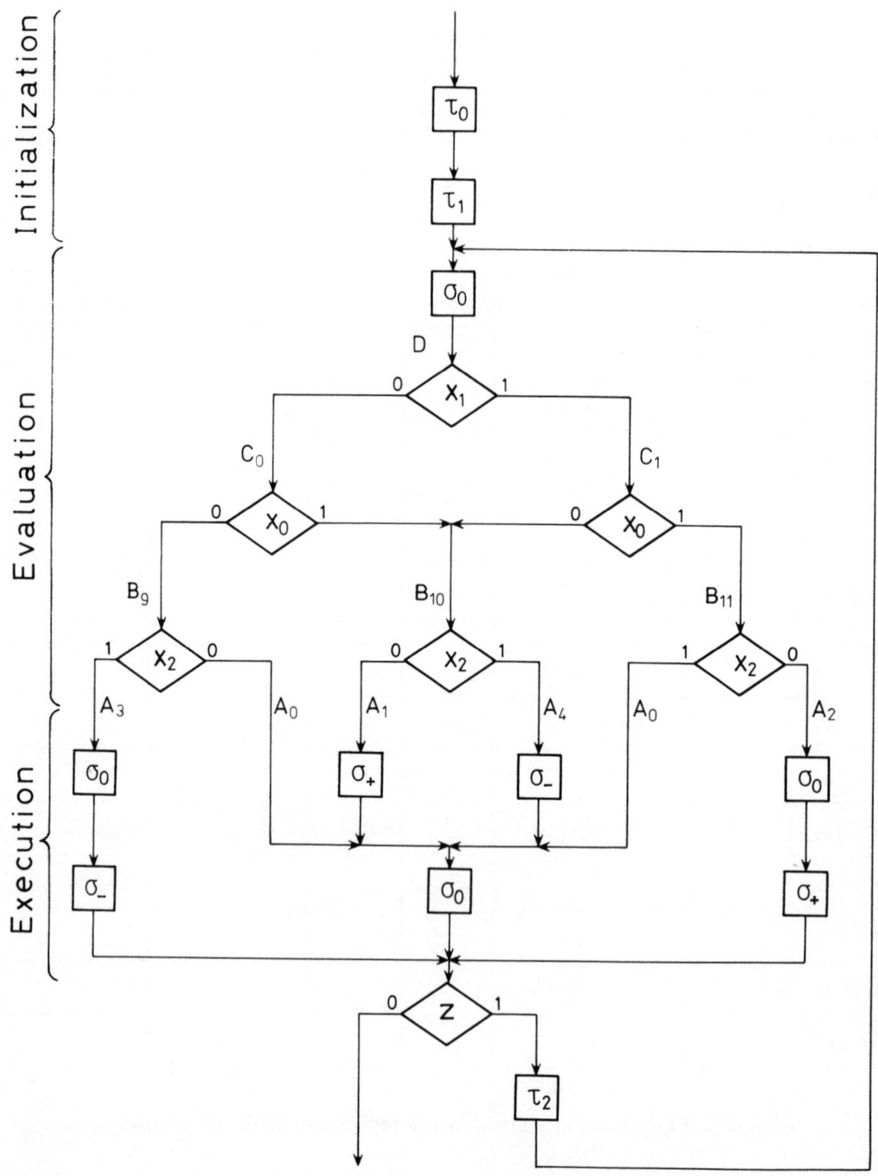

Figure 8. Optimal program for the modified Booth algorithm

From a practical point of view it may be recognized at once in Figure 7 that the optimal programs are those including the instruction B_{10}. Indeed we have seen that, from a cost point of view, it is important to recognize the reconvergent instructions in a program. These instructions are characterized by domain functions which are a sum of cubes. Between all the domain functions B_i, B_{10} is the only one which is a sum of cubes; it is thus the only reconvergent branching instruction in any conditional program (remember that the A_j's are domain functions associated with execution instructions).

Comments. The above algorithm assumes that the program is partitioned into two parts, namely an evaluation part followed by an execution part. If we assume, for example, that execution instructions may appear between

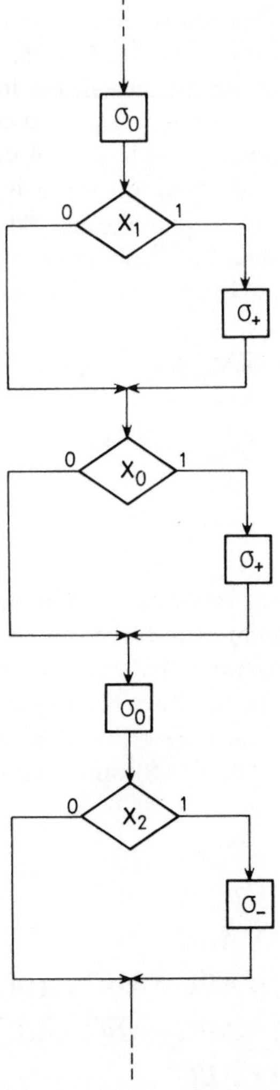

Figure 9. Modified program configuration for the modified Booth algorithm

two branching instructions, this more general algorithm scheme may in some cases result in a better program from a cost point of view.

In Example 2 we verify that the flow charts of Figures 8 and 9 correspond to programs which are (semantically) equivalent. The program of Figure 9 has a better cost than the program of Figure 8; however, the latter remains optimal as long as the computation duration is considered.

2.5 Transformation of a program into an equivalent simple conditional program

In the preceding sections we have considered the transformation of an instruction into an optimal conditional program. In the present section we shall consider the transformation of a program into an optimal conditional program. It is indeed clear that the optimization of a program will generally lead to a better result than the optimization of each of its instructions.

The program optimization algorithm will be described through the intermediate of the concept of P-function associated with a program. The concept of P-function associated with a program and the algorithm for generating optimal conditional programs equivalent to a program are both immediate generalizations of corresponding concept and algorithm introduced in Sections 2.2, 2.3, and 2.4, respectively.

Remember that a program is an indexed sequence of instructions of the type (1), i.e.:

$$P = N_0, N_1, \ldots, N_{(2^p-1)}$$

with

$$\begin{array}{llll} N_k & f_{0k} & \lambda & N_{0k} \\ & f_{1k} & \lambda & N_{1k} \\ & \cdots & \cdots & \cdots \\ & f_{(q-1)k} & \lambda & N_{(q-1)k}, \qquad 0 \leq k \leq 2^p - 1 \end{array} \tag{29}$$

Example 2 (continuation of Example 1, Chapter XII). The instructions (19, 20) of Chapter XII evaluate the functions S_0, C_0, S_1, C_1 of a controlled Add/Subtract cell (CAS). Assume that, according to the value of a binary parameter y, we want to evaluate either the functions S_0, C_0 or the functions S_1, C_1. This corresponds to the possibility of using the CAS cell either for addition or the subtraction. The CAS input/output relationship is specified by the following program:

$$\begin{array}{llll} N & \bar{y} & \lambda & N_0 \\ & y & \lambda & N_1 \end{array}$$

$$\begin{array}{llll} N_0 & f_{00}(ABC) = \bar{A}\bar{B}\bar{C} & \lambda & M_0 \\ & f_{10}(ABC) = A\bar{B}C \vee AB\bar{C} \vee \bar{A}BC & \lambda & M_1 \\ & f_{20}(ABC) = \bar{A}\bar{B}C \vee A\bar{B}\bar{C} \vee \bar{A}B\bar{C} & \lambda & M_2 \\ & f_{30}(ABC) = ABC & \lambda & M_3 \end{array}$$

$$
\begin{array}{lllll}
N_1 & f_{01}(ABC)=\bar{A}B\bar{C} & & \lambda & M_0 \\
 & f_{11}(ABC)=\bar{A}\bar{B}C \vee A\bar{B}\bar{C} \vee ABC & & \lambda & M_1 \\
 & f_{21}(ABC)=\bar{A}\bar{B}\bar{C} \vee AB\bar{C} \vee \bar{A}BC & & \lambda & M_2 \\
 & f_{31}(ABC)=A\bar{B}C & & \lambda & M_3
\end{array}
\tag{30}
$$

$$
\begin{array}{lllll}
M_0 & \mathrm{t} & \text{output}=00 & M & M \\
M_1 & \mathrm{t} & \text{output}=01 & M & M \\
M_2 & \mathrm{t} & \text{output}=10 & M & \\
M_3 & \mathrm{t} & \text{output}=11 & M &
\end{array}
$$

$M=$ next instruction

In view of the form of the two instructions N_0, N_1 (identical next instructions M_i) it is clear that a simultaneous synthesis of both instructions will always be preferable to a separate synthesis (e.g. in view of an optimization process). The simultaneous synthesis of N_0, N_1 will be obtained by introducing a system of P-functions associated with the pair of instructions N_0, N_1; it is defined as follows:

$$
\underbrace{\langle f_{k0},\ f_{k1}\quad ;}_{\text{domain function}}\quad \underbrace{\quad M_k\quad \rangle}_{\text{codomain function}}, \qquad 0 \leqslant k \leqslant 3,
$$

$$
\left.
\begin{array}{l}
A_0=\langle \bar{A}\bar{B}\bar{C},\ \bar{A}B\bar{C};\ M_0\rangle \\
A_1=\langle A\bar{B}C \vee AB\bar{C} \vee \bar{A}BC,\ \bar{A}\bar{B}C \vee A\bar{B}\bar{C} \vee ABC;\ M_1\rangle \\
A_2=\langle \bar{A}\bar{B}C \vee A\bar{B}\bar{C} \vee \bar{A}B\bar{C},\ \bar{A}\bar{B}\bar{C} \vee AB\bar{C} \vee \bar{A}BC;\ M_2\rangle \\
A_3=\langle ABC,\ A\bar{B}C;\ M_3\rangle
\end{array}
\right\}
\begin{array}{l}\text{initial system}\\ \text{of } P\text{-functions.}\end{array}
\tag{31}
$$

The domain functions being in this example pairs of functions, we define the composition law T_i as the componentwise extension of T_i^1. Starting with the initial system (31) of P-functions, we apply iteratively the law T_i^1 in order to obtain an ultimate system of two P-functions have $(1, -)$ and $(-, 1)$ as domain functions respectively ($-$ means an indeterminate value):

$$
\begin{array}{l}
B_0=A_2T_C^1A_1=\langle A\bar{B} \vee \bar{A}B,\ \bar{A}\bar{B} \vee AB;\ \bar{C}M_2 \vee CM_1\rangle, \\
B_1=A_0T_C^1A_2=\langle \bar{A}\bar{B},\ \bar{A}B;\ \bar{C}M_0 \vee CM_2\rangle, \\
B_2=A_1T_C^1A_3=\langle AB,\ A\bar{B};\ \bar{C}M_1 \vee CM_3\rangle, \\
C_0=B_0T_A^1B_2=\langle B,\ \bar{B};\ \bar{A}\bar{C}M_2 \vee (\bar{A}C \vee A\bar{C})M_1 \vee ACM_3\rangle, \\
C_1=B_1T_A^1B_0=\langle \bar{B},\ B;\ \bar{A}\bar{C}M_0 \vee (\bar{A}C \vee A\bar{C})M_2 \vee ACM_1\rangle, \\
\left.
\begin{array}{l}
D_0=C_1T_B^1C_0=\langle 1, 0;\ \Sigma f_{j_0}M_j\rangle,\ 0 \leqslant j \leqslant 3 \\
D_1=C_0T_B^1C_1=\langle 0, 1;\ \Sigma f_{j_1}M_j\rangle,\ 0 \leqslant j \leqslant 3
\end{array}
\right\}
\begin{array}{l}\text{ultimate sytem}\\ \text{of } P\text{-functions.}\end{array}
\end{array}
\tag{32}
$$

To the domain functions $(1, 0)$ and $(0, 1)$ correspond the codomain functions $\Sigma f_{j_0}M_j$ and $\Sigma f_{j_1}M_j$, respectively; this derives from (31) and from the componentwise application of the laws T_i^1. The conditional program

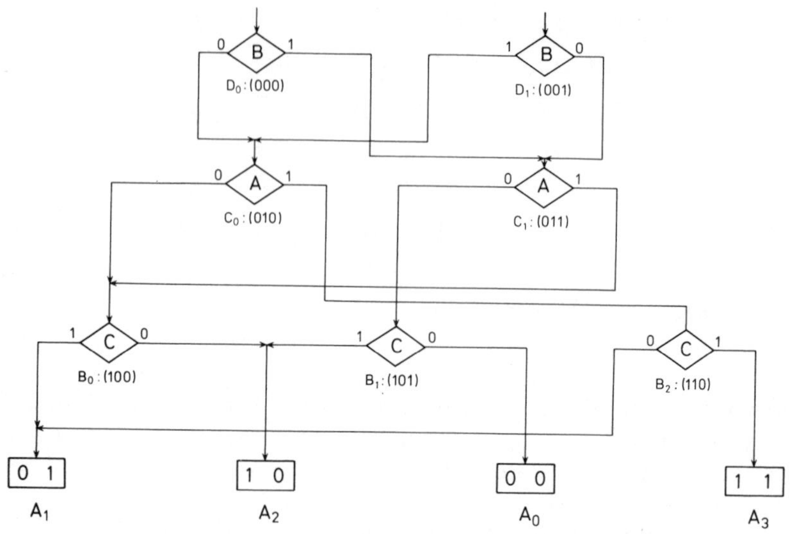

Figure 10. Conditional program describing a CAS cell

corresponding to the tranformation: system (31) → system (32) is depicted in Figure 10. If we start this program with the instruction labelled D_0 it computes (S_0, C_0) while if we start it with the instruction labelled D_1 it computes (S_1, C_1). Finally it can be verified that the program of Figure 10 is optimal both in compuation time and in cost. This program was derived by means of a tabular method which is the componentwise extension of the method proposed in Section 2.4. The computations are gathered in Figure 11.

It is shown in Baranov and Keevallik (1981) that the program synthesis and optimization may be divided into several independent subproblems by partitioning the set of instructions into subsets. The optimal conditional program for the set of instructions results from the optimal conditional programs for all the subset of instructions.

To partition the set of instructions (29) write in a left-hand column and in a right-hand column the instruction labels $\{N_k\}$ and $\{N_{jk}\}$ which appear at the left and the right of the instructions (29) respectively (see Figure 12). The arrow from N_k to N_{jk} is marked f_{jk}; $f_{jk}(\mathbf{x})$ is the Boolean function in (29) which indicates that to the instruction N_k is associated the next instruction N_{jk} for $f_{jk}(\mathbf{x}) = 1$. Draw arrows $\forall\{k, jk\}$ with the convention that the absence of an arrow between two labels corresponds to a Boolean function identically zero. This construction defines a partition on the set of instructions into blocks; these blocks are shown in Figure 12. Now it is not difficult to prove that if $P = \{N_k\}$ is partitioned into blocks the synthesis and optimization of P by means of a conditional program reduces to a separate

	$\bar{A}$	A	$\bar{B}$	B	$\bar{C}$	C
$A_0 = \bar{A}\bar{B}\bar{C}, \bar{A}B\bar{C}$	$\bar{B}\bar{C}, B\bar{C}$		$\bar{A}\bar{C}, -$	$-, \bar{A}\bar{C}$:	$\bar{A}\bar{B}, \bar{A}B; B_1$	
$A_1 = A\bar{B}C \vee AB\bar{C} \vee \bar{A}BC, \bar{A}\bar{B}C \vee A\bar{B}\bar{C} \vee ABC$	$BC, \bar{B}C$	$\bar{B}C \vee C\bar{C}, \bar{B}\bar{C} \vee BC$	$\bar{A}C, \bar{A} \vee A\bar{C}$	$A\bar{C} \vee \bar{A}C, AC$	$AB, A\bar{B}; B_2$	$A\bar{B} \vee \bar{A}B, \bar{A}\bar{B} \vee AB; B_0$
$A_2 = \bar{A}\bar{B}C \vee A\bar{B}\bar{C} \vee \bar{A}B\bar{C}, \bar{A}\bar{B}\bar{C} \vee AB\bar{C} \vee \bar{A}BC$	$\bar{B}C, B\bar{C}, \bar{B}\bar{C}, BC$	$\bar{B}\bar{C}, B\bar{C}$	$\bar{A}C \vee A\bar{C}, \bar{A}\bar{C}$	$\bar{A}\bar{C}, A\bar{C} \vee \bar{A}C$	$A\bar{B} \vee \bar{A}B, \bar{A}\bar{B} \vee AB; B_0$	$\bar{A}\bar{B}, \bar{A}B; B_1$
$A_3 = ABC, A\bar{B}C$		$BC, \bar{B}C$	$-, AC$	$AC, -$		$AB, A\bar{B}; B_2$
$B_0 = A\bar{B} \vee \bar{A}B, \bar{A}\bar{B} \vee AB$	$B, \bar{B}; C_0$	$\bar{B}, B; C_1$				
$B_1 = \bar{A}\bar{B}, \bar{A}B$	$\bar{B}, B; C_1$					
$B_2 = AB, A\bar{B}$		$B, \bar{B}; C_0$				
$C_0 = \bar{B}, B$			$1, -; D_0$	$-, 1; D_1$		
$C_1 = B, \bar{B}$			$-, 1; D_1$	$1, -; D_0$		
$D_0 = 1, -$						
$D_1 = -, 1$						

Figure 11. Optimization algorithm for the CAS cell

synthesis and optimization of each of its blocks (see Baranov and Keevallik, 1981).

Consider the block A of Figure 12; to the set of instructions $N_0, N_1, \ldots, N_{m-1}$ we associate the following set of P-functions:

$$\left.\begin{array}{l}\langle f_{00}, \quad f_{10}, \quad \ldots, f_{(m-1)0}; \quad N_0\rangle \\ \langle f_{01}, \quad f_{11}, \quad \ldots, f_{(m-1)1}; \quad N_1\rangle \\ \ldots \\ \langle \underbrace{f_{0(r-1)}, f_{1(r-1)}, \ldots, f_{(m-1)(r-1)}}_{\text{domain functions}}; \underbrace{N_{r-1}}_{\text{codomain functions}}\rangle \end{array}\right\} \begin{array}{l}\text{initial system}\\ \text{of } P\text{-functions}\end{array} \qquad (33)$$

To a block of m instructions $\{N_k\}$ with r next instruction labels $\{N_\ell\}$ we associate thus a system of r P-functions with m domain functions. The

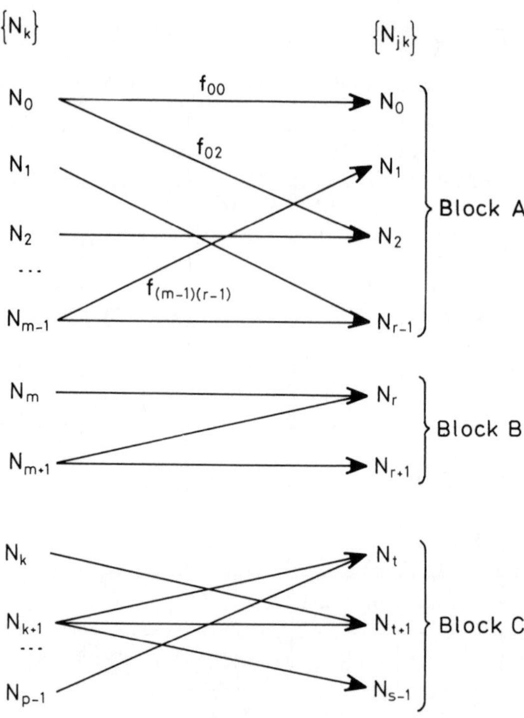

Figure 12. Partition on the set of instructions into blocks

domain functions $\{f_{k\ell}\}$ satisfy the relations:

$$\bigvee_{\ell} f_{k\ell} \equiv 1, \qquad 0 \leq \ell \leq r-1, \quad \forall k,$$
$$f_{k\ell} f_{k\ell'} = 0 \; \forall \ell \neq \ell'. \tag{34}$$

We extend componentwise the composition laws T_i for P-functions having m domain functions

Theorem 4. *Starting with the initial system* (33) *of r P-functions, the iterative generation of P-functions by means of laws T_i produces in at least one way the ultimate system* (35) *of m P-functions:*

$$\left.\begin{array}{l} \langle 1, -, \ldots, -; \vee f_{0\ell} N_\ell \rangle \\ \langle -, 1, \ldots, -; \vee f_{1\ell} N_\ell \rangle \\ \ldots \quad \ldots \\ \langle -, -, \ldots, 1; \vee f_{(m-1)\ell} N_\ell \rangle \end{array}\right\} \text{ultimate system of } P\text{-functions} \tag{35}$$

To the transformation (33) → (35) *is associated a conditional program performing the set of instructions* $\{N_0, N_1, \ldots, N_{m-1}\}$.

Proof. Since the functions $\{f_{k\ell}\}$ satisfy the relations (32) it is always possible to obtain, by means of an iterative application of the laws T_i, a P-function having a 1 in its jth domain component:

$$\langle -, -, \ldots, 1_j, \ldots, -; \vee f_{j\ell} N_\ell (0 \leqslant \ell \leqslant r-1) \rangle. \tag{36}$$

The corresponding codomain function is $\vee f_{j\ell} N_\ell$ which describes the input–output relation of instruction N_j. In view of Theorem 2, to the transformation (33) → (35) is associated a conditional program performing the instruction N_j. Hence to the transformation (33) → (35) is associated a conditional program performing the program $P = \{N_0, N_1, \ldots, N_{m-1}\}$.

To the transformation (33) → (35), obtained with a minimum number of application of the laws T_i, corresponds the generation of an optimal conditional program. □

Example 2 (continued). Assume that, according to the value of two binary parameters x, y, we want to evaluate one of the functions: S_0, C_0, S_1, C_1 of the CAS. Its input–output relationship is specified by the following program:

$$\begin{array}{llll}
N & \bar{x}\bar{y} & \lambda & N_0 \\
 & \bar{x}y & \lambda & N_1 \\
 & x\bar{y} & \lambda & N_2 \\
 & xy & \lambda & N_3
\end{array}$$

$$\begin{array}{llll}
N_0 & f_{00}(ABC) = \bar{A} \oplus B \oplus C & \lambda & M_0 \\
 & f_{10}(ABC) = A \oplus B \oplus C & \lambda & M_1 \\[1ex]
N_1 & f_{01}(ABC) = \overline{AB \vee BC \vee AC} & \lambda & M_0 \\
 & f_{11}(ABC) = AB \vee BC \vee AC & \lambda & M_1 \\
N_2 & f_{02}(ABC) = A \oplus B \oplus C & \lambda & M_0 \\
 & f_{12}(ABC) = \bar{A} \oplus B \oplus C & \lambda & M_1 \\[1ex]
N_3 & f_{03}(ABC) = \overline{A\bar{B} \vee AC \vee \bar{B}C} & \lambda & M_0 \\
 & f_{13}(ABC) = A\bar{B} \vee AC \vee \bar{B}C & \lambda & M_1 \\
M_0 & \text{t} \;\; \text{output} = 0 \;\; M & & \\
M_1 & \text{t} \;\; \text{output} = 1 \;\; M & & \\
\multicolumn{4}{l}{M = \text{next instruction}}
\end{array} \tag{37}$$

The conditional program which simultaneously implements the instructions N_0, N_1, N_2, N_3 is obtained from the transformation between systems

(38) and (39) of P-functions:

$$\left.\begin{aligned} A_0&=\langle f_{00}, f_{01}, f_{02}, f_{03}; M_0\rangle \\ A_1&=\langle f_{10}, f_{11}, f_{12}, f_{13}; M_1\rangle \end{aligned}\right\} \text{initial system of } P\text{-functions;} \qquad (38)$$

$$\begin{aligned}
B_0&=A_0T_CA_1=\langle AB\vee\bar{A}\bar{B},\ A\bar{B}\vee\bar{A}B,\ A\bar{B}\vee\bar{A}B,\ AB\vee\bar{A}\bar{B};\ \bar{C}M_0\vee CM_1\rangle,\\
B_1&=A_1T_CA_0=\langle A\bar{B}\vee\bar{A}B,\ -,\ AB\vee\bar{A}\bar{B},\ -;\ \bar{C}M_1\vee CM_0\rangle;\\
C_0&=B_0T_AB_1=\langle \bar{B},\ -,\ B,\ -;\ (\bar{A}\bar{C}\vee AC)M_0\vee(\bar{A}C\vee A\bar{C})M_1\rangle,\\
C_1&=B_1T_AB_0=\langle B,\ -,\ \bar{B},\ -;\ (A\bar{C}\vee\bar{A}C)M_0\vee(AC\vee\bar{A}\bar{C})M_1\rangle,\\
C_2&=B_0T_AA_1=\langle \bar{B}\bar{C},\ B,\ B\bar{C},\ \bar{B};\ \bar{A}\bar{C}M_0\vee(A\vee C)M_1\rangle,\\
C_3&=A_0T_AB_0=\langle BC,\ \bar{B},\ \bar{B}C,\ B;\ (\bar{A}\vee\bar{C})M_0\vee ACM_1\rangle;
\end{aligned}$$

$$\left.\begin{aligned}
D_0&=C_0T_BC_1=\langle 1, 0, 0, 0;\ f_{00}M_0\vee f_{10}M_1\rangle\\
D_1&=C_3T_BC_2=\langle 0, 1, 0, 0;\ f_{01}M_0\vee f_{11}M_1\rangle\\
D_2&=C_1T_BC_0=\langle 0, 0, 0, 1;\ f_{02}M_0\vee f_{12}M_1\rangle\\
D_3&=C_2T_BC_3=\langle 0, 0, 0, 1;\ f_{03}M_0\vee f_{13}M_1\rangle
\end{aligned}\right\} \begin{array}{l}\text{ultimate system}\\ \text{of } P\text{-functions}\end{array} \qquad (39)$$

This program is depicted by the flow chart of Figure 13. To the entries D_0, D_1, D_2, and D_3 of this program correspond the evaluation of the functions S_0, C_0, S_1, and C_1 respectively. It could easily be shown that this conditional program is optimal in both the cost and the computation time.

Remark. Hardware realization of evaluation programs. As a conclusion of this section we present briefly some relations that exist between the software

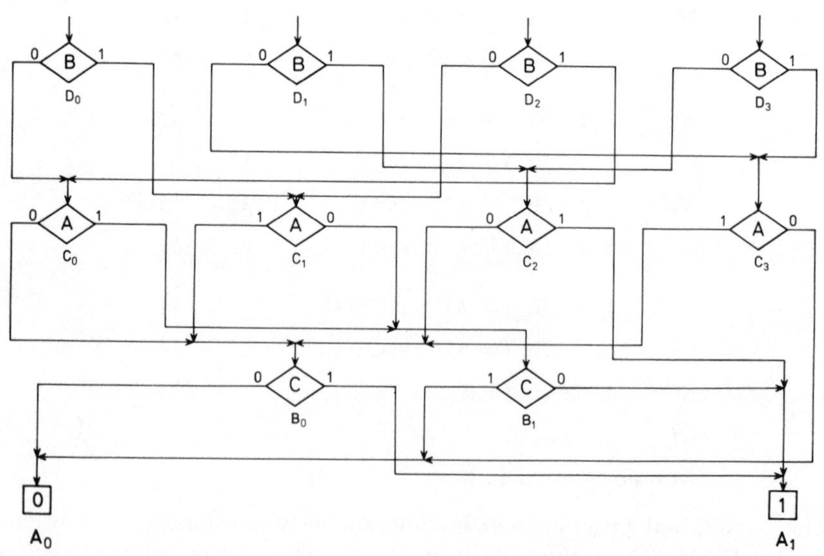

Figure 13. Conditional program describing the functions of a CAS cell

description (as depicted, for example, by the flow chart of Figure 1) of an evaluation program and its hardware realization by means of either multiplexers of demultiplexers (see also Chapter III).

Remember that a binary demultiplexer of one control variable x is a logic gate with one input terminal y and two output terminals z_0, z_1 realizing the functions:

$$z_0 = \bar{x}y \quad \text{and} \quad z_1 = xy.$$

A multiplexer of one binary control variable x is a logic gate with two input terminals y_0, y_1 and one output terminal z realizing the function:

$$z = \bar{x}y_0 \vee xy_1.$$

Theorem 5 (Hardware implementation of programs by means of multiplexers). *If a logical network is deduced from the conditional program of Figure* 1

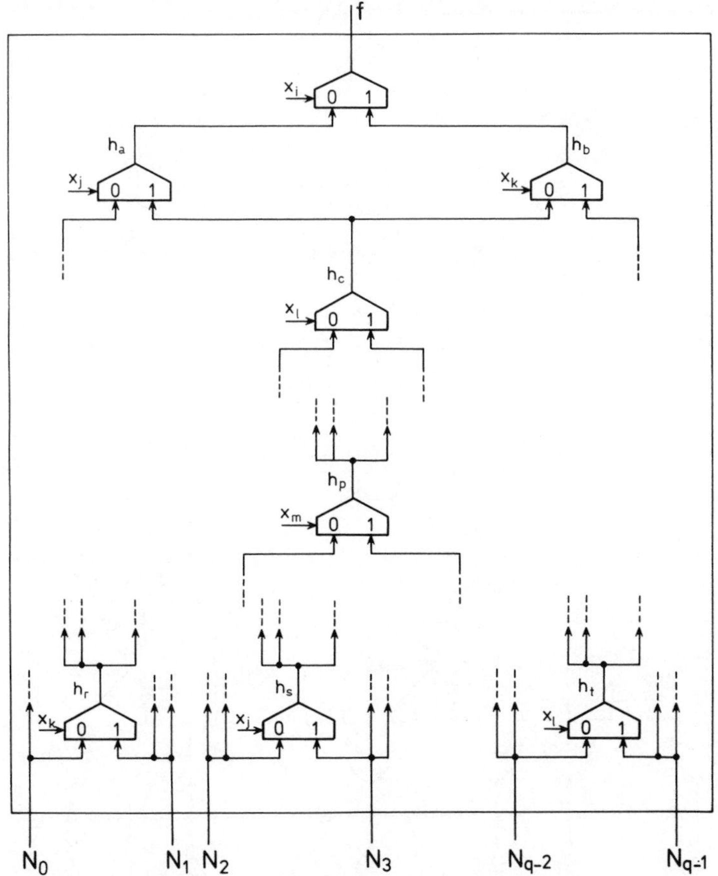

Figure 14. Hardware realization of a program by means of multiplexers

in the following way:

(a) *to each instruction M of the program corresponds a multiplexer;*

(b) *the next instructions to M corresponding to the values* 0, 1 *of the control variable* x_i *are connected to the inputs of the multiplexer (with identical control variable* x_i*); and*

(c) *the inputs of the initial multiplexers (corresponding to the instructions whose P-functions are* $\langle g_i\,;N_i\rangle$*) are connected to the instruction labels* N_i *interpreted here as discrete constants, then it realizes the function* $f_N = \bigvee g_i N_i$.

Moreover, the multiplexer corresponding to the instruction M having $\langle g; h\rangle$ *as P-function realizes the function h.*

Proof. The proof immediately derives from the fact that the functions of the

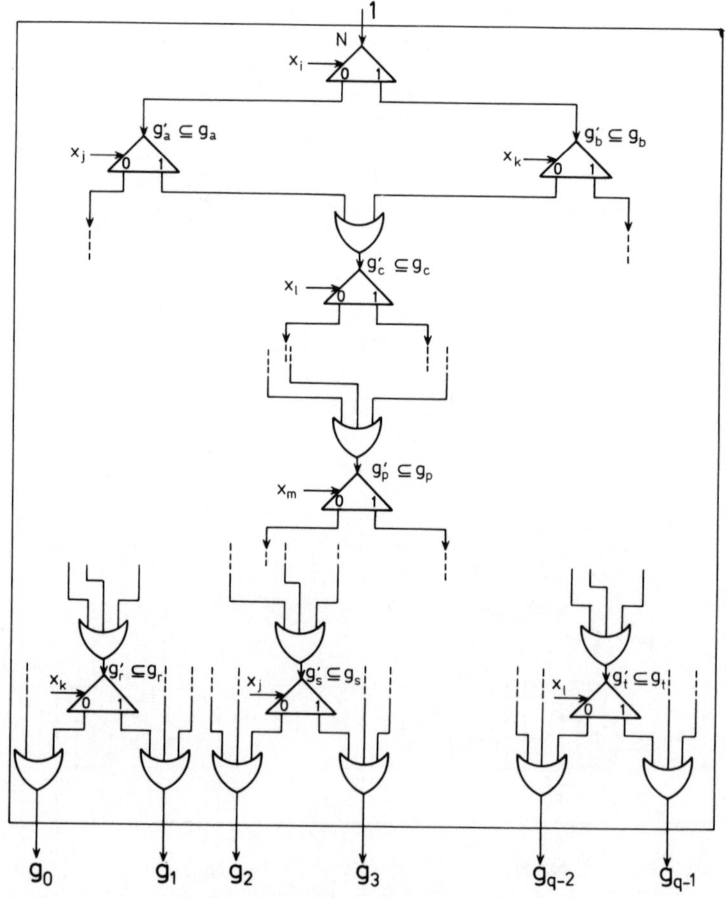

Figure 15. Hardware realization of a program by means of demultiplexers and OR-gates

P-functions are generated by the composition law:

$$h = h_0\bar{x} \vee h_1 x,$$

which corresponds to the output function z of a multiplexer. □

The hardware realization by means of multiplexers of the program of Figure 1 is depicted in Figure 14.

Theorem 6 (Hardward implementation of programs by means of demultiplexers and OR-gates; see Figures 1 and 15). *If a logical network is deduced from the conditional program of Figure* 1 *in the following way:*

(a) *to each instruction M of the program corresponds a system: demultiplexer-OR-gate; the demultiplexer input terminal is connected to the OR-gate output terminal;*

(b) *the next instructions to M corresponding to the values* 0, 1, *of the control variable x_i are the outputs of the demultiplexer respectively: the inputs to the OR-gate correspond to the instructions preceding M (if there is a single instruction preceding M, the OR-gate may be dropped and this instruction is connected to the demultiplexer input), and*

(c) *the inputs of the initial demultiplexer (corresponding to the instruction whose P-function is $\langle 1; f_N\rangle$) is connected to the Boolean constant* 1, *then it realizes the functions $\{g_i\}$ and the OR-gate corresponding to the P-function $\langle g_\ell, h_\ell\rangle$ realizes g'_ℓ, $g'_\ell \subseteq g_\ell$ as output function.*

Proof. Identical to that of Theorem 5. □

The hardware realization by means of demultiplexers and OR-gates of the program of Figure 1 is depicted in Figure 15.

BIBLIOGRAPHICAL REMARKS

Much advanced research on program transformation and on implementation and transformation of algorithms based on automata has been published by Cerny, Davio, Deschamps, Mange, Sanchez, Stauffer, and Thayse: Cerny, Mange and Sanchez (1979), Mange (1979, 1980, 1981), Stauffer (1980), Mange, Sanchez and Stauffer (1982), Deschamps (1981), Thayse (1980, 1981a, 1981b, 1981c, 1982), Sanchez and Thayse (1981), and Davio and Thayse (1978, 1980). Observe finally that the Gluskhov model is generally known in the English literature as the 'Algorithmic state machine' (Clare, 1973).

EXERCISES

E1. Amplitude comparision has been studied in Chapter IV and in Exercise E2 of Chapter XII. Use the optimization techniques described in this chapter to obtain an optimal binary program for amplitude comparison. Derive from that program

combinational realizations of amplitude comparison using
(i) multiplexers, and
(ii) demultiplexers and OR gates.

E2. It is possible to generalize the Booth algorithm for multiplication. Consider indeed the integer A having the np-bit representation:

$$a_{np-1}\ldots a_{(n-1)p}\vdots\cdots\vdots\, a_{2p-1}\ldots a_{p+1}a_p \vdots\, a_{p-1}\ldots a_1 a_0.$$

Thus

$$A=-a_{np-1}2^{np-1}+\sum_{i=0}^{np-2} a_i 2^i.$$

Put $a_{-1}=0$ and

$$\alpha_j=-a_{(j+1)p-1}2^{p-1}+\sum_{i=0}^{p-2} a_{jp+i}2^i+a_{jp-1}. \tag{2}$$

Hence $\alpha_j\in\{0,\pm1,\pm2,\ldots,\pm2^{p-1}\}$ and

$$A=\sum_{j=0}^{n-1}\alpha_j 2^{pj}.$$

(For $p=1$ and 2, the above formulae describe the Booth algorithm and the modified Booth algorithm, respectively.) The Horner scheme describing the multiplication $A\cdot B$ is then described by

$$A\cdot B=(\ldots((0\cdot 2^p+\alpha_{n-1}B)2^p+\alpha_{n-2}B)2^p+\ldots+\alpha_1 B)2^p+\alpha_0 B,$$

i.e. its typical step is of the form:

$$R:=R\cdot 2^p+\alpha_j B. \tag{1}$$

Once again, in a context where simple shifts only are available, it is possible to intermesh the left shifts of R and the additions of B to reach each the appropriate result. Let indeed

$$\sigma: R:=2R,$$
$$\tau: R:=R+B.$$

Then, the product transfer

$$\sigma\tau^{r_{p-1}}\sigma\tau^{r_{p-2}}\ldots\sigma\tau^{r_i}\ldots\sigma\tau^{r_1}\sigma\tau^{r_0}.$$

realizes the operation:

$$\begin{aligned}R&:=(((R\cdot 2+r_{p-1}B)2+r_{p-2}B)2+\ldots+r_1B)2+r_0B,\\ R&:=2^pR+(2^{p-1}r_{p-1}+\ldots+2r_1+r_0)B.\end{aligned} \tag{3}$$

Using the optimization techniques developed in this chapter, obtain optimal binary decision algorithms for $p=3$ and $p=4$. Can you infer a general rule? The above presentation, however, suggests an important remark: it becomes obvious that, while the transfer $R:=R\cdot 2^p+\alpha_j B$ by an equivalent conditional program, it is not mandatory to maintain a separation between the evaluation and the excecution phases. Replace in (1) the quantity α_j by its value (2) and express your result under a form similar to (3). Describe the associated conditional program and compare it to the previous program.

Chapter XIV
Special purpose control units

1 INTRODUCTION

In the preceding chapters we introduced the concept of *type of instruction.* We encountered a situation in which a specific property of the type of instruction, namely its conditional character, was reflected in a modification of the control automaton hardware structure so as to become an effective parameter of a speed–size trade-off.

The concept of instruction type will play a key role in the present chapter; we first give an interpretation of this concept in terms of finite automata. Until now we have identified the concepts of instruction and of internal state. In this terminology, to classify the instructions into types restricts us to defining a partition on the set of internal states of the finite automaton. The membership of an internal state to a given block of this partition, i.e. the knowledge of the instruction type, determines some restrictions on the next state function and on the output function. We know, for example, that in the case of conditional programs in the restricted sense the branching instructions are characterized by a command signal λ (λ = empty command). It is the introduction of the concept of instruction type which allows us to reduce the number of fields in the ROM when implementing a conditional program in the restricted sense (field merging technique).

In this chapter our purpose is to take advantage of the notion of instruction type in order to decompose the combinatorial network corresponding to the transition function of the control automaton.

This principle is first illustrated in Section 2 by the introduction of the concept of the incremental program; it is a program whose branching instructions are of the form:

$$\begin{array}{llll} N_i & \bar{x} & \lambda & N_{i+1} \\ & x & \lambda & N_j \end{array}$$

Clearly, a specialized combinatorial network allows us to compute the next instruction N_{i+1} and it is no longer necessary to memorize this address in the ROM. Figure 1 roughly illustrates this view by introducing a possible type of architecture corresponding to the above instruction.

Section 3 is devoted to programs in which the branching conditions are

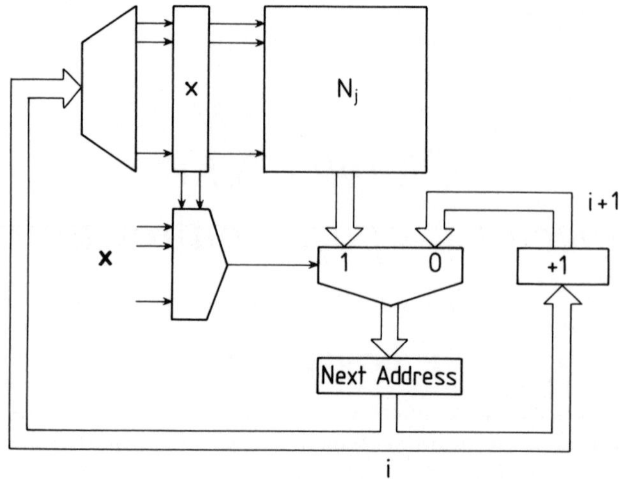

Figure 1. Type of program implementation

fixed by a reduced number of condition variables. Program transformations may be performed which yield several types of implementation. Cost upper bounds for these implementations are derived.

Subroutines are programs used by other programs to accomplish a particular task. Frequently, many programs contain identical sections of partial computations. Instructions can be saved by employing subroutines which perform these partial computations. Programs that use subroutines are described in Section 4.

2 INCREMENTAL PROGRAMS

2.1 The instruction set and its associated architecture

We call an *incremental program* a program written using only one of the three types of instructions:

(a) Conditional jump:

$$\begin{array}{llll} N_i & \bar{x} & \lambda & N_{i+1} \\ & x & \lambda & N_j; \end{array} \tag{1}$$

(b) Unconditional jump:

$$N_i \quad t \quad \lambda \quad N_j; \tag{2}$$

(c) Execution:

$$N_i \quad t \quad \sigma \quad N_{i+1}. \tag{3}$$

In order to state the universality of incremental programs it is sufficient to show that the typical branching instruction ((6a) of Chapter XII) of a conditional program may be replaced by an equivalent incremental program.

Let $\{N_i\}$ be the set of instructions of the initial conditional program and $\{M_j\}$ be the set of instructions of the equivalent incremental program. In order to construct this equivalent incremental program we first associate to each instruction N_i one and only one instruction $M_j = \psi(N_i)$: it is the initial instruction of the incremental program equivalent to N_i. Putting $M_i = \psi(N_i)$, $M_j = \psi(N_j)$ and $M_k = \psi(N_k)$, we may replace the instruction:

$$\begin{array}{llll} N_i & \bar{x} & \sigma_0 & N_j \\ & x & \sigma_1 & N_k, \end{array} \tag{4}$$

by the equivalent subprogram:

$$\begin{array}{llll} M_i & \bar{x} & \lambda & M_{i+1} \\ & x & \lambda & M_\ell \\ M_{i+1} & \mathrm{t} & \sigma_0 & M_{i+2} \\ M_{i+2} & \mathrm{t} & \lambda & M_j \\ M_\ell & \mathrm{t} & \sigma_1 & M_{\ell+1} \\ M_{\ell+1} & \mathrm{t} & \lambda & M_k \end{array} \tag{5}$$

We could describe in a similar way the transformation of a conditional program in the restricted sense into an incremental program. We observe that in both cases the number of instructions of the equivalent incremental program increases in a relatively important way with respect to the number of instructions of the initial program.

Consider again an initial program, with disjoint predicates and one instruction per row, whose typical instruction is (see also (5), Chapter XII)

$$\begin{array}{llll} N & g_0(\mathbf{x}) & \sigma_0 & N_0 \\ & g_1(\mathbf{x}) & \sigma_1 & N_1 \\ & \cdots & \cdots & \cdots \\ & g_{q-1}(\mathbf{x}) & \sigma_{q-1} & N_{q-1} \end{array} \tag{6}$$

Let us transform this program into an incremental program. We first separate the evaluation phase from the execution phase (see also (2), Chapter XII) and obtain subprograms (7) and (8):

$$\begin{array}{llll} N & g_0(\mathbf{x}) & \lambda & N'_0 \\ & g_1(\mathbf{x}) & \lambda & N'_1 \\ & \cdots & \cdots & \cdots \\ & g_{q-1}(\mathbf{x}) & \lambda & N'_{q-1} \end{array} \tag{7}$$

$$\begin{array}{llll} N'_0 & \mathrm{t} & \sigma_0 & N_0 \\ N'_1 & \mathrm{t} & \sigma_1 & N_1 \\ & \cdots & \cdots & \cdots \\ N'_{q-1} & \mathrm{t} & \sigma_{q-1} & N_{q-1} \end{array} \tag{8}$$

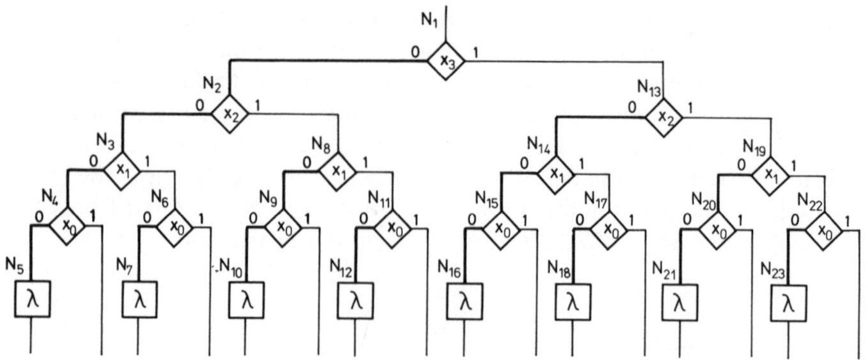

Figure 2. Evaluation phase of an incremental program

The transformation of the program (7) into an incremental program is illustrated by Figure 2 for the case $n=4$. The construction of Figure 2 is clearly of a general nature: it shows that the evaluation instruction (7) is replaced by an incremental subprogram containing $(3 \cdot 2^{n-1})$ instructions. Moreover, every instruction of (8) is replaced by two incremental instructions. If the initial program includes 2^p instructions, and if q is the upper bound of predicates per row, then the number of instructions of the incremental program is at most

$$3 \cdot 2^{n-1} + 2q - 1. \tag{9}$$

Another upper bound can be deduced from the Muller decomposition (see Section 3 of Chapter XII):

(1) The evaluation of the 2^{n-k} minterms in $x_0, x_1, \ldots, x_{n-k-1}$ requires

$$\frac{3}{2} 2^{n-k} - 1$$

instructions.

(2) The evaluation of all the q-tuples of k variable functions requires

$$2q^{2^k}$$

instructions.

(3) The execution phase is performed in $2q$ instructions.

Hence, every instruction of the initial program is replaced by at most:

$$\frac{3}{2} 2^{n-k} - 1 + 2q^{2^k} + 2q \tag{10}$$

incremental instructions.

Let us define $\tilde{k}$ and s as follows:

$$\tilde{k} = k + \log \log q,$$

$$s = n + \log \frac{3}{4} + \log \log q.$$

The number of instructions (10) may be written in the following form:

$$2(2^{s-\tilde{k}}+2^{(2^{\tilde{k}})})-1+2q. \tag{11}$$

It remains to choose $\tilde{k}$ in order to minimize (11); as in Section 3.1 of Chapter XII. (see equation (14), Chapter XII) we choose:

$$\tilde{k}_{\text{opt}}=\log(s-\log s)$$

and we obtain the following upper bound for (11):

$$2K\frac{2^s}{s}-1+2q.$$

The upper bound of the number of instructions of the whole incremental program is equal to

$$\left(2K\frac{2^s}{s}+2q-1\right)2^p=\frac{3}{2}\left(\frac{K2^n\log q}{n+\log\frac{3}{4}+\log\log q}+2q-1\right)2^p. \tag{12}$$

The practical implementation of an incremental program which contains 2^{p_3} instructions is based on the use of two type variables v_0 and v_1:

conditional jump: $v_0=0, v_1=1$;
unconditional jump: $v_0=1, v_1=0$;
execution: $v_0=1$.

Figure 3 shows a realization using a read only memory divided into three fields:

(1) The first field contains $\lceil\log n\rceil$ columns; it indicates the index of the condition variable under test.

(2) A two column field defines the instruction type.

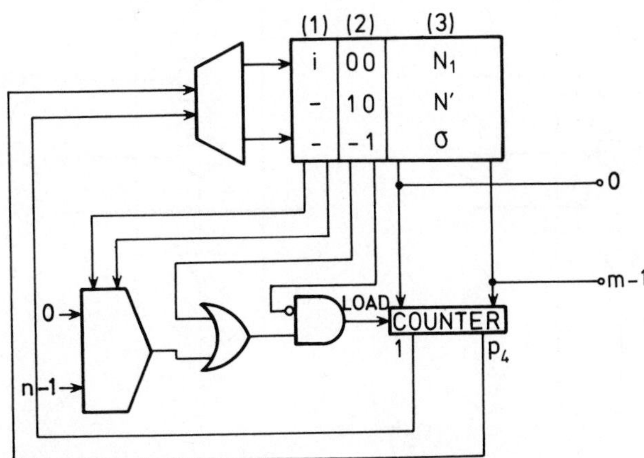

Figure 3. Implementation of an incremental program

(3) The last field defines either the next instruction number or the command.
The circuit contains a programmable counter working under the control of the variable LOAD: when LOAD = 0, it works as a counter; when LOAD = 1 it receives its contents from the read only memory.

Figure 4 describes the realization of a programmable counter by means of an half adder (HA ≃ half adder).

Figure 5 shows an equivalent realization based on the use of a programmable logic array.

Example. Addition of two integers in the representation sign and magnitude. When two operands are given in sign-magnitude form, one may wish to build a circuit to add the two numbers directly, resulting in a sign-magnitude sum.

Remember that an integer X belonging to the interval $[-(2^n-1), 2^n-1]$ may be represented by the binary vector:

$$x_s, x_{n-1}, \ldots, x_1, x_0;$$

the sign bit x_s is 0 if $X \geqslant 0$ and is 1 if $X < 0$; the vector $x_{n-1}, \ldots, x_1, x_0$ is the binary representation of the absolute value $|X|$ of X.

Two numbers X and Y being given by means of their representations

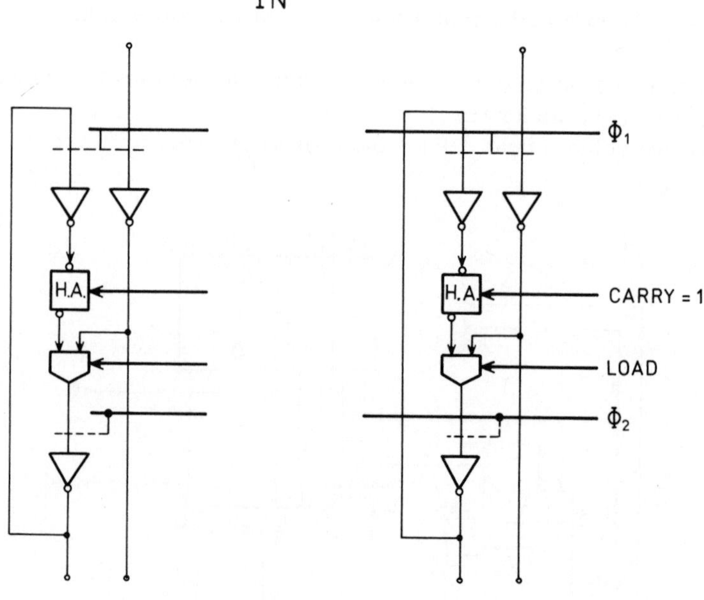

Figure 4. Programmable counter

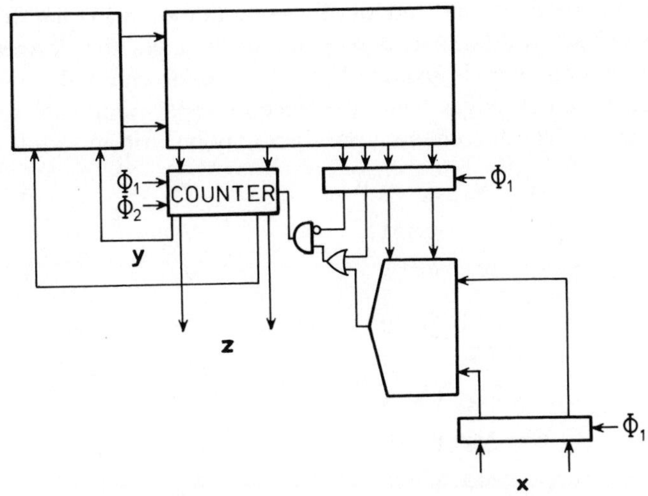

Figure 5. Incremental program implementation

$(x_s, |X|)$ and $(y_s, |Y|)$ respectively, we want to compute the representation $(z_s, |Z|)$ of their sum $Z = X + Y$. In order to realize this objective we use an operational automaton whose architecture is described by Figure 6. This automaton is formed, among others, by the registers containing the representations of X, Y, and Z and by an add–subtract unit controlled by the variable t ($t = 0$: addition, $t = 1$: subtraction); this unit generates the binary variable w which represents either the carry bit or the borrow according to the value of t. The variable w is memorized in the bistable flip-flop OV

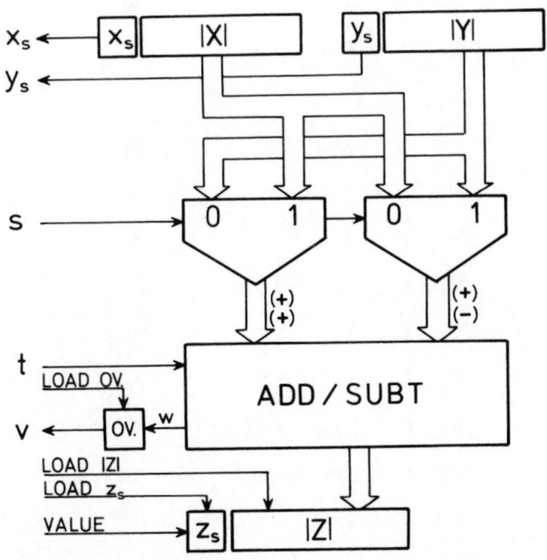

Figure 6. Addition of two numbers (sign and magnitude): hardware organization

which will be used at the end of the computation in order to detect a possible overflow. A flow chart description of the algorithm for realizing the addition of two numbers is given in Figure 7; the incremental constraints are represented by block edges while the unconditional jumps are represented by block circles. The incremental program may be written as follows:

$$
\begin{array}{llll}
N_0 & \bar{x}_s & \lambda & N_1 \\
 & x_s & \lambda & N_{12} \\
N_1 & \bar{y}_s & \lambda & N_2 \\
 & y_s & \lambda & N_4 \\
N_2 & \mathrm{t} & z_s := 0 & N_3 \\
N_3 & \mathrm{t} & \lambda & N_{20} \\
N_4 & \mathrm{t} & \mathrm{OV} := w & N_5 \\
N_5 & \bar{v} & \lambda & N_6 \\
 & v & \lambda & N_9 \\
N_6 & \mathrm{t} & z_s := 0 & N_7 \\
N_7 & \mathrm{t} & |Z| := |X| - |Y| & N_8 \\
N_8 & \mathrm{t} & \lambda & F \\
N_9 & \mathrm{t} & z_s := 1 & N_{10} \\
N_{10} & \mathrm{t} & |Z| := |X| - |Y| & N_{11} \\
N_{11} & \mathrm{t} & \lambda & F \\
N_{12} & \bar{y}_s & \lambda & N_{13} \\
 & y_s & \lambda & N_{19} \\
N_{13} & \mathrm{t} & \mathrm{OV} := w & N_{14} \\
N_{14} & \bar{v} & \lambda & N_{15} \\
 & v & \lambda & N_{17} \\
N_{15} & \mathrm{t} & z_s := 1 & N_{16} \\
N_{16} & \mathrm{t} & \lambda & N_7 \\
N_{17} & \mathrm{t} & z_s := 0 & N_{18} \\
N_{18} & \mathrm{t} & \lambda & N_{10} \\
N_{19} & \mathrm{t} & z_s := 1 & N_{20} \\
N_{20} & \mathrm{t} & |Z| := |X| + |Y| & N_{21} \\
N_{21} & \mathrm{t} & \mathrm{OV} := w & N_{22} \\
N_{22} & \mathrm{t} & \lambda & F
\end{array}
\qquad (13)
$$

The corresponding control automaton is represented by Figure 8. In this representation the condition variables x_s, y_s, and v and the constant 1 are

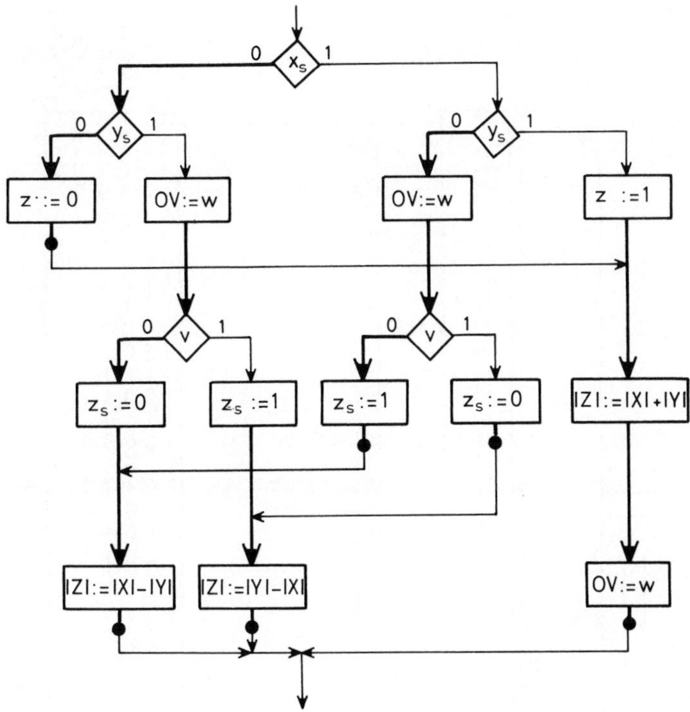

Figure 7. Addition of two numbers (sign and magnitude): flow chart

coded (0, 0), (0, 1), (1, 0), and (1, 1) respectively. The validation signal v is used in order to inhibit the loading lines LOAD z_s, LOAD OV and LOAD $|Z|$. Note also that the signals s and t must be specified only for the commands LOAD OV and LOAD $|Z|$; moreover s must be specified only when t is 0 while the value to be loaded in z_s must only be specified for the command: LOAD z_s.

2.2 Cost estimates

Let us estimate the cost of an incremental program implementation by means of a programmable logic array as shown by Figure 5.

(1) The programmable logic array realizes

$$T = \lceil \log n \rceil + 2 + \max(p_3, m)$$

functions of $N = p_3$ variables. According to equation (15), Chapter III, its cost is equal to

$$C_4(\lceil \log n \rceil + 2 + \max(p_3, m))2^{p_3}.$$

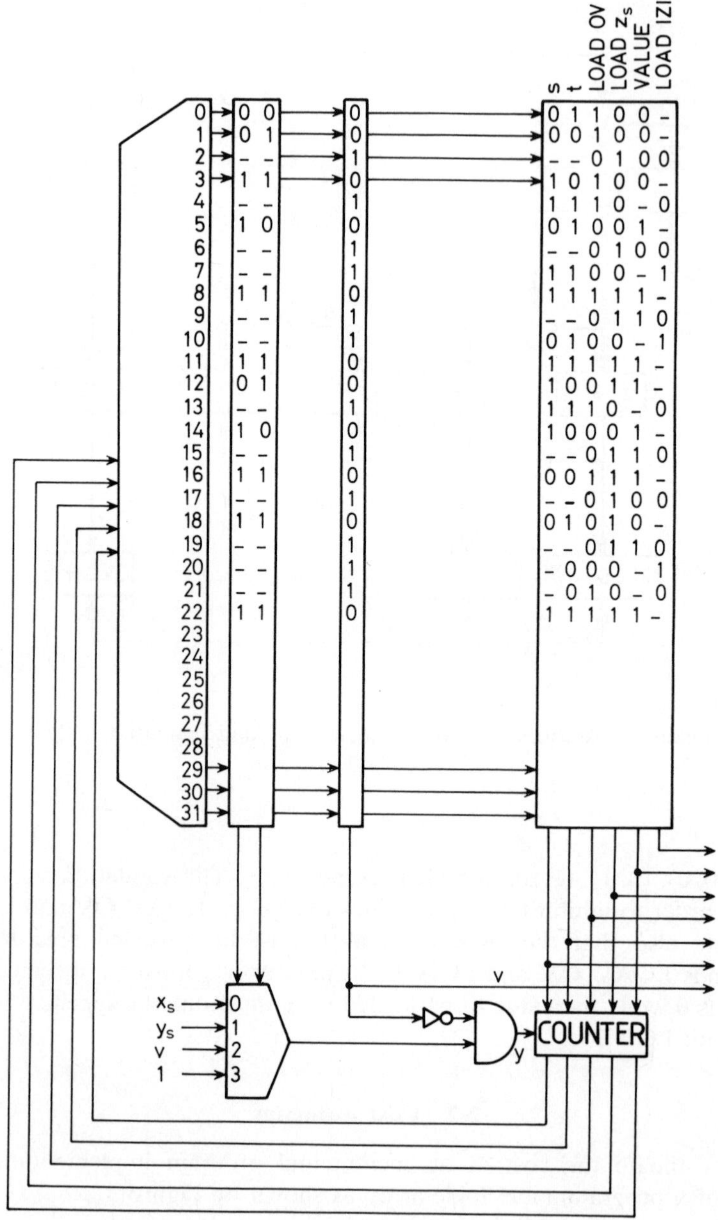

Figure 8. Addition of two numbers (sign and magnitude): control organization

(2) According to Section 2 of Chapter III, the cost of the multiplexer is equal to

$$C_1^*\lceil \log n\rceil 2^{(\lceil \log n\rceil -1)}$$

and it is bounded by:

$$C_1^* n\lceil \log n\rceil .$$

(3) The cost of the counter is proportional to the number of stored bits; we estimate the counter cost by:

$$C_6 p_3.$$

(4) As in Section 4.4 of Chapter XII, we estimate the interconnection cost by:

$$C_5(p_3^2+n^2+m^2).$$

The cost of the whole circuit is equal to:

$$\begin{aligned} C_4(\lceil \log n\rceil +2+\max (p_3, m))2^{p_3}+C_1 n\lceil \log n\rceil \\ +C_6 p_3+C_5(p_3^2+n^2+m^2). \end{aligned} \tag{14}$$

Consider again an initial program which contains 2^p instructions and let us transform it into an incremental program. According to (12) it contains at most:

$$2^{p_3}=\frac{3}{2}\left(\frac{K2^n \log q}{n+\log \frac{3}{4}+\log \log q}+2q-1\right)2^p$$

incremental instructions. A series of approximations similar to those performed in Chapter XII for conditional programs (see Section 4.4 of Chapter XII) yields:

$$2^{p_3}\cong \tfrac{3}{2}K \log q\frac{2^{n+p}}{n}$$

and thus:

$$p_3\cong n+p+\log \left(\frac{3K \log q}{2n}\right).$$

If $n \gg \log q$ we may choose the following upper bound for p_3:

$$p_3\cong n+p.$$

We finally deduce the following cost estimate c_4 for the incremental program:

$$c_4=C_4\tfrac{3}{2}K \log q\frac{2^{n+p}}{n}\max (n+p, m)+C_5((n+p)^2+n^2+m^2). \tag{15}$$

Let us compare c_4 with the cost c_3 of a conditional program in the restricted sense (see Section 4.4 of Chapter XII).

$$c_3 - c_4 = C_4 \frac{K \log q}{n} 2^{n+p}(n+p-\tfrac{1}{2}\max(n+p, m)).$$

If $n+p \geqslant m$, then $c_3 > c_4$ and:

$$c_4 = C_4 \tfrac{3}{2} K \log q \frac{2^{n+p}}{n}(n+p) + C_5((n+p)^2 + n^2 + m^2). \tag{16}$$

If $n+p < m$, then $c_3 < c_4$ and

$$c_3 = C_4 \frac{K \log q}{n} 2^{n+p}(n+p+m). \tag{17}$$

Completing the cost comparisons made in Section 4.4 of Chapter XII with the above results, we may state that:

(1) If $m \gg p$ we use the solution of cost c_2, that is, the direct implementation with two programmable logic arrays.

(2) If $m \ll p$ we use the solution of cost c_4, that is, the implementation of an equivalent incremental program.

(3) The implementation of the equivalent conditional program in the restricted sense could be the best solution when p and m have the same magnitude order.

3 PROGRAM IMPLEMENTATION

3.1 Introduction

Chapter XII and Section 2 of this chapter were devoted to the implementation of programs with disjoint predicates whose typical instruction is (6).

In many practical situations, the functions g_i do not depend on all the condition variables $\mathbf{x} = x_{n-1}, \ldots, x_1, x_0$, but on a subset of the variables. Let us denote by

$$\mathbf{x}_N \subseteq \mathbf{x}$$

the subset of variables on which do actually depend the predicates g_i of instruction (6). Instruction (6) may be replaced by

$$\begin{array}{llll} N & h_0(\mathbf{x}_N) & \sigma_0 & N_0 \\ & h_1(\mathbf{x}_N) & \sigma_1 & N_1 \\ \hline & h_{q-1}(\mathbf{x}_N) & \sigma_{q-1} & N_{q-1} \end{array} \tag{18}$$

where

$$\forall \mathbf{x} \quad \text{and} \quad \forall i: h_i(\mathbf{x}_N) = g_i(\mathbf{x}).$$

Let ν_N denote the number of variables in $\mathbf{x}_N$, and ν the maximum of ν_N over all N.

In Chapter XII we studied programs having the following characteristics:

(1) there are p address variables $\mathbf{v}$;

(2) there are m output variables $\mathbf{z}$;

(3) there are n condition variables $\mathbf{x}$; and

(4) every instruction contains at most q rows.

The programs we study in this section contain instructions of type (18) and thus have the additional characteristic:

(5) the predicates of any given instruction depend on at most ν condition variables.

As a result of the disjoint nature of functions h_i it is obvious that

$$q \leqslant 2^{\nu}.$$

In order to derive complexity upper bounds we shall suppose that

$$q = 2^{\nu}. \tag{19}$$

Furthermore, this assumption often holds in practical situations.

An upper bound of the number of rows in the initial program is

$$2^{p+\nu}. \tag{20}$$

Several implementation methods are described in the next sections: direct implementation, implementation based on bit replacements, conditional implementation, and incremental implementation.

3.2 Direct implementation

The corresponding circuit is shown in Figure 9a. Every row of the read only memory contains the next address and the current command. The row is selected by the current address and the value of $\mathbf{x}$. The read only memory stores

$$2^{n+p}(p+m) \tag{21}$$

bits.

Figure 9b shows an alternative solution based on the fact that the number of different rows in the read only memory cannot exceed the number of different rows in the initial program. The last number is bounded by (20). The two read only memories store

$$2^{n+p}(p+\nu)+2^{p+\nu}(p+m)=2^{n+p}(p+m)\left(\frac{p+\nu}{p+m}+2^{\nu-n}\right) \tag{22}$$

bits.

Solution (22) is better than solution (21) if

$$p+\nu \ll p+m \quad \text{and} \quad \nu \ll n.$$

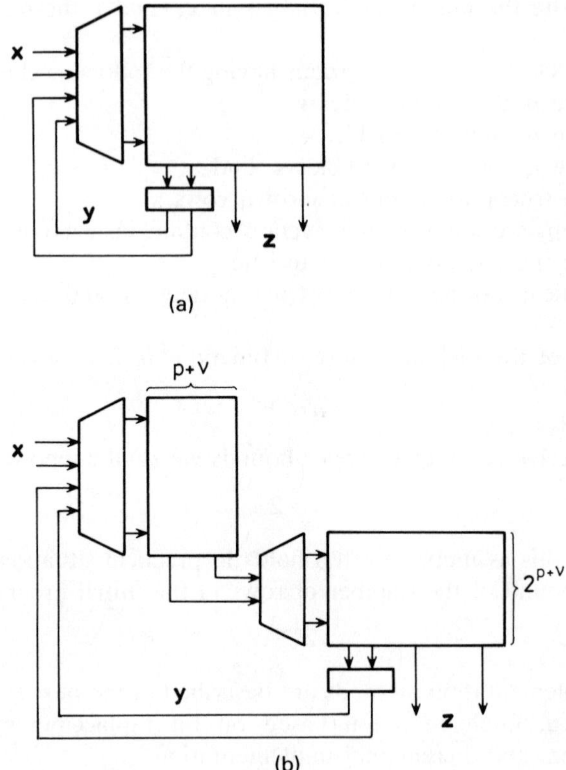

Figure 9. Program implementation: (a) direct implementation; (b) implementation with two read only memories

3.3 Bit replacement

Let us consider a program whose instructions belong to either one of the following two types.

(1) *Branching instruction*:

$$
\begin{array}{llll}
N & m_0(\mathbf{x}_N) & \lambda & k_N 2^\nu \\
 & m_1(\mathbf{x}_N) & \lambda & k_N 2^\nu + 1 \\
\hline
 & m_{b_N-1}(\mathbf{x}_N) & \lambda & k_N 2^\nu + b_N - 1
\end{array}
\tag{23}
$$

where

(a) $\mathbf{x}_N$ is a subset of $\mathbf{x}$ including ν_N variables,
(b) $b_N = 2^{\nu_N}$,
(c) k_N is a non-negative integer, and
(d) $m_i(\mathbf{x}_N)$ is the ith minterm in $\mathbf{x}_N$.

Example: $n=6$, $\nu=3$; $\mathbf{x}_N = x_1, x_5$, and thus $\nu_N = 2$; $k_N = 2$:

$$\begin{array}{llll} N & \bar{x}_5\bar{x}_1 & \lambda & 16 \\ & \bar{x}_5 x_1 & \lambda & 17 \\ & x_5\bar{x}_1 & \lambda & 18 \\ & x_5 x_1 & \lambda & 19 \end{array} \tag{24}$$

(2) *Execution instruction*:

$$N \quad \mathrm{t} \quad \sigma \quad N'. \tag{25}$$

Figure 11 shows a circuit which implements the program. The read only memory is divided into four fields:

(1) the $(P-\nu)$ most significant bits of the next address;
(2) the ν least significant bits of the next address;
(3) a type variable t, and
(4) a field containing either an n-bit masking vector $\mathbf{M}$, or a command.

The ith component of the masking vector associated with branching instruction number N is equal to 1 if x_i belongs to $\mathbf{x}_N$, and to 0 otherwise. For instance, to instruction (24) corresponds the next row in the read only

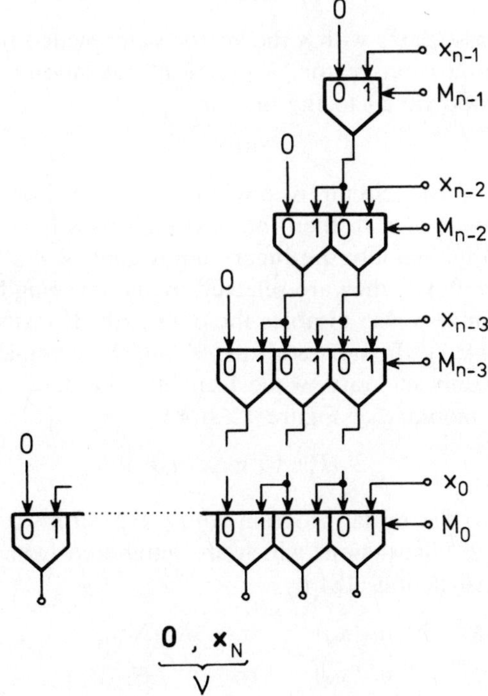

Figure 10. Shift circuit

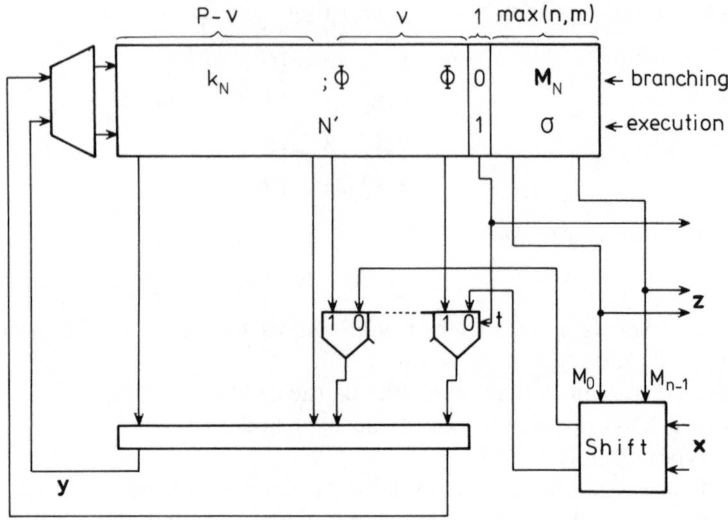

Figure 11. Bit replacement

memory:

$$00\ldots010;\quad \emptyset\emptyset\emptyset;\quad 0;\quad 010001.$$

The shift circuit associates with $\mathbf{x}$ the vector $\mathbf{x}_N$ preceded by $(\nu-\nu_N)$ zeroes, in order to obtain a ν-bit vector. A practical realization is shown in Figure 10. Its cost is proportional to the product

$$n\nu. \tag{26}$$

The working of the circuit is obvious. If instruction number N is a branching instruction ($t=0$), then the next addresses have the same $(P-\nu)$ most significant bits, namely the binary equivalent of k_N, while the ν least significant bits are $\mathbf{0}, \mathbf{x}_N$; they are selected by the masking field and the shift circuit. Type variable t also disables the data path. If instruction number N is an execution ($t=1$), the next address N' and the command σ are directly read in the read only memory, while t enables the data path.

The read only memory of Figure 11 stores

$$2^P(P+1+\max(n, m)) \tag{27}$$

bits.

Let us return to the initial program whose typical instruction is given by (18). It contains 2^p instructions which are numbered from 0 to (2^p-1).

First replace instruction (18) by

$$\begin{array}{llll} N & m_0(\mathbf{x}_N) & \sigma_{\ell(0)} & N_{\ell(0)} \\ & m_1(\mathbf{x}_N) & \sigma_{\ell(1)} & N_{\ell(1)} \\ \hline & m_{b_N-1}(\mathbf{x}_N) & \sigma_{\ell(b_N-1)} & N_{\ell(b_N-1)} \end{array} \tag{28}$$

where $\ell(i)$ is the only index j (if any) such that

$$m_i(\mathbf{x}_N) \leqslant h_j(\mathbf{x}_N). \tag{29}$$

if $(\Sigma_i h_i) \neq 1$, some rows are not specified.

With instruction number N we associate a series of unused instruction numbers:

$$k_N 2^\nu, k_N 2^\nu + 1, k_N 2^\nu + 2, \ldots, k_N 2^\nu + b_N - 1,$$

where

$$\begin{aligned} k_N &= 2^{p-\nu} + N, \quad \text{if } p \geqslant \nu, \\ &= 1 + N, \qquad \text{if } p < \nu. \end{aligned} \tag{30}$$

Notice that

$$k_0 2^\nu \geqslant 2^p \quad \text{and} \quad k_N 2^\nu \geqslant k_{N-1} 2^\nu + b_{N-1}.$$

Example: $p = 3$; $\nu_0 = \nu_1 = \nu_2 = \nu_3 = 2$; $\nu_4 = \nu_5 = \nu_6 = \nu_7 = 1$, and thus $\nu = 2$; $k_N = 2 + N$; $2^\nu = 4$:

$$\begin{aligned} &0 \to 8, 9, 10, 11; \qquad &4 \to 24, 25; \\ &1 \to 12, 13, 14, 15; \qquad &5 \to 28, 29; \\ &2 \to 16, 17, 18, 19; \qquad &6 \to 32, 33; \\ &3 \to 20, 21, 22, 23; \qquad &7 \to 36, 37. \end{aligned}$$

Instruction (28) is replaced by the next subprogram:

$$\begin{array}{llll} N & m_0(\mathbf{x}_N) & \lambda & k_N 2^\nu \\ & m_1(\mathbf{x}_N) & \lambda & k_N 2^\nu + 1 \\ \hline & m_{b_N - 1}(\mathbf{x}_N) & \lambda & k_N 2^\nu + b_N - 1 \end{array} \tag{31}$$

$$\begin{array}{llll} k_N 2^\nu & \mathrm{t} & \sigma_{\ell(0)} & N_{\ell(0)} \\ k_N 2^\nu + 1 & \mathrm{t} & \sigma_{\ell(1)} & N_{\ell(1)} \\ \hline k_N 2^\nu + b_N - 1 & \mathrm{t} & \sigma_{\ell(b_N - 1)} & N_{\ell(b_N - 1)} \end{array} \tag{32}$$

It contains one branching instruction (23) and b_N execution instructions.

Hence, the initial program is replaced by an equivalent program which can be implemented by the circuit of Figure 11. The number 2^p of instructions of the transformed program is bounded by

$$\begin{aligned} k_{2^p - 1} 2^\nu + b_{2^p - 1} &\leqslant (2^{p-\nu} + 2^p - 1) 2^\nu + 2^\nu = 2^p (2^\nu + 1) \quad \text{if} \quad p \geqslant \nu, \\ &\leqslant (1 + 2^p - 1) 2^\nu + 2^\nu = 2^\nu (2^p + 1) \quad \text{if} \quad p \leqslant \nu. \end{aligned}$$

Further on, we shall only consider the case where $p \geqslant \nu$. By taking into account the fact that

$$2^{p+\nu} < 2p(2^\nu + 1) \leqslant 2^{p+\nu+1},$$

one deduces that

$$P = p + \nu + 1. \tag{33}$$

According to (27), the read only memory stores at most

$$2^p(2^\nu + 1)(p + \nu + 2 + \max(n, m)) \tag{34}$$

bits.

An alternative solution, based on the fact that the number of different commands does not exceed the number (20) of rows in the initial program, uses two read only memories storing

$$2^p(2^{\nu+1})(p + \nu + 2 + \max(n, p + \nu)) + m2^{p+\nu}. \tag{35}$$

Notice that the number of different masking vectors **M** can be much smaller than the maximum theoretical number

$$n + \binom{n}{2} + \ldots + \binom{n}{\nu},$$

that is the number of non-zero n-bit vectors whose weight does not exceed ν. In that case we may encode the masking vectors and either use a second read only memory acting as a decoder or replace the shift circuit by a specialized circuit.

Example 2, Chapter XIII (continued). Modified Booth algorithm. Expression (27) of chapter XIII represents a Hörner scheme allowing us to perform the multiplication of two fractional numbers A and B. This Hörner scheme yields a step by step algorithm whose elementary step consists in replacing a partial result r by $r2^{-2} + \alpha_i b$. We use an operational automaton including register A which initially contains A, register B which initially contains B, accumulator register R, counting register I, and an arithmetic unit. The operational automaton accepts the following commands:

$$\sigma_0 : [A] := A; \quad [B] := B; \quad [R] := 0; \quad [I] := n - 1,$$
$$\sigma_1 : [I] := [I] - 1; \quad [A] := [A]/4,$$
$$\sigma_2 : [R] := [R]/2,$$
$$\sigma_3 : [R] := [R] + [B],$$
$$\sigma_4 : [R] := [R] - [B].$$

There are four condition variables: x_0, x_1, and x_2 are the three most significant bits of register A; x_3 is the 'zero condition' of register I:

$$x_3 = 1, \quad \text{if } [I] = 0,$$
$$= 0, \quad \text{otherwise.}$$

The elementary step is performed according to the rules indicated in Figure 12:

a_{2j+2}	a_{2j+1}	a_{2j}	α_j	$r2^{-2}+\alpha_j b$
0	0	0	0	$(r/2)/2$
0	0	1	-1	$((r/2)/2)-b$
0	1	0	$1/2$	$((r/2)+b)/2$
0	1	1	$-1/2$	$((r/2)-b)/2$
1	0	0	$1/2$	$((r/2)+b)/2$
1	0	1	$-1/2$	$((r/2)-b)/2$
1	1	0	1	$((r/2)/2)+b$
1	1	1	0	$(r/2)/2$

Figure 12. Elementary step for the modified Booth algorithm

The following program defines the sequence of operations (see also Figure 13):

0	t	σ_0	1
1	t	σ_2	2
2	$\bar{x}_2\bar{x}_1\bar{x}_0$	σ_2	6
	$\bar{x}_2\bar{x}_1x_0$	σ_2	3
	$\bar{x}_2x_1\bar{x}_0$	σ_3	4
	$\bar{x}_2x_1x_0$	σ_4	4
	$x_2\bar{x}_1\bar{x}_0$	σ_3	4
	$x_2\bar{x}_1x_0$	σ_4	4
	$x_2x_1\bar{x}_0$	σ_2	5
	$x_2x_1x_0$	σ_2	6
3	t	σ_4	6
4	t	σ_2	6
5	t	σ_3	6
6	$\bar{x}_3$	σ_1	1
	x_3	λ	7
7	END		

(36)

The program has the following parameters: $p=3$, $n=4$, and $\nu=3$. It can be transformed into an equivalent program which can be implemented by the

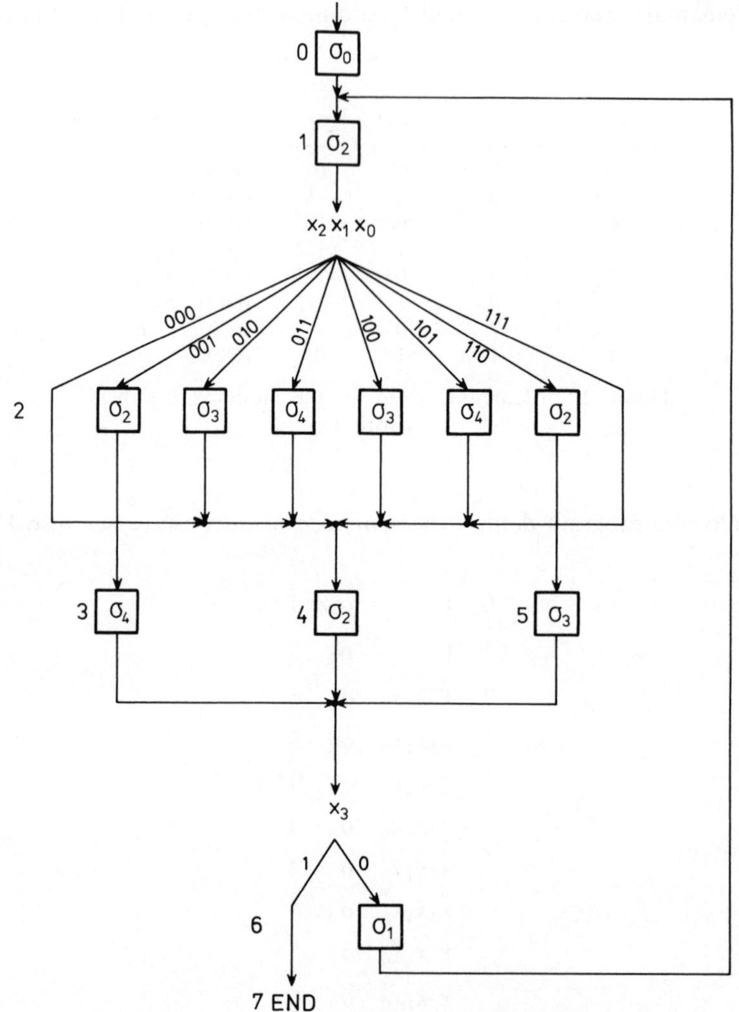

Figure 13. Multiplication algorithm

circuit of Figure 11:

0	t	σ_0	1		$x_2x_1\bar{x}_0$	λ	14
1	t	σ_2	2		$x_2x_1x_0$	λ	15
2	$\bar{x}_2\bar{x}_1\bar{x}_0$	λ	8	3	t	σ_4	6
	$\bar{x}_2\bar{x}_1x_0$	λ	9	4	t	σ_2	6
	$\bar{x}_2x_1\bar{x}_0$	λ	10	5	t	σ_3	6
	$\bar{x}_2x_1x_0$	λ	11	6	$\bar{x}_3$	λ	16
	$x_2\bar{x}_1\bar{x}_0$	λ	12		x_3	λ	17
	$x_2\bar{x}_1x_0$	λ	13	7	END		

8	t	σ_2	6	13	t	σ_4	4
9	t	σ_2	3	14	t	σ_2	5
10	t	σ_3	4	15	t	σ_2	6
11	t	σ_4	4	16	t	σ_1	1
12	t	σ_3	4	17	t	λ	7 or END

The corresponding control automaton is shown in Figure 14. The commands have been encoded:

$$\sigma_0 \to 000, \qquad \sigma_1 \to 001, \qquad \sigma_2 \to 010, \qquad \sigma_3 \to 011, \qquad \sigma_4 \to 100.$$

Two unused conditions are used instead of a true masking vector:

$$\begin{aligned} &\text{test of } x_0, x_1, x_2 \to 110, \\ &\text{test of } x_3 \qquad\quad\; \to 111. \end{aligned}$$

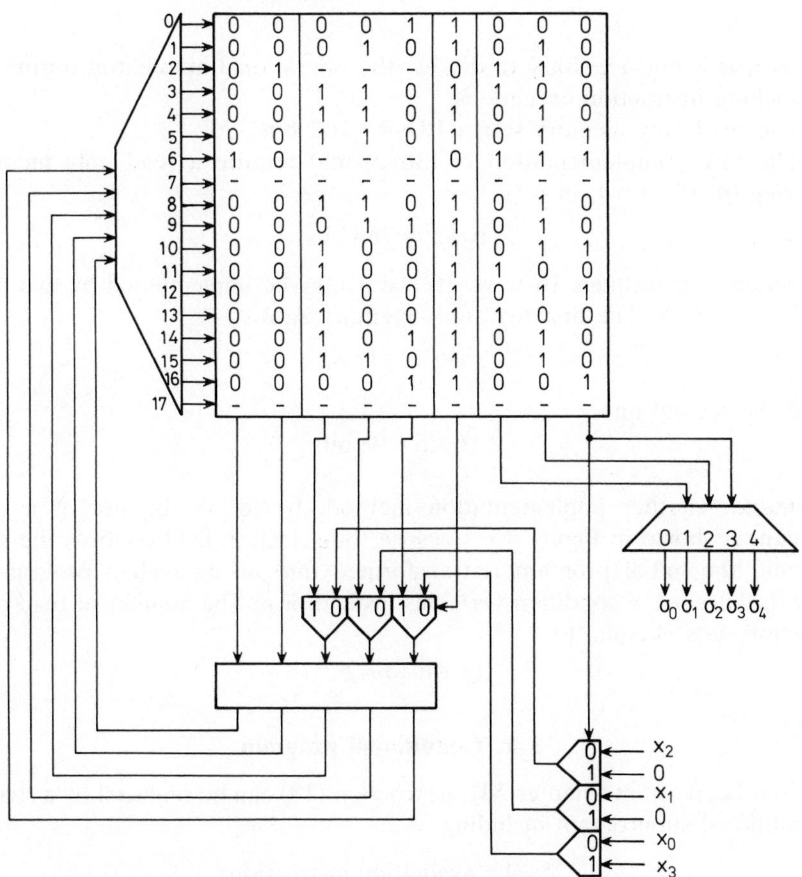

Figure 14. Control automaton

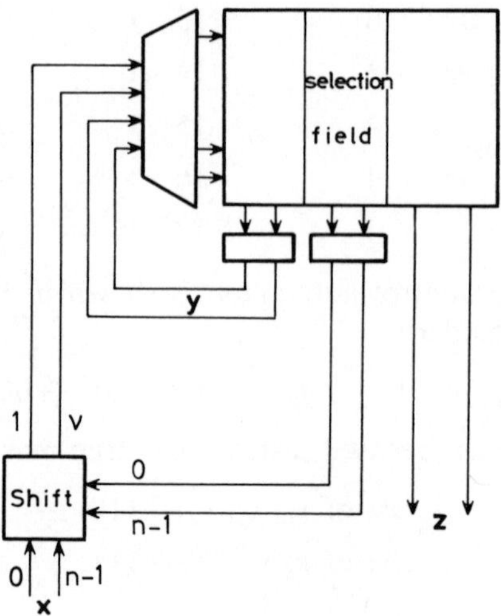

Figure 15. Use of a shift circuit

Hence, it is not necessary to disable the operational automaton during the branching instruction executions.

The read only memory stores $18 \times 9 = 162$ bits.

The direct implementation of (36) would require a read only memory storing ($p = 3$, $n = 4$, $m = 3$)

$$6 \times 2^7 = 768 \text{ bits.}$$

Program (36) includes 16 rows. Hence it may be implemented by two read only memories. The first read only memory stores

$$4 \times 2^7 = 512 \text{ bits,}$$

and the second one

$$16 \times 6 = 96 \text{ bits.}$$

Remark. Another implementation method, based on the use of a shift circuit, is shown in Figure 15. Because the selection field controls the shift circuit, the initial program is transformed into an equivalent program in which there are ν condition variables instead of n. The number of read only memory bits is equal to

$$(p + n + m)2^{p+\nu}.$$

3.4 Conditional program

As has been sen in Chapter XII, instruction (18) can be replaced by a strictly conditional subprogram including

$$2^{\nu} - 1 \quad \text{evaluation instructions}$$

and

$$q \quad \text{execution instructions.}$$

Notice that when $q = 2^\nu$ the Muller decomposition is useless: the program evaluating all q-tuples of k variable functions would contain

$$(q)^{2k} = (2^\nu)^{2k} > 2^\nu - 1$$

evaluation instructions.

The initial program can be replaced by an equivalent strictly conditional program containing at most

$$2^p(2^\nu - 1 + 2^\nu) = 2^p(2^{\nu+1} - 1)$$

instructions.

Notice that

$$2^\nu \leqslant 2^{\nu+1} - 1 < 2^{\nu+1}.$$

The addresses of the conditional program include $(p+\nu+1)$ bits (p if $\nu = 0$). The read only memory stores

$$2^p(2^{\nu+1} - 1)(\lceil \log n \rceil + p + \nu + 2 + \max(p+\nu+1, m)) \tag{37}$$

bits (see Figure 8, Chapter XII).

There is another solution using two read only memories storing

$$2^p(2^{\nu+1} - 1)(\lceil \log n \rceil + 1 + (p+\nu+1) + \max(p+\nu+1, p+\nu)) + m2^{p+\nu}$$
$$= 2^p(2^{\nu+1} - 1)(\lceil \log n \rceil + 2p + 2\nu + 3) + m2^{p+\nu} \tag{38}$$

bits.

3.5 Incremental program

Instruction (18) can also be replaced by an incremental subprogram containing

$$\tfrac{3}{2}2^\nu - 1 \quad \text{conditional jump instructions}$$

and

$$2q \quad \text{execution instructions.}$$

The initial program can be replaced by an incremental program containing at most

$$2^p(\tfrac{3}{2}2^\nu - 1 + 2^{\nu+1}) = 2^p(\tfrac{7}{2}2^\nu - 1)$$

instructions. The addresses of the incremental program contain at most $(p+\nu+2)$ bits since

$$2^{\nu+1} < \tfrac{7}{2}2^\nu - 1 < 2^{\nu+2}.$$

The read only memory stores

$$2^p(\tfrac{7}{2}2^\nu - 1)(\lceil \log n \rceil + 2 + \max(p+\nu+2, m)) \tag{39}$$

bits (see Figure 3).

The alternative solution with two read only memories would require the storing of

$$2^p(\tfrac{7}{2}2^\nu - 1)(\lceil \log n \rceil + p + \nu + 4) + m2^{p+\nu}. \tag{40}$$

bits.

3.6 Cost comparisons

The various solutions are compared under the assumption that both p and m are the dominant parameters. The comparisons are based on the number of stored bits: it has been seen in Chapter XII that this number appears in the dominant term of the cost formulae:

$$\begin{aligned}
&(21) \Rightarrow n_1 = 2^{n+p}(p+m),\\
&(22) \Rightarrow n_2 = 2^{n+p}(p+\nu) + 2^{p+\nu}(p+m),\\
&(34) \Rightarrow n_3 = 2^p(2^\nu+1)(p+\nu+2+m),\\
&(35) \Rightarrow n_4 = 2^p(2^\nu+1)(2p+2\nu+2) + m2^{p+\nu},\\
&(37) \Rightarrow n_5 = 2^p(2^{\nu+1}-1)(\lceil \log n \rceil + p + \nu + 2 + \max(p+\nu+1, m)),\\
&(38) \Rightarrow n_6 = 2^p(2^{\nu+1}-1)(\lceil \log n \rceil + 2p + 2\nu + 3) + m2^{p+\nu},\\
&(39) \Rightarrow n_7 = 2^p(\tfrac{7}{2}2^\nu - 1)(\lceil \log n \rceil + 2 + \max(p+\nu+2, m)),\\
&(40) \Rightarrow n_8 = 2^p(\tfrac{7}{2}2^\nu - 1)(\lceil \log n \rceil + p + \nu + 4) + m2^{p+\nu}.
\end{aligned}$$

If $\nu = 0$, the direct implementation of Figure 9a can be simplified: the condition variables may be deleted and the number of stored bits is

$$n_1 = 2^p(p+m).$$

It is smaller than $n_2, \ldots, n_8$.

If $\nu = n$, then

$$2^\nu + 1 > 2^n, \qquad 2^{\nu+1} - 1 > 2^n, \qquad \tfrac{7}{2}2^\nu - 1 > 2^n$$

and thus n_1 is smaller than $n_2, \ldots, n_8$.

If $\nu \geq 1$, then

$$2^{\nu+1} - 1 \geq 2^\nu + 1 \quad \text{and} \quad \tfrac{7}{2}2^\nu - 1 \geq 2(2^\nu + 1).$$

Hence,

$$n_5 \geq n_3, \qquad n_6 \geq n_4, \qquad n_7 \geq n_3, \qquad n_8 \geq n_4.$$

Let us compare n_1, n_2, n_3, and n_4 under the following assumptions: $\nu \ll p$ and $1 \leq \nu \leq n-1$:

$$\begin{aligned}
n_1 &= (2^n)p2^p + (2^n)m2^p,\\
n_2 &\cong (2^n + 2^\nu)p2^p + (2^\nu)m2^p,\\
n_3 &\cong (2^\nu + 1)p2^p + (2^\nu + 1)m2^p,\\
n_4 &\cong 2(2^\nu + 1)p2^p + (2^\nu)m2^p;
\end{aligned}$$

$$1 \leq \nu \leq n-1 \Rightarrow 2^\nu + 1 < 2^n \quad \text{and} \quad 2(2^\nu + 1) \leq 2^n + 2 \leq 2^n + 2^\nu.$$

Hence, n_3 is smaller than n_1, and n_4 is smaller than or equal to n_2.

In order to compare n_3 and n_4, one should make assumptions on the relative values of p and m. Notice that if $m \gg p$, then

$$n_2, n_4, n_6, n_8 \cong m2^{p+\nu}.$$

On the other hand, if $p \gg m$, $n_4 \cong 2n_3$.

4 SUBROUTINES

4.1 Introduction

We have seen the usefulness of detecting in a combinatorial algorithm identical partial computations; in a total sequentialization process we may use only one copy of the resource performing this partial computation. A similar situation may appear at a program level where it is possible to recognize several occurrences of an identical subprogram. In this case also we may want to use in each of these occurrences a single copy of the subprogram. We shall restrict ourselves to the particular case where the subprograms have a single initial instruction and a single ultimate instruction. This type of subprogram will be called a *subroutine*. A subroutine P having initial and final instructions N_{ip} and N_{fP} respectively is symbolized by the scheme of Figure 16. A program in which we detect multiple occurrences of an identical subroutine is called *main program*.

To enter into a subroutine does not state any particular problem: we have access to the subroutine P by means of an unconditional jump to the instruction N_{ip}. However, the return from the subroutine to the main program requires additional information: the different occurrences of an identical subroutine generally have different return addresses to the main program.

A solution to this practical problem is suggested by the scheme of Figure 17 where two different address registers are used. The *next address register* contains the address of an instruction of the main program or of a subroutine: it is the instruction which is presently executed. The second register, called the *return address register* contains the appropriate return address during a particular execution of the subroutine. Note that the internal state

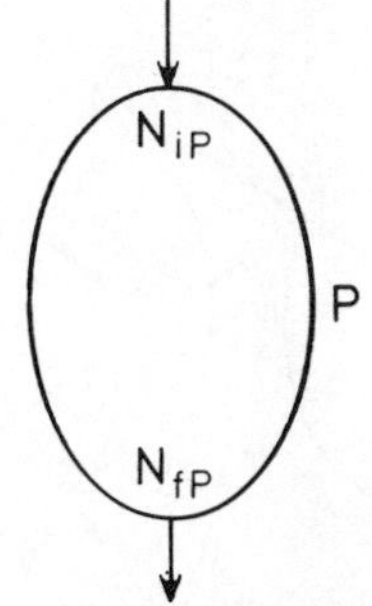

Figure 16. Subroutine

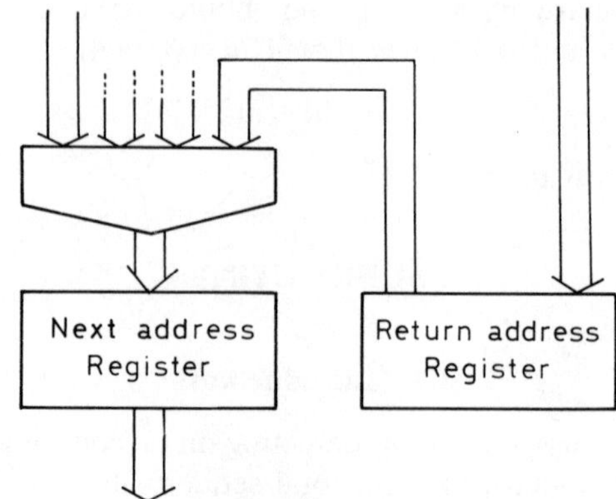

Figure 17. Enter- and return-addresses implementation

of the control automaton contains in the present case two components, namely the internal state N_i of the standard address register and the internal state B_j of the return address. Accordingly, the internal state of the control automaton will be denoted (N_i, B_j). This suggests that the input and output instructions of a subroutine (CALL and RETURN instructions) should be written under the following forms respectively:

$$\text{subroutine input:} \quad (N_i, B_j) \;\; t \;\; \lambda(N_{iP}, N_r), \tag{41}$$

$$\text{subroutine output:} \quad (N_{fP}, N_r) \;\; t \;\; \lambda(N_r, N_r), \tag{42}$$

where N_r is the return address. We assume that the return address N_r remains memorized during the execution time of the subroutine P.

It is easy to generalize this addressing process for the case of subroutines linked according to the scheme of Figure 18. In this case a return address register is no longer sufficient for gathering the set of information necessary for computing the return addresses. For example, for the linked subroutines schemes of Figure 18 we must have a memory of return addresses which will

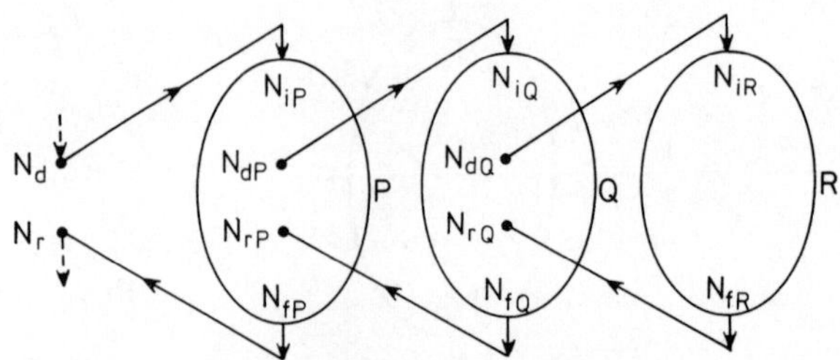

Figure 18. Linked subroutines

contain the successive information:

$$(N_r), (N_{rP}, N_r), (N_{rQ}, N_{rP}, N_r), (N_{rP}, N_r), (N_r).$$

A possible way to organize a register file that stores addresses for subroutine calls and returns is to use a first-in, last-out stack (FILO). The stack organization and its use in subroutine calls and returns are explained in Section 4.2.

We shall first adapt our notations (41, 42) to the case of linked subroutines by describing the internal state of the control automaton by a pair $(N_i, \mathbf{B}_j)$, where $\mathbf{B}_j$ represents the vector of return addresses. In order to simplify the notations we shall mention in the vector $\mathbf{B}_j$ only the return addresses which will effectively be used. Besides this, the first left component in the writing of $\mathbf{B}_j$ will represent the last memorized return address; this address will thus be the first one that will be used, and after that it will be dropped.

The input and output instructions of the subroutine will thus be written in the form:

$$\text{Subroutine input:} \quad (N_i, \mathbf{B}_j) \quad t \quad \lambda \quad (N_{i\mathbf{p}}, N_r\mathbf{B}_j), \tag{43}$$

$$\text{Subroutine output:} \quad (N_{fP}, N_{r\mathbf{B}_j}) \quad t \quad \lambda \quad (N_t, \mathbf{B}_j). \tag{44}$$

We illustrate this type of instruction by writing the corresponding set of input–output instructions relative to the linked subroutines of Figure 18:

$(N_d, -)$	t	λ	(N_{iP}, N_r)
(N_{iP}, N_r)	...	...	...
(N_{dP}, N_r)	t	λ	$(N_{iQ}, N_{rP}N_r)$
$(N_{iQ}, N_{rP}N_r)$	...	...	...
$(N_{dQ}, N_{rP}N_r)$	t	λ	$(N_{iR}, N_{rQ}N_{rP}N_r)$
$(N_{iR}, N_{rQ}N_{rP}N_r)$	...	...	...
$(N_{fR}, N_{rQ}N_{rP}N_r)$	t	λ	$(N_{rQ}, N_{rP}N_r)$
$(N_{rQ}, N_{rP}N_r)$	...	...	...
$(N_{fQ}, N_{rP}N_r)$	t	λ	(N_{rP}, N_r)
(N_{rP}, N_r)	...	...	...
(N_{fP}, N_r)	t	λ	$(N_r, -)$

4.2 Stacks

A stack is a storage device that stores information in such a manner that the item stored last is the first item retrieved. It is essentially a portion of a memory unit accessed by an address that is always incremented or decremented after the memory access. The register that holds the address for a stack is called a *stack pointer* because its value always points to the top item in the stack. The two generations of a stack are the insertion and deletion of items. The operation of insertion is called *push* and it can be thought of as the result of pushing a new item onto the top of the stack. The operation of deletion is called *pop* and it can be thought of as the result of removing one item so that the stack pops out. Those operations are represented by incrementing or decrementing the stack pointer register.

We shall adopt further on the classical terminology which refers to the organization of the return addresses of the subroutines. We call the *top* of the stack the memory location where the last return address has been written: it is the address which will be read first. In this terminology the first return address that has been written in the stack occupies the lowest position. Let us call $R_0, R_1, \ldots, R_{P-1}$ the registers of the stack: if R_0 is the *top* of the stack, R_{P-1} is its *bottom* and P is its *depth*. The depth of a stack clearly represents the maximum number of subroutines which can be linked, each to others. From a practical point of view the stacks will contain a mechanism which detects that its bottom has been reached and that the further return addresses will be lost. It will also frequently be useful in the synthesis process of a program to have the possibility of reaching any register of the stack without loss of the information it contains. The above two practical remarks are essential in the choice of the stack materialization.

It is easy to build a stack of P words of n bits by using n two-dimensional shift registers of length P. We write an address in the stack by presenting this address on the input lines and by shifting the contents from one position to the final one. During the read process of a stack, the return address of the subroutine is at the top (of the stack); this return address is then transferred in the next instruction register. As quoted above these two operations and the associated logic signals are called PUSH and POP respectively.

Figure 19a illustrates the building up of a shift register: it must be able to perform two operations:

(a) a bottom-shift operation ((PUSH-POP) = (1, 0)),

(b) a top-shift operation ((PUSH-POP) = (0, 1)).

To the command (PUSH-POP) = (0, 0) is associated the absence of shift. From the external variables PUSH and POP an elementary combinatorial circuit allows us to generate an activation signal a and a directional signal b (see Figure 19a). Note that for the POP operation the register is connected as a ring which allows us to read the register contents without losing this information. The register symbol is given in Figure 19b. Figure 19c illustrates the general stack organization. It is basically made up of registers $R_0, R_1, \ldots, R_{n-1}$ and with a system allowing us to detect that the bottom of the stack has been reached. This kind of organization is used for low and medium sized applications.

A second building up scheme for a shift register is illustrated by Figure 20; the counter used in this realization is called the *stack pointer*. This type of realization has basically identical properties to the former one so far as the read operation without contents destruction and the stack bottom

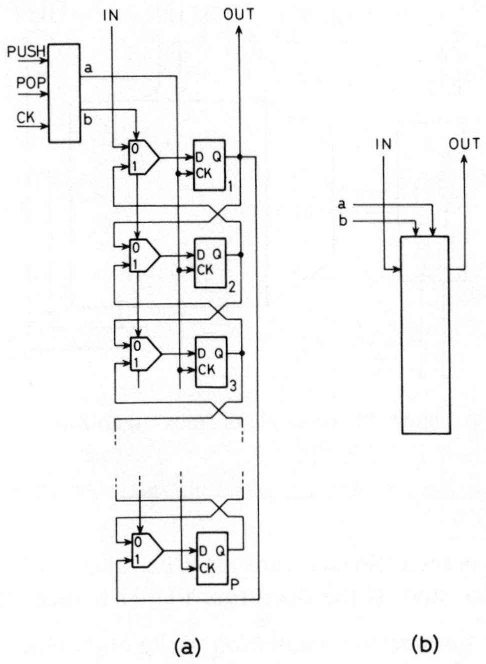

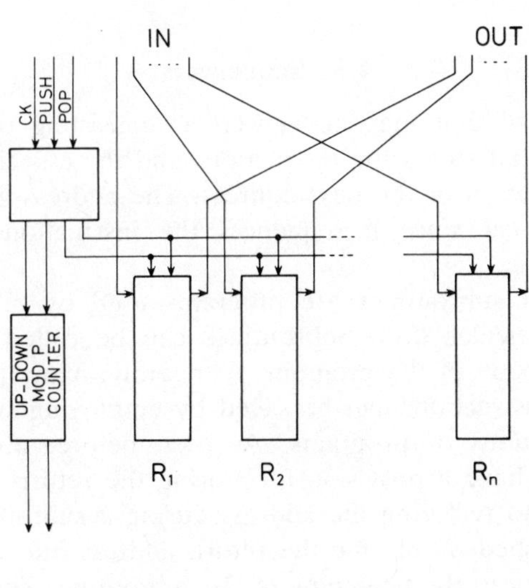

Figure 19. (a) Building-up of a shift-register. (b) Register symbol. (c) General stack organization

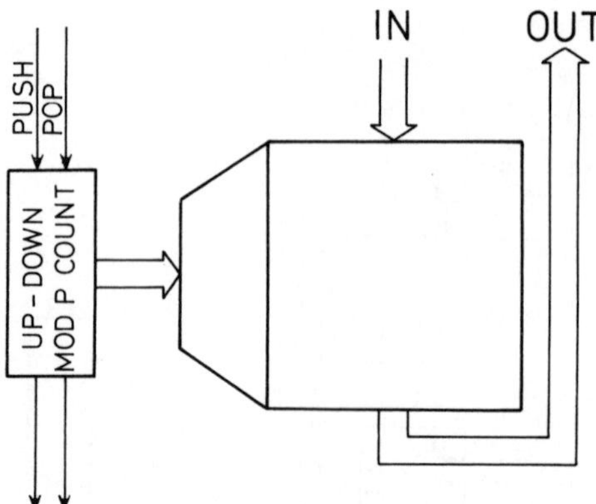

Figure 20. Example of stack organization

detection are concerned. Special care must be taken so far as the timing of operations is concerned. If the operation PUSH realizes the sequence:

pointer incrementation; write operation;

then the operation POP must realize the sequence:

read operation; Pointer decrementation

4.3 Sequencers

A program control unit may be viewed as consisting of two parts: the control memory that stores the instructions and the associated circuits that control the generation of the next address. The address-generation part is called the *sequencer* since it sequences the instructions in the control memory.

Remember that subroutines are programs used by other programs to accomplish a particular task. Subroutines can be called from any point within the main body of the program. Frequently, many programs contain identical parts. Instructions can be saved by employing subroutines which use common sections of programs. As seen before, programs that use subroutines must have a provision for storing the return address during a subroutine call and restoring the address during a subroutine return. This may be accomplished by placing the return address into a special register and then branching to the beginning of the subroutine. This special register can then become the address source for setting the address register for the return to the main program.

The above requirements allow us to write down an elementary block diagram of a program sequencer. Assume first that the sequencer must be able to perform the operations:

$$\begin{aligned} &\text{PUSH: } (N_i, \mathbf{B}_j) \quad &t \quad \lambda \quad &(N_{iP}, N_{i+1}\mathbf{B}_j), \\ &\text{POP: } \quad (N_{fP}, N_{i+1}\mathbf{B}_j) \quad &t \quad \lambda \quad &(N_{i+1}, \mathbf{B}_j). \end{aligned} \tag{45}$$

It must also be able to perform the basic instructions of incremental programs which are written in the formalism adopted here:

$$\begin{array}{llccl} \text{JUMP INC} & (N_i, \mathbf{B}_j) & t & \lambda & (N_k, \mathbf{B}_j) \\ \text{JUMP COND} & (N_i, \mathbf{B}_j) & \bar{x} & \lambda & (N_{i+1}, \mathbf{B}_j) \\ & & x & \lambda & (N_k, \mathbf{B}_j) \\ \text{EXEC} & (N_i, \mathbf{B}_j) & t & \sigma & (N_{i+1}, \mathbf{B}_j) \end{array} \tag{46}$$

Figure 21 illustrates the materialisation of a sequencer which implements the five types of instructions (45, 46); let us code these instructions by means of the three type variables (t_2, t_1, t_0) according to the scheme:

PUSH $\simeq$ (1, 0, −); POP $\simeq$ (1, 1, −); JUMP COND $\simeq$ (0, 1, 0);
JUMP INC $\simeq$ (0, 1, 1), EXEC $\simeq$ (0, 0, −).

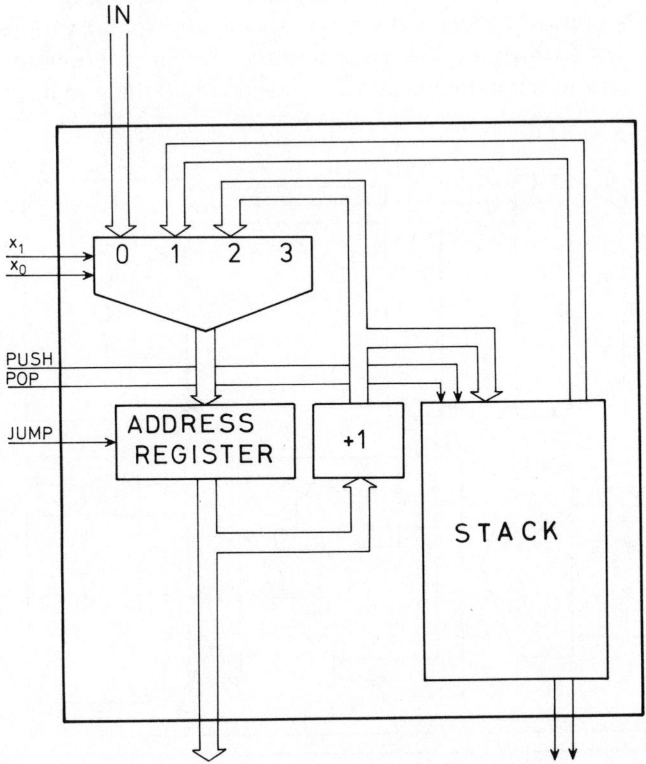

Figure 21. Example of sequencer materialization

An elementary computation shows us that the commands x_1, x_0, PUSH, POP and the validation signal v (see Figure 21) are given by the expressions:

$$\begin{aligned} x_1 &= \bar{t}_2(\bar{t}_1 \vee \bar{t}_0 \bar{x}), \\ x_0 &= \text{POP} = t_2 t_1, \\ \text{PUSH} &= t_2 \bar{t}_1, \\ \text{VALID} &= \bar{t}_2 \bar{t}_1. \end{aligned} \tag{47}$$

A general type of organization of the memory part and of the sequencer of the control is depicted in Figure 22. The scheme of Figure 22 is quite general and the procedure followed for synthesizing a sequencer may be detailed in the following way:

(a) Analysis of the computation to be performed and of the sequences of addresses; synthesis of the sequencer.

(b) Analysis of the information contained in each of the types of instructions and definition of the size and of the organization of the read only memory.

(c) Coding of the instructions types and synthesis of the combinatorial circuit which generates the input variables to the sequencer.

To the theoretical concepts discussed above correspond various commercial components which are known as a sequencer, microprogram controller, etc. These various components may be classified into three main categories.

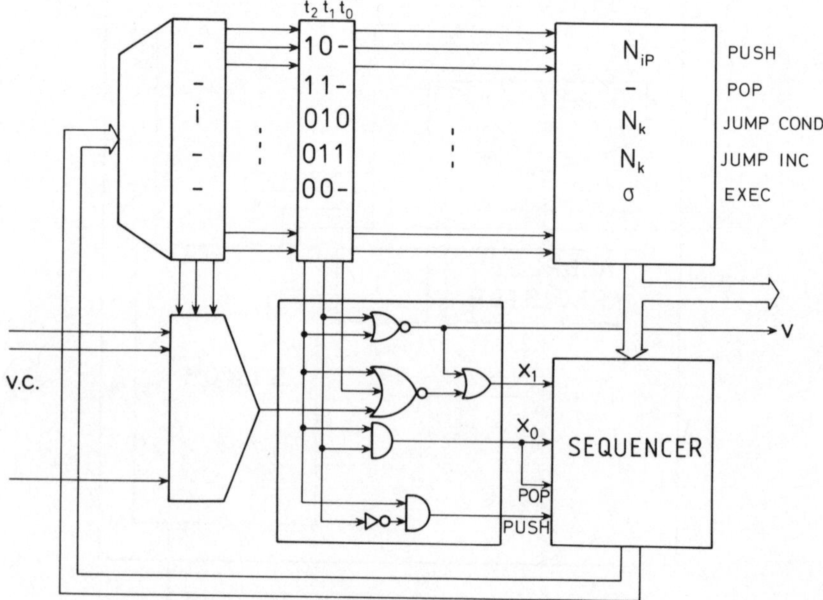

Figure 22. Sequencer use in circuit design

Category I. These components correspond to what was called a sequencer above; they are thus exclusively components which determine the next address of the control memory.

Category II. These components correspond to the sequencer of Category I and to its combinatorial input circuit; they are planned for the branching execution (see the 8X02, Signetics, detailed below).

Category III. These components correspond to the sequencer of Category II and to its input multiplexer (see Figure 22).

The interest of these types of sequencer lies in the fact that the control unit is made up of only two components, namely the read only memory and the sequencer.

We give a description of a sequencer of Category II produced by Signetics under the code number 8X02; its instruction set is given in Figure 23. A first analysis of this type of instruction set shows that the only instructions which refer to a new address denoted N_k are BSR (4) and BRT (6); the address N_k must therefore be written in the program memory. We thus normally associate to a sequencer a program memory which contains, besides the usual instruction type field and variable instruction field, only one information field. All the remaining instructions (other than (4) and (6)) may contain a command component.

We also observe that the next address register may receive five kinds of information: address coming from the main memory (thus outside to the sequencer), present address +1, present address +2, address of the top of the stack, and address 0. This allows us to clarify the architecture of the sequencer as indicated by Figure 24. A decoder network which received the indication of the type of instruction under the coded form (AC2, AC1,

N° Code	Signal	Description	Instruction			
0	TSK	TEST & SKIP	$(N_i, \mathbf{B}_j)$	$\bar{x}$	...	$(N_{i+1}, \mathbf{B}_j)$
				x	...	$(N_{i+2}, \mathbf{B}_j)$
1	INC	INCREMENT	$(N_i, \mathbf{B}_j)$	t	...	$(N_{i+1}, \mathbf{B}_j)$
2	BLT	BRANCH TO LOOP	$(N_i, N_k\mathbf{B}_j)$	$\bar{x}$	...	$(N_{i+1}, \mathbf{B}_j)$
		IF TEST TRUE		x	...	$(N_k, \mathbf{B}_j)$
3	POP	POP STACK	$(N_i, N_k\mathbf{B}_j)$	t	...	$(N_k, \mathbf{B}_j)$
4	BSR	BRANCH TO SUBR.	$(N_i, \mathbf{B}_j)$	$\bar{x}$	...	$(N_{i+1}, \mathbf{B}_j)$
		IF TEST TRUE		x	...	$(N_k, N_{i+1}\mathbf{B}_j)$
5	PLP	PUSH FOR LOOPING	$(N_i, \mathbf{B}_j)$	t	...	$(N_{i+1}, N_i\mathbf{B}_j)$
6	BRT	BRANCH IF TEST	$(N_i, \mathbf{B}_j)$	$\bar{x}$	...	$(N_{i+1}, \mathbf{B}_j)$
		TRUE		x	...	$(N_k, \mathbf{B}_j)$
7	RST	SET ADDRESS TO ZERO	$(N_i, \mathbf{B}_j)$	t	...	$(0, \mathbf{B}_j)$

Figure 23. Instructions of the sequencer 8X02

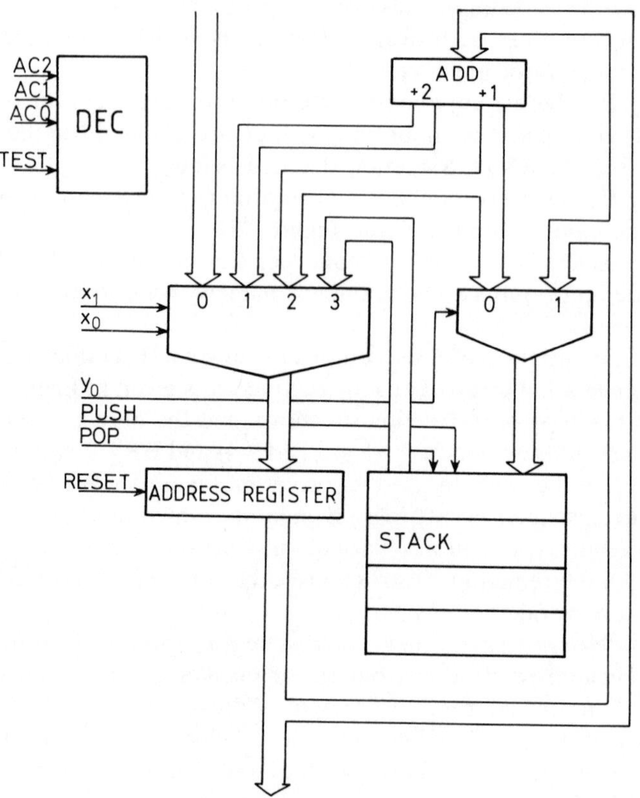

Figure 24. Sequencer 8X02. (Reproduced with the permission of Signetics Corp. Copyright 1981)

AC0) and the condition variable to be tested (TES) will produce the different command signals x_1, x_0, y_0, PUSH, POP, and RESET.

Let us examine in a more detailed way the set of instructions gathered in Figure 23. First observe that the instructions INC(1) and BRT(6) correspond to our usual execution instructions and conditional jump respectively. These two instructions constitute thus a universal set (of instructions) for the writing of a program. Two other instructions TSK(0) and RST(7) do not request any intervention of the stack: they may roughly be considered as facilities for the program designer. Finally, four instructions are concerned with the stack: two of them BSR(4) and PLP(5) correspond to writings in the stack while the two others POP(3) and BLT(2) correspond to readings in the stack. In each group of two instructions we find an unconditional instruction and a conditional instruction.

The conditional instructions which request the intervention of the stack

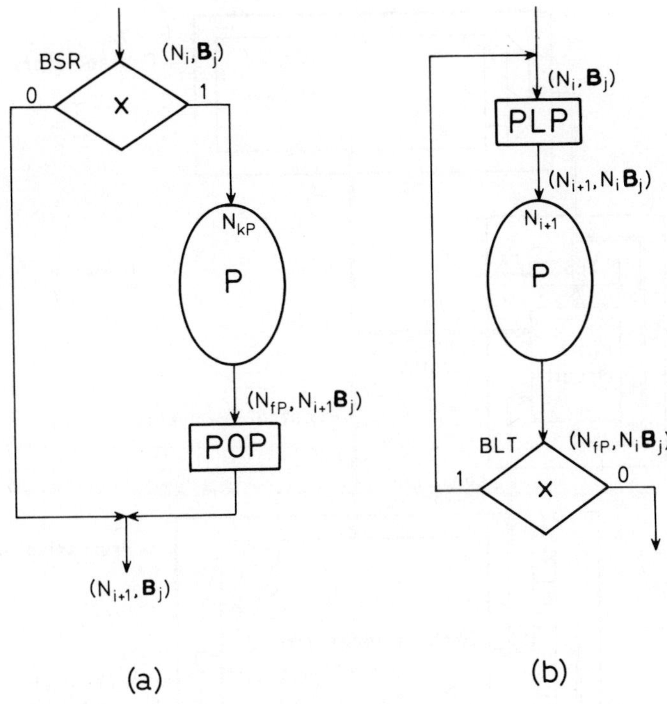

Figure 25. Use of the set of instructions of the sequencer 8X02

are encountered for the first time. We study these instructions in a more detailed way by giving two examples for their use.

The pair (BSR and POP) may be used according to the scheme of Figure 25a which does not require any particular comment, besides the fact that it is one of the most used in practice. The pair (PLP and BLT) may be used as indicated by Figure 25b. We observe that the generation of a program feedback loop by this means (instead of by means of a conditional jump) allows us to save a field of the memory. Observe, moreover, that this feedback loop corresponds to the classical structure *repeat . . . until.*

5 PIPELINE

The pipeline technique was described in Chapter VIII, within the context of the throughput increase of combinational circuits. It can be used within the more general context of systems which implement non-combinatorial algorithms, if the condition variable values are correctly interpreted.

Figure 26 describes a general circuit architecture which contains a control store, a processing unit, and a sequencer represented as a finite automaton.

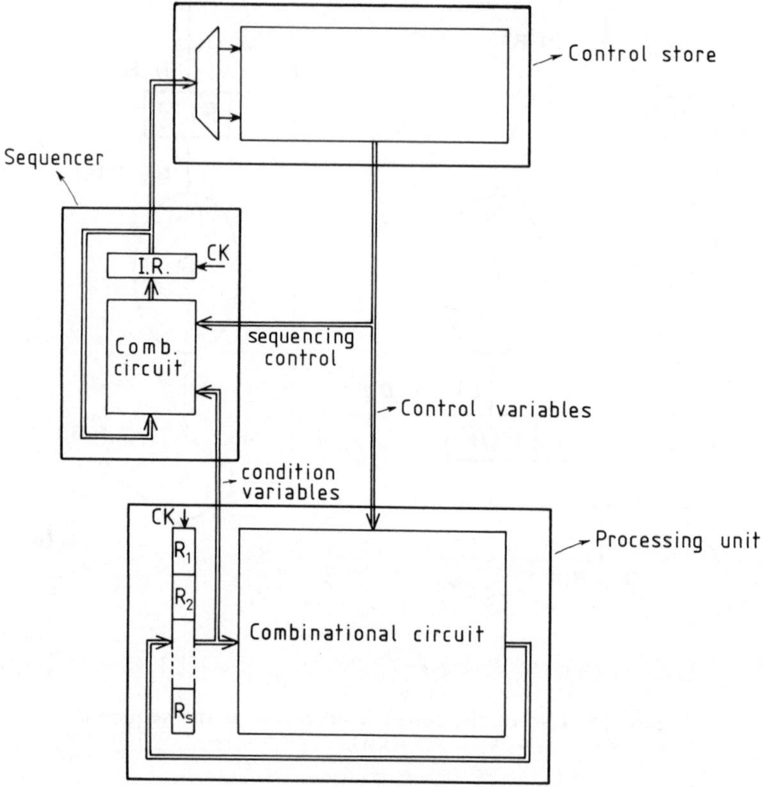

Figure 26. Pipeline technique

Let us introduce some notations:

$T_{\text{c.s.}}$ is the delay introduced by the control store,

$T_{\text{p.u.}}$ is the delay introduced by the processing unit,

$T_{\text{sequ.}}$ is the delay introduced by the sequencer,

T is the clock signal period.

During one clock period the following operations are performed:

(a) the instruction address is latched in register *I.R.*:

(b) the preceding computation results are latched in registers R_i, as well as the corresponding values of the condition variables;

(c) the instruction is read: this takes a time $T_{\text{c.s.}}$;

(d) the instruction is executed by the processing unit: this takes a time $T_{\text{p.u.}}$, and

(e) at the same time, the next address instruction is computed by the sequencer: this takes a time $T_{\text{sequ.}}$.

Hence,

$$T > T_{\text{c.s.}} + \max(T_{\text{p.u.}}, T_{\text{sequ.}}). \tag{48}$$

If the register loading times are negligible, then (48) gives a lower bound of the clock signal period.

Remark. In the instructions defined above ((14) of Chapter X), the sequencing is based on results of the computation performed during the last instruction execution. An alternative instruction type is

$$\begin{array}{lll} N\ \sigma & f_0(\mathbf{x}) & N_0 \\ & f_1(\mathbf{x}) & N_1 \\ & \text{-------} & \text{-------} \\ & f_{r-1}(\mathbf{x}) & N_{r-1} \end{array} \tag{49}$$

The sequencing is based on the results of the computation performed during the present instruction execution. The following operations are successively performed in one clock signal period: reading of an instruction, execution of the command, and computation of the next instruction address. Thus

$$T > T_{\text{c.s.}} + T_{\text{p.u.}} + T_{\text{sequ.}} \tag{50}$$

Furthermore, a piece of program such as

$$\begin{array}{llll} N & \sigma & f_0(\mathbf{x}) & N_0 \\ & & f_1(\mathbf{x}) & N_1 \\ & & \text{-------} & \text{-------} \\ & & f_{r-1}(\mathbf{x}) & N_{r-1} \\ N_0 & \sigma_0 & \text{-------} & \text{-------} \\ N_1 & \sigma_1 & \text{-------} & \text{-------} \\ \text{-------} & \text{-------} & \text{-------} & \text{-------} \\ N_{r-1} & \sigma_{r-1} & \text{-------} & \text{-------} \end{array}$$

is equivalent to

$$\begin{array}{llll} \text{-------} & & \sigma & N \\ N & f_0(\mathbf{x}) & \sigma_0 & N_0 \\ & f_1(\mathbf{x}) & \sigma_1 & N_1 \\ & \text{-------} & \text{-------} & \text{-------} \\ & f_{r-1}(\mathbf{x}) & \sigma_{r-1} & N_{r-1} \\ \text{-------} & \text{-------} & \text{-------} & \text{-------} \end{array}$$

Hence the chosen formalism naturally leads to the simultaneous working of the processing unit and of the sequencer, and is equivalent to the one we would deduce from (49).

The pipeline technique allows to lower the bound (48), that is, to make the control store and the processing unit work simultaneously. For this

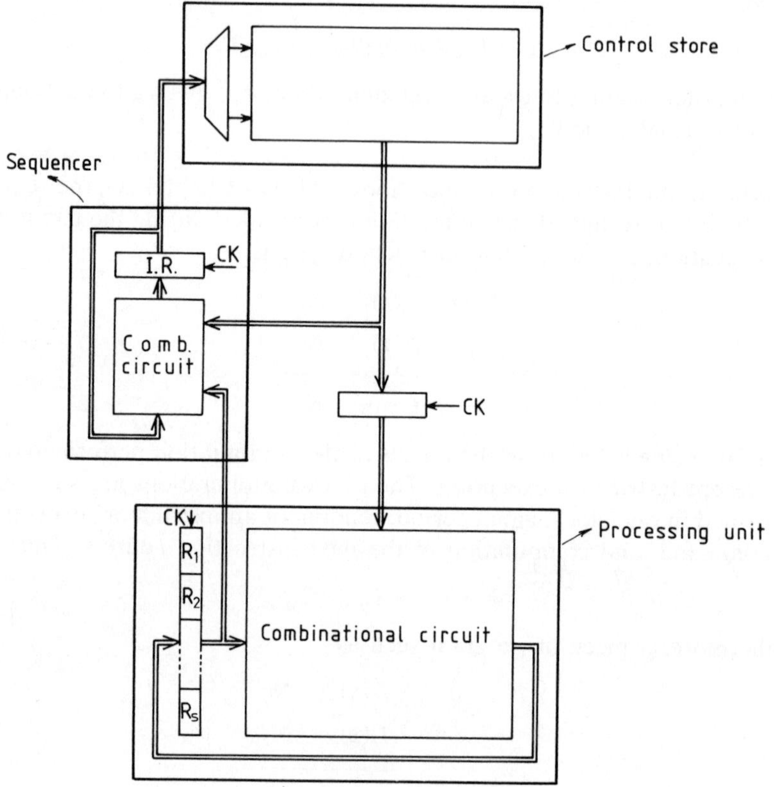

Figure 27. Pipeline technique

purpose we add a register, as shown in Figure 27. The execution of each instruction is delayed for one clock signal period. As a consequence, the value taken by the condition variables at the time when the next instruction address is computed, depends on the penultimate computation results and not on the ultimate computation results. Hence, it is necessary to take a lot of precautions. If the case arises, it may be necessary to introduce fictitious instructions such as

$$N \quad t \quad \lambda \quad N'.$$

EXERCISES

E1. Consider the following instruction type sets

(i) (a) $N_i \quad \bar{x} \quad \lambda \quad N_{i+1}$

$\qquad x \quad \lambda \quad N_j$

(b) $N_i \quad \bar{x} \quad \lambda \quad N_j$

$\qquad x \quad \lambda \quad N_{i+1}$

(c) N_i t σ N_{i+1}

(d) N_i t σ N_j

(ii) (a) N_i t σ N_j

(b) N_i $\bar{x}$ λ N_j

x λ N_k

(c) N_i $\bar{x}$ λ N_j

x σ N_j

Discuss their universality. Derive appropriate control unit. Propose other types of universal instruction type sets.

E2. It has been shown in Exercise E4 of Chapter XII that every program may be replaced by an equivalent regular program. Regular programs do not contain any more explicit jumps: the latter are actually replaced by subroutine calls. Design a sequencer appropriate to handle regular programs.

E3. As an alternative to E2, develop a method for translating a regular program into a conditional program. The translation program should itself be a conditional program.

Chapter XV
Microprocessors

1 INTRODUCTION

According to the viewpoint throughout this work, a digital system implements an algorithm. This last is characterized by the computation resources it requires, and by the way the various operations are linked together. This leads to the distinction between the processing unit and the control unit.

The implementation of a new algorithm can be performed in several ways. One of these consists in designing an entirely new circuit by using, for instance, some of the methods developed above. In particular, the designer will devise a processing unit able to perform the computations included in the algorithm; he will list the control variables necessary to control the processing unit, and choose a series of test points, i.e. condition variables, useful for the sequencing.

It can occur that the algorithm only requires standard computation resources; if the case arises, the algorithm could be transformed with this objective. Then it is possible to use a previously designed processing unit, whether it has been devised for another application or whether it is a standard processing unit, for instance a series of bit slices. The control unit can be made up of standard circuits (sequencer, PLA, ROM, . . .). The task of the designer includes the writing of the control program, the implementation of the program on the physical medium (PLA, ROM, . . .), and the connection of the various parts. The designing of digital systems is thus broken down into two tasks: the realization of reliable circuits, whose behaviour is well known and which are produced in great numbers, and the writing of a correct program, that is a program which actually corresponds to the algorithm.

The advent of a third type of design is due to the following observation: the processing unit is made up of registers, combinational computation circuits, and connection circuits; the amount of computations and transfers it can perform in one time unit is relatively small: prominence was given to this fact in the preceding chapters, among others in the introduction of Chapter X. Hence, the number of steps necessary to execute the algorithm is large. Furthermore, it is necessary, at every step, to specify the state of a series of control wires: register loading, connection network positioning, and computation resource programming. Eventually, the writing of the control program can be a hard and tedious task, and thus a process which generates errors.

These points led to the conception of systems which externally accept a restricted number of useful commands under encoded form; in fact, the execution of every command is controlled by a subprogram generating the values which must successively be applied on the control wires of the processing unit. As an example, let us examine a processing unit based on Figure 17 of Chapter XI. The functional transfer

$$[M_i] := [M_i] + [M_j] \tag{1}$$

requires two clock periods, each of which is decomposed into two phases:

$$\begin{aligned}
&\phi_1 = 1, \phi_2 = 0\text{: ADDR.} := j, [S_A] := [M_j],\\
&\phi_1 = 0, \phi_2 = 1\text{: } [M_{ACC}] := [S_A];\\
&\phi_1 = 1, \phi_2 = 0\text{: ADDR.} := i, [S_A] := [M_i], [S_{ACC}] := [M_{ACC}],\\
&\phi_1 = 0, \phi_2 = 1\text{; ADDR.} := i, \text{EN} = 1, \text{ OPER.} = \text{addition.}
\end{aligned} \tag{2}$$

This functional transfer frequently occurs during the execution of a program. This leads to the devising of a processing unit which, apparently, can directly execute transfer (1), expressed under encoded form; in fact, it includes a circuit which associates with the encoded command the sequence of instructions (2) which the processing unit actually can execute.

If we generalize this basic idea, we get a system including two instruction levels: those which include encoded commands and form the *main program* and those which can be executed directly by the processing unit. The translation of every instruction of the first type into a series of instructions of the second type, concomitantly with the execution of these last instructions by the processing unit, is often called *interpretation* of the main program. The interpretation algorithm is in turn implemented by a digital system including a processing unit, namely the initial processing unit, and a control circuit defined by an *interpretation program.* We so obtain the general architecture of Figure 1. Every instruction of the initial program contains sequencing information and commands which can be executed by the virtual processing unit made up of the interpretation program memory, the interpretation program sequencer, and the processing unit. Every command initiates a particular interpretation subprogram. This obviously raises the problem of synchronizing the main program and the interpretation program.

Figure 1 suggests several comments.

(1) It gives prominence to a recurrent construction: the processing unit of Figure 1 could also be replaced by a virtual processing unit containing an interpretation program memory, an interpretation program sequencer, and a processing unit, and so on. We obtain a series of programs related by interpretation programs; the translation of main program instructions into series of instructions which the last processing unit can execute is broken down into a series of interpretation processes.

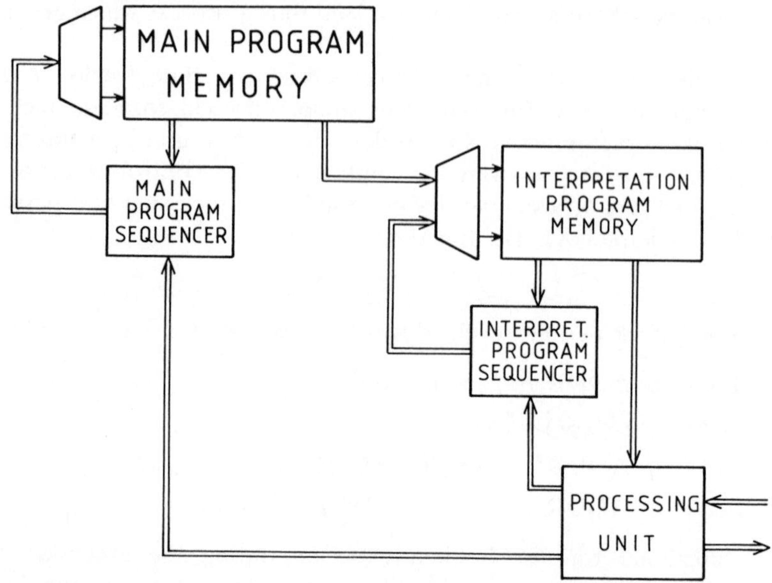

Figure 1. Two program structure

(2) The main program sequencer is a kind of processing unit specialized in the computing of the main program next instruction address. A frequent alternative consists in integrating the main program sequencer into the processing unit; the main program address counter then becomes a particular register of the processing unit, and the interpretation program controls both the operations and the sequencing of the main program. This increases the amount of structured circuit parts (ROM, PLA, . . .) and resolves the problem of synchronizing the main program and the interpretation program since all operations are controlled by the interpretation program.

(3) The addition of input and output circuits gives to the circuit of Figure 1 the structure of the *von Neumann machine* (see, for instance, Baer, 1980) The *control unit* includes the interpretation program memory, the interpretation program sequencer, and the main program address counter. The *memory* includes the main program memory and some of the registers contained in the processing unit and in the main program sequencer; hence the memory can store instructions of the main program, data, and addresses. The *arithmetic and logic unit* consists of the combinational circuits of the processing unit and of the remaining registers.

(4) The implementation of the interpretation algorithm by means of a memory and a sequencer is not a general rule; in some cases, the control automaton is a few structured circuits whose area has been minimized; it is a *hardwired control* system.

Figure 1 gives an example of *microprocessor architecture*; it is a standard circuit, able to interpret and sequence the instructions of a main program. The interpretation program is written once for all; the designer's task is limited to the writing of the main program, its implementation in the main program memory, and the connection of a few of chips: the microprocessor, the memories if they are separate from the microprocessor, and some auxiliary circuits.

We have just described three types of design. In the first case, that is the creation of an entirely new circuit, the design effort is considerable. This choice is made only in very special cases; numerous circuit production, special functions, high performances, etc. The second type of design consists in assembling existing circuits (bit slice processing units, sequencers, memories, etc.), and in writing and implementing the control program. The design effort is reduced while the performances may remain comparable: the designer then has the option to somehow fit the system architecture to the application under consideration; furthermore, if necessary, high performance standard circuits can be chosen. The third type of design is based on the use of microprocessors; in this case, the design effort is minimal. Every time the obtained performances are sufficient for the application, we should consider this third solution.

2 Example of two instruction level system

Let us consider the processing unit of Figure 2 of Chapter X; we shall now define an interpretation algorithm able to transform it into a virtual processing unit which can execute the following instructions (we suppose that $n = r$):

(1) *Transfers*:

N_ℓ t $[R_S]:=[R_P]$ $N_{\ell+1}$; (MOV R_P, R_S)

N_ℓ t $[R_S]:=[R_X]$ $N_{\ell+1}$; (MOV R_X, R_S)

N_ℓ t $[R_P]:=[R_Q]$ $N_{\ell+1}$; (MOV R_Q, R_P)

N_ℓ t $[R_P]:=[R_A]$ N_{l+1}; (MOV R_A, R_P)

(2) *Multiplication in* $GF(2^n)$:

N_ℓ t $[R_p]:=[R_P][R_S]$ $N_{\ell+1}$; (MUL)

(3) *Counting*:

N_ℓ t $[C(i)]:=[C(i)]-1$ $N_{\ell+1}$; (DEC)

(4) *Initializations*:

N_ℓ t $[R_P]:=1$ $N_{\ell+1}$; (RESET R_P)

N_ℓ t $[C(i)]:=n$ $N_{\ell+1}$; (RESET C)

(5) *Shift*:

N_ℓ t $[R_A]:=2[R_A]$ $N_{\ell+1}$; (SHIFT)

(6) *Jumps* (x stands for $\bar{x}_2$ or x_3):

$N_\ell \quad \bar{x} \quad \lambda \quad N_{\ell+1}$

$\quad\quad x \quad \lambda \quad N_s$; (JNZ x, N_S)

$N_\ell \quad t \quad \lambda \quad N_s$; (JMP N_s)

For every instruction type, a *mnemonic* expression is written between brackets.

These instructions allow incremental programs to be constructed, in which appear two condition variables: $\bar{x}_2$ and x_3.

With every command we associate a subprogram that the processing unit can execute:

(1) MOV R_P, $R_S : \sigma_6$.
(2) MOV R_X, $R_S : \sigma_8$.
(3) MOV R_Q, $R_P : \sigma_7$.
(4) MOV R_A, R_P; this transfer may be performed by the following series of operations:

$[R_Q] := 0, [R_P] := 1;$

for $j := n-1$ *downto* 0 *do*
begin $[R_Q] := 2[R_Q]$;

$[R_Q] := [R_Q] \oplus a_j[R_p]$

end;

$[R_P] := [R_Q]$.

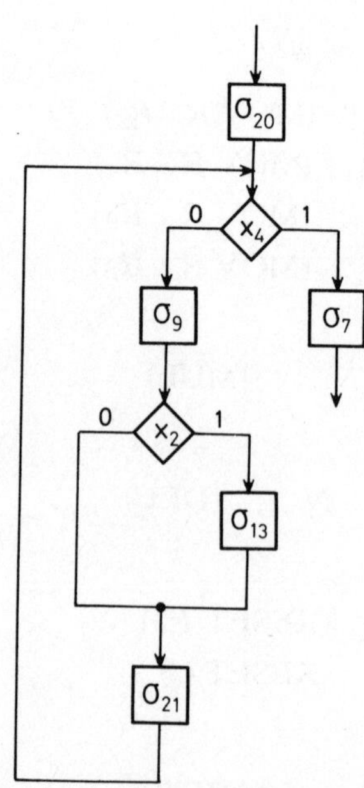

Figure 2. Subprogram

The corresponding subprogram is shown in Figure 2; we have introduced two new composite commands:

$$\sigma_{20}=\sigma_0\sigma_1\sigma_3, \qquad \sigma_{21}=\sigma_5\sigma_{11}.$$

Let us observe that, at the end of the computation $[R_Q]=[R_P]$ and $[R_A]=0$.

(5) MUL: the corresponding subprogram is given in Figure 10 of Chapter X.

(6) DEC: σ_4.

(7) RESET R_P: σ_0.

(8) RESET C: σ_2.

(9) SHIFT: σ_{11}.

(10) JNZ x, N_s and JMP N_s: λ.

Every command may be encoded by four binary variables w_0, w_1, w_2, w_3 as shown in Figure 3.

	w_3	w_2	w_1	w_0
JNZ x, N_s and JMP N_s	0	0	0	0
MOV R_P, R_S	0	0	0	1
MOV R_X, R_S	0	0	1	0
MOV R_Q, R_P	0	0	1	1
MOV R_A, R_P	0	1	0	0
MUL	0	1	0	1
DEC	0	1	1	0
RESET R_P	0	1	1	1
RESET C	1	0	0	0
SHIFT	1	0	0	1

Figure 3. Command encoding

The interpretation program is described in Figure 4. The synchronization of the main program and of the interpretation program is performed by the auxiliary variable S which enables the main program counter at the beginning of every loop in the interpretation program. The corresponding circuit is shown in Figure 5 (clock signals are not represented and timing problems are not discussed).

The following main program controls the functional transfer

$$[R_P]:=[R_X]^{[R_A]}:$$

No.	Mnemonic	i	v_0	v_1	w_3	w_2	w_1	w_0
0	RESET R_P	–	–	1	0	1	1	1
1	RESET C	–	–	1	1	0	0	0
2	JNZ x_3, N_{11}	1	0	0	1	0	1	1
3	MOV R_P, R_S	–	–	1	0	0	0	1
4	MUL	–	–	1	0	1	0	1
5	JNZ $\bar{x}_2$, N_8	0	0	0	1	0	0	0
6	MOV R_X, R_S	–	–	1	0	0	1	0
7	MUL	–	–	1	0	1	0	1
8	DEC	–	–	1	0	1	1	0
9	SHIFT	–	–	1	1	0	0	1
10	JMP N_2	–	1	0	0	0	1	0
11	END							

(3)

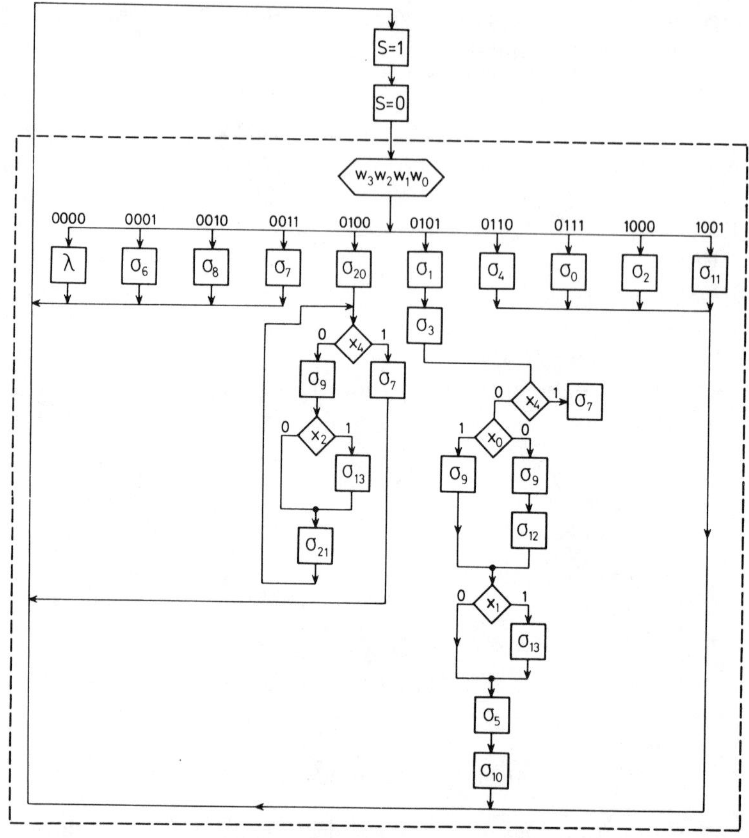

Figure 4. Interpretation program

Another algorithm may be implemented by the same circuit, with a different main program. As an example, the following main program controls the transfer

$$[R_P]:=[R_A][R_X]:$$

No.	Mnemonic	i	v_0	v_1	w_3	w_2	w_1	w_0
0	MOV R_A, R_P	–	–	1	0	1	0	0
1	MOV R_X, R_S	–	–	1	0	0	1	0
2	MUL	–	–	1	0	1	0	1
3	END							

Let us recall that $[R_A]=0$ at the end of the execution.

It has been seen in the introductory section that the main program sequencer may be deleted if the main program sequencing is controlled by the interpretation program. The main program counter PC then becomes

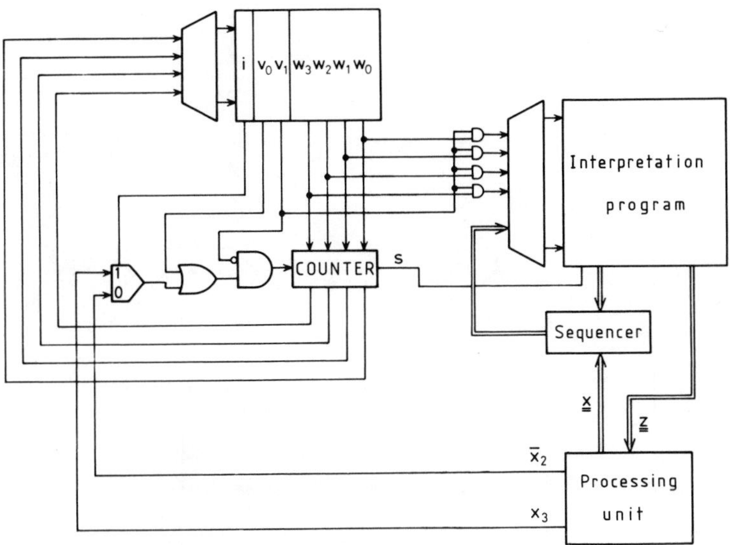

Figure 5. Complete circuit

part of the processing unit. As an example, the exponentiation program (3), which computes

$$[R_p]:=[R_X]^{[R_A]}$$

then becomes

$$
\begin{array}{llll}
N & ([PC]=0) & [R_P]:=1, [PC]:=[PC]+1 & N \\
 & ([PC]=1) & [C(i)]:=n, [PC]:=[PC]+1 & N \\
 & ([PC]=2) & [PC]:=([PC]+1)\bar{x}_3+11x_3 & N \\
 & ([PC]=3) & [R_S]:=[R_P], [PC]:=[PC]+1 & N \\
 & ([PC]=4) & [R_P]:=[R_P][R_S], [PC]:=[PC]+1 & N \\
 & ([PC]=5) & [PC]:=8\bar{x}_2+([PC]+1)x_2 & N \\
 & ([PC]=6) & [R_S]:=[R_X], [PC]:=[PC]+1 & N \\
 & ([PC]=7) & [R_P]:=[R_P][R_S], [PC]:=[PC]+1 & N \\
 & ([PC]=8) & [C(i)]:=[C(i)]-1, [PC]:=[PC]+1 & N \\
 & ([PC]=9) & [R_A]:=2[R_A], [PC]:=[PC]+1 & N \\
 & ([PC]=10) & [PC]:=2 & N \\
 & ([PC]=11) & \text{END} &
\end{array} \tag{4}
$$

The transfer of the main program counter *PC* from the control unit to the processing unit allows the exponentiation program (3) to be transformed into the program (4) which includes only one instruction; this program transformation was mentioned at the end of Chapter X.

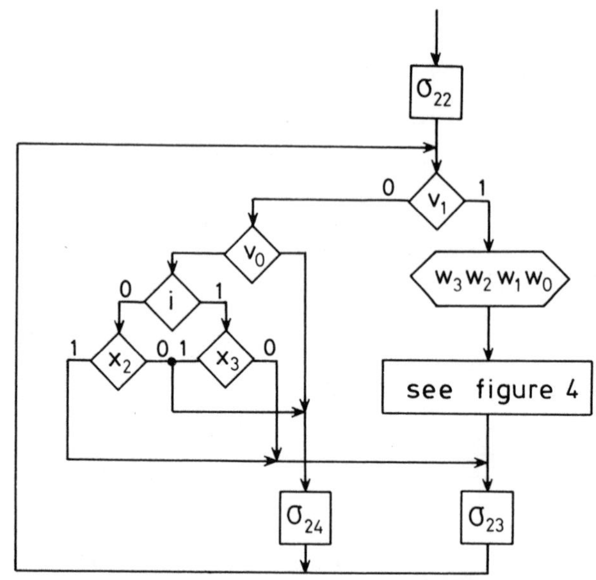

Figure 6. Interpretation program

As above in (3), the various commands may be encoded by means of seven variables i, v_0, v_1, w_3, w_2, w_1, w_0. The new interpretation program is shown in Figure 6; it includes three new commands:

$$\sigma_{22}:[PC]:=0,$$

$$\sigma_{23}:[PC]:=PC+1,$$

$$\sigma_{24}:[PC]:=w_3w_2w_1w_0.$$

The circuit is shown in Figure 7. Its processing unit includes the initial processing unit of Figure 2 of Chapter X. and the program counter PC controlled by three variables z_{14}, z_{15}, and z_{16} which allow the commands σ_{22}, σ_{23} and σ_{24} to be activated.

Comments.

(1) With regard to the main programs, there is no difference between the circuits of Figures 5 and 7.

(2) In the circuit of Figure 7, the main program sequencer has been deleted and replaced by a few more instructions in the interpretation program. The proportion of well structured circuit parts has been increased.

(3) If the processing unit contains an adder, necessary to execute the commands of the main program, it could also be used to increment the main program counter PC, under the control of the interpretation program. The program counter then becomes a simple register. Nevertheless, this type of

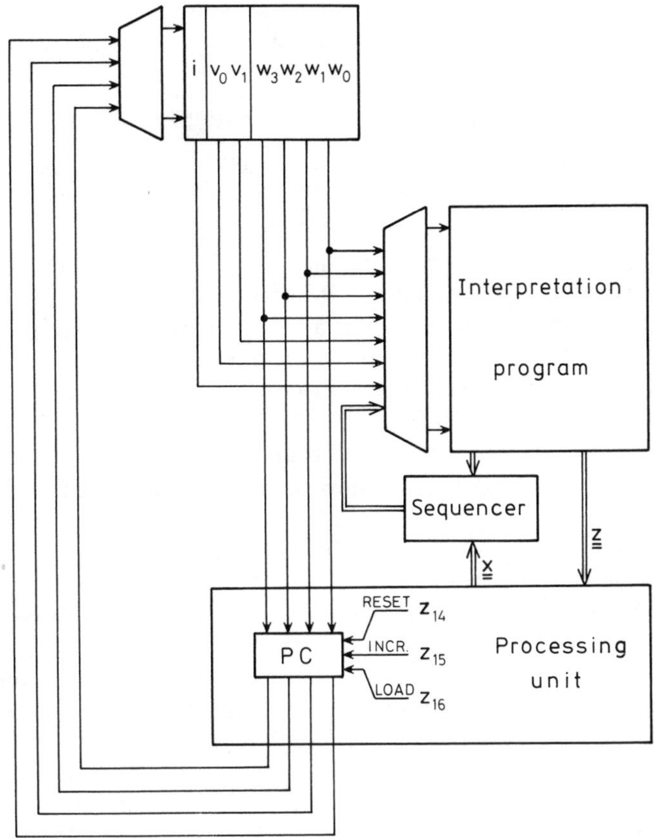

Figure 7. Complete circuit

organization prevents one from simultaneously managing the program counter and executing commands: once more we are faced with a time–complexity trade-off.

(4) Instead of using the processing unit of Figure 10 of Chapter X, we could use the very general processing unit of Figure 17 of Chapter XI, provided its arithmetic and logic unit can perform the following operations: value transfer, incrementation, decrementation, left shift, and exclusive or. The registers R_X, R_A, R_M, R_P, R_Q, R_S, and PC are chosen from among the registers $M_1, M_2, \ldots$; they are addressed by the interpretation program which confers on every register its particular function. The constant values 0, 1, and n must also be stored. As the arithmetic and logic unit can also be designed as a stack of identical slices, the whole circuit becomes extremely simple and structured. Obviously, this type of circuit would be very slow.

(5) The interpretation program can be transformed and implemented

according to the methods presented in Chapters XII, XIII, and XIV: conditional form, incremental form, sequencing by bit replacement, etc.

3 SECOND EXAMPLE: A VON NEUMANN MACHINE

Throughout the course of this section we will construct a system able to execute a given instruction set.

A series of choices must be made before starting the design.

(1) As regards the *instruction set*, that is the list of instruction types the system is able to execute, we choose it in a rather arbitrary way; it is a very simple general purpose instruction set allowing main programs, to be written with the minimum of ease. It includes the following basic operations: clearing (CLR), complementation (CPL), AND, OR, addition (ADD), and data transfers (MOV, IN, OUT). The sequencing can be controlled by jump instructions (JMP, JC, JZ) and by subroutine calls (CALL, RET).

(2) In order to simplify the example, all handled values will be *bytes*, that is eight bit vectors.

(3) Data and instructions are stored in a random access memory which, according to choice (2), contains 256 bytes: $R0, R1, \ldots, R255$.

(4) As regards the instruction *format*, there are one byte, two byte, and three byte instructions. The first byte is the *operation code*; the second and third bytes give memory addresses or data necessary to execute the instruction. This is a rather uneconomical format, but it respects choice (2) and, again, simplifies the example.

(5) The functions of the arithmetic and logic unit will be clarified below, but we, *a priori*, know that it can generate a carry and store it in a carry flip-flop CY.

(6) Data input and output is performed by way of port P.

Here is the instruction set.

One byte instructions.

(1) *CLR CY*:	N_ℓ t $[CY]:=0$	$N_{\ell+1}$	;
(2) *CPL CY*:	N_ℓ t $[CY]:=[\overline{CY}]$	$N_{\ell+1}$	;
(3) *RET*:	subroutine return;		
(4) *NOP*:	N_ℓ t λ	$N_{\ell+1}$	.

Two byte instructions.

(5) *CPL Ri*:	N_ℓ t $[Ri]:=[\overline{Ri}]$	$N_{\ell+2}$	;
	$N_{\ell+1}$ is used to store address i		;
(6) *CLR Ri*:	N_ℓ t $[Ri]:=0$	$N_{\ell+2}$	;
	$N_{\ell+1}$ is used to store address i		;
(7) *IN Ri*:	N_ℓ t $[Ri]:=[P]$	$N_{\ell+2}$	;
	$N_{\ell+1}$ is used to store address i		;
(8) *OUT Ri*:	N_ℓ t $[P]:=[Ri]$	$N_{\ell+2}$	;
	$N_{\ell+1}$ is used to store address i		;
(9) *JMP Ns*:	N_ℓ t λ	N_s	;
	$N_{\ell+1}$ is used to store address s		;

(10) *JC Ns*: N_ℓ $[\overline{CY}]$ λ $N_{\ell+2}$;
$[CY]$ λ N_s ,
$N_{\ell+1}$ is used to store address s ;

(11) *CALL Ns*: call of the subroutine whose first instruction label is N_s;
$N_{\ell+1}$ is used to store address s.

Three byte instructions.

(12) *MOV Ri, Rj*: N_ℓ t $[Rj]:=[Ri]+[Rj], [CY]:=\text{carry}$ $N_{\ell+3}$
$N_{\ell+1}$ is to store address i;
$N_{\ell+2}$ is used to store address j;

(13) *AND Ri, Rj*: N_ℓ t $[Rj]:=[Ri]\wedge[Rj]$ $N_{\ell+3}$;
$N_{\ell+1}$ is used to store address i;
$N_{\ell+3}$ is used to store address j;

(14) *OR Ri, Rj*: N_ℓ t $[Rj]:=[Ri]\vee[Rj]$ $N_{\ell+3}$;
$N_{\ell+1}$ is used to store address i;
$N_{\ell+2}$ is used to store address j;

(15) *JZ Ri, Ns*: N_ℓ $\overline{([Ri]=0)}$ λ $N_{\ell+3}$;
$([Ri]=0)$ λ N_s ;
$N_{\ell+1}$ is used to store address i;
$N_{\ell+2}$ is used to store address s;

(16) *MOV Ri, Rj*: N_ℓ t $[Rj]:=[Ri]$ $N_{\ell+3}$;
$N_{\ell+1}$ is used to store address i;
$N_{\ell+2}$ is used to store address j;

(17) *MOV D, Ri*: N_ℓ t $[Ri]:=D$ $N_{\ell+3}$;
$N_{\ell+1}$ is used to store datum D;
$N_{\ell+2}$ is used to store address i.

Let us now design a processing unit able to execute these instructions. We again make a series of choices:

(1) Two bus structure: this allows a functional transfer, $[Rj]:=F([Ri], [Rj])$, to be performed in two clock periods (see Section 3.3 of Chapter XI). The processing unit thus includes an *accumulator register ACC*.

(2) Two instruction level system: the existence of two and three byte instructions, and the fact that there are only two buses naturally lead to this choice.

(3) The main program sequencing will be performed by the processing unit under the control of the interpretation program. Hence, the processing unit includes a *program counter PC*.

(4) The memory contains both instructions and numerical data. Instructions are addressed by the program counter, while data are often addressed by values contained in the main program. Hence, the program counter may not be permanently connected to the memory address decoder. This leads to the inclusion of a *memory address register AD*.

(5) The operation codes stored in the memory are supplied to the control unit, usually to the interpretation program sequencer, while the numerical data are supplied to the processing unit. The memory output may not be

permanently connected to the control unit. Hence, we include an *instruction register IR* which stores the current instruction operation code.

(6) The possibility to call subroutines imposes the presence of a stack. It will be made up of a part of the memory and of a *stack pointer*; it is a register whose initial content is 255, and whose content is decremented at each subroutine call and incremented at each subroutine return.

(7) The conditional jump JZ is based on the value, zero or not, of $[Ri]$, for some $i=0, 1, \ldots, 255$. Hence, we include a NOR gate able to test the value of $[Ri]$ and a flip-flop Z which stores the test result. Flip-flops such as CY or Z are sometimes called *flags*. In some cases, the various flags are gathered and form a *status word register* which, sometimes, also includes the stack pointer.

These various choices yield the processing unit of Figure 8. The clock signals are not represented. The wires z_0 to z_8 control the register loading; the wires z_9 to z_{13} control the bus contents; the wires z_{14} to z_{16} control the arithmetic and logic unit, the working of which will be specified below.

The condition variables provided to the control unit, are the flag contents (x_0 and x_1) and the instruction register contents (w_0 to w_7).

The general structure of the interpretation program is shown in Figure 9. The RESET operation consists in driving the program counter to state 0, which is by hypothesis the main program's first instruction address, and the stack pointer to state 255. Figure 10 describes the interpretation subprogram of every main program instruction. We deduce from these various interpretation subprograms the list if operations the arithmetic and logic unit must be able to perform:

(1) Reset to zero: $[CY]:=0$, $[R[AD]]:=0$.

(2) Complementation: $[CY]:=[\overline{CY}]$, $R[AD]:=\overline{R[AD]}$.

(3) Incrementation: $[SP]:=[SP]+1, [PC]:=[PC]+1, [PC]:=R[AD]+1$.

(4) Decrementation: $[SP]:=[SP]-1$.

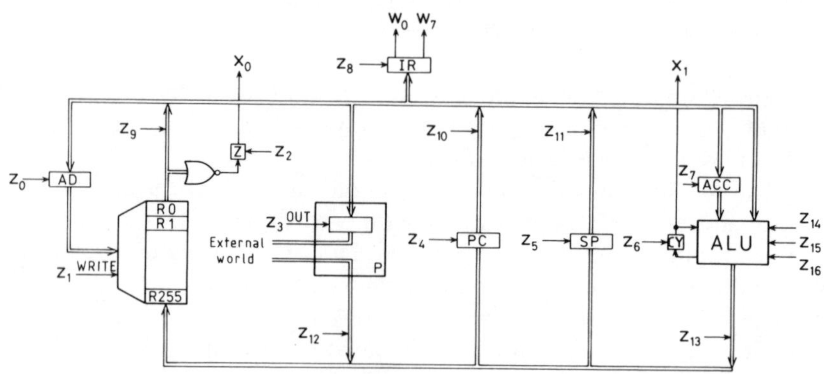

Figure 8. Processing unit

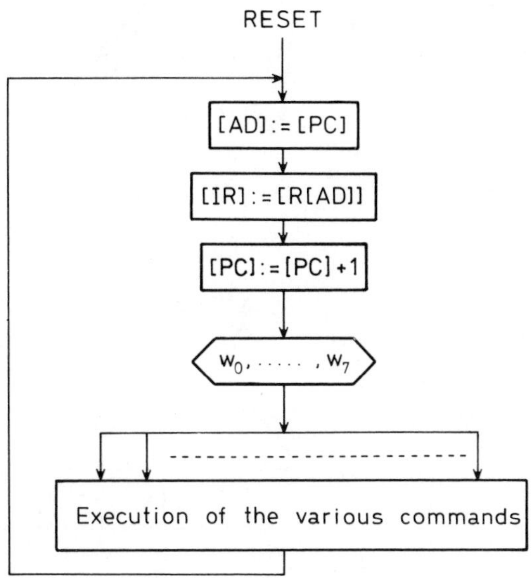

Figure 9. General structure of the interpretation program

(5) Operand composition: $[R[AD]]:=[ACC]+R[AD]$, $[R[AD]]:=[ACC]\wedge R[AD]$ $[R[AD]]:=[ACC]\vee R[AD]$.

(6) Data transfers: some transfers cannot be directly performed by the bus structure of Figure 8, namely:

$$[R[AD]]:=[ACC], [PC]:=[ACC], [PC]:=[R[AD]], [R[AD]]:=[PC].$$

Instead of adding connections to the buses, they can be made through the arithmetic and logic unit. Furthermore, the two last transfers are performed in two steps, via *ACC*.

Hence, there are essentially eight different input–output behaviours of the arithmetic and logic unit; they are encoded by the control variables z_{14}, z_{15}, and z_{16} according to Figure 11.

Figure 12 gives the list of commands which are actually used, and the corresponding values of the control variables z_0 to z_{16}.

A first version of the interpretation algorithm, before encoding of the main program instructions, is given in Figure 13; identical program parts have been merged in order to minimize the number of instructions. We deduce the instruction encoding of Figure 14 which is conceived in order to simplify the branching condition evaluation: the first branching, in the interpretation program of Figure 13, divides the seventeen instructions in two groups

$$\text{ADD } Ri, Rj; \quad \ldots; \quad JZ\, Ri, N_s$$

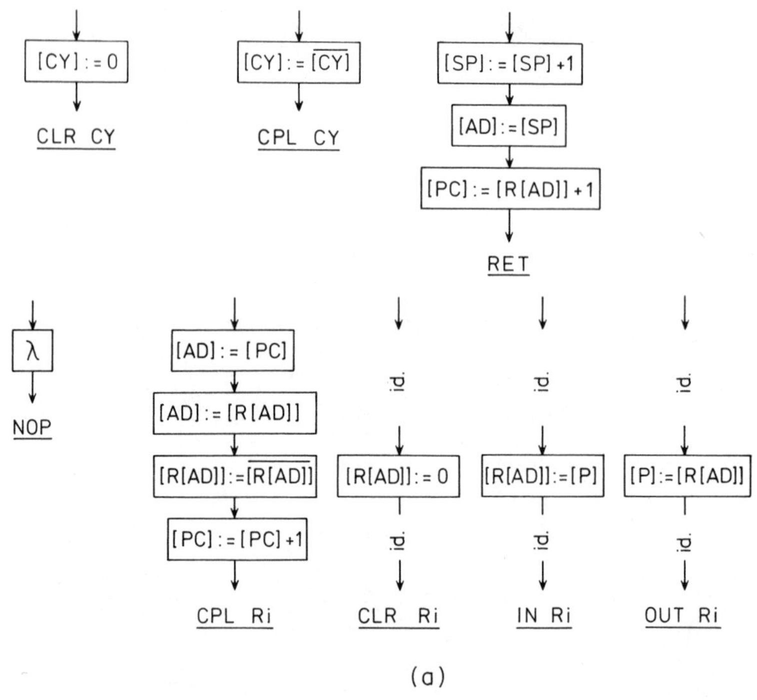

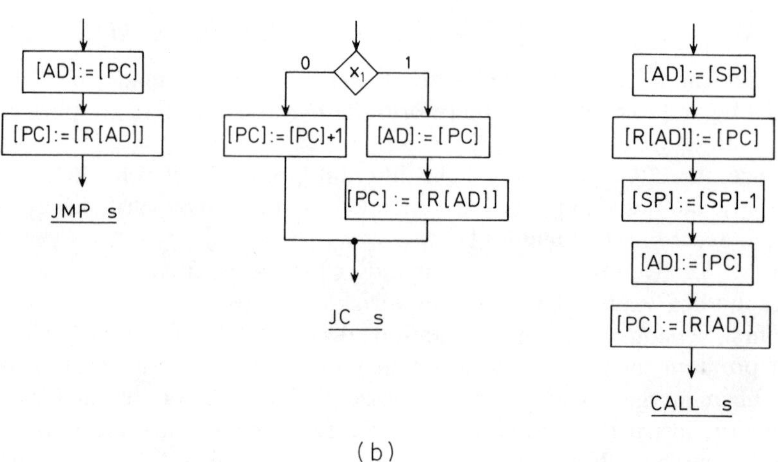

Figure 10. Interpretation of every main program instruction

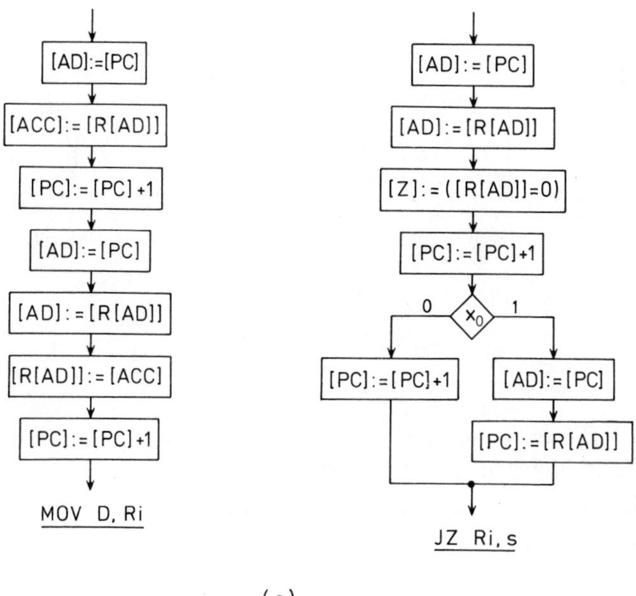

(c)

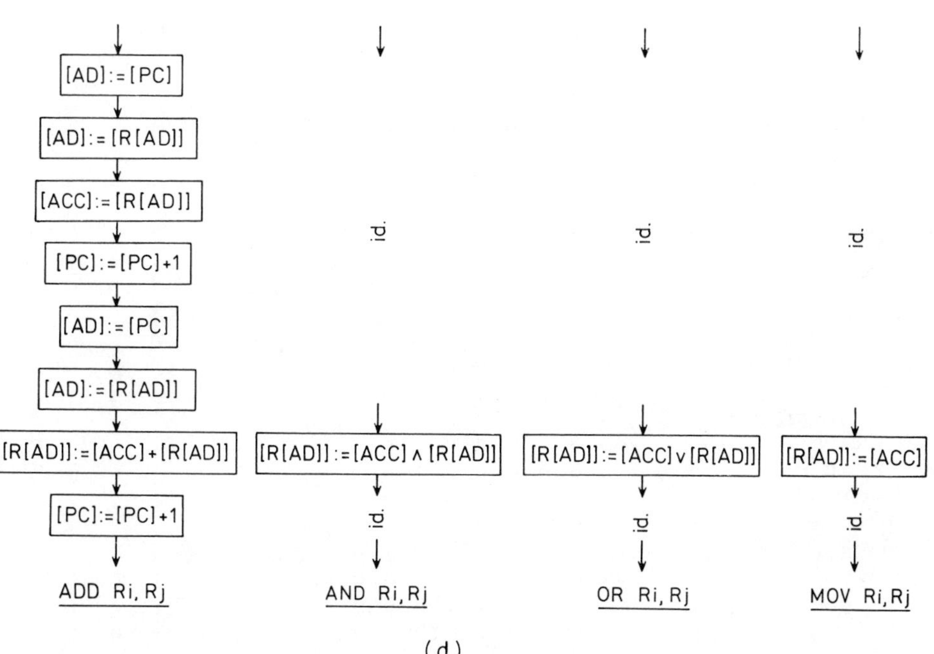

(d)

Figure 10 (*Continued*)

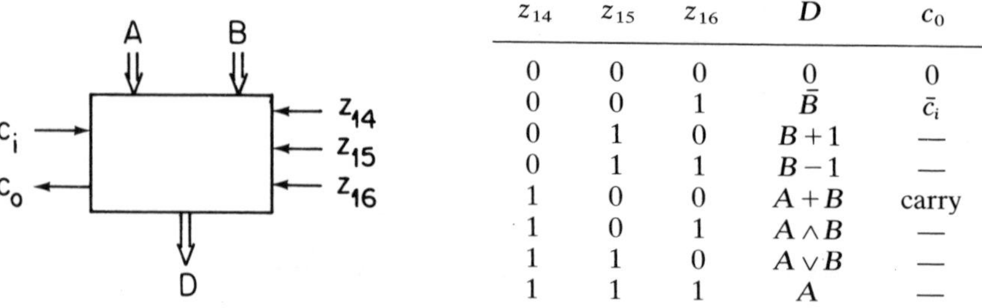

z_{14}	z_{15}	z_{16}	D	c_0
0	0	0	0	0
0	0	1	$\bar{B}$	$\bar{c}_i$
0	1	0	$B+1$	—
0	1	1	$B-1$	—
1	0	0	$A+B$	carry
1	0	1	$A \wedge B$	—
1	1	0	$A \vee B$	—
1	1	1	A	—

Figure 11. Arithmetic and logic unit

and

MOV *D*, *Ri*; . . . ; NOP.

In Figure 14, these two groups are distinguished by w_0. Similarly, a second branching divides the first group in three subgroups

ADD; AND; OR; MOV

CPL; CLR; IN; OUT

JZ,

which are distinguished by w_1 and w_2. Finally, inside every subgroup, the actual instruction is encoded by w_4 and w_5. The eight instructions of the

	z_0	z_1	z_2	z_3	z_4	z_5	z_6	z_7	z_8	z_9	z_{10}	z_{11}	z_{12}	z_{13}	z_{14}	z_{15}	z_{16}
σ_0 $[AD]:=[PC]$	1	0	0	0	0	0	0	0	0	0	1	0	—	—	—	—	—
σ_1 $[IR]:=[R[AD]]$	0	0	0	0	0	0	0	0	1	1	0	0	—	—	—	—	—
σ_2 $[PC]:=[PC]+1$	0	0	0	0	1	0	0	0	0	0	1	0	0	1	0	1	0
σ_3 $[AD]:=[R[AD]]$	1	0	0	0	0	0	0	0	0	1	0	0	—	—	—	—	—
σ_4 $[ACC]:=[R[AD]]$	0	0	0	0	0	0	0	1	0	1	0	0	—	—	—	—	—
σ_5 $[R[AD]]:=[ACC]+[R[AD]]$	0	1	0	0	0	0	1	0	0	1	0	0	0	1	1	0	0
σ_6 $[R[AD]]:=[ACC]\wedge[R[AD]]$	0	1	0	0	0	0	0	0	0	1	0	0	0	1	1	0	1
σ_7 $[R[AD]]:=[ACC]\vee[R[AD]]$	0	1	0	0	0	0	0	0	0	1	0	0	0	1	1	1	0
σ_8 $[R[AD]]:=\overline{[R[AD]]}$	0	1	0	0	0	0	0	0	0	1	0	0	0	1	0	0	1
σ_9 $[R[AD]]:=0$	0	1	0	0	0	0	0	0	0	—	—	—	0	1	0	0	0
$\sigma_{10}[R[AD]]:=[P]$	0	1	0	0	0	0	0	0	0	—	—	—	1	0	—	—	—
$\sigma_{11}[P]:=[R[AD]]$	0	0	0	1	0	0	0	0	0	1	0	0	—	—	—	—	—
$\sigma_{12}[PC]:=[ACC]$	0	0	0	0	1	0	0	0	0	—	—	—	0	1	1	1	1
$\sigma_{13}[Z]:=([R[AD]]=0)$	0	0	1	0	0	0	0	0	0	—	—	—	—	—	—	—	—
$\sigma_{14}[CY]:=0$	0	0	0	0	0	0	1	0	0	—	—	—	—	—	0	0	0
$\sigma_{15}[CY]:=\overline{[CY]}$	0	0	0	0	0	0	1	0	0	—	—	—	—	—	0	0	1
$\sigma_{16}[R[AD]]:=[ACC]$	0	1	0	0	0	0	0	0	0	—	—	—	0	1	1	1	1
$\sigma_{17}[AD]:=[SP]$	1	0	0	0	0	0	0	0	0	0	0	1	—	—	—	—	—
$\sigma_{18}[ACC]:=[PC]$	0	0	0	0	0	0	0	1	0	0	1	0	—	—	—	—	—
$\sigma_{19}[SP]:=[SP]-1$	0	0	0	0	0	1	0	0	0	0	0	1	0	1	0	1	1
$\sigma_{20}[SP]:=[SP]+1$	0	0	0	0	0	1	0	0	0	0	0	1	0	1	0	1	0
$\sigma_{21}[PC]:=[R[AD]]+1$	0	0	0	0	1	0	0	0	0	1	0	0	0	1	0	1	0
λ	0	0	0	0	0	0	0	0	0	—	—	—	—	—	—	—	—

Figure 12. Encoding of the commands

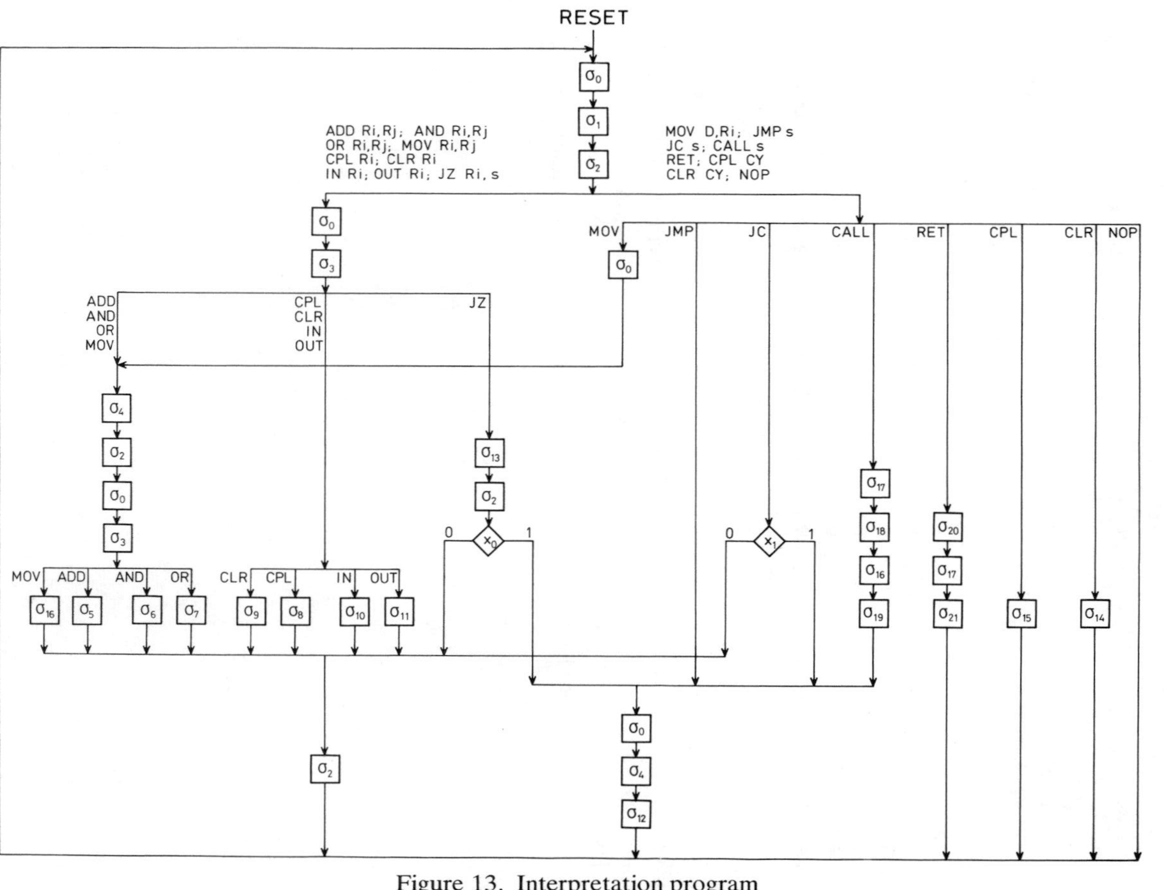

Figure 13. Interpretation program

	w_0	w_1	w_2	w_3	w_4	w_5	w_6	w_7
MOV *Ri*, *Rj*	1	1	0	1	0	0	—	—
ADD *Ri*, *Rj*	1	1	0	0	1	0	—	—
AND *Ri*, *Rj*	1	1	0	0	0	1	—	—
OR *Ri*, *Rj*	1	1	0	0	0	0	—	—
CLR *Ri*	1	0	1	1	0	0	—	—
CPL *Ri*	1	0	1	0	1	0	—	—
IN *Ri*	1	0	1	0	0	1	—	—
OUT *Ri*	1	0	1	0	0	0	—	—
JZ *Ri*, N_s	1	0	0	—	—	—	—	—
JMP N_s	0	1	0	0	0	0	0	0
JC N_s	0	0	1	0	0	0	0	0
MOV *D*, *Ri*	0	0	0	1	0	0	0	0
CALL N_s	0	0	0	0	1	0	0	0
RET	0	0	0	0	0	1	0	0
CPL *CY*	0	0	0	0	0	0	1	0
CLR *CY*	0	0	0	0	0	0	0	1
NOP	0	0	0	0	0	0	0	0

Figure 14. Encoding of the instructions

second group are encoded by a 0 or 1 out of 7 code, yielding a rather simple branching subprogram including seven incremental instructions.

The interpretation program is shown in Figure 15. It can be transformed into an incremented program including 64 instructions by adding some unconditional jumps:

N_0	t	σ_0	N_1
N_1	t	σ_1	N_2
N_2	t	σ_2	N_3
N_3	$\bar{w}_0$	λ	N_4
	w_0	λ	N_{38}
N_4	$\bar{w}_1$	λ	N_5
	w_1	λ	N_{22}
N_5	$\bar{w}_2$	λ	N_6
	w_2	λ	N_{36}
N_6	$\bar{w}_3$	λ	N_7
	w_3	λ	N_{26}
N_7	$\bar{w}_4$	λ	N_8
	w_4	λ	N_{18}
N_8	$\bar{w}_5$	λ	N_9
	w_5	λ	N_{14}
N_9	$\bar{w}_6$	λ	N_{10}
	w_6	λ	N_{12}
N_{10}	$\bar{w}_7$	λ	N_{11}
	w_7	λ	N_{63}
N_{11}	t	λ	N_0

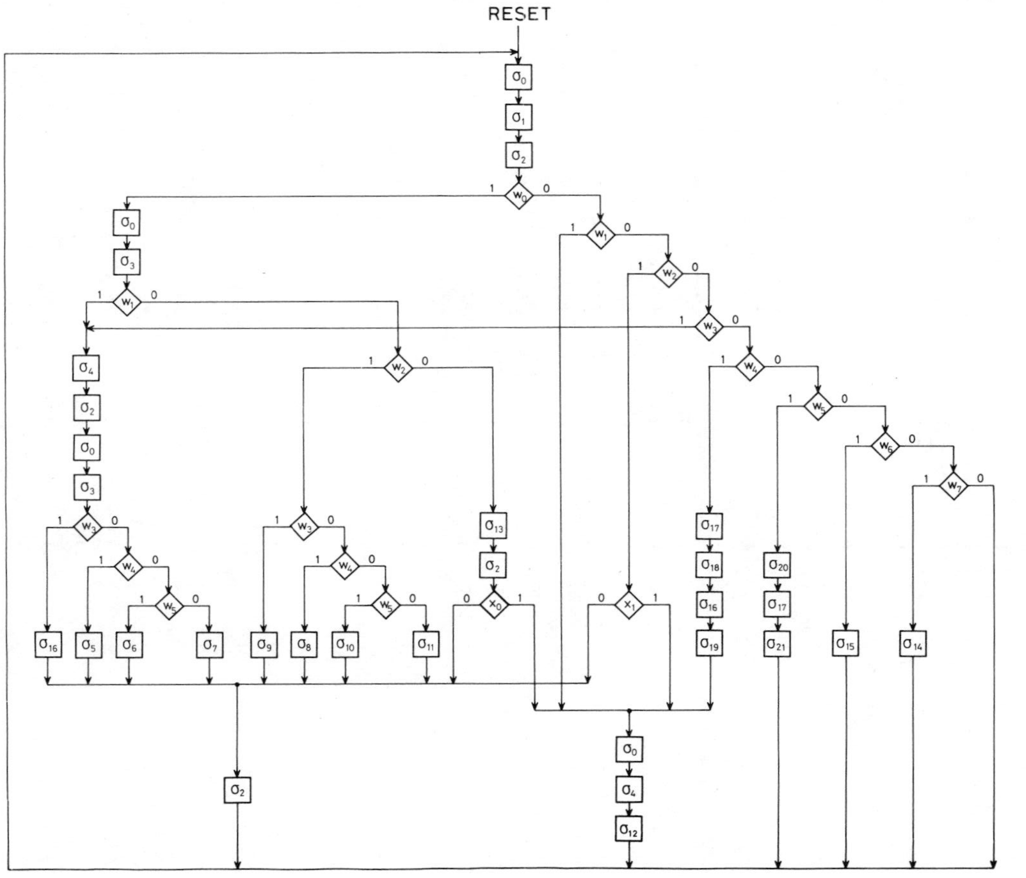

Figure 15. Conditional version of the interpretation program

N_{12}	t	σ_{15}	N_{23}
N_{13}	t	λ	N_0
N_{14}	t	σ_{20}	N_{25}
N_{15}	t	σ_{17}	N_{16}
N_{16}	t	σ_{21}	N_{17}
N_{17}	t	λ	N_0
N_{18}	t	σ_{17}	N_{19}
N_{19}	t	σ_{18}	N_{20}
N_{20}	t	σ_{16}	N_{21}
N_{21}	t	σ_{19}	N_{22}
N_{22}	t	σ_0	N_{23}
N_{23}	t	σ_4	N_{24}
N_{24}	t	σ_{12}	N_{25}
N_{25}	t	λ	N_0
N_{26}	t	σ_4	N_{27}
N_{27}	t	σ_2	N_{28}
N_{28}	t	σ_0	N_{29}
N_{29}	t	σ_3	N_{30}
N_{30}	$\bar{w}_3$	λ	N_{31}
	w_3	λ	N_{57}
N_{31}	$\bar{w}_4$	λ	N_{32}
	w_4	λ	N_{59}
N_{32}	$\bar{w}_5$	λ	N_{33}
	w_5	λ	N_{61}
N_{33}	t	σ_7	N_{34}
N_{34}	t	σ_2	N_{35}
N_{35}	t	λ	N_0
N_{36}	$\bar{x}_1$	λ	N_{37}
	x_1	λ	N_{22}
N_{37}	t	λ	N_{34}
N_{38}	t	σ_0	N_{39}
N_{39}	t	σ_3	N_{40}
N_{40}	$\bar{w}_1$	λ	N_{41}
	w_1	λ	N_{26}
N_{41}	$\bar{w}_2$	λ	N_{42}
	w_2	λ	N_{46}
N_{42}	t	σ_{13}	N_{43}
N_{43}	t	σ_2	N_{44}
N_{44}	$\bar{x}_0$	λ	N_{45}
	x_0	λ	N_{22}
N_{45}	t	λ	N_{34}
N_{46}	$\bar{w}_3$	λ	N_{47}
	w_3	λ	N_{51}
N_{47}	$\bar{w}_4$	λ	N_{48}
	w_4	λ	N_{53}

N_{48}	$\bar{w}_5$	λ	N_{49}
	w_5	λ	N_{55}
N_{49}	t	σ_{11}	N_{50}
N_{50}	t	λ	N_{34}
N_{51}	t	σ_9	N_{52}
N_{52}	t	λ	N_{34}
N_{53}	t	σ_8	N_{54}
N_{54}	t	λ	N_{34}
N_{55}	t	σ_{10}	N_{56}
N_{56}	t	λ	N_{34}
N_{57}	t	σ_{16}	N_{58}
N_{58}	t	λ	N_{34}
N_{59}	t	σ_5	N_{60}
N_{60}	t	λ	N_{34}
N_{61}	t	σ_6	N_{62}
N_{62}	t	λ	N_{34}
N_{63}	t	σ_{14}	N_0

This incremental program may be implemented by the circuit of Figure 16 (see Section 2 of Chapter XIV). The *control store*, that is the memory which stores the interpretation program, contains commands under encoded form; they are further decoded by the AND and OR arrays.

Comments.

(1) The main program is made up of one, two, and three byte instructions. The first byte gives the operation and the two other bytes contain numerical values. This is uneconomical since the operation code uses eight bits for encoding seventeen instructions. It is probably possible to use the first byte for both the operation code and, if necessary, a register address, by reducing the number of directly addressable registers and using indirect addressing (see Section 4.3) of the registers. Anyway, the fact of defining instructions with different formats can be compared with the decomposition of complex operations into simpler ones, which are executed in one time unit: instructions are decomposed into parts which can be stored in one memory unit.

(2) The choice of a two bus architecture was entirely arbitrary. Some indications concerning one, two, and three bus architectures are given in Chapter XI. As a matter of fact, we must find, for a given application, a good compromise between circuit simplicity and computation speed.

(3) We decided that all handled values would be bytes and this implied that the memory only contains 256 bytes. This is a very unrealistic choice. In practice, the program counter would include more than eight bits, and the stack management would be somewhat more complicated.

(4) Once the instruction set and the processing unit have been defined, it remains to write an interpretation algorithm, to encode the main program

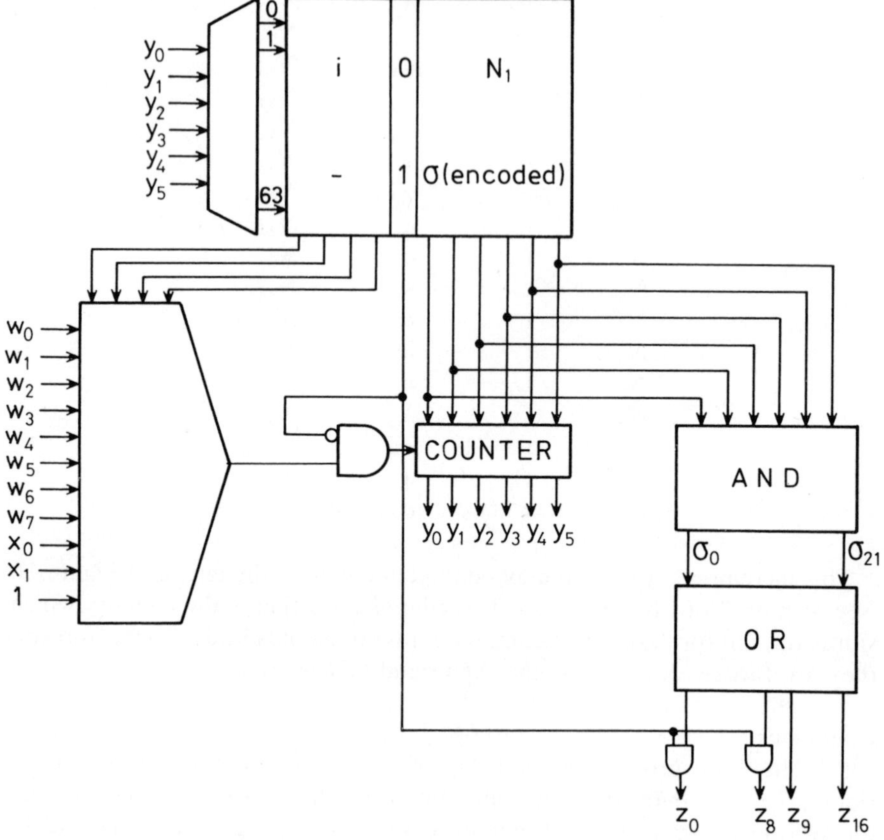

Figure 16. Control unit

instructions, and to synthesize the control unit. These two last points are related: in the example, we chose an encoding making the branching evaluations relatively simple, in the case of a conditional or incremental implementation. Other types of implementations, for example by bit replacement, would lead to another choice. Hence, there is an optimization problem, which is not dealt with in this work and is very hard; it is a generalized version of the input state assignment problem of a finite automation: the circuit structure is partially known in advance, but the condition variables can be tested simultaneously, or sequentially in a chosen order, or by small groups, etc.

As regards the choice of the encoding and of the interpretation program, it is interesting to notice that in practical cases some main program instructions appear much more often than other ones. As an example, instructions MOV appear more often than instruction JZ, and the latter is probably used more frequently than instruction OR. In practical cases it is necessary to take this fact into account so as to maximize the circuit performances at the main program level.

There is thus a very hard optimization problem. Systematic methods allow the solution of partial problems such as the prime implicant computation, the state number minimization of a sequential machine, or the optimization of a binary tree; there are no such methods for designing a complete control unit with optimality criteria. The examination of a large number of solutions aided by designer experience (which is not simple tradition or conformity), seems the only present solution.

(5) As regards the control unit, there are many variants which are not described in the previous chapters. One of them consists in defining cycles and phases. The transmission of one command from the control unit to the processing unit constitutes a *phase*; according to our model it corresponds to one time unit. *Cycles* are sets of phases forming logic entities such as: next instruction fetching, operand fetching, execution, etc. This leads to a natural decomposition of the control automation into a *cycle generator* and a *phase generator*: every instruction of the control program is characterized by the cycle to which it belongs and by the particular phase within the cycle.

(6) We did not take into account the fact that some operations, such as memory accesses, are slower than other ones, such as register to register transfers. In order to get an optimized circuit, the unit execution time hypothesis should be forsaken: the timing characteristics of every resource must be studied so as to define a precedence relation between the operations and to deduce an acceptable labelling; in this way, the intrinsic algorithm *parellelism* can be exploited (see Chapters VIII and XI). Another frequent variant consists in implementing circuits with some degree of *pipelining* (see Chapters VIII and XIV).

(7) The whole circuit (Figures 8 and 16) essentially includes a main memory, a control memory with a decoding programmable logic array, an arithmetic and logic unit which may be realized by a series of bit slices, buses, and some registers and gates. It contains very few random logic parts, while forming a rather general architecture. It can be fitted to a particular application, or to a particular set of applications, by implementing an interpretation program and a main program, and by defining the functions performed by the arithmetic and logic unit. The main program defines an algorithm by means of instructions which are interpreted by the processing unit controlled by the interpretation program. The latter and the set of functions performed by the arithmetic and logic unit define an *instruction set,* while the main program defines a particular *algorithm.*

4 GENERALITIES ABOUT MICROPROCESSORS

This is the last stage of our study. We started from elementary components (Chapter I) and now arrive at the description of the most popular family of LSI circuits. Our aim consists in giving some general informations, in proving the feasibility of these circuits in view of the preceding sections and, eventually, in establishing the link between digital systems and computer science.

4.1 Definition. Typical architecture

The term microprocessor covers a large number of circuits, with rather different architectures, but with some common features; among others:

(1) They are used as basic components for building Von Neumann machines. In particular, they include, at the least, an arithmetic and logic unit and a control unit, the set of these two units being often called the *central processing unit* (CPU).

(2) They are implemented as single LSI chips which are presently most often realized in MOS technology.

In summary, a microprocessor is a central processing unit on a chip.

The realization of a Von Neumann machine by means of a microprocessor obviously requires the addition of other circuits, such as memories and input–output parts, and the connection of the various parts.

It is not possible to describe a standard architecture, which corresponds to all existing microprocessors. Nevertheless, we may give prominence to some important points, which have already been mentioned in the preceding section and will lead to an example of architecture.

Let us return to the Von Neumann model. The memory contains a series of instructions, which constitute the main program, and data which must be processed or interpreted according to the instructions. The arithmetic and logic unit must be able to perform the data transformations. There is again a compromise between the circuit complexity and the computation time: if we only look for functional completeness, the arithmetic and logic unit can be made up of a two-input NAND gate; most often, the data are stored under the form of n bit vectors, and the arithmetic and logic unit will be able to perform 'useful' operations, such as the addition of two n bit numbers, with possible facilities for decimal and 2's complement operations, and n bit componentwise Boolean operations. However, it is not an absolute rule that the arithmetic and logic unit width is equal to the memory word size.

The central processing unit controls the main program sequencing and thus includes a *program counter*. It also includes *flags* which store information deduced from intermediate results.

The memory contains instructions and numerical values. Hence, it is not always addressed by the program counter and, as in the example of Section 3, we must include an *address register*. More generally, in order to allow address computations when particular addressing modes are used (see Section 4.3), we may include a set of registers, selected by the control unit, able to store addresses.

As it is a single chip circuit, the connection complexity must be minimized, and thus a bus technique is used. Many microprocessors contain

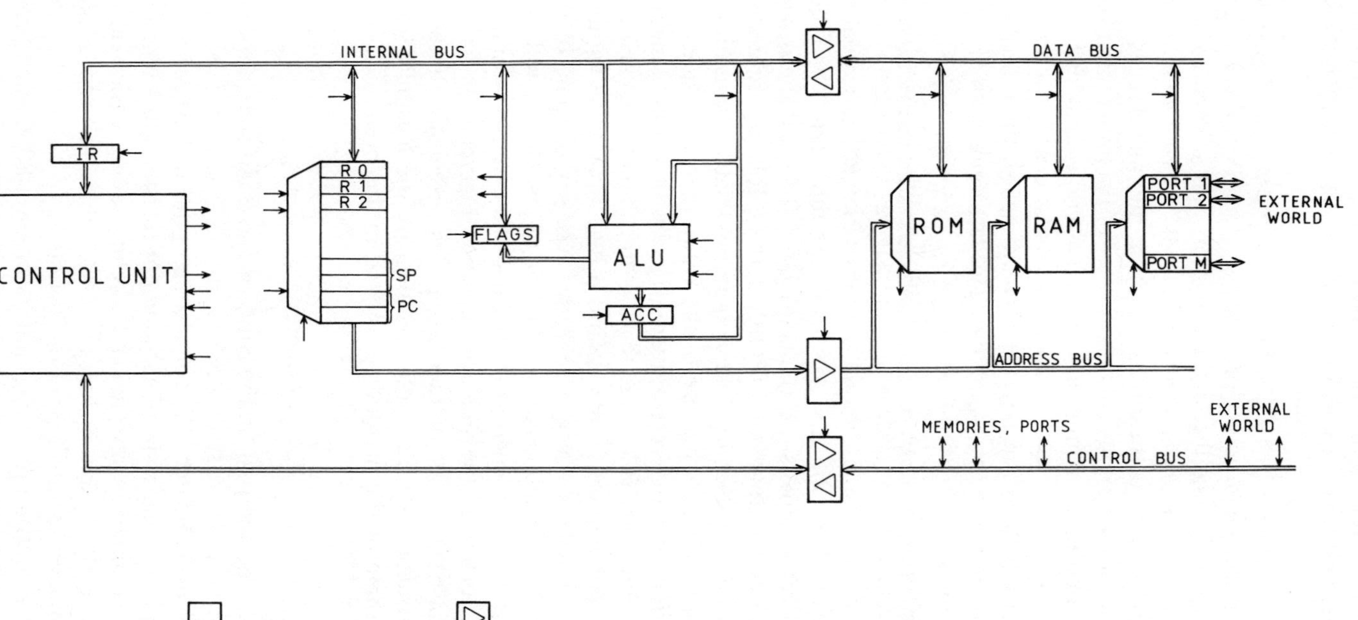

Figure 17. Example of microprocessor architecture

a single internal bus. Hence we include an *accumulator register*. Let us add some *working registers* able to store data and intermediate results temporarily.

The following hypotheses are realistic but do not correspond to a general rule; they help to make the structure precise: the memory is divided into a read only memory containing the main program, and a random access memory which contains data and intermediate results. These two memories, as well as the input–output ports, are distinct from the microprocessor.

It remains to connect the various circuits. In view of the limited number of circuit pins, it is important to minimize the connection complexity. This leads to the linking of the various circuits by means of buses. According to the type of information they transmit, we may distinguish three buses: the *address bus*, the *data bus*, and the *control bus*. This last is made up of control and condition variables.

The previous discussion leads us to the circuit of Figure 17. The registers are stacked and selected by an address decoder. It has been supposed that the memory word size and the arithmetic and logic unit width are equal to n, and that the addresses have $2n$ bits. Addresses are thus stored by register pairs. The two last register pairs are the stack point (*SP*) and the program counter (*PC*). The input–output ports (PORT 1 to PORT M) are gathered and selected by an address decoder. The condition variables transmitted to the control unit via the control bus are of the type: interrupt request, hold request before a direct memory access, ready signal issued from a memory, etc. Examples of control variables are: read, write, acknowledgement, bus control, etc. We do not emphasize these points, which already fall within the province of computer science, even though the necessity of some of these signal comes from the dynamic characteristics of the memories and of the interface circuits between the microprocessor and the external world.

The example of Section 3, the similarity between Figures 8 and 17, and the fact that the instruction set of the example is a realistic subset of a microprocessor instruction set, are sufficient to prove the feasibility of this type of circuit by some of the methods presented in this book.

4.2 Example

Figure 18 describes the internal architecture of the 8080 microprocessor. The program counter and the stack pointer are managed by a particular circuit: incrementer/decrementer. The connection of the arithmetic and logic unit to the internal bus is made via an accumulator register and a temporary register; furthermore there is an auxiliary circuit making easier decimal computations. The control unit includes a cycle generator and a phase generator (see Comment (5) at the end of Section 3).

Figure 19 shows a central processing unit based on a 8080 microprocessor. As the pin number is equal to forty, data and control informations are time multiplexed and transmitted on pins 3 to 10. The time multiplexing and

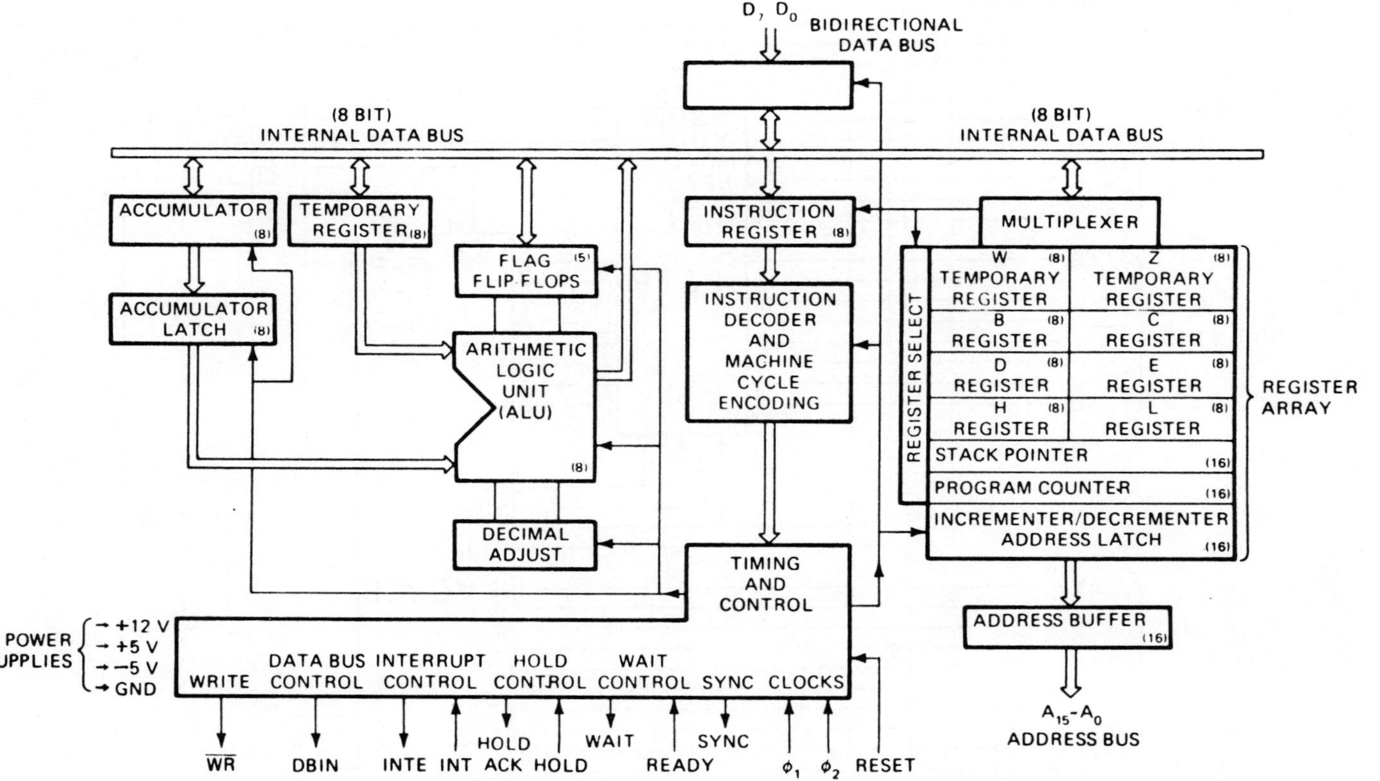

Figure 18. Intel 8080 block diagram. (Reprinted by permission of Intel Corporation)

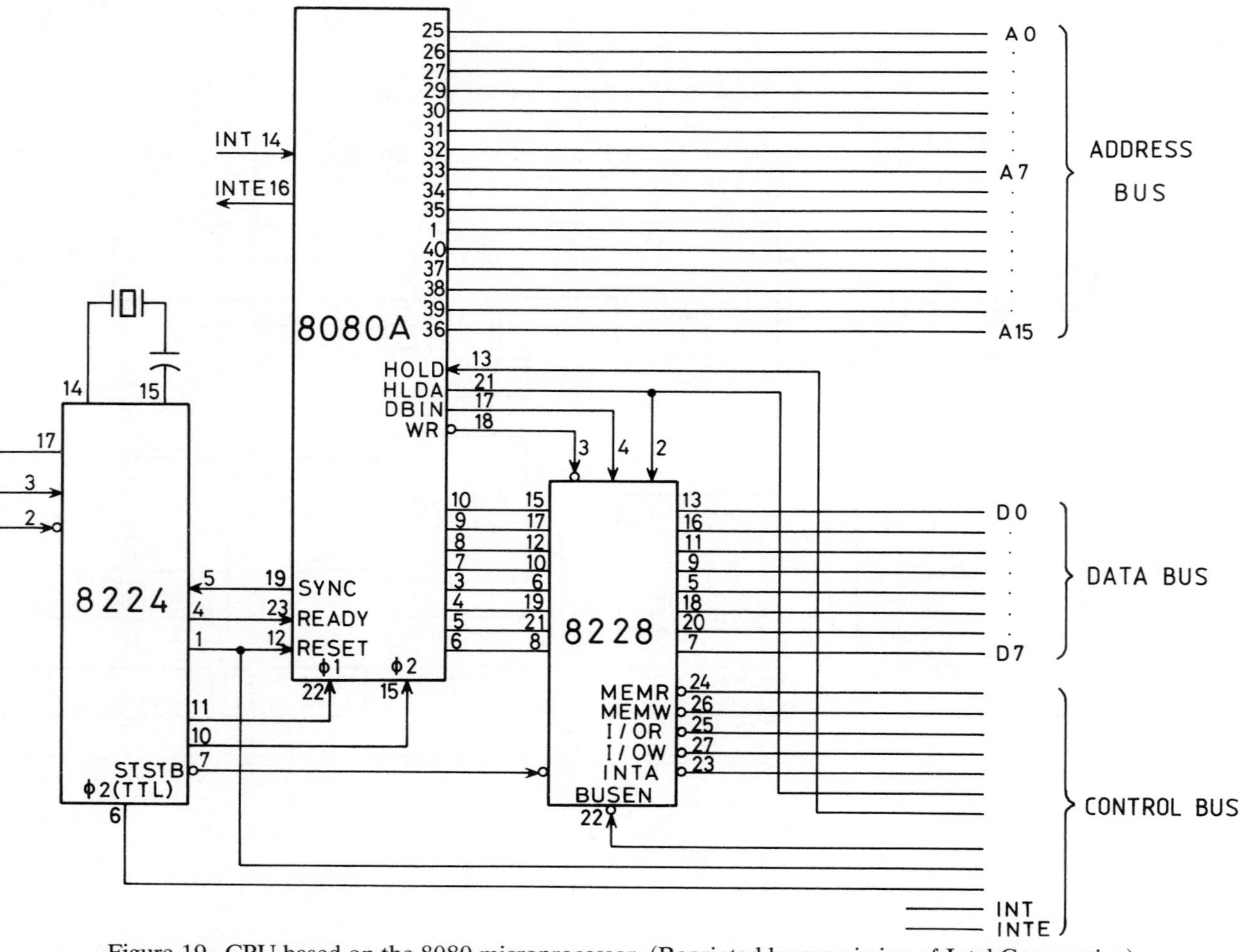

Figure 19. CPU based on the 8080 microprocessor. (Reprinted by permission of Intel Corporation)

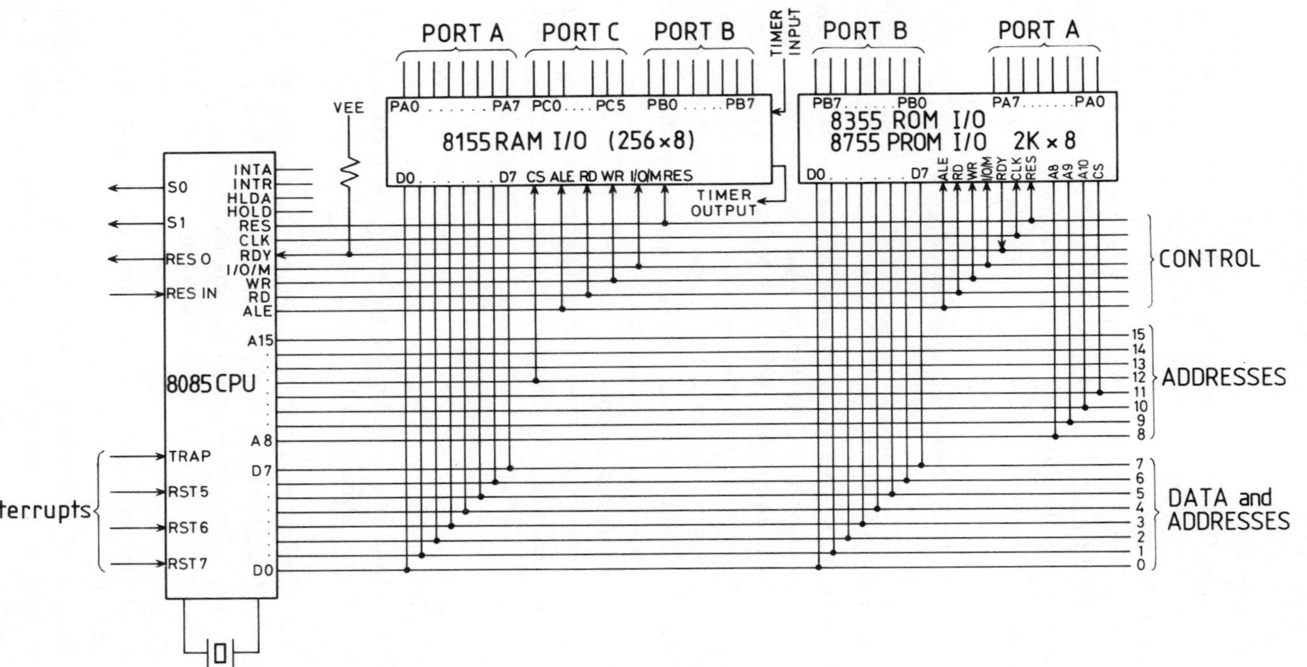

Figure 20. Von Neumann machine based on the 8085 microprocessor. (Reprinted by permission of Intel Corporation)

demultiplexing is performed by the 8228 auxiliary circuit. The synchronization signals are generated by the 8224 auxiliary circuit.

The 8085 microprocessor is an improved version of the 8080; it includes the synchronization signal generator. Nevertheless, the pin number is still insufficient; the eight least significant address bits are time multiplexed with the eight data bits. As shown in Figure 20, the memory and input–output circuits perform the time demultiplexing under the control of the variable ALE (address latching enable). These circuits combine RAM and I/O PORTS (8155), ROM and I/O PORTS (8355), or PROM and I/O PORTS (8755). The circuit of Figure 20 thus forms a complete Von Neumann machine, whose size is small, but which only includes three chips.

4.3 Some comparison points

We conclude this section with a list of some degrees of freedom given to the microprocessor designer.

(1) The *number of bits.* We mean by this the size of the data which are stored in the memories or in the working registers of the processing unit. In some cases, in order to lower the circuit complexity, the operations may be sequentialized, and the arithmetic and logic unit width is not equal to the memory and register size. As an example, the CP 1600 microprocessor handles sixteen bit numbers by means of an eight bit arithmetic and logic unit.

(2) The *number of internal buses.*

(3) The *type of control.* We have presented control units, realized by means of read only memories or programmable logic arrays, implementing a control program. In some cases, the control is hardwired. Another variant could consist in using more than two program levels, that is to allow several interpretation levels. Let us also observe that in our examples, the execution times of the various instructions are not identical; nevertheless we could design fixed execution time systems. Many control unit examples are described by Anceau and Etienne (1979) and Anceau (1980).

(4) *Pipelining.*

(5) *Instruction set.* The designer must choose the operations, transfers, branching conditions, etc. He also has to define the *addressing modes.* The example of Section 3 described two addressing modes.

(i) In CPL *Ri*, the address *i* of the datum is part of the instruction; it is a *direct addressing.*

(ii) In MOV *D*, *Ri*, the datum *D* itself is part of the instruction; it is an *immediate addressing.*

There are other possible addressing modes. Let us quote three of them:

(iii) *Indirect addressing*: the address of the address, rather than the address itself is part of the instruction.

(iv) *Indexed addressing*: the central processing unit adds the contents of an index register to the address supplied with the instruction, in order to find the effective address.

(v) *Relative addressing*: the central processing unit adds the contents of the program counter to the address supplied with the instruction in order to find the effective address.

(6) *Number of chips.* As an example, Figure 19 describes a three chip central processing unit, and Figure 20 a three chip complete system. There are also single chip complete systems (single chip microcomputers).

(7) *Memory organization, number of registers, stack implementation,* etc.

BIBLIOGRAPHICAL REMARKS

There are many books devoted to microprocessors, e.g. Aspinall (1978), Aspinall and Dagless (1977), Leventhal (1978), Zaks (1977). These writers are mainly concerned with the use of microprocessors rather than with their design. The well-known book of Mead and Conway (1980) describes two chips which are basic parts of a whole computer system, namely the processing unit and the interpretation program sequencer. Information on the design of processing and control units may also be found in Mano (1979). So far as microprogramming is concerned, let us quote Andrews (1980). For additional discussions, refer also to Patterson and Sequin (1980).

References

Akers, S. (1972) 'A Rectangular Logic Array', *IEEE Trans.* **C-21,** 848–857.

Amarel, S., Cook G. and Winder R. (1964) 'Majority Gate Networks', *IEEE Trans.* **EC-13,** 4–13.

Anceau, F. (1980) *Architecture and Design of Von Neumann Microprocessors,* NATO Advanced Study Institute on Design Methodologies for VLSI, Louvain La Neuve.

Anceau, F. and Etienne, R. (1979) *Conception des parties controles de microprocesseurs,* Ecole Nationale Superieure d'Informatique et de Mathematiques Appliquees, Grenoble.

Andrews, M. (1980) *Principles of Firmware Engineering in Microprogram Control,* Pitman, London.

Ashcroft, E. and Manna, Z. (1972) 'The translation of GOTO programs to WHILE programs', *Information Processing,* 250–255, North Holland, Amsterdam.

Aspinall, D. (1978) *The Microprocessor and its Application,* Cambridge University Press.

Aspinall, D. and Dagless, E. L. (1977) *Introduction to Microprocessors,* Pitman, London.

Atrubin, A. J. (1965) 'A One-Dimensional Real-Time Iterative Multipler', *IEEE Trans.* **EC-14,** 394–399.

Avizienis, A. (1961) 'Signed Digit Number Representation for Fast Parallel Arithmetic', *IRE Trans.* **EC-10,** 389–399.

Baer, J. L. (1980) *Computer Systems Architecture,* Pitman, London.

Baranov, S. and Keevallik, A. (1981) 'Transformations of Graph-schemes of Algorithms', *Digital Processes* **6,** 127–147.

Baugh, C. R. and Wooley, B. A. (1973) 'A Two's Complement Parallel Array Multiplication Algortihm', *IEEE Trans.* **C-22,** 1045–1047.

Bell, C. G. and Newell, A. (1971) *Computer Structures: Readings and Examples,* McGraw-Hill, New York.

Benedek, M. (1977) 'Developing Large Binary to BCD Conversion Structures', *IEEE Trans. on Computers* **C-26,** 688–699.

Bioul, G., Davio, M. and Quisquater, J. J. (1975) 'A Computation Scheme for an Adder Modulo $(2**n)-1$', *Digital Processes* **1,** 309–318.

Birkhoff, G. and Bartee, T. (1970) *Modern Applied Algebra,* McGraw-Hill, New York.

Birkoff, G. (1967) 'Lattice Theory', *Am. Math. Soc. Colloquium Publications* **25,** 1.

Böhm, C. and Jacopini, G. (1966) 'Flow diagrams, Turing Machines and languages with only two formation rules', *Com. ACM,* **9,** 5, 366–371.

Boole, G. (1847) *The Mathematical Analysis of Logic,* Cambridge.

Boole, G. (1854) *The Laws of Thought,* Cambridge.

Booth, T. L. (1967) *Sequential Machines and Automata Theory,* John Wiley, New York.

Booth, T. L. (1971) *Digital Networks and Computer Systems*, John Wiley, New York.

Boute, R. (1976) 'The Binary Decision Machine as a Programmable Controller', *Euromicro Newsletter* **2,** 16–22.

Burks, A. W., Goldstine, H. H. and Von Newmann (1946) 'Preliminary Discussion of the Logical Design of an Electronic Computing Instrument', in Bell & Newell (1971).

Canaday, R. (1964) *Two Dimensional Iterative Logic*, Report ESL-R-210, MIT Electronic Systems, Mass.

Cappa, M. and Hamacher, V. C. (1973) 'An Augmented Iterative Array for High-Speed Binary Division', *IEEE Trans.* **C-22,** 172–175.

Cerny, E., Mange, D. and Sanchez, E. (1979) 'Synthesis of Minimal Binary Decision Trees', *IEEE Trans.* **C-28,** 472–482.

Chen, I. and Willoner, R. (1979) 'An O(n) Parallel Multiplier with Bit-Sequential Input and Output', *IEEE Trans.* **C-28,** 721–727.

Clare, C. C. (1973) *Designing Logic Systems Using State Machines*, McGraw-Hill, New York.

Coffman, E. J. (1976) *Computer and Job Shop Scheduling Theory*, John Wiley, New York.

Coffman, E. and Graham, R. (1972) 'Optimal scheduling for two-processor system', *Acta Informatica,* **1,** 200–213.

Cohn, M. and Lindaman R. (1961) 'Axiomatic Majority Decision Logic', *IRE Trans.* **EC-10,** 17–21.

Cotten, L. W. 'Circuit Implementation of High Speed Pipeline Systems', *Proc. Fall Joint Computer Conference*, 489–504.

Cotten, L. W. (1969) 'Maximum Rate Pipeline Systems', *Proc. Spring Joint Computer Conference.* 581–586.

Couleur, J. F. (1958) 'A Binary to Decimal or Decimal to Binary Converter', *IRE Trans. Electr. Computers* **EC-7,** 313.

Dadda, L. (1965) 'Some Schemes for Parallel Multipliers', *Alta Frequenza* **34,** 349–356.

Dao, T. T. (1977) 'Threshold I2L and its Applications to Binary Symmetric Functions and Multivalued Logic', *IEEE Trans.* **SC-12,** 532–534.

Dao, T. T., McCluskey E. J. and Russel, L. K. (1977) 'Multivalued Integrated Injection Logic', *IEEE Trans* **C-26,** 1233–1241.

Davio, M. and Bioul, G. (1977) 'Fast Parallel Multiplication', *Philips Research Reports* **32,** 1977.

Davio, M. and Bioul, G. (1978) 'Efficiency of Pipelined Combinational Circuits', *Digital Processes* **4,** 3–16.

Davio, M. and De Lit, C. (1970) *State Minimal Covers of Deterministic Mealy Automata*, MBLE Tech. Note N.69.

Davio, M. and Deschamps, J. P. (1977) 'Synthesis of some Partially Symmetric Boolean Functions by Ternary Algebra', *Electronic Letters* **13,** 16, 473–474.

Davio, M. and Deschamps, J. P. (1978) 'Addition in signed-digit number systems', *Proc. 8th Symp. M.V.L.*, pp. 104–112.

Davio, M. and Deschamps, J. P. (1981) 'Synthesis of Discrete Functions Using I2L Technology', *IEEE Trans* **C-30,** 653–661.

Davio, M. and Quisquater, J. J. (1973a) 'Rectangular Universal Iterative Arrays', *Electronic Letters* **9,** 485–486.

Davio, M. and Quisquater, J. J. (1973b) 'Iterative Universal Logical Modules', *Philips Research Reports,* **28,** 265–293.

Davio, M. and Quisquater, J. J. (1975) *Complexity of Discrete Functions*, MBLE Internal Report.

Davio, M. and Ronse, C. (1980) *Rotator Design*, Philips Res. Lab., Brussels Report 422. To appear in *Discrete Applied Mathematics.*

Davio, M. and Thayse, A. (1978) 'Optimization of Multivalued Decision Algorithms, *Philips Journal of Research* **33,** 31–65.

Davio, M. and Thayse, A. (1979) 'Algorithms for Minimal Length Schedules', *Philips Journal of Research* **34,** 26–47.

Davio, M. and Thayse, A. (1980) 'Implementation and Transformation of Algorithms Based on Automata, Part I. Introduction and Elementary Optimization Problems', *Philips Journal of Research* **35,** 122–144.

Davio, M., Deschamps, J. P. and Liénard, J. C. (1977) 'Optimization of Strictly Non-Blocking and Rearrangeable Interconnections', *Philips Research Reports* **32,** 266–296.

Davio, M., Deschamps, J. P. and Thayse, A. (1978) *Discrete and Switching Functions*, McGraw-Hill, New York.

Davis, R. L. (1974) 'Uniform Shift Networks', *Computer* **1974,** 60–71.

De Mori, R. (1969) 'Suggestion for an IC Fast Parallel Multiplier', *Electronic Letters* **5,** 50–51.

Dean, K. J. (1968) 'Binary Division Using a Data Dependent Iterative Array', *Electronic Letters* **4,** 283–284.

Deschamps, J. P. (1980) *Hardware Implementation of Algorithms*, Philips Internal Report, Brussels.

Deschamps, J. P. (1981) 'Hardware Implementation of Algorithms: Coast Estimates of the Control Units', to appear.

Deverell, J. (1975) 'Pipeline Iterative Arrays', *IEEE Trans.* **C-24,** 317–322.

Dietmeyer, D. (1971) *Logic Design of Digital Systems*, Allyn and Bacon, Boston.

Dijkstra, E. W. (1968) 'GOTO statement considered harmful', *Comm. ACM*, pp. 147–148.

Etiemble, D. (1979) 'Contribution a l'etude des circuits multivalues et de leur utilisation', PhD. Dissertation. Univ. Paris VI.

Feng, T. S. (1981) 'A Survey of Interconnection Networks', *Computer* **14,** 12, 12–30.

Flores, I. (1963) *The Logic of Computer Arithmetic*, Prentice-hall, Englewood Cliffs, N.J.

Forbes, B. E. (1977) 'Silicon on Sapphire Technology Produces High-Speed Single Chip Processor', *Hewlett Packard Journal* **April,** 1–8.

Gajski, T. D. and Tulpule, B. R. (1978) 'High-speed Masking Rotator', *Digital Processes* **4,** 67–81.

Gerace, G. B. (1968) 'Digital System Design Automation. A Method for Designing a Digital System as a Sequential Network System', *IEEE Trans.* **C-17,** 1044–1061.

Gerace, G. B., Vanneschi, M. and Casaglia, G. F. (1974) 'Models and Comparisons of Microprogrammed Systems', *Note Scientifiche*, **S-74-18,** Univ. de Pisa, Istit. di Scienze dell'Informazione.

Gill. A. (1966) *Linear Sequential Circuit Analysis*, McGraw-Hill, New York.

Gimpel, J. (1967) 'The Minimization of TANT Networks', *IEEE Trans.* **EC-16,** 18–38.

Ginzburg, A. (1968) *Algebraic Theory of Automata*, Academic Press, New York.

Glaser, A. and Subak-Sharpe, G. (1979) *Integrated Circuit Engineering*, Addison Wesley, Reading, Mass.

Glushkov, V. M. (1966) *Introduction to Cybernetics*, Academic Press, New York.

Glushkov, V. M. (1970) 'Some Problems in the Theories of Automata and Arrificial Intelligence', *Kibernetica* **6,** 2.

Glushkov, V. M. (1965) 'Automata Theory and Formal Microprogram Transformation', *Kibernetica* **1,** 1–9.

Golomb, S. (1967) *Shift Register Sequences*, Holden Day, San Francisco.

Grasselli, A. and Gimpel, J. (1966) *The Synthesis of Three-Level Logic Networks*, Internal Report R-49, Princeton University, Digital Systems Laboratory.

Grätzer, G. (1971) *Lattice Theory*, Freeman, San Francisco.

Greenfield, J. D. (1977) *Practical Design Using IC's*, John Wiley, New York.

Guild, H. H. (1969) 'Fully Iterative Fast Array for Binary Multiplication and Fast Addition', *Electronic Letters* **5,** 263.

Guild, H. H. (1970) 'Some Cellular Logic Arrays for Non-restoring Binary Division', *The Radio and Elec. Eng.* **39,** 345–348.

Hallyn, T. G. and Flynn, M. J. (1972) 'Pipelining of Arithmetic Functions', *IEEE Trans.* **C-21,** 880–886.

Hamacher, V. C., Vranesic, Z. G. and Zaky, S. G. (1978) *Computer Organization*, McGraw-Hill, New York.

Hammer, P. and Rudeanu, S. (1968) *Boolean Methods in Operations Research*, Springer, Berlin.

Harrison, M. A. (1965) *Introduction to Switching and Automata Theory*, McGraw-Hill, New York.

Hartmanis, J. and Stearns, R. E. (1966) *Algebraic Structure Theory of Sequential Machines*, Prentice-Hall, Englewood Cliffs, N.J.

Hennie, F. C. (1961) *Iterative Arrays of Logical Circuits*, MIT Press & John Wiley, New York.

Hennie, F. C. (1968) *Finite State Models for Logical Machines*, John Wiley, New York.

Hill, F. and Peterson, G. (1968) (2nd edn. 1974) *Introduction to Switching Theory and Logical Design*, John Wiley, New York.

Hohn, F. E. (1966) (2nd edn. 1970) *Applied Boolean Algebra*, Macmillian New York.

Howard, B. (1975a) 'Determinacy of Computation Schemata for Both Parallel and Simultaneous Operation', *Electronic Letters* **11,** 485–487.

Howard, B. (1975b) 'Parallel Computation Schemata and their Hardware Implementation', *Digital Processes* **1,** 183–206.

Huffman, D. A. (1959) 'Canonical Forms for Information Lossless Finite-state Logical Machines', *IRE Trans.* **CT-6,** Spec. Suppl., 41–59.

Hwang, K. (1979) *Computer Arithmetic*, John Wiley, New York.

Ito, T. (1973) 'A Theory of Formal Microprograms', *SIG MICRO Newsletter* **4,** 5–17.

Kambayashi, Y. (1979) 'Logic Design of Programmable Logic Arrays', *IEEE Trans.* **C-28,** 609–616.

Karnaugh, M. (1953) 'The map method for synthesis of combinational logic circuits', *Trans AIEE, Part I, Communications and Electronics,* **72,** 593–599.

Kautz, W. (1971) 'Programmable Cellular Logic', in *Recent Developments in Switching Theory* (A. Mukhopadhyay, ed.), Academic Press, New York.

Kelley, J. E. (1961) 'Critical path Planning and Scheduling Mathematical Basis', *Operations Research* **9,** 296–320.

Kelley, J. E. and Walker, M. R. (1959) 'Critical Path Planning and Scheduling', *Proc. Eastern Joint Computer Conf.*, 160–173.

Kleene, S. C. (1956) 'Representation of Events in Nerve Nets and Finite Automata', in *Automata Studies*, Princeton University Press.

Knuth, D. E. (1969) *The Art of Computer Programming*, vol. 2, *Seminumerical Algorithms*, Addison Wesley, Reading, Mass.

Knuth, D. E. (1973) *The Art of Computer Programming*, vol. 3. *Sorting and Searching*, Addison Wesley, Reading, Mass.

Knuth, D. E. and Floyd, R. W. (1961) 'Notes on avoiding GOTO statements', *Information Processing Letters*, **1,** 23–31.

Kohavi, Z. (1970) *Switching and Finite Automata Theory*, McGraw-Hill, New York.

Kostopoulos, G. K. (1975) *Digital Engineering*, John Wiley, New York.

Kuntzmann, J. (1965) *Algebre de Boole,* Dunod, Paris.

Kurmit, A. A. (1974) *Information Lossless Automata of Finite Order,* John Wiley, New York.

Lai, H. C., and Muroga, S. (1979) 'Minimum Parallel Binary Adders with NOR (NAND) gates', *IEEE Trans.* **C-28,** 648–659.

Lanning, W. C. (1975) 'General Algorithms for Direct Radix Conversion', *Computer Design,* **1975,** 61–73.

Ledgard, H. J. and Marcotti, M. (1975) 'A geneology of control structures', *Comm. ACM,* **18,** 629–639.

Lee, C. 'Representation of Switching Circuits by Binary Decision Programs', *BSTJ* **38,** 985–999.

Lee, S. (1976) *Digital Circuits and Logic Design,* Prentice-Hall, Englewood Cliffs, N.J.

Lee, S. (1978) *Modern Switching Theory and Logical Design,* Prentice-Hall, Englewood Cliffs, N.J.

Leventhal, L. A. (1978) *Introduction to Microprocessors,* Prentice-Hall, Englewood Cliffs, N.J.

Liu, T. K., Hohulin, K. R. *et al.* (1974) *Optimal One-Bit Full Adders with Different Types of Gates. IEEE Trans.* **C-23,** 63–70.

Lupanov, O. (1958) 'Ob Odnom Metode Sinteza Skhem', *Isvestiya VUZ (Radiofizika),* 120–140.

Majithia, J. C. (1976) 'Some Comments Concerning the Design of Pipeline Arithmetic Arrays', *IEEE Trans.* **C-25,** 1132–1134.

Maley, G. and Earle, J. *The Logic Design of Transistor Digital Computers,* Prentice-Hall, Englewood Cliffs, N.J.

Mange, D. (1978) *Analyse et Synthese des Systemes Logiques,* Georgi, St Saphorin.

Mange, D. (1979) 'Compteurs Microprogrammes', *Bulletin de l'Association Suisse des Electriciens,* **70,** 1087–1095.

Mange, D. (1980) 'Microprogrammation Structuree', *Le Nouvel Automatisme,* **25,** 45–54.

Mange, D. (1981) 'Programmation Structuree', *Bulletin de l'Association Suisse des Electriciens,* **72,** 339–345.

Mange, D., Sanchez, E. and Stauffer, A. (1982) *Systemes Logiques Programmes,* Presses Polytechniques Romandes, Lausanne.

Manna, Z. (1974) *Mathematical Theory of Computation,* McGraw-Hill, New York.

Manning, F. (1972) *Autonomous Synchronous Counters Constructed only of JK Flip-Flops,* MAC TR-96, MIT Press, Cambridge, Mass.

Mano, M. M. (1972) *Computer Logic Design,* Prentice-Hall, Englewood Cliffs, N.J.

Mano, M. M. (1979) *Digital Logic and Computer Design,* Prentice-Hall, Englewood Cliffs, N.J.

Marcus, M. (1962) (2nd edn. 1967) *Switching Circuits for Engineers,* Prentice-Hall, Englewood Cliffs, N.J.

Martin, R. L. (1969) *Studies in Feedback Shift Registers,* MIT Press, Cambridge, Mass.

McCluskey, E. J. (1956) *Introduction to the Theory of Switching Circuits,* McGraw-Hill, New York.

MacLane S., and Birkhoff, G. (1967) *Algebra,* Macmillan, New York.

McSorley, O. L. (1961) 'High Speed Arithmetic in Binary Computers', *Proc. IRE* **49,** 67–91.

Mead, C., and Conway, L. (1980) *Introduction to VLSI Systems,* Addison-Wesley, Reading, Mass.

Mealy, G. H. (1955) 'A Method for Synthesizing Sequential Circuits', *BSTJ* **34,** 1045–1079.

Meisel, W. (1967) 'A Note on Internal State Minimization in Incompletely Specified Sequential Networks; *IEEE Trans.* **EC-16,** 508–509.

Mishchenko, A. (1968a) 'The Formal Synthesis of an Automaton by a Microprogram; *Cybernetios* **4,** 20–26.

Mishchenko, A. T. (1968b) 'Formal Synthesis of an Automaton by a Microprogram', part II, *Cybernetics* **4,** 17–22.

Mishchenko, A. T. (1967) 'Transformations of Microprograms', *Kibernetica* **3,** 7–13.

Moore, E. F. (1956) 'Gedanken Experiments on Sequential Machines', in *Automata Studies* (C. E. Shannon and E. J. McCarthy, eds.), Princeton University Press.

Mukhopadhyay, A. (1971) *Recent Developments in Switching Theory*, Academic Press, New York.

Muller, D. (1956) 'Complexity in Electronic Switching Circuits', *IRE Trans.* **EC-5,** 15–19.

Myers, G. J. (1980) *Digital System Design with LSI Bit-Slice Logic*, John Wiley, New York.

Neiman, V. J. (1969) 'Structure et Commande Optimale des reseaux de Connexion sans Bloc age', *Annales des Telecommunications* **24,** 232–238.

Nicoud, J. D. (1969) 'Decimal Binary Conversion', *Electronic Letters* **5,** 26 June.

Oberman, R. M. (1970) *Disciplines in Combinational and Sequential Circuit Design*, McGraw-Hill, New York.

Patterson, D. A. and Sequin, C. H. (1980) 'Design Considerations for Single Chip Computers of the Future', *IEEE Trans.* **C-29,** 108–115.

Paull, M. and Unger, H. (1959) 'Minimizing the Number of States in Incompletely Specified Sequential Circuits', *IEEE Trans.* **EC-8,** 356–367.

Penney, W. M. and Lau, L. (1972) *MOS Integrated Circuits*, Van Nostrand, New York.

Pezaris, S. D. (1971) 'A 40 ns 17-bit-by-17-bit Array Multiplier', *IEEE Trans.* **C-20,** 422–447.

Pippenger, N. (1980) *Pebbling*, IBM Research Report RC 8258 (No. 35937).

Preparata, F. and Yeh, R. (1973) *Introduction to Discrete Structures*, Addison-Wesley, Reading, Mass.

Quine, W. (1952) 'The problem of simplifying truth functions', *Am. Math. Monthly*, **59,** 521–537.

Quine, W. (1953) 'Two theorems about truth functions', *Bul. Soc. Mat. Mexicana*, **10,** 64–70.

Quine, W. (1955) 'A way to simplify truth-functions', *Am. Math. Monthly*. **62,** 627–631.

Rabin, M. O. and Scott, D. (1959) 'Finite Automata and their Decision Problems', *IBM Journal of Research and Development*, **3,** 114–125.

Rauscher, T. G., and Adams, P. N. (1980) 'Microprogramming: a Tutorial and Survey of Recent Developments', *IEEE Trans.* **C-29,** 2–19.

Robertson, J. E. (1958) 'A new Class of Division Methods', *IEEE Trans.* **C-7,** 218–222.

Ronse, C. (1980) *Non-linear Shift-Registers: a Survey*, MBLE Research Report R 430.

Rudeanu, S. (1965) 'On Tohma's Decomposition of Boolean Functions', *IEEE Trans.* **EC-14,** 929–931.

Rudeanu, S. (1974) *Boolean Functions and Equations*, North-Holland, Amsterdam.

Rutherford, D. E. (1965) *Introduction to Lattice Theory*, Hafnes, New York.

Sanchez, E. and Stauffer, A. (1979) *Horloge Microprogrammee*, 10eme Congres International de Chronometrie, Geneve, 279–284.

Sanchez, E. and Thayse, A. (1981) 'Implementation and Transformation of Algorithms Based on Automata', Part III, 'Optimization of Evaluation Programs', *Philips Journal of Research* **36,** 159–172.

Sasao, (1981) 'Boolean Functions and Programmable Logic Arrays', *IEEE Trans.* **C-30,** 635–643.

Savage, J. E. (1972) 'Computational Work and Time on Finite Machines', *Journal of the ACM* **19,** 660–674.

Segovia, T. (1980) *Optimisation en Surface des PLA*, Rapport de l'Institut National Polytechnique de Grenoble.

Séquin, C. H. and Tompsett, M. F. (1975) *Charge Transfer Devices*, Academic Press, New York.

Sethi, R. (1976) 'Algorithms for Minimal Length Schedules', in *Computer and Job-Shop Scheduling* (Coffman E. J. ed), John Wiley, New York.

Shannon, C. E. (1949) 'A Mathematical Theory of Communication', *BSTJ*, Part I, **27,** 379–423; Part II, **27,** 623–656.

Sikorski, R. (1964) *Boolean Algebras*, Springer, Heidelberg.

Skokan, Z. E. (1973) 'Emitter Function Logic. Logic Family for LSI', *IEEE Journal of Solid State Physics*, **SC-8,** 356–361.

Slansky, J. (1960a) 'Conditional Sum Addition Logic', *IRE Trans.* **EC-9,** 226–231.

Slansky, J. (1960b) 'An Evaluation of Several Two Summand Binary Adders', *IRE Trans.* **EC-9,** 213–226.

Spira, P. (1971a) 'On Time–Hardware Trade-offs for Boolean Functions', *Proc. 4th Hawaii Int. Conf. on System Sciences*, pp. 525–527.

Spira, P. (1971b) 'On the Time Necessary to Compute Switching Functions', *IEEE Trans.* **C-20,** 104–105.

Spira, P. (1973) 'Computation Times of Arithmetic and Boolean Functions in (d, r)-circuits', *IEEE Trans.* **C-22,** 552–555.

Stabler, E. P. (1970) 'Microprogram Transformations', *IEEE Trans.* **C-19,** 908–916.

Stauffer, A. (1980) 'Methode de Synthese des Systemes Digitaux', *Bulletin de l'Association Suisse des Electriciens* **71,** 143–150.

Stefanelli, R. (1972) 'A Suggestion for High-Speed Parallel Binary Divider', *IEEE Trans.* **C-21,** 42–55.

Stone, H. S. *et al.* (1975) *Introduction to Computer Architecture*, Science Research Associates, Chicago.

Stone, M. (1935) *Subsumption of Boolean Algebras under the Theory of Rings*, Proc. Nat. Acad. Sci. USA **29,** 103–105.

Swartzlander, E. E. (1973) 'Parallel Counters', *IEEE Trans.* **C-22,** 1021–1024.

Swartzlander, E. E. (1979) 'Microprogrammed Control for Specialized Processors', *IEEE Trans.* **C-28,** 930–933.

Taub, H. and Schilling, D. (1977) *Digital Integrated Electronics*, McGraw-Hill and Kogakusha, Tokyo.

Thayse, A. (1980) 'Implementation and Transformation of Algorithms Based on Automata. Part II, Synthesis of Evaluation Programs', *Philips Journal of Research* **35,** 190–216.

Thayse, A. (1981a) 'Programmable and Hardwired Synthesis of Discrete Functions', Part I, 'One Level Addressing Networks', *Philips Journal of Research*, **36,** 40–73.

Thayse, A. (1981b) 'Programmable and Hardwired Synthesis of Discrete Functions, Part II, Two level Addressing Networks', *Philips Journal of Research* **36,** 140–158.

Thayse, A. (1981c) 'P-Functions: A New Tool for the Analysis and Synthesis of Binary Programs', *IEEE Trans.* **C-30,** 698–705.

Thayse, A. (1982) 'Synthesis and Optimization of Programs by Means of P-Functions', *IEEE Trans.* **C-31,** 34–40.

Tohma, Y. (1964) 'Decomposition of Logical Functions Using Majority Decision Elements', *IEEE Trans.* **EC-13,** 698–705.

Torng, H. C. and Wilhelm, N. C. (1977) 'The Optimal Interconnection of Circuit Modules in Microprocessor and Digital System Design', *IEEE Trans.* **C-26,** 450–457.

Unger, S. H. (1977) 'Tree Realizations of Iterative Circuits', *IEEE Trans.* **C-26,** 4, 365–383.

Veitch, E. (1952) 'A chart method for simplifying truth functions', *Proc. ACM,* 127–133.

Venn, J. (1876) 'Boole's logical system', *Mind,* **1,** 479–491.

Venn, J. (1894) *Symbolic Logic,* Macmillan, London.

Vuillemin, J. (1980) 'A combinatorial limit to the computing power of VLSI circuits', *Proc. 21st Annual Symposium on Foundations of Computer Science, IEEE,* NY.

Waksman, A. (1968) 'A Permutation Network', *Journal of the ACM,* **15,** 159–163.

Wallace, C. S. (1964) 'A Suggestion for Parallel Multiplier', *IEEE Trans.* **EC-13,** 14–17.

Wilkes, M. V. (1951) *The Best Way to Design an Automatic Calculating Machine,* Manchester University Inaugural Conference, pp. 14–18.

Winograd, S. (1965) 'On the Time Required to Perform Addition', *Journal of the ACM,* **12,** 277–285.

Winograd, S. (1967) 'On the Time Required to Perform Multiplication', *Journal of the ACM* **14,** 793–802.

Wood, R. (1979) 'A High Density Programmable Logic Array Chip', *IEEE Trans.* **C-28,** 602–607.

Zahnd, J. (1969) 'Homomorphismes et recouvrements de machines de Mealy', *Systemes Logiques* **1,** 16–18.

Zhand, J. (1980) *Machines Sequentielles,* Georgi, St Saphorin, CH 1813.

Zhand, J. and Mange, D. (1975) 'Reduction des tables d'etat Incompletement Defines', *Systemes Logiques* **6**.

Zaks, R. (1977) 'Microprocessors: from Chips to Systems', *SYBEX.*

Author index

Subject index